Environmental Science and Ecology

Environmental Science and Ecology

Editor: Molly Ismay

Callisto Reference
www.callistoreference.com

Callisto Reference,
118-35 Queens Blvd., Suite 400,
Forest Hills, NY 11375, USA

Visit us on the World Wide Web at:
www.callistoreference.com

ISBN: 978-1-63239-994-6 (Hardback)

Printed in China.

Cataloging-in-Publication Data

Environmental science and ecology / edited by Molly Ismay.
p. cm.
Includes bibliographical references and index.
ISBN 978-1-63239-994-6
1. Environmental sciences. 2. Ecology. I. Ismay, Molly.
GE105 .E58 2018
363.7--dc23

Table of Contents

Preface

The world is advancing at a fast pace like never before. Therefore, the need is to keep up with the latest developments. This book was an idea that came to fruition when the specialists in the area realized the need to coordinate together and document essential themes in the subject. That's when I was requested to be the editor. Editing this book has been an honour as it brings together diverse authors researching on different streams of the field. The book collates essential materials contributed by veterans in the area which can be utilized by students and researchers alike.

Ecology primarily studies the diverse interactions that occur within an ecosystem. These interactions can be amongst the organisms or between the organisms and their environment. The major areas of concern in ecology are biodiversity, habitat conservation, food webs, population ecology, etc. Ecology has various practical applications in areas like environmental science, natural resource management, city planning, etc. This book provides comprehensive insights into the field of ecology. Through this book, we attempt to further enlighten the readers about the new concepts in this field.

Each chapter is a sole-standing publication that reflects each author´s interpretation. Thus, the book displays a multi-facetted picture of our current understanding of application, resources and aspects of the field. I would like to thank the contributors of this book and my family for their endless support.

Editor

1

Traditions in Spider Monkeys Are Biased towards the Social Domain

Claire J. Santorelli[1]*, Colleen M. Schaffner[1], Christina J. Campbell[2], Hugh Notman[3], Mary S. Pavelka[4], Jennifer A. Weghorst[5], Filippo Aureli[6]

1 Department of Psychology, University of Chester, Chester, United Kingdom, **2** Department of Anthropology, California State University, Northridge, California, United States of America, **3** Anthropology, Centre for Work and Community Studies, Athabasca University, Athabasca, Alberta, Canada, **4** Department of Anthropology, University of Calgary, Calgary, Alberta, Canada, **5** Natural History Museum and Biodiversity Research Center, University of Kansas, Lawrence, Kansas, United States of America, **6** Research Centre in Evolutionary Anthropology and Palaeoecology, School of Natural Sciences and Psychology, Liverpool John Moores University, Liverpool, United Kingdom

Abstract

Cross-site comparison studies of behavioral variation can provide evidence for traditions in wild species once ecological and genetic factors are excluded as causes for cross-site differences. These studies ensure behavior variants are considered within the context of a species' ecology and evolutionary adaptations. We examined wide-scale geographic variation in the behavior of spider monkeys (*Ateles geoffroyi*) across five long-term field sites in Central America using a well established ethnographic cross-site survey method. Spider monkeys possess a relatively rare social system with a high degree of fission-fusion dynamics, also typical of chimpanzees (*Pan troglodytes*) and humans (*Homo sapiens*). From the initial 62 behaviors surveyed 65% failed to meet the necessary criteria for traditions. The remaining 22 behaviors showed cross-site variation in occurrence ranging from absent through to customary, representing to our knowledge, the first documented cases of traditions in this taxon and only the second case of multiple traditions in a New World monkey species. Of the 22 behavioral variants recorded across all sites, on average 57% occurred in the social domain, 19% in food-related domains and 24% in other domains. This social bias contrasts with the food-related bias reported in great ape cross-site comparison studies and has implications for the evolution of human culture. No pattern of geographical radiation was found in relation to distance across sites. Our findings promote *A. geoffroyi* as a model species to investigate traditions with field and captive based experiments and emphasize the importance of the social domain for the study of animal traditions.

Editor: Tapio Mappes, University of Jyväskylä, Finland

Funding: Research at Barro Colorado Island was supported by grants from the National Science Foundation (SBR-9711161), the Leakey Foundation, the Department of Anthropology, University of California, Berkeley (www.berkeley.edu) and a Short-term Fellowship from the Smithsonian Tropical Research Institute (www.stri.org). Research at Corcovado National Park's Sirena Biological Station was supported by NSF award 0233248 (with R. Sussman), the Wenner-Gren Foundation, the Leakey Foundation, the American Society of Primatologists (www.asp.org), and Washington University in St. Louis (www.wustl.edu). Funds for Sirena's field lab facility were provided to L. E. Gilbert (Univ. of Texas at Austin) by NSF BSR 8315399 and a matching WWF grant, and funds for updating Sirena's trail system and installation of spatial reference system were provided by the Mellon Foundation through the Institute of Latin American Studies at UT Austin. Research at Santa Rosa and Punta Laguna was supported by The British Academy (www.britac.ac.uk), the Wenner-Gren Foundation (www.wennergren.org), the Leakey Foundation (www.leakeyfoundation.org) and the North of England Zoological Society (www.chesterzoo.org). CJS was supported by a Gladstone bursary from the University of Chester (www.chester.ac.uk) and by the Santander University Scheme (www.santander.co.uk). Research at Runaway Creek was supported by the Natural Sciences and Engineering Research Council of Canada. The funders had no role in study design, data collection and analysis, decision to publish, or preparation of the manuscript.

Competing Interests: The authors have declared that no competing interests exist.

* E-mail: c.santorelli@chester.ac.uk

Introduction

Traditions in wild populations are defined as "enduring behavior patterns shared among members of a group that depend to a measurable degree on social contributions to individual learning, resulting in shared practices among members of a group" [1] (p. 3). Evidence for traditions is often initially achieved through the documentation of between-group behavioral variation, once ecological and genetic explanations are excluded [2–9]. A crucial aspect of a tradition is that it derives from socially learned information (i.e. the ability to extract information from observing, or interacting with, another individual or its products [10]), and that it is not genetically inherited or individually learned information [11,12]. Under the appropriate circumstances, adaptations as a result of social learning can be more rapid than those resulting from natural selection and less risky than those obtained through individual trial and error learning [13,14]. Although criticized for being unable to definitively rule out the influence of genes or ecology [15–17], cross-site studies have the advantage of ensuring that behaviors are considered within the context of a species' ecology and evolutionary adaptations [2].

Spider monkeys (*Ateles* spp.) are well suited for a study of traditions because they possess several characteristics thought to promote social learning. Firstly, infants and juveniles are slow to develop compared to monkeys of a similar size and lifespan [18], providing prolonged exposure to maternal skills. Secondly, spider monkeys are socially tolerant [19], a feature predicted to facilitate social learning [20,21]. Thirdly, spider monkeys live in communities characterized by a high degree of fission-fusion dynamics, in which individuals split and merge into subgroups of variable composition [19]. This social system is relatively rare among mammals, but it is shared with chimpanzees (*Pan troglodytes*), bonobos (*P. paniscus*) and humans (*Homo sapiens*) [22–24]. Milton [25,26] pointed out that the foraging patterns of species with such fluid fission-fusion dynamics would also place them under great pressure to develop key skills, including enhanced communication systems for rapid recognition and greeting behaviors to facilitate reunions cf. [24]. The behavioral repertoire of *A. geoffroyi* comprises a number of gestures including embracing and pectoral sniffing, that likely function as greetings [19,27,28]. In addition, spider monkeys use a range of substrate marking behaviors for delayed olfactory communication [29–31], which may convey information between community members that visit the same location in separate subgroups at different times. Variation across communities in greeting and marking behaviors may occur and additionally serve to convey community identity, making them ideal potential behaviors for traditions.

With the exception of Perry et al. 's [5] study on capuchin monkeys (*Cebus capucinus*), documenting variation in behaviors involving extractive foraging and tool-use has been a main focus of previous cross-site primate studies [3,6,32–34]. There is a discrepancy between this focus and the awareness that many human traditions involve social behavior [35]. This discrepancy may be due to tool use or object manipulation being clearly identifiable [36], and object function being immediately apparent [37].

Of the three features that are fundamental for material culture (*sensu* McGrew [38]), including extractive foraging, dexterous manipulation and tolerant gregariousness [21], spider monkeys rarely show the first two. They are ripe fruit specialists [39,40] and their nutritional needs are largely met by plant substrates easily accessed in the canopy [41]. Dexterous manipulation in spider monkeys is likely limited due to a dramatic reduction of the pollex or opposable thumb [42] and to them not having separate control of individual fingers [43,44] (Figure S1), which would make the firm gripping of objects problematic. These two hand adaptations are thought to afford *Ateles* with the skills needed for their highly arboreal lifestyle and specialized locomotion [42,43]. Consequently, these anatomical and dietary adaptations indicate that spider monkeys would be unlikely to engage in many behaviors relating to extractive foraging or tool use.

To our knowledge, no systematic study of traditions across different populations of spider monkeys has been carried out, although several publications suggest potential behaviors that could show inter-community variation and patterns of transmission via social learning, including meliponid bee (*Scaptotrigona* spp.) eating [45,46], self-anointing behavior with plant substrates [30,47], terrestrialism [48] and self-scratching using sticks [49]. These reports all document potential community variation in *Ateles* behavioral repertoire, but offer no indication that these behaviors are either learned socially or, in the case of tool-use, are being successfully transmitted between individuals. The aim of our study was to provide the first systematic evidence for traditions in spider monkeys, using a large sample of candidate behaviors across five distantly located populations of the same species with special emphasis on the domains in which the traditions occur. The similarity of their social system with that of humans [19,23] makes the investigation of traditions in the social domain particularly relevant. First, we predicted that evidence for traditions within the social domain of spider monkeys would be more prevalent than in other domains including material traditions. Second, we predicted that candidate behaviors for traditions would likely incorporate behaviors related to community identity, such as greeting and marking behaviors.

Results

A survey list of 62 behaviors (Table S1) was compiled and used by the authors to document the occurrence and prevalence of each behavior within each monkey community at the five long-term field sites. This method allowed for a comparison of behavioral variance across sites while minimizing ecological and genetic differences (see Methods). In keeping with the original methodology used by Whiten et al. [3,32], and subsequently followed by Panger et al. [33] and van Schaik et al. [6], each behavior was classified into one of the following categories: customary, habitual, present, absent, ecological explanation and unknown (see Method for definitions).

Behaviors that failed criteria for traditions

Forty of the proposed 62 behavior variants failed to meet the necessary criteria for traditions [3] for four reasons presented in bands A–D of Table S2. Ten of these behaviors were absent across all five sites (band A, Table S2). Three such behaviors were related to the consumption of non-vegetative matter and were included in the original questionnaire as other *Ateles* species consume them [39,45]. The remaining seven behavior variants in this band were included in the survey list as they occurred in at least one site, but did not meet the 'present' criteria.

Six behavior variants were absent from the majority of sites, but clearly present at one or two, although not to the extent of being habitual or customary (band B, Table S2). It is possible that these behaviors are examples of current innovations at these sites; however, before social transmission can be inferred it seems reasonable that more than two individuals are required to exhibit such behaviors [33]. Four further behaviors were shown to be habitual or customary at some sites, but their absence at the other sites could be explained by ecological factors, or the existence of substrates used to perform the behavior at a site was unknown (band C, Table S2). Although social learning of these four behaviors cannot be ruled out, the currently available data are inconclusive for their inclusion as traditions. For example, one of these four behaviors, 'raiding', involves males walking on the ground single file in silence into the territory of neighboring communities [50]. Raiding has been observed by all males of the Eastern community multiple times at the Punta Laguna site and therefore deemed customary. It is unclear whether raiding occurs at the Corcovado site because although the subjects' actions were similarly described, no inter-community encounter was observed, and it is unknown how deep into the neighboring territory these incursions were. More importantly, as raiding likely occurs as a response to key socioecological conditions, such as reduced mating opportunities and strong male-male coalitions [50], its absence at other sites could be due to these conditions not being met, rather than an absence due to lack of social transmission. Thus, without clear evidence of how other communities respond to similar socioecological conditions, it would be premature to categorize raiding as a tradition.

Twenty behavior variants were observed across all sites (i.e., were 'universal' behaviors [3,32]), although with differing degrees of prevalence among community members (band D, Table S2). These universal behaviors included a number of greetings, which are characteristic of spider monkey repertoires [19]. In addition, there were also a number of behaviors used for threats or aggressive escalation. As Whiten et al. [32] suggested from observing similar behavior patterns across chimpanzee communities, there is no way of knowing if these are genetically-based species-specific behaviors or traditions that have arisen independently at each site. They could be examples of traditions that have become homogenized within communities due to conformity and led to reduced inter-community variation [51,52]. However, the ethnographic record cannot establish the origin of this pattern and, as a consequence of the absence of variation across sites, there was a lack of direct evidence for traditions.

Behaviors that met criteria for traditions

The remaining 22 behaviors showed variation in their occurrence across the study sites ranging from absent through to customary with absence in at least one site not due to an ecological explanation (Table 1; Figure 1). These patterns of occurrence across sites provide evidence that these behaviors are not species-specific or absent due to ecological reasons, and they best fit the criteria for traditions. The number of traditions was slightly greater within the three Southern sites than the two Northern sites. Spider monkeys at the Northern sites of Runaway Creek and Punta Laguna showed six and seven traditions respectively, whereas individuals at the Southern sites of Barro Colorado and Corcovado showed nine each, and Santa Rosa the most with thirteen.

Three of these behaviors were variants of substrate marking, which provide delayed olfactory information to conspecifics: 'chest rub', 'ano-genital rub', and 'rub with *Ficus* root'. One was a greeting variant, 'kiss'. Two behaviors were variants of aggressive behavior: 'false branch shake' and 'overlording'. There were two variants of rare affiliative behavior, 'interspecies grooming' and 'play on the ground', and one variant of a locomotive behavior, 'special bridging for an infant'. There were also seven variants related to food consumption choices. The remaining six variants included four drinking techniques, 'bipedal locomotion' and a potential thermoregulatory behavior, 'wind catching'. Very few traits showed a similar distribution across multiple sites; however, wind catching and the consumption of rocks, mushrooms, and *Phoradendron* leaves all reached a habitual level at Santa Rosa and were absent from the other sites.

Table 1. Prevalence of the 22 behaviors which met criteria for traditions across the five sites.

Domain	Behavior†	Prevalence				
		BCI	CV	SR	RC	PL
Social	Chest rub	H	H	C	H	A
	Ano-genital rub	H	A	H	H	H
	Rub with *Ficus* root	A	A	H	P	A
	Kiss	A	H	C	P	H
	False branch shake	H	C	P	H	A
	Overlording	A	C	H	H	P
	Interspecies grooming	A	H	H	A	A
	Play on ground	H	H	A*	A	P
	Special bridging for an infant	H	P	H	A	H
Food	Consume *Enterolobium cyclocarpum* fruit	A	C	A	A	C
	Consume *Anacardium excelsum* pith	A	C	A	E	E
	Consume *Phoradendron* leaves	A	A	H	A	E
	Consume *Phoradendron* fruit	A	A	A	H	E
	Consume caterpillars/larvae	H	A	H	A	P
	Consume mushrooms	A	A	H	A	A
	Consume rocks	A	A	H	A	A
Other	Drink from ground waterhole/lake	A	E	P	A	C
	Drink using head only	+	+	H	A	C
	Dribble water into mouth	H	C	A	C	H
	Lick water off palm	H	A	A	A	A
	Bipedal walking	C	P	P	A	P
	Wind catching	A	A	H	A	A

C = customary; H = habitual; P = present; A = absent; E = ecological explanation; + = behavior occurs but detailed information was not collected.
†For full explanation of behaviors see Table S1 in supporting information.
*play on the ground was observed with capuchin monkeys.
BCI = Barro Colorado Island, Panama; CV = Corcovado National Park, Costa Rica; SR = Santa Rosa National Park, Costa Rica; RC = Runaway Creek Nature Reserve, Belize; PL = Punta Laguna Reserve, Mexico.

Tradition domains

The occurrence of the 22 identified traditions varied across sites (Table 1). On average 57% of the identified traditions were in the social domain (Table 2). The observed bias of traditions toward the social domain is not surprising given the relative prevalence of social behaviors in the spider monkey repertoire, reflected by over half (53%) of the 62 behavior variants examined in our survey belonging to the social domain. However, this bias is still relevant from a comparative perspective when evaluating the relative occurrence of traditions in previous primate studies, where, unlike for spider monkeys, the majority belonged to the food-related domain [3,6]. When the percentage of traditions in the social domain was calculated out of the identified number of traditions at each site, it ranged from 43% at Punta Laguna to 67% at Corcovado and Runaway Creek (Figure 2). Similar classifications across chimpanzee and orangutan (*Pongo* spp.) study sites further highlights species differences in the distribution of traditions across domains. The mean percentage of traditions in the social domain across the nine chimpanzee study sites and across the six orangutan study sites was 42% and 34% respectively, which is lower than the mean value across the five spider monkey sites (Table 2). There was, however, high variability especially across chimpanzee sites with the percentage of traditions in the social domain ranging from 0% to 64%.

No traditions relating to aggressive interactions were recorded at the Punta Laguna site, while no traditions relating to affiliative behaviors were found at the Runaway Creek site (Figure 3). Behaviors relating to locomotion did not meet the criteria for tradition at the Corcovado and Runaway Creek sites. Candidate behaviors for traditions related to feeding, drinking and substrate marking were present across all five sites.

Geographic distribution

Geographic distances between study sites range from 350 km between Runaway Creek and Punta Laguna to 2,010 km between Barro Colorado Island and Punta Laguna (Table S3). There was no significant correlation between the distance and the number of habitual or customary behaviors [r (10) = 0.04, p = 0.914], or the number of absent behaviors shared between each pair of sites [r (10) = 0.311, p = 0.282] (Figure 4).

Discussion

We documented behavioral variation across five populations of spider monkeys providing evidence for traditions, which is relevant for a better understanding of human cultural evolution [53]. Our findings place *A. geoffroyi* alongside other species for which multiple traditions have been documented, such as humans, chimpanzees, orangutans, capuchin monkeys, bottlenose dolphins (*Tursiops* spp.), killer whales (*Orcinus orca*) and guppies (*Poecilia reticulata*) [3–

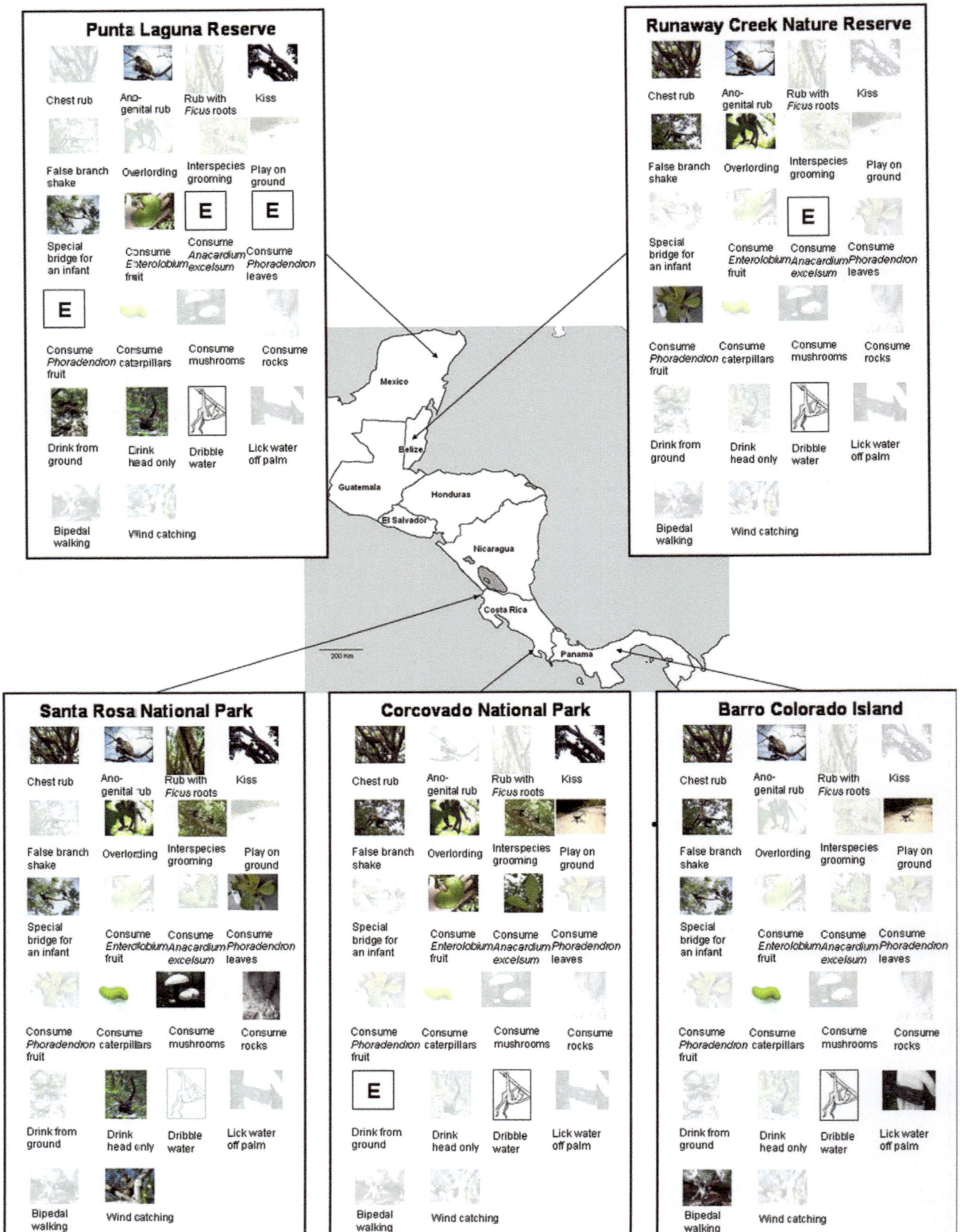
Punta Laguna Reserve
Chest rub
Ano-genital rub
Rub with Ficus roots
Kiss
False branch shake
Overlording
Interspecies grooming
Play on ground
Special bridge for an infant
Consume Enterolobium fruit
E
Consume Anacardium excelsum
E
Consume Phoradendron leaves
E
Consume Phoradendron fruit
Consume caterpillars
Consume mushrooms
Consume rocks
Drink from ground
Drink head only
Dribble water
Lick water off palm
Bipedal walking
Wind catching
Runaway Creek Nature Reserve
Chest rub
Ano-genital rub
Rub with Ficus roots
Kiss
False branch shake
Overlording
Interspecies grooming
Play on ground
Special bridge for an infant
Consume Enterolobium fruit
E
Consume Anacardium excelsum
Consume Phoradendron leaves
Consume Phoradendron fruit
Consume caterpillars
Consume mushrooms
Consume rocks
Drink from ground
Drink head only
Dribble water
Lick water off palm
Bipedal walking
Wind catching
Mexico
Belize
Guatemala
Honduras
El Salvador
Nicaragua
Costa Rica
Panama
200 Km
Santa Rosa National Park
Chest rub
Ano-genital rub
Rub with Ficus roots
Kiss
False branch shake
Overlording
Interspecies grooming
Play on ground
Special bridge for an infant
Consume Enterolobium fruit
Consume Anacardium excelsum
Consume Phoradendron leaves
Consume Phoradendron fruit
Consume caterpillars
Consume mushrooms
Consume rocks
Drink from ground
Drink head only
Dribble water
Lick water off palm
Bipedal walking
Wind catching
Corcovado National Park
Chest rub
Ano-genital rub
Rub with Ficus roots
Kiss
False branch shake
Overlording
Interspecies grooming
Play on ground
Special bridge for an infant
Consume Enterolobium fruit
Consume Anacardium excelsum
Consume Phoradendron leaves
Consume Phoradendron fruit
Consume caterpillars
Consume mushrooms
Consume rocks
E
Drink from ground
Drink head only
Dribble water
Lick water off palm
Bipedal walking
Wind catching
Barro Colorado Island
Chest rub
Ano-genital rub
Rub with Ficus roots
Kiss
False branch shake
Overlording
Interspecies grooming
Play on ground
Special bridge for an infant
Consume Enterolobium fruit
Consume Anacardium excelsum
Consume Phoradendron leaves
Consume Phoradendron fruit
Consume caterpillars
Consume mushrooms
Consume rocks
Drink from ground
Drink head only
Dribble water
Lick water off palm
Bipedal walking
Wind catching

Figure 1. Distribution of traditions observed across the five study sites. The photograph of a behavior indicates its presence at the customary or habitual level at the field site. A faded photograph of a behavior indicates its absence or occurrence only at the present level at the field site. A missing photograph with an 'E' indicates the absence of the behavior at a field site due to ecological reasons (Photographs by Claire J. Santorelli and Luisa Rebecchini and drawing by Claire J. Santorelli).

6,33,35,54]. In addition, the findings supported our two predictions. The identified traditions were more prevalent in the social domain than in other domains and included key behaviors for community identity.

Characteristics of spider monkey traditions

A larger proportion of the 22 identified traditions belonged to the social domain. This may partially be a reflection of the species' behavioral repertoire, which includes a large number of social behaviors [39] and relatively few food processing behaviors due to their limited dexterous manipulation [42]. The overall bias of spider monkey behavior toward the social domain is also confirmed by the independent observations of each research team at the five field sites, which contributed to the selection of the 62 behaviors as appropriate candidates for traditions, over half of which belonged to the social domain.

Watson and Caldwell [55] pointed out that a large number of tradition studies have focused on food-related behaviors, either food-consumption processes during cross-site comparisons or food-rewarded behaviors in experimental procedures. This is in contrast with the bias of traditions toward the social domain in spider monkeys, which does not match the pattern reported in similar studies on orangutans and chimpanzees in which the majority belonged to the food-related domain. The 22 traditions documented here are similar in number to the 24 documented across six orangutan sites [6], but considerably fewer than the 39 documented across nine chimpanzee sites [3,32]. It is interesting to note that the number of traditions documented per number of examined sites was relatively consistent across the three studies: 4.3 (39/9) for chimpanzees, 4 (24/6) for orangutans, and 4.4 (22/5) for spider monkeys. Thus, the inclusion of additional sites in a study seems to yield further cases of traditions, reflecting the unique repertoire of each community. A direct comparison with capuchin monkey cross-site data was not possible as separate studies focused on foraging and a sub-set of social interactions [5,33] but it is probable that a pattern similar to that of the two great apes may apply to capuchin monkeys given that 20 variants in food processing techniques were reported across three sites [33]. Another contrast with these studies is that none of the seven food-related traditions of spider monkeys were food processing techniques.

Social behaviors have, by their very nature, a shared and often public quality, which is not necessarily the case for food-related or other subsistence-based behaviors as they are not reliant on the presence of at least one other individual to be performed [5]. This characteristic may facilitate social learning and, as a consequence, the development of traditions in the social domain. In addition, the social demands faced by species with a high degree of fission-fusion dynamics may place a greater emphasis on the functional importance of social behavior variants that are linked to group or community identity [23,56,57]. Support for this prediction was found as the greeting variant 'kiss' was identified among the traditions in our study. Additionally, variants of olfactory communication in the form of three marking behaviors were identified as traditions, further illustrating how behaviors incorporating signals of community identity might be used to convey information between frequently dispersed individuals (i.e., when in different subgroups). In some cases, such as 'rub with *Ficus* root' performed by multiple individuals only in the Santa Rosa community, the selection for community identity through this marking variant may have led to a form of ritual, where most subgroup members are simultaneously involved. There was variation in the percentage of traditions in the social domain (43–67%) across the five study sites. Higher variation (0–64%) occurred across chimpanzee communities. It would be interesting to assess whether such variation is associated with variation in the degree of fission-fusion dynamics across populations. Additionally, a within-population examination of behavioral variation, where genetic and ecological differences between communities are likely to be negligible, would provide more detailed evidence for such social learning opportunities and potential traditions (Santorelli et al. *in prep.*).

The community identity hypothesis to partially explain the bias of traditions toward the social domain in spider monkeys does not however account for the large number of greeting behaviors that did not meet the criteria for traditions and were listed as 'universal'. It is noteworthy that several of these greeting variants may involve high risk, such as embraces and pectoral sniffs, which involve close body-contact and leave the recipient vulnerable to bites [27]. It is possible that the performance of these behaviors partly functions to test and strengthen relationships between community members [5]. This is in contrast to the relatively low risk 'kiss', reported to have reached habitual or customary status at three of the study sites, which involves less intimate contact with another individual. It is therefore possible that any variation in a high-risk greeting behavior could cause confusion, with potentially injurious consequences, especially when immigrating individuals are in the process of integrating into a new community. Innovation and dissemination of variants of low-risk social behaviors may be less problematic than innovation and dissemination of high-risk

Table 2. Mean (±SE) percentages of traditions across field sites in different domains for three primate species identified through cross-site surveys.

	Chimpanzee[3]	Orangutan[6]	Spider monkey*
Food-related domain	45% (±9.88)	41% (±5.66)	19% (±3.46)
Social domain	42% (±6.59)	34% (±9.28)	57% (±4.47)
Other domain	13% (±4.37)	25% (±11.50)	24% (±6.07)

*This study.

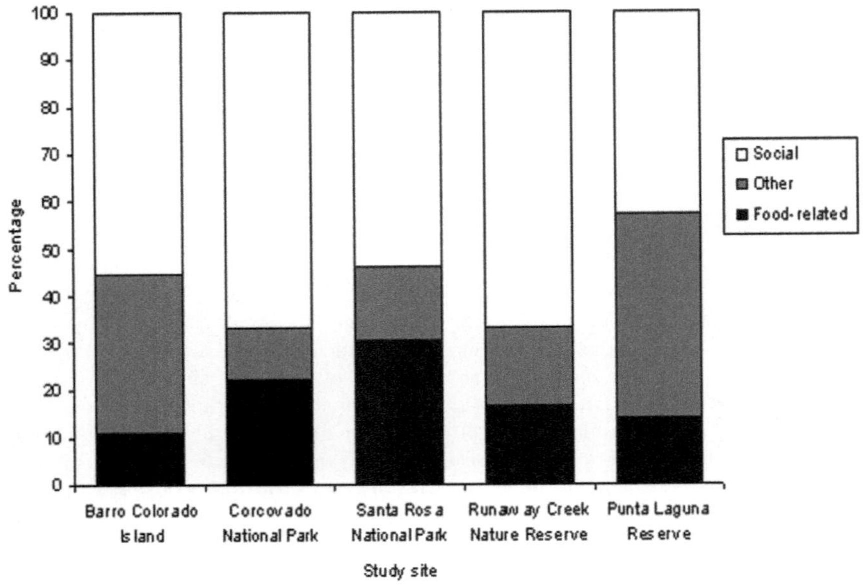

Figure 2. Percentage of behaviors, which met criteria for traditions, belonging to the three domain categories (social, food-related and other) at each site. See Table S1 for the division of the 22 traditions into the three domains and Figure 1 for the traditions at each site.

social behaviors. Thus, it might be expected that for species with a high degree of fission-fusion dynamics, for whom rapid community identity is particularly valuable, the emergence of traditions within their repertoire depends on this risk-based distinction.

Geographic distribution pattern

Genetic variation across the site populations invariably exists. Whilst it is impossible to eliminate genetic variation in wild populations, it was minimized in this study by only examining individuals of one of the four *Ateles* species, *A. geoffroyi* [58] (see Methods). Across large geographic spaces, it is likely that inter-community genetic variation would be greatest between communities that were more geographically distant from one another. This might lead to the expectation that if genetic differences alone are responsible for explaining behavioral variation across sites, patterns of shared traditions would diminish the further apart the communities were, yet we found that geographic distance did not correlate with number of shared traditions. The failure to find a correlation, while providing no evidence of a link between genetic variation and behavioral variation, does not rule out behavioral variation due to innovation and transmission by social learning. Given that there are features of transmission processes that might affect the dispersal of socially learned behaviors between populations (i.e. immigrants as poor demonstrators of a behavior, or the transmission of a behavior performed by peers), which do not affect behaviors based on a proximate genetic cause.

The geographic distribution of traditions may reflect patterns of innovation, diffusion and transmission, which can be affected by factors such as the dispersal of individuals between communities and restrictions imposed by geographical features [6]. A loss of knowledgeable individuals, through habitat loss or hunting pressures may also affect the distribution of reported absent behaviors at a particular site over time [59]. Accordingly, a positive correlation between geographic distance and cultural difference (i.e., the percentage of shared customary and habitual variants) was found across six orangutan field sites [6]. We found no such correlation across our five field sites. Van Schaik et al. [6] suggested that the possible cause of such a correlation was a result of emigrating orangutans spreading new variants easily as they move from a site of origin to new localities. In contrast, the lack of a correlation in spider monkeys suggests strong conformity of behaviors within populations and a low likelihood of immigrants spreading new variants, as was also suggested for chimpanzee traditions [32]. Patterns of similar behavior variants are likely to emerge if behavior variants either, originate independently at multiple sites, or are introduced via immigration and then consequently spread when the costs of acquiring a particular new behavior are low [32]. However, migration does not always result in the transmission of socially learned information. In some species, such as vervet monkeys (*Chlorocebus aethiops*), individuals of the philopatric sex are preferred demonstrators of behavior than individuals of the migrating sex, and as a result of this selection highly localized traditions can emerge [60]. The social model hypothesis, which predicts primates living in structured social groups are most likely to pick knowledgeable, older or high ranking group members as demonstrators of a behavior [61], might explain such motivation. If this is the case for spider monkeys, emigrating females are unlikely to be chosen as behavior modelers and would be unlikely to transmit novel behavior variants. Although spider monkeys do not live in rigidly hierarchical social groups but in more socially tolerant communities [19], the pattern of traditions reported here similarly suggests that conformity for community specific behaviors maintain variants and, over large distances migrating females may be poor dispersers of behavior variants. The average migration distance for an emigrating female spider monkey remains unknown, but is thought to be a considerable distance (i.e., greater than four neighboring communities away [62]). Not knowing how many communities an emigrating individual passes by before settling makes it difficult to predict how closely patterns of geographic variation in traditions might reflect patterns of dispersing individuals. Consequently, limited dispersion of behavior variants between sites is more likely to result in the maintenance of site-specific behavior patterns. Finally, the lack

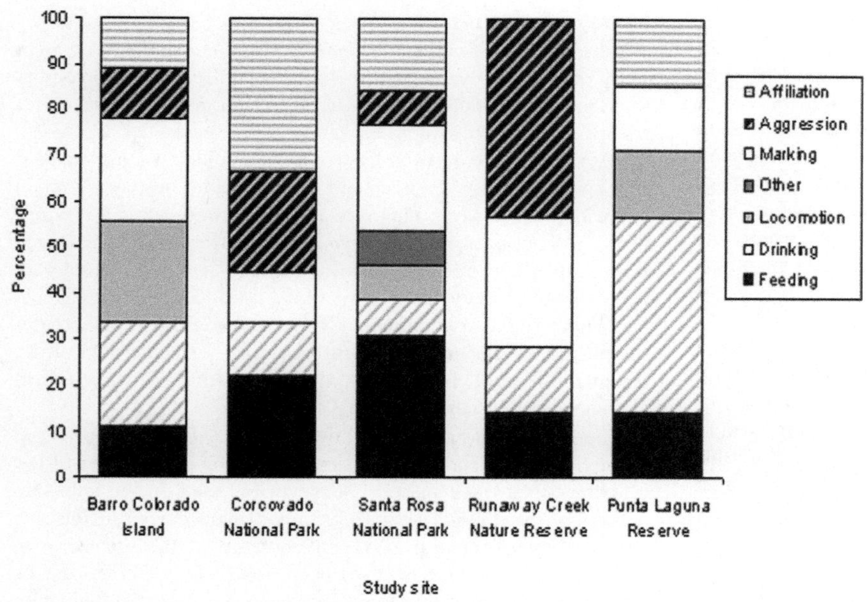

Figure 3. Percentage of behaviors showing evidence of tradition at each site by sub-domain category. See Table S1 for the division of behaviors into sub-domains and Figure 1 for the traditions at each site. Note that the only behavior in the sub-domain 'Other' was wind catching at the Santa Rosa site.

of continuous forest across Central America limits opportunities for wide-scale dispersion of individuals and, therefore, behavior variants between communities [63,64].

Whiten [65] suggested that cases where behavioral variants identified in one or more sites are common, but absent in at least one other site, imply that animals at the latter site are at a disadvantage. Potential self-medicating traditions may offer opportunities for case studies. For example, further cross-site research can help identify whether *Phoradendron* leaf or rock feeding conveys a selective advantage for spider monkeys in communities where it is practiced over individuals in communities where it is not practiced. The consumption of the widely available *Enterolobium cyclocarpum* fruit by individuals only in Corcovado and Punta Laguna suggests another example of a feeding variant which conveys nutritional advantages for individuals that feed upon it over individuals that do not, although, it is always possible that other communities compensate by eating another food resource [66].

Site specific consumption of widely available food resources illustrates how ecological conditions may impact to promote or hinder innovation, or the subsequent maintenance of socially transmitted behaviors. Opportunities to innovate may be influenced by the accessibility of associated substrates or social resources which may be seasonal, rare or highly unpredictable in their availability [67,68]. For example, the fruiting cycle of *Enterolobium cyclocarpum* can be unpredictable [69]. During years

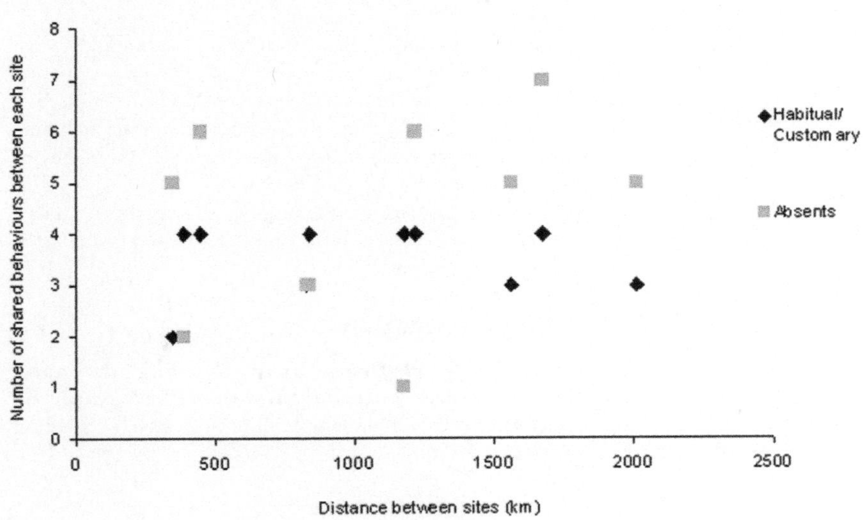

Figure 4. Association between the number of customary/habitual behaviors or absence of behavior each pair of study sites shared and the distance between each pair of sites.

when this tree does not fruit, individuals have no opportunity to innovate food processing techniques, or socially learn how to consume it. Similarly, the adaptive value of a variant or risks associated with exploration may affect innovation or social learning opportunities for some individuals more than others [68]. Therefore, despite occurrences in which social learning may account for behavioral variation across communities, the subtle interactions of ecology and personal genetic predisposition may still affect individuals' likelihood for innovation and transmission processes, contributing in part, to the establishment of traditions [67].

In other species, captive and field based experiments have been instrumental in complementing findings from cross-site comparison studies and are invaluable for exploring social learning mechanisms and transmission processes [15,70–77]. The use of spider monkeys as a focal species for similar experiments would help provide evidence for the social learning mechanisms commonly used by this species and may explain the differential development of traditions across sites.

Materials and Methods

Ethics statement

The study was carried out in the field with free-ranging monkeys and was completely observational. Research was conducted at all times in accordance with the laws of participating countries. Approval and permission to conduct research was granted by the University of California IACUC committee, the Animal Studies Committee of Washington University # 20020071, the University of Chester Psychology Department Ethics Committee and approved by the University of Chester Animal Ethics Committee, the Animal Care Certification in compliance with the Canadian Council on Animal Care, the Costa Rica Ministry of Environment and Energy (MINAE) permit #s 418-2001-OFAU, 226-2002-OFAU and ACG-PL-030-2006, the Belize Forest Department permit # CD/60/3/09(05) and the Mexican government under the auspices of Pronatura, Peninsula de Yucatan, A.C. (PPY) # 1577105.

Study site selection

Only field sites where research on spider monkeys of the species *A. geoffroyi* was carried out were considered in our study, in order to minimize genetic influences on any behavioral variations observed. Five sites were selected for the study (Figure S2 and Table S4), which met the following two criteria: 1) behavioral data were collected for a minimum of 12 months, in order to have a reasonable amount of observation time to document behavioral variations; and 2) the monkeys were individually recognized, so that assessment of whether individuals engaged in behavioral variants multiple times could be made, allowing for the categorization of each behavior into categories based on its prevalence at each site.

The five sites included in the survey were Barro Colorado Island, Panama (hereafter Barro Colorado [78,79]); Corcovado National Park, Costa Rica (hereafter Corcovado [80]); Santa Rosa National Park, Costa Rica (hereafter Santa Rosa [66,81,82]); Runaway Creek Nature Reserve, Belize (hereafter Runaway Creek; Pavelka & Notman, unpublished data); and Otoch Ma'ax Yetel Kooh Reserve, Mexico, also known as Punta Laguna Reserve (hereafter Punta Laguna [83,84]).

Data on two monkey communities were available at each of three sites: Corcovado, Runaway Creek and Punta Laguna. Since the aim of the survey was to examine behavioral variation across a large geographical area, responses from the two communities at each of these sites were merged.

Survey procedure

An initial list of candidate behaviors was collated from a pilot study carried out over a two year period on the spider monkeys at the Santa Rosa and Punta Laguna field sites, as well as behaviors reported from the literature on various *Ateles* species (31,39,45–49). Then, the list was reviewed by researchers at all participating field sites and care was taken to ensure researchers accurately identified behaviors across sites. This was achieved using detailed descriptions, photographs and video clips to clarify behaviors nuances. Based on the joint feedback, additions or consolidations of behaviors were made, leading to a final list of 62 behaviors (Table S1). Survey data were compiled from data originally collected for the purpose of various behavioral studies by retrieving them from systematic records. In addition, all researchers used detailed field notes to identify patterns of rare behaviors.

The survey consisted of two phases. Phase I required researchers to document the presence or absence of each of the 62 behaviors at their field site. Categories based on the following definitions were used: *present* – behavior has occurred at the site; *absent* – behavior has never been observed at the site; *ecological explanation* - behavior has never been observed at the site but its absence is explicable by site ecology (e.g., if a particular substrate was not present at the field site, thereby removing the opportunity for behaviors associated with that substrate to occur); and *unknown* - insufficient opportunity to observe a behavior to reliably know if it was present or absent. This last category was especially relevant for behaviors that require rare conditions or might be less likely to occur in the presence of observers, despite habituation.

Phase II required researchers to classify each observed behavior at their field site using one of the categories based on the following definitions derived from Whiten et al. [3], depending on how often and by whom the behavior was performed: *customary* - behavior occurs in all or most able-bodied members of at least one age-sex class (e.g. all adult males); *habitual* - behavior is not customary but has occurred repeatedly in several individuals, consistent with some degree of social transmission; and *present* - behavior is neither customary nor habitual, but is performed multiple times by at least two individuals. Thus, in phase II performance of a behavior by only one individual at one site was classified as 'absent'. A behavior variant was considered to be a tradition when it occurred at a habitual or customary level in at least one site while being absent in at least one other site without an ecological explanation [3,32]. For the three sites where data from two communities were collected, the more prevalent occurrence of a behavior in each community was used for the overall site record. For example, if a behavior was 'present' within one community and 'habitual' within the other community at the same field site, the behavior was recorded as 'habitual' for that field site.

Pearson correlations between the number of shared customary/habitual or absent behaviors and the distance between each pair of sites were run using SPSS v.15.0.

Supporting Information

Figure S1 *Ateles geoffroyi* hand showing dramatic reduction in pollex (external thumb) (Photograph by Claire J. Santorelli). Photograph illustrates area of reduced pollex on the left hand.

Figure S2 Map of Central America showing locations of the five field sites participating in the study. Arrows illustrate location of participating field sites within their host country.

Table S1 Definitions and domains of the 62 behaviors considered in the study.

Table S2 Prevalence of behavior variants across study sites.

Table S3 Distance (kilometers) between sites (using Google Earth ruler, http://earth.google.com).

Table S4 Site information.

Acknowledgments

Authors would like to thank Antigone Thomas and Thomas A. Wake (Barro Colorado Island); Sabrina Adleman, Delanie Hurst, Robert Lessnau and his team of assistants, Andres Vega, Larry Gilbert, Erika Deinert, Reinaldo Aguilar, the administration of ACOSA and MINAE, particularly Javier Guevara and Eliecer Arce, and Paulino Valverde and the staff of Sirena Biological Station, Corcovado National Park (Corcovado); Elvin Murillo, Roberto Espinoza, Roger Blanco and María Marta Chavarría at Area Conservation Guanacaste (Santa Rosa); Athabasca University, Kayla Hartwell, Brittany Dean and Kayley Evans (Runaway Creek); Eulogio Canul-Aban, Macedonio Canul-Chan, Juan Canul-Chan, Augusto Canul-Aban, Gabriel Ramos-Fernández, Laura Vick, Luisa Rebecchini, Kathy Slater, and staff at Pronatura (Punta Laguna).

Author Contributions

Conceived and designed the experiments: CJS CMS FA. Performed the experiments: CJS CJC CMS FA HN MSP JAW. Analyzed the data: CJS CMS FA. Wrote the paper: CJS CJC CMS FA HN MSP JAW.

References

1. Fragaszy DM, Perry S (2003) Towards a biology of traditions. In: Fragaszy DM, Perry S, eds. The biology of traditions: Models and evidence. Cambridge: Cambridge University Press. pp 1–32.
2. McGrew WC (1998) Culture in nonhuman primates? Annual Review of Anthropology 27: 301–328.
3. Whiten A, Goodall J, McGrew WC, Nishida T, Reynolds V, et al. (1999) Cultures in chimpanzees. Nature 399: 682–685.
4. Rendell L, Whitehead H (2001) Culture in whales and dolphins. Behavioral and Brain Sciences 24: 309–382.
5. Perry S, Baker M, Fedigan L, Gros-Louis J, Jack K, et al. (2003) Social conventions in wild White-faced Capuchin monkeys: Evidence for traditions in a neotropical primate. Current Anthropology 44: 241–268.
6. van Schaik C, Ancrenaz M, Borgen G, Galdikas B, Knott CD, et al. (2003) Orangutan cultures and the evolution of material culture. Science 299: 102–105.
7. Krützen M, Mann J, Heithaus MR, Connor RC, Bejder L, et al. (2005) Cultural transmission of tool use in bottlenose dolphins. Proceeding of the National Academy of Sciences 102: 8939–8943.
8. Krützen M, van Schaik CR, Whiten A (2006) The animal cultures debate: response to Laland and Janik. TRENDS in Ecology and Evolution 22: 6.
9. van der Post DJ, Hogeweg P (2009) Cultural inheritance and diversification of diet in variable environments. Animal Behaviour 78: 155–166.
10. Laland KN (2002) Culture in nonhuman organisms. In: Smelser NJ, Baltes PB, eds. International Encyclopedia of the Social & Behavioral Sciences. Oxford: Elsevier. pp 3164–3167.
11. Boesch C (2003) Is culture a golden barrier between human and chimpanzee? Evolutionary Anthropology 12: 82–91.
12. Laland KN, Hoppitt W (2003) Do animals have culture? Evolutionary Anthropology 12: 150–159.
13. Galef BG, Laland KN (2005) Social learning in animals: Empirical studies and theoretical models. BioScience 55: 489–499.
14. Borenstein E, Feldman MW, Aoki K (2008) Evolution of learning in fluctuating environments: When selection favors both social and exploratory individual learning. Evolution 62: 586–602.
15. Laland KN, Janik VM (2006) The animal cultures debate. Trends in Ecology and Evolution 21: 542–547.
16. Byrne RW (2007) Culture in great apes: using intricate complexity in feeding skills to trace the evolutionary origin of human technical prowess. Philosophical Transactions of the Royal Society B 362: 577–585.
17. Laland KN (2008) Animal cultures. Current Biology 18: 366–370.
18. Vick L (2008) Immaturity in spider monkeys: A risky business. In: Campbell CJ, ed. Spider monkeys: Behavior, ecology and evolution of the genus *Ateles*. Cambridge: Cambridge University Press. pp 288–328.
19. Aureli F, Schaffner CM (2008) Social interactions, social relationships and the social system of spider monkeys. In: Campbell CJ, ed. Spider monkeys: Behavior, ecology and evolution of the genus Ateles. Cambridge: Cambridge University Press. pp 236–265.
20. Coussi-Korbel S, Fragaszy DM (1995) On the relation between social dynamics and social learning. Animal Behaviour 50: 1441–1453.
21. van Schaik C, Deaner RO, Merrill MY (1999) The conditions for tool use in primates: implications for the evolution of material culture. Journal of Human Evolution 36: 719–741.
22. Nishida T, Hiraiwa-Hasegawa M (1987) Chimpanzees and Bonobos: Cooperative relationships among males. In: Smuts BB, Cheney DL, Seyfarth RM, Wrangham RW, Struhsaker TT, eds. Primate Societies. Chicago: University of Chicago Press. pp 165–177.
23. Rodseth L, Wrangham RW, Harrigan AM, Smuts BB (1991) The human community as a primate society. Current Anthropology 32: 221–254.
24. Aureli F, Schaffner CM, Boesch C, Bearder SK, Call J, et al. (2008) Fission-fusion dynamics: New research frameworks. Current Anthropology 49: 627–654.
25. Milton K (2000) Quo vadis? Tactics of food search and group movement in primates and other animals. In: Boinski S, Garber PA, eds. One the move: How and why animals travel in groups. Chicago: University of Chicago Press. pp 375–417.
26. Milton K (2006) Analyzing nutritional ecology: Introduction. In: Hohmann G, Robbins, MM, Boesch C, eds. Feeding ecology in apes and other primates. Ecological, physical and behavioral aspects. Cambridge: Cambridge University Press. pp 381–396.
27. Schaffner CM, Aureli F (2005) Embraces and grooming in captive spider monkeys. International Journal of Primatology 26: 1093–1106.
28. Aureli F, Schaffner CM (2007) Aggression and conflict management at fusion in spider monkeys. Biology Letters 3: 147–149.
29. Klein LL (1972) The ecology and social organization of the spider monkey, *Ateles belzebuth*. PhD. Thesis. Berkeley: University of California.
30. Campbell CJ (2000) Fur Rubbing Behavior in Free-Ranging Black-Handed Spider Monkeys (*Ateles geoffroyi*) in Panama. American Journal of Primatology 51: 205–208.
31. Heymann EW (2006) Scent marking strategies of New World primates. American Journal of Primatology 68: 650–661.
32. Whiten A, Goodall J, McGrew WC, Nishida T, Reynolds V, et al. (2001) Charting cultural variation in chimpanzees. Behaviour 138: 1481–1516.
33. Panger M, Perry S, Rose L, Gros-Louis J, Vogel E, et al. (2002) Cross-site difference in foraging behavior of white-face capuchins (*Cebus capucinus*). American Journal of Physical Anthropology 119: 52–66.
34. Leca J.-B, Gunst N, Huffman MA (2008) Food provisioning and stone handling tradition in Japanese Macaques: A comparative study of ten troops. American Journal of Primatology 70: 803–813.
35. Richerson PJ, Boyd R (2005) Not by genes alone: How culture transformed human evolution. Chicago: The University of Chicago Press.
36. McGrew WC (2004) The cultured chimpanzee: Reflections on cultural primatology. Cambridge: Cambridge University Press.
37. van Schaik C (2003) Local traditions in orangutans and chimpanzees: social learning and social tolerance. In: Fragaszy DM, Perry S, eds. The biology of traditions: Models and evidence. Cambridge: Cambridge University Press. pp 297–328.
38. McGrew WC (1992) Chimpanzee material culture: Implications for human evolution. Cambridge: Cambridge University Press.
39. van Roosmalen MGM, Klein LL (1988) The spider monkeys, genus *Ateles*. In: Mittermeier RA, Rylands AB, Coimbra-Filho AF, da Fonseca GBA, eds. Ecology and behavior of Neotropical primates. Washington DC: World Wide Fund for Nature. pp 455–537.
40. Di Fiore A, Link A, Dew JL (2008) Diets of wild spider monkeys. In: Campbell CJ, ed. Spider monkeys: Behavior, ecology and evolution of the Genus *Ateles*. Cambridge: Cambridge University Press. pp 81–137.
41. Chapman CA, Chapman LJ (1991) The foraging itinerary of spider monkeys: When to eat leaves? Folia Primatologica 56: 162–166.
42. Rosenberger AL, Halenar L, Cooke SB, Hartwig WC (2008) Morphology and evolution of the spider monkey, genus *Ateles*. In: Campbell CJ, ed. Spider monkeys: Behavior, ecology and evolution of the Genus *Ateles*. Cambridge: Cambridge University Press. pp 19–49.
43. Erikson GE (1963) Brachiation in New World monkeys and in anthropoid apes. Symposia of the Zoological Society of London 10: 135–164.
44. Aziz MA, Dunlap SS (1986) The human extensor digitorum profundus muscle with comments on the evolution of the primate hand. Primates 27: 293–319.

45. Link A (2003) Insect-eating by spider monkeys. Neotropical Primates 11: 104–107.
46. Pozo RWE (2004) Agrupación y dieta de Ateles belzebuth belzebuth en el Parque Nacional Yasuní, Ecuador. Anuar Investigación Científica 2: 77–102.
47. Laska M, Bauer V, Hernandez Salazar LT (2007) Self-anointing behavior in free-ranging spider monkeys (*Ateles geoffroyi*) in Mexico. Primates 48: 160–163.
48. Campbell CJ, Aureli F, Chapman CA, Matthews K, Ramos-Fernández G, et al. (2005) Terrestrial behavior of *Ateles* spp. International Journal of Primatology 26: 1039–1051.
49. Lindshield SM, Rodrigues MA (2009) Tool use in wild spider monkeys (*Ateles geoffroyi*). Primates 50: 269–272.
50. Aureli F, Schaffner CM, Verpooten J, Slater K, Ramos-Fernández G (2006) Raiding parties of male spider monkeys: Insights into Human warfare? American Journal of Physical Anthropology 131: 486–497.
51. Henrich J, Boyd R (1998) The evolution of conformist transmission and the emergence of between-group differences. Evolution and Human Behavior 19: 215–241.
52. Efferson C, Lalive R, Richerson PJ, McElreath R, Lubell M (2008) Conformists and mavericks: the empirics of frequency-dependent cultural transmission. Evolution and Human Behavior 29: 56–64.
53. Heyes CM, Galef BG (1996) Social learning in animals: The roots of culture. London: Academic Press.
54. Lachlan RF, Crooks L, Laland KN (1998) Who follows whom? Shoaling preferences and social learning of foraging information in guppies. Animal Behaviour 56: 181–190.
55. Watson CFI, Caldwell CA (2009) Understanding behavioral traditions in primates: Are current experimental approaches too focused on food? International Journal of Primatology 30: 143–167.
56. Boughman JW, Wilkinson GS (1998) Greater spear-nosed bats discriminate group mates by vocalizations. Animal Behaviour 55: 1717–1732.
57. Datta S, Sturtivant C (2002) Dolphin whistle classification for determining group identities. Signal Processing 82: 251–258.
58. Collins AC (2008) The taxonomic status of spider monkeys in the twenty-first century. In: Campbell, CJ, eds. Spider monkeys: Behavior, ecology and evolution of the Genus Ateles. Cambridge: Cambridge University Press. pp 50–78.
59. van Schaik C (2002) Fragility of traditions: the disturbance hypothesis for the loss of local traditions in Orangutans. International Journal of Primatology 23: 527–537.
60. van de Waal E, Renevey N, Favre CM, Bshary R (2010) Selective attention to philopatric models causes directed social learning in wild vervet monkeys. Proceedings of the Royal Society B 277: 2105–2111.
61. de Waal FBM (2001) The ape and the sushi master: cultural reflections of a primatologist. New York: Basic Books.
62. Shimooka Y, Campbell CJ, Di Fiore A, Felton AM, Izawa K, et al. (2008) Demography and group composition of Ateles. In: Campbell CJ, ed. Spider monkeys: Behavior, ecology and evolution of the Genus *Ateles*. Cambridge: Cambridge University Press. pp 329–348.
63. Estrada A, Coates-Estrada R (1988) Tropical rain forest conversion and perspectives in the conservation of wild primates (*Alouatta* and *Ateles*) in Mexico. American Journal of Primatology 14: 315–327.
64. Sader SA, Hayes DJ, Irwin DE, Saatchi SS (2001) Preliminary forest cover change estimates for Central America (1990's), with reference to the proposed Mesoamerican biological corridor. St LouisMissouri: ASPRS 2001 Annual meeting.
65. Whiten A (2006) The significance of socially transmitted information for nutrition and health in the great ape clade. In: Wells JCK, Strickland S, Laland KN, eds. Social information transmission and human biology. London: Taylor and Francis. pp 119–136.
66. Chapman CA (1987) Flexibility in diets in three species of Costa Rican primates. Folia Primatologica 49: 90–105.
67. van Schaik CP (2009) Geographic variation in the behaviour of wild great apes: Is it really cultural? In: Laland KN, Galef BG, eds. The question of animal culture. Cambridge: Harvard University Press. pp 70–98.
68. Humle T (2010) How are army ants shedding new light on culture in chimpanzees? In: Lonsdorf EV, Ross SR, Matsuzawa T, eds. The mind of the chimpanzee. Chicago: Chicago University Press. pp 116–126.
69. Rocha OJ, Aguilar G (2001) Reproductive biology of the dry forest tree *Enterolobium cyclocarpum* (Guanacaste) in Costa Rica: A comparison between the trees left in pastures and trees in continuous forest. American Journal of Botany 88: 1607–1614.
70. Brosnan SF, de Waal F (2004) Socially learned preferences for differentially rewarded tokens in the brown capuchin monkey (*Cebus apella*). Journal of Comparative Psychology 118: 133–139.
71. Galef BG (2004) Approaches to the study of traditional behaviors of free-living animals. Learning and Behavior 32: 53–61.
72. Horner V, Whiten A, Flynn E, de Waal FBM (2006) Faithful replication of foraging techniques along cultural transmission chains by chimpanzees and children. Proceeding of the National Academy of Sciences 103: 13878–13883.
73. Whiten A, Spiteri A, Horner V, Bonnie KE, Lambeth SP, et al. (2007) Transmission of multiple traditions within and between Chimpanzee groups. Current Biology 17: 1038–1043.
74. Dindo M, Thierry B, Whiten A (2008) Social diffusion of novel foraging methods in brown capuchin monkeys (*Cebus apella*). Proceedings of the Royal Society B 275: 187–193.
75. Humle T, Snowdon CT (2008) Socially biased learning in the acquisition of a complex foraging task in juvenile cottontop tamarins, *Saguinus oedipus*. Animal Behaviour 75: 267–277.
76. Thornton A (2008) Social learning about novel foods in young meerkats. Animal Behaviour 76: 1411–1421.
77. Whiten A, Mesoudi A (2008) Establishing an experimental science of culture: animal social diffusion experiments. Philosophical Transactions of the Royal Society B 363: 3477–3488.
78. Campbell CJ (2003) Female directed aggression in free-ranging *Ateles geoffroyi*. International Journal of Primatology 24: 223–238.
79. Milton K, Hopkins ME (2005) Growth of a reintroduced spider monkey population on Barro Colorado Island, Panama. In: Estrada A, Garber P, Pavelka M, Luecke L, eds. New Perspectives in the Study of Mesoamerican Primates: Distribution, Ecology, Behavior and Conservation. New York: Springer. pp 417–436.
80. Weghorst J (2007) High population density of black-handed spider monkeys (*Ateles geoffroyi*) in Costa Rican lowland wet forest. Primates 48: 108–116.
81. Chapman CA, Wrangham RW, Chapman LJ (1995) Ecological constraints on group size: an analysis of spider monkey and chimpanzee subgroups. Behavioral Ecology and Sociobiology 36: 59–70.
82. Asensio N, Korstjens A H, Aureli F (2009) Fissioning minimises ranging costs in spider monkeys: a multiple-level approach. Behavioral Ecology and Sociobiology 63: 649–654.
83. Ramos-Fernández G, Vick LG, Aureli F, Schaffner CM, Taub DM (2003) Behavioral ecology and conservation status of spider monkeys in the Otoch Ma'ax Yetel Kooh protected area. Neotropical Primates 11: 115–158.
84. Ramos-Fernández G, Ayala-Orozco B (2002) Population size and habitat use of spider monkeys in Punta Laguna, Mexico. In: Marsh LK, ed. Primates in Fragments: Ecology and Conservation. New York: Kluwer/Plenum Press. pp 191–209.

2

Towards Operational Modeling and Forecasting of the Iberian Shelves Ecosystem

Martinho Marta-Almeida[1]*, Rosa Reboreda[1], Carlos Rocha[1], Jesus Dubert[1], Rita Nolasco[1], Nuno Cordeiro[1], Tiago Luna[2], Alfredo Rocha[1], João D. Lencart e Silva[1], Henrique Queiroga[3], Alvaro Peliz[4], Manuel Ruiz-Villarreal[5]

1 CESAM-Dep Física, Universidade de Aveiro, Aveiro, Portugal, **2** Instituto do Ambiente e Desenvolvimento, Universidade de Aveiro, Aveiro, Portugal, **3** CESAM-Dep Biologia, Universidade de Aveiro, Aveiro, Portugal, **4** Centro de Oceanografia, Universidade de Lisboa, Lisboa, Portugal, **5** Instituto Español de Oceanografía, Centro Oceanográfico A Coruña, Galicia, Spain

Abstract

There is a growing interest on physical and biogeochemical oceanic hindcasts and forecasts from a wide range of users and businesses. In this contribution we present an operational biogeochemical forecast system for the Portuguese and Galician oceanographic regions, where atmospheric, hydrodynamic and biogeochemical variables are integrated. The ocean model ROMS, with a horizontal resolution of 3 km, is forced by the atmospheric model WRF and includes a Nutrients-Phytoplankton-Zooplankton-Detritus biogeochemical module (NPZD). In addition to oceanographic variables, the system predicts the concentration of nitrate, phytoplankton, zooplankton and detritus (mmol N m^{-3}). Model results are compared against radar currents and remote sensed SST and chlorophyll. Quantitative skill assessment during a summer upwelling period shows that our modelling system adequately represents the surface circulation over the shelf including the observed spatial variability and trends of temperature and chlorophyll concentration. Additionally, the skill assessment also shows some deficiencies like the overestimation of upwelling circulation and consequently, of the duration and intensity of the phytoplankton blooms. These and other departures from the observations are discussed, their origins identified and future improvements suggested. The forecast system is the first of its kind in the region and provides free online distribution of model input and output, as well as comparisons of model results with satellite imagery for qualitative operational assessment of model skill.

Editor: Juan A. Añel, University of Oxford, United Kingdom

Funding: The authors CR and NC are supported by project RAIA funded by Programa Operativo de Cooperacion Transfronteriza Espa~a-Portugal, POCTEP 2007–2013 (0313-RAIA-1-E). R.Reboreda is supported by the FCT grant SFRH/BD/33388/2008. The funders had no role in study design, data collection and analysis, decision to publish, or preparation of the manuscript.

Competing Interests: The authors have declared that no competing interests exist.

* E-mail: mma@ua.pt

Introduction

Providing operational oceanographic data on biological and chemical variables has become an issue of concern over the last years. The demand for this kind of information arises from a range of fields and applications such as scientific research on marine ecosystems, monitoring of seawater quality and decision-making support for marine and coastal management. A recent questionnaire conducted by ICES-WGOOFE (International Council for the Exploration of the Sea, Working Group on Operational Oceanographic Products for Fisheries and Environment) showed that temperature, currents, salinity, chlorophyll standing stock and primary production were the most requested products among ocean sciences scientific community, who scored several biological parameters in the top 10 rankings of products on demand [1]. There is a well known increasing interest in combined physical, chemical and biological operational products, including near-real time and forecast, which are currently possible due to facilitated access to computational resources, development of numerical solutions and improvement of modelling algorithms and performance.

The marine policies implemented in many countries to protect the sea from increasing environmental pressures has urged the need for monitoring systems of seawater quality. In the European Union, the Water Framework Directive and the Marine Strategy Framework Directive launched in 2000 and 2008, commit the member states to the evaluation and monitoring of the ecological status of their river basins/coastal waters and marine waters, respectively (e.g., [2,3]). The setting up of the European Earth Monitoring Programme GMES (Global Monitoring for Environment and Security) by the European Commission, which includes an important ocean component and aims to be fully operational by 2014, highlights the interest on reliable and up-to-date marine environmental information [4].

Biogeochemical models constitute a valuable tool for operational oceanography when coupled with circulation models. They can complement the time and space limitation of observations and offer the possibility to help explain processes and variability. The simplest version of these models are the NPZD (Nutrients-Phytoplankton-Zooplankton-Detritus) models. They can give information on concentration of the biological state variables over time, and have strong potentialities for analysis and prediction.

Nevertheless, modelling the ocean biogeochemical properties has a number of difficulties that make it a challenging task. The parameters used in biogeochemical models are often not well constrained in the literature and finding an appropriate set of parameters through trial runs is seldom straightforward. In addition, independently of the level of complexity of a biogeochemical model, it is still a great simplification of the reality.

The strong dependency of biology on hydrodynamics requires an ocean circulation model able to reproduce adequately the main features and variability in the modelling domain, at both seasonal and event time scales. In regional coastal domains, sufficiently high resolution is required to resolve the shelf mesoscale eddy processes, which depends upon high performance computational facilities.

Operational modelling products of ocean biogeochemical variables are rarely provided by online operational systems, presumable because of the difficulties mentioned above. In the western Iberia, an early experience of chlorophyll forecast in 1998 in the Gulf of Cadiz is referred in Pinard and Woods [5], Chap 11. The GMES ocean data server MyOcean (http://www.myocean.eu) provides modeled biogeochemical data for the European northwest shelf since April 2011, including the northernmost part of the Iberian maritime region. The model used has been implemented operationally over the last years in the seas around the UK and Ireland [6]. The Project EasyCO (model products in several European maritime regions, including the western Iberia) foresees to distribute biogeochemical data, although this type of information is currently available only for the western French coast. Other regions in Europe with operational modelling products for biology are the Baltic Sea, developed by the Baltic ocean community [7] and the Mediterranean Sea, developed by the Istituto Nazionale di Oceanografia e di Geofisica Sperimentale, OGS, in collaboration with other Italian Institutions (INGV, CINECA and GOS-ISAC-CNR) [8,9]. Both available through MyOcean.

In this contribution we present an operational biogeochemical forecast system implemented for the Atlantic Iberian shelf ecosystem: Portuguese and Galician shelf and slope and nearby oceanic regions. This product is the first of its kind in the region and provides free online access to all ocean state and biogeochemical simulated variables. The hydrodynamic forecast system is working since 1st November 2008 and the biogeochemical module is available operationally since 10th June 2011.

The next section describes the ocean model setup, the atmospheric model providing surface data for the ocean model, the biogeochemical model and the operational system setup. Then, results are presented and compared with observations. In the last section we present the summary and outlook of upcoming improvements.

Models

Ocean Model

The numerical model implemented is the Regional Ocean Modelling System, ROMS [10,11,12]. ROMS is a free-surface terrain-following primitive equation hydrostatic model with Boussinesq approximations and configurable for realistic applications. ROMS has been used in a variety of time and space scales, from very small regions, like harbours, to the coastal, large and global scale.

The model configuration in use has been successfully applied in the western Iberian region during roughly the last ten years, to many different studies, like dispersion of larvae [13,14,15], river plumes [16,17,18,19] and pollution transport [20,21]. The Gibraltar exchange flow and the large scale circulation associated with the Azores Current are included in the modelling configuration and were based on [22,23].

The ocean model uses two different domains, illustrated in Figure 1 as OCEAN-0 and OCEAN-1. The large domain, OCEAN-0, has horizontal resolution of 1/10°, able to resolve the

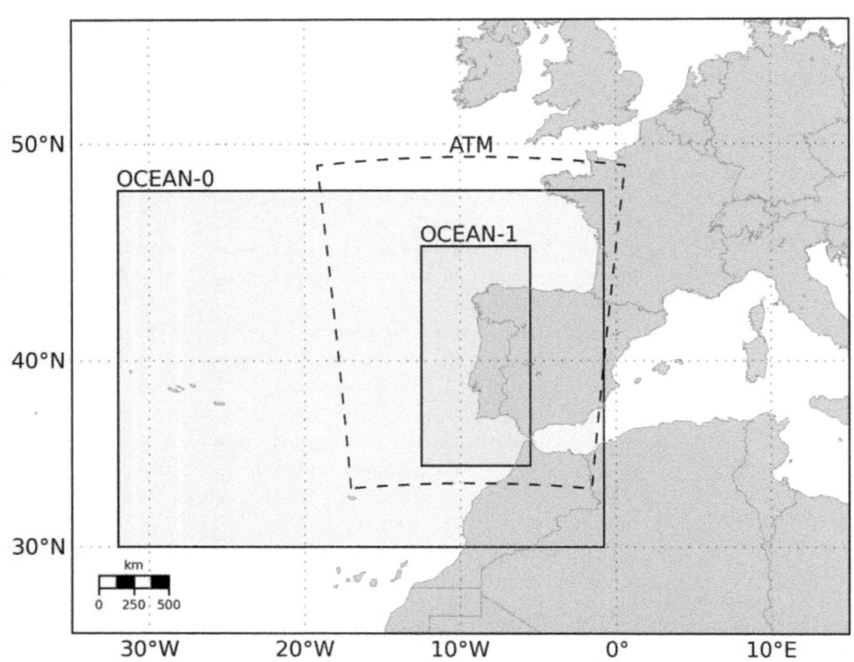

Figure 1. Study region with nested grids configuration of the atmospheric model (ATM) and ocean model (OCEAN-0 and target domain OCEAN-1).

large scale circulation features, and vertical resolution of 30 s-levels with enhanced surface and bottom resolution. The simulations of OCEAN-0 started from rest with temperature and salinity climatologies from [24] and [25], used to initialise and constrain the open boundaries of this domain. The surface was forced with fluxes from the Atlas of Surface Marine Data [26]. Monthly geostrophic velocities (referenced to 1200 m) and Ekman velocities were calculated from the climatologies and applied along the lateral boundaries. The Mediterranean inflow/outflow and spreading was parametrised following [22]. This configuration ran for several years until a stable kinetic solution was attained by year five. The monthly mean of this last year is considered a good resolution climatology and thus used to provide initial and boundary data to the high resolution configuration OCEAN-1.

The target domain OCEAN-1 (Figures 1 and 2) has a horizontal resolution of about 3 km extending from Gibraltar to North Galicia (~1200 km). The domain width is about 600 km. In the vertical 60 s-levels are used with increased near bottom resolution, in order to deal with the Mediterranean exchange (parametrised in the same way as in OCEAN-0). This model was initialised at 1st November 2008 and forced with realistic high resolution surface fluxes from a local solution of the atmospheric model WRF (described in the next section).

The main riverine fresh water inputs from Portugal and Galicia were included in both domains as monthly climatologies. All rivers used are indicated in Figure 2 together with the corresponding percentage of freshwater input in the model domain. The river discharges were obtained from the Portuguese Water Institute (INAG, http://www.inag.pt) and from estimations presented in [27]. A fixed value of 12°C was used for river temperatures.

Atmospheric Model

The model which provides surface forcing data to the ocean domain OCEAN-1 (Figure 2), is an operational implementation of the Weather Research and Forecasting Advance Research model (WRF-ARW) version 3.2.1 [28]. The WRF model is driven by GFS forecasts [29] with an horizontal resolution of 0.5° x 0.5°, and vertical domain extending from a surface pressure of 1000 hPa to 0.27 hPa, discretized in 64 vertical unequally-spaced levels (15 levels are located below 800 hPa and 24 levels above 100 hPa).

The atmospheric forecasts are performed daily by the Group of Meteorology and Climatology of Aveiro University (CliM@UA, http://climetua.fis.ua.pt) and encompass the Portuguese mainland, Madeira and Azores archipelagos. The oceanic operational system uses the weather forecasts for Portuguese mainland domain with spatial resolution of 25 km (Figure 1).

The atmospheric model in use by CliM@UA was configured after a set of numerical experiments described by Ferreira [30]. Model outputs were tested against observations and the best combination of model parametrisations was found for the region.

Biological Model

The biogeochemical model contains a 4-component nitrogen based ecosystem NPZD model (based on [31]), computing 4 state variables: nutrients (nitrate), phytoplankton, zooplankton and detritus, all expressed in mmol N m^{-3}. Chlorophyll *a* (mg m^{-3}) is derived from phytoplankton concentration using a variable chlorophyll:C ratio (θ) that is a function of light and nutrients availability, and a C:N ratio of 6.625 (mmol C/mmol N), i.e., a Redfield ratio. The equation for θ describes the proportion of photosynthetically fixed carbon that is used for chlorophyll a biosynthesis considering the model of Geider et al. [32], which implementation in the ocean model is described in Gruber et al. [33]. The 3D time evolution of the concentration of any of the biogeochemical variables (B_i) follows the general equation:

$$\frac{\delta B_i}{\delta t} = \nabla \cdot K \nabla B_i - u \cdot \nabla_h B_i - (w - w_{\sin k}) \frac{\delta B_i}{\delta z} + SMS(B_i)$$

where the terms in the right hand side account for diffusion, horizontal advection, vertical mixing and sink minus source (*SMS*) biological processes, respectively. K is the eddy kinematic diffusivity tensor, $\boldsymbol{u}$ is the horizontal velocity of the fluid, w and w_{sink} are the vertical velocity of the fluid and the vertical sinking rate of the biogeochemical tracer, respectively, with the exception of zooplankton and nitrate, to which no sinking rate is attributed.

The biogeochemical processes included in *SMS* are specific for each variable and represent the following conceptual description: phytoplankton uptakes nitrate at a rate that is dependent on the instantaneous nitrate concentration; temperature and light intensity/PAR (photosynthetically available radiation). PAR at the surface is calculated as 43% of the incident radiation and attenuated with depth as it is absorbed by water and chlorophyll. Phytoplankton dies at a constant linear rate, been automatically incorporated to the detritus pool. Zooplankton growth relies on its grazing on phytoplankton, which rate depends on prey concentration, and its assimilation efficiency. A constant excretion rate is attributed to zooplankton, providing a source of nutrients to the nitrate pool. Zooplankton incorporates to the detritus pool at a constant linear mortality rate. Mineralization of detritus is formulated as direct transformation to nitrate at a constant nitrification rate. Model parameters for the sink/source terms selected to represent our region of study are listed in Table 1. A detailed description of the biogeochemical model equations, including sink-source terms, can be found in Koné et al. [34].

The concentration of nitrate and chlorophyll *a* for the model initial and boundary conditions were supplied by the climatological data sets World Ocean Atlas 2005 [35] and SeaWiFS, respectively. The initial and boundary data of phytoplankton and zooplankton were derived from chlorophyll *a*. Detritus were introduced constant with the value 0.02 mmol N m^{-3}. Boundary conditions were supplied seasonally. The riverine inputs of nitrate and chlorophyll were used constant along the year and were obtained from Ferreira et al. [36] and from the European Environment Agency (http://www.eea.europa.eu). The values used are listed in Table 2.

The biological model runs as a module integrated in the ocean model. It became operational at 10th June 2011.

Operational System

The ocean and biological modelling system was implemented operationally with OOFε (Operational Ocean Forecast Python Engine, [37]). OOFε is a comprehensive set of tools, written in the Python programming language, which creates and operates the model input/output and controls all the required tasks for the operationality of the ocean model. The operational engine executes daily analysis and forecast (currently three days), whose cycles are repeated continuously in a robust and fully automated manner. The operational engine also includes a visualisation module able to produce many types of slices and plots from the ocean model inputs/outputs. Some graphical outputs of the current implementation can be seen at the homepage of the forecast system, http://neptuno.fis.ua.pt/oof. Among the several outputs are horizontal slices at several depths of temperature, salinity and chlorophyll, as well as forcing wind, sea surface height, currents, etc. For a quick comparison to observational data sets, the web site also shows satellite observations, namely sea surface temperature from O&SI-SAF (Ocean & Sea Ice Satellite

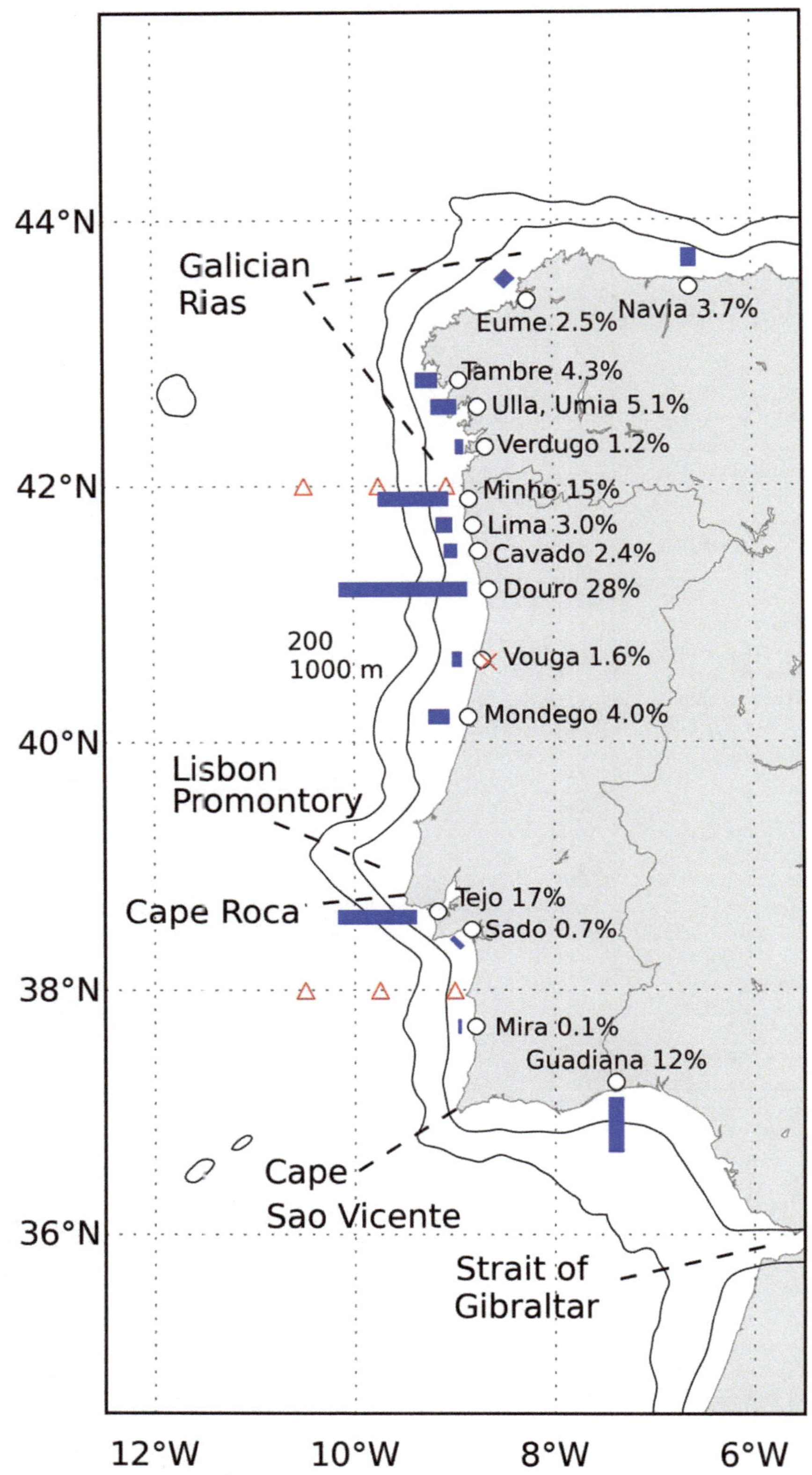

Figure 2. Operational ocean model domain with the location of geographical features and landmarks referenced through the text. The model grid covers the Portuguese and Galician coasts. The main riverine sources of the model are indicated together with the corresponding percentages of fresh water inputs (rivers were introduced as monthly climatologies). The three triangles at the latitudes 42°N and 38°N indicate the locations of surface currents (42° only), temperature and chlorophyll comparisons, between model and observations, done in this work. The cross near the Vouga river mouth is the location of wind comparison.

Table 1. Parameter values of the NPZD model.

Parameter	Value	Units
Light attenuation in seawater	0.04	m^{-1}
Light attenuation by chlorophyll	0.024	m^{-1} (mg Chla $m^{-3})^{-1}$
Initial slope of the P-I curve	1.0	mg C (mg Chla W m^{-2} d$)^{-1}$
C:N ratio for phytoplankton	6.625	mol C (mol N$)^{-1}$
Maximum Cellular chlorophyll:C ratio	0.03	mg Chla (mg C$)^{-1}$
Half-saturation for phytoplankton NO_3 uptake	1.5	mmol N m^{-3}
Zooplankton half-saturation constant for ingestion	1.0	mmol N m^{-3}
Maximum zooplankton growth rate	0.9	d^{-1}
Zooplankton assimilation coefficient	0.75	n.d.
Phytoplankton mortality (to detritus) rate	0.03	d^{-1}
Zooplankton mortality (to detritus) rate	0.1	d^{-1}
Zooplankton specific excretion rate	0.1	d^{-1}
Detrital mineralisation to NO_3 rate	0.05	d^{-1}
Sinking velocity for phytoplankton	0.5	m d^{-1}
Sinking velocity for detritus	5	m d^{-1}

Application Facility, www.osi-saf.org) and from OSTIA (Operational Sea Surface Temperature and Sea Ice Analysis, http://ghrsst-pp.metoffice.com), and high resolution chlorophyll processed and distributed by IFREMER (http://cersat.ifremer.fr/science/ocean color).

The full model data sets are freely available through the Python OPenDAP server Pydap (http://pydap.org). OPeNDAP is an efficient protocol for allowing the remote access of data sets (e.g [38,39]). Data usually available as a set of different individual large data files is made available in an efficient and consistent way by OPeNDAP servers.

The system implemented with OOFε has been successfully operating the ROMS model in other locations, like the Brazilian region [40] and the Northern Gulf of Mexico [41].

Table 2. River inputs of nitrate and chlorophyll.

River	Nitrate (μg l^{-1})	Chlorophyll (μg l^{-1})
Navia	0.1	0.1
Eume	6.4	0.1
Tambre	23.2	0.1
Ulla, Ulmia	13.3	0.1
Verdugo	4.4	0.1
Minho	34.7	2.0
Lima	11.8	2.4
Cavado	33.2	0.9
Douro	88.5	5.4
Vouga	44.4	0.1
Mondego	2.3	5.0
Tejo	21.0	8.5
Sado	21.3	9.6
Mira	10.6	0.9
Guadiana	5.2	0.1

Results

The results presented give an overview of model ability to reproduce the hydrodynamic (surface currents and surface temperature) and chlorophyll fields for the four months following the operational implementation of the biogeochemical model (10th June 2011 to 10th October 2011). These preliminary results are thus addressed as a general reliability assessment rather than a thorough validation, which would need more comparisons with observations and for a longer time period. The period studied corresponds to summer and early autumn, which is characterised in the western coastal region by recurrent upwelling conditions. Therefore, the model is evaluated during one of the most dynamic and biologically productive seasons of the year.

Surface model currents are compared with a high frequency radar data, owned by the Spanish Merchant Navy, at three locations off Vigo (latitude 42°N and longitudes 9°W, 9.75°W and 10.5°W). Given the importance of atmospheric forcing for the ocean model to adequately reproduce the circulation, the surface wind provided by the WRF atmospheric model is compared with wind measurements at University of Aveiro weather station. Then, in order to assess model ability to reproduce coastal blooms, the onset and evolution of a strong upwelling/coastal bloom event (13th to 26th of July) is shown, comparing modeled sea surface temperature and chlorophyll concentration fields in the model with satellite observations (OSISAF-EUMETSAT Ocean & Sea Ice Satellite Application Facility and IFREMER optimised interpolation of MODIS, SeaWiFS and MERIS observations, respectively). To better evaluate the model skill for surface temperature and chlorophyll along the four months, and at different locations within the domain, time series comparing model results and satellite data are presented for two latitudes (42°N and 38°N) in three locations for each latitude, corresponding to middle shelf, off the shelf break and open-sea (same longitudes as used for currents comparison at 42°N). Finally, the model skill in the whole domain is further explored by plotting model bias of SST and chlorophyll which allow to evaluate whether the model is over-or underestimating the observations.

Due to the lack of reliable observational data other than satellite chlorophyll and SST for the time period under study, we do not further explore model skill assessment methods [42,43]. For the same reason, neither concentration of nitrate or the vertical structure of the variables was analysed in the current study.

Surface Wind and Currents

The surface wind measured at Aveiro University weather station (the location is indicated in Figure 2 as x, near the Vouga river mouth) is displayed in Figure 3A. The wind field, upwelling favourable, is typical of the summer season for Western Iberian region, with predominance of northerly winds [44]. The most intense event is observed between 13 and 28 of July, with wind intensity higher than 5 m s^{-1} during most of the days. Figure 3B shows the ocean model forcing wind, generated by the WRF atmospheric model, interpolated at the same location. In general, the modeled wind is more intense than the observed one and the wind inversion after the relaxation is also more prominent.

The comparison of modeled and observed surface currents at the latitude 42°N is shown in Figure 4. (the three comparison sites are indicated in Figure 2). The data shown has been filtered with a low-pass filter with sub-diurnal cutoff frequency. In the shelf (Figure 4A, 4B), the model reproduces correctly the intensity and direction of the current field, which is dominated by an equatorward flow in response to the upwelling favourable wind forcing, with presence of typical velocities of 0.2 m s^{-1}, and maximum values of about 0.35 m s^{-1} associated to the peak of wind forcing by day 22nd of July. In the vicinity of the shelf break (Figure 4C, 4D), the southward tendency is also present in the observations and model results, but the model presents in general higher intensities. In the open-ocean (Figure 4E, 4F) the mesoscale variability is very high and the model cannot reproduce the observed eddy activity.

Coastal Bloom Event, 13th to 26th July 2011

The general trend on chlorophyll concentration simulated along the considered period was that expected for a typical summer situation in the Iberian upwelling region. Successive coastal blooms appeared associated with upwelling pulses, and were more persistent in the northwest coast around Galicia and in the proximity of Cape Roca. In several occasions the bloom occupied the coast from North to South.

We selected the period 13th to 26th July, characterised by strong northerly wind, upwelling favourable, that led to the onset and evolution of a conspicuous coastal bloom, to evaluate model performance reproducing this kind of blooms. Surface temperature and chlorophyll in the model were compared with satellite observations (Figures 5 and 6). Surface temperature shows a distinctive band of cold, upwelled water near the coast which was narrower and more restricted to the North coast in the satellite images than in the model. Two filaments with origin in Cape Roca and Cape S. Vicente appeared in the model, and some evidence of their presence was also observable in the satellite (e.g. days 23rd and 24th). As for the phytoplankton bloom, the increase in chlorophyll concentration along the coast (up to about 6 mg m^{-3}) appeared at about the same time in the model and in the satellite, around 13th of July. The trend was particularly well reproduced in the region between 40°N and 42°N, where the concentration reached 10 mg m^{-3}, whereas in the southwest coast (37°N to 39°N), the concentration in the model was higher than in observations (highest values around 6 and 2 mg m^{-3}, respectively). In contrast, the model was not able to reproduce the high inner shelf chlorophyll concentration along the South coast of Portugal and the Gulf of Cadiz (36°N to 37°N). The bloom situation continued on the next days, spreading offshore with increasing chlorophyll concentration. By day 23th, the bloom seemed to decrease on satellite observations, and then it intensified again (24th to 26th). This decrease was not observed in the model and the bloom continued extending offshore and along the coast. As a consequence, at the end of July, the bloom was more intense and spread farther offshore in the model than in the satellite observations. This is probably related with a slower relaxation of the upwelling winds in the model than in the real ocean.

SST and Chlorophyll Time Series

The model skill for sea surface temperature and chlorophyll was investigated comparing time series between the model and the satellite at six points within the domain (Figures 7 and 8). The comparison sites are indicated in Figure 2 and represent different zones: middle shelf; offshore vicinity of shelf brake; and offshore zone, for two latitudes, in the northern and southern part of the domain (42°N and 38°N). At both latitudes, the modeled shelf surface temperature was lower than temperature from satellite during most part of the analysed period, with typical bias higher than 1°C (Figure 7A and Figure 8A). At the offshore sites the model had a better skill reproducing both the low and high (event

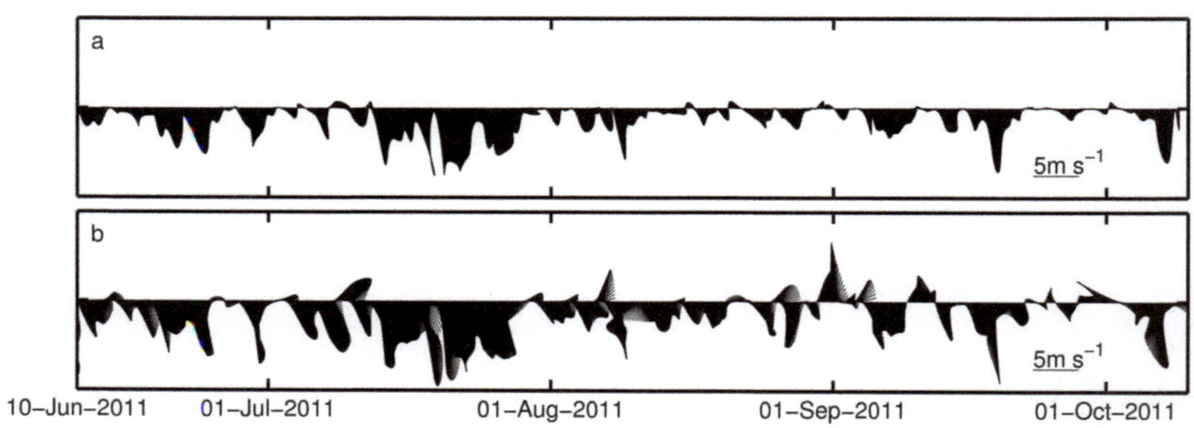

Figure 3. Wind measured by the Aveiro University automatic weather station (a) and ocean model forcing wind at the same location (b). The site is indicated in Figure 2 as x close to the Vouga river mouth.

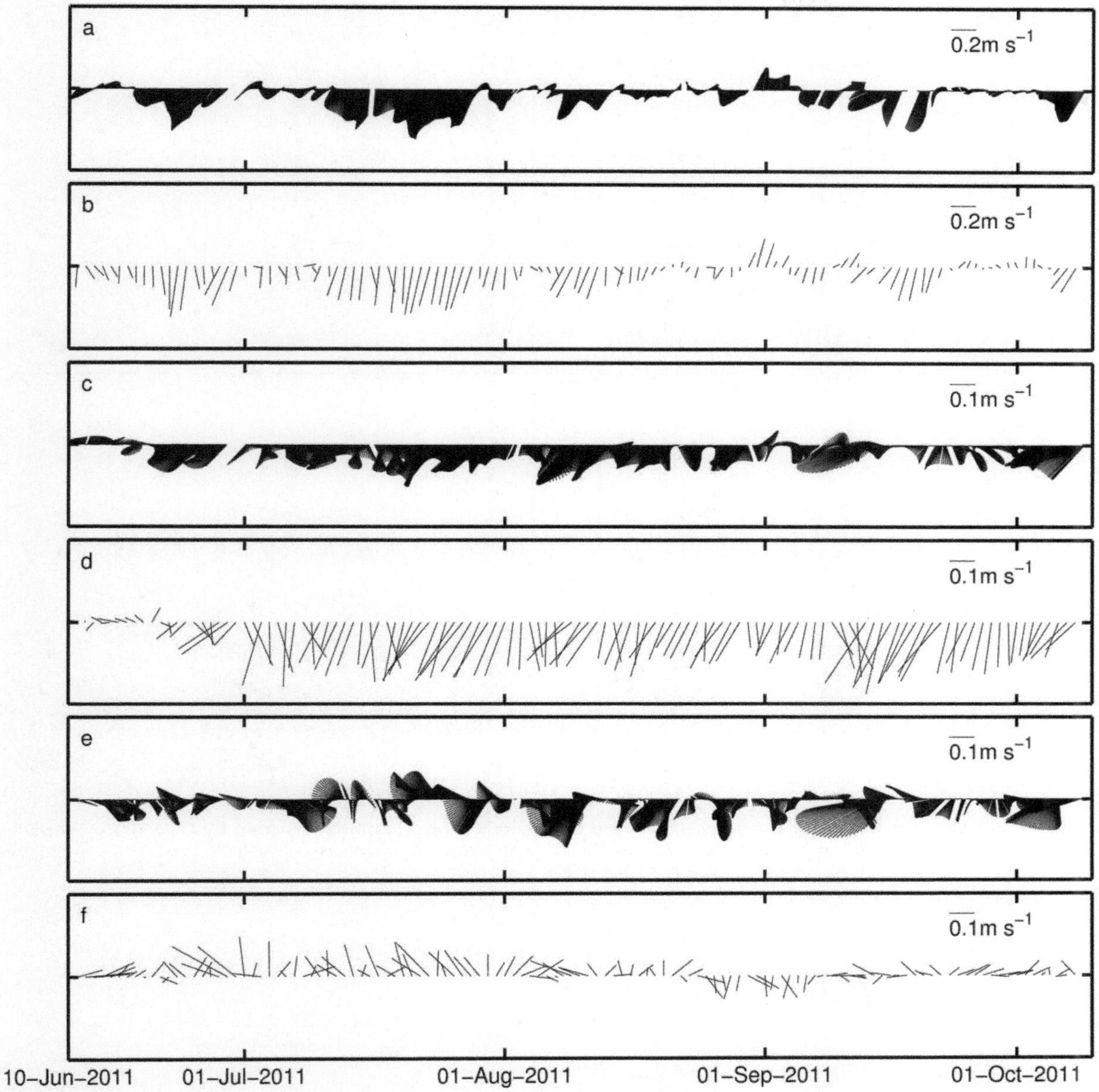

Figure 4. Observed and modeled surface currents at three locations along the latitude 42°N at the longitudes 9°W (a and b), 9.75°W (c and d) and 10.5°W (e and f). These locations are shown in Figure 2 as triangles.

scale) variability of the surface temperature (Figure 7C, 7E and Figure 8C, 8E). The relatively colder waters near the coast may be associated with several factors including over-mixing and inability of the model to re-stratify after strong wind events, deficient parametrisation of air-sea heat fluxes and wind-stress. However, the mixing parametrisations we have used are the ones that are standard and recommended in the majority of ROMS applications. The K-Profile Parameterization, KPP model, described in detail in Large et al. [45], is fully tested for most of the mesoscale applications. It can be argued that near the coast other mixing schemes could perform better, but they usually require additional local information. On the other hand, the colder bias near the coast have been reported in several upwelling systems studies, and attributed to the lack of resolution of the wind stress field near the coast, in which the wind curl is not properly resolved (e.g., [46,47]). A recent comparison across different upwelling systems [48] shows that the bias is common to all upwelling systems and that in part there is a warm tendency of SST data bases.

The surface chlorophyll comparisons show that the relatively simple biological model is able to reproduce many of the observed features, mainly in terms of variability. At 42°N the model simulated the main episodes of chlorophyll concentration variability, namely the shelf blooms by the end of July and beginning of August (Figure 7B). Chlorophyll concentration in both peaks was similar in the model and in satellite observations. The model peak in August lasted longer, most probably as a consequence of the model difficulty to represent the relaxation of the upwelling (Figure 3). A third bloom appeared in early autumn (end of September) that was not captured by the model in the shelf (Figure 7B), but it was captured next to the shelf brake (Figure 7D).

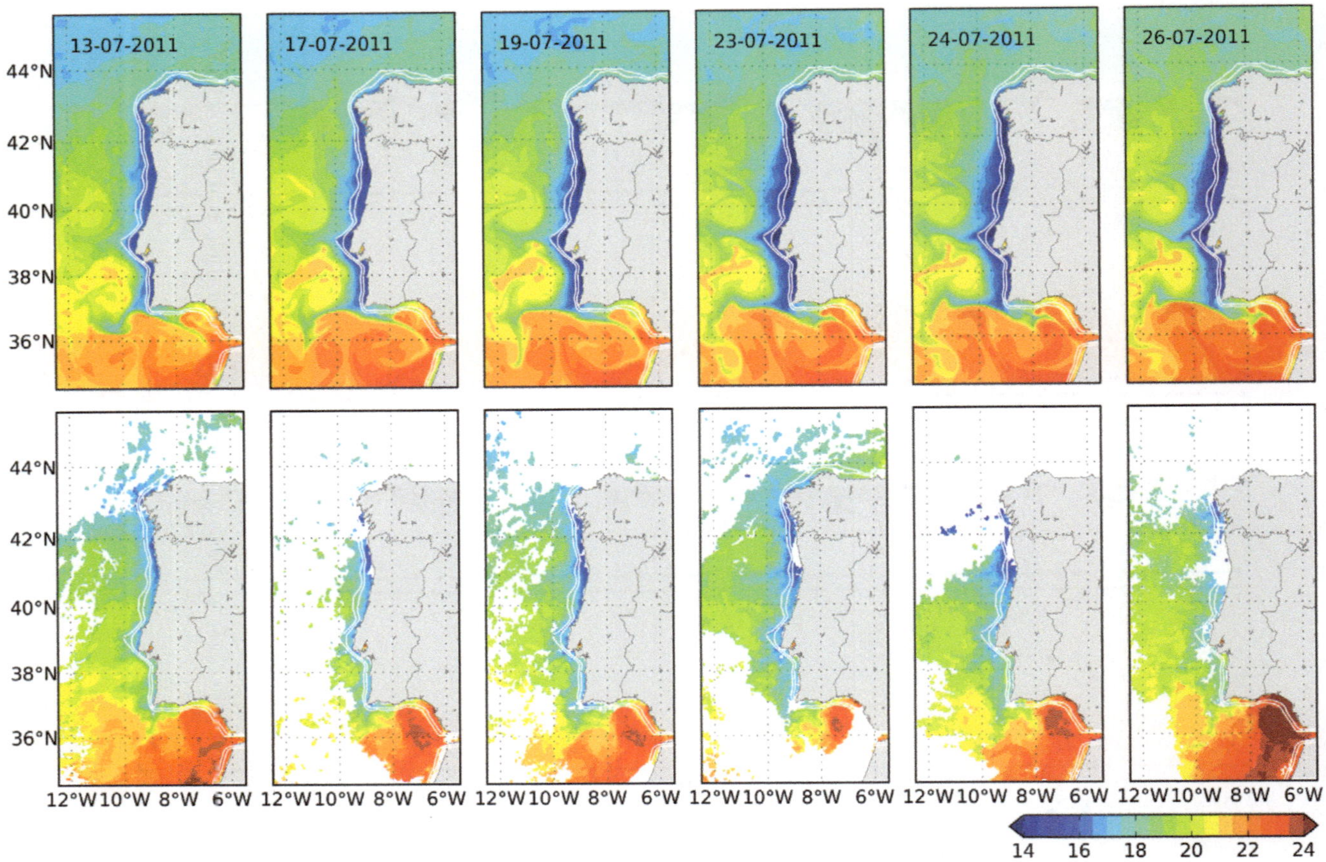

Figure 5. Sea surface temperature from model (upper panel) and from satellite observations provided by OSISAF (lower panel, processed by Meteo-France/CMS-Lannion in the framework of the OSISAF project). Snapshots between 13th and 26th July 2011 are depicted, illustrating one episode of upwelling intensification and coastal bloom (next Figure).

In general, chlorophyll concentration in the model was of the same order as satellite values, except for the most offshore location (10.5°W) at the end of the period, when model values tended to be systematically lower (Figure 7F).

For the latitude 38°N (Figure 8) the model also shows some sensitivity to the event at the end of July, but giving much higher chlorophyll concentrations than the observed ones. The model simulated blooms in the shelf (Figure 8B) and near the shelf brake (Figure 8D) that were considerably higher than in the satellite observations (or even non-existent in the satellite). This behaviour is believed to be a consequence of the stronger modeled upwelling in the southern coast, which is reflected in lower temperatures with a bias of near 2°C at the shelf until middle of August (Figure 8A). It is worth noting that the southern shelf is considerably narrower than the northern shelf (Figure 2), which may play a role in this difference due to the lower numerical discretization of the shelf in this region, and hence higher sensitivity to wind forcing accuracy. Conversely, the chlorophyll peaks observed offshore in early autumn were not properly captured in the model (Figure 8D 8F). Still, the model shows evidences of being able to reproduce the slowly varying base concentration, around 0.2 mg m^{-3}.

SST and Chlorophyll Model Bias

The model bias (MB) indicates whether the model is overestimating or underestimating the observations, being the model results as better as MB is closest to zero. For each grid point, it was calculated as:

$$MB = \frac{\Sigma_{t=1}^{nt}(M_t - D_t)}{\Sigma_{t=1}^{nt} D_t}$$

where M_t and D_t are the modeled and observed values at time index t, respectively. Figure 9 shows the model bias of sea surface temperature and chlorophyll for the domain. There is a systematic negative bias of temperature over the western shelf, as expected, because of the higher intensity of modeled wind than measured (Figure 3), leading to higher upwelling and lower surface temperatures extending offshore, beyond the shelf break. The bias of chlorophyll was lowest in the inner shelf and the offshore region around Galicia. Higher differences are found throughout the western outer shelf and in general near the model domain boundaries. In some regions, like the shelf off Asturias (northern coast) and the outer shelf and slope around 40.5°N, modeled chlorophyll concentration was three times higher than satellite values. The chlorophyll bias in the outer shelf is supposed to be related with the negative bias in temperature, i.e., higher upwelling. Near the boundaries, the differences reflect the usage of climatological boundary conditions biogeochemical variables, which vary very slowly and are very often in disagreement with reality.

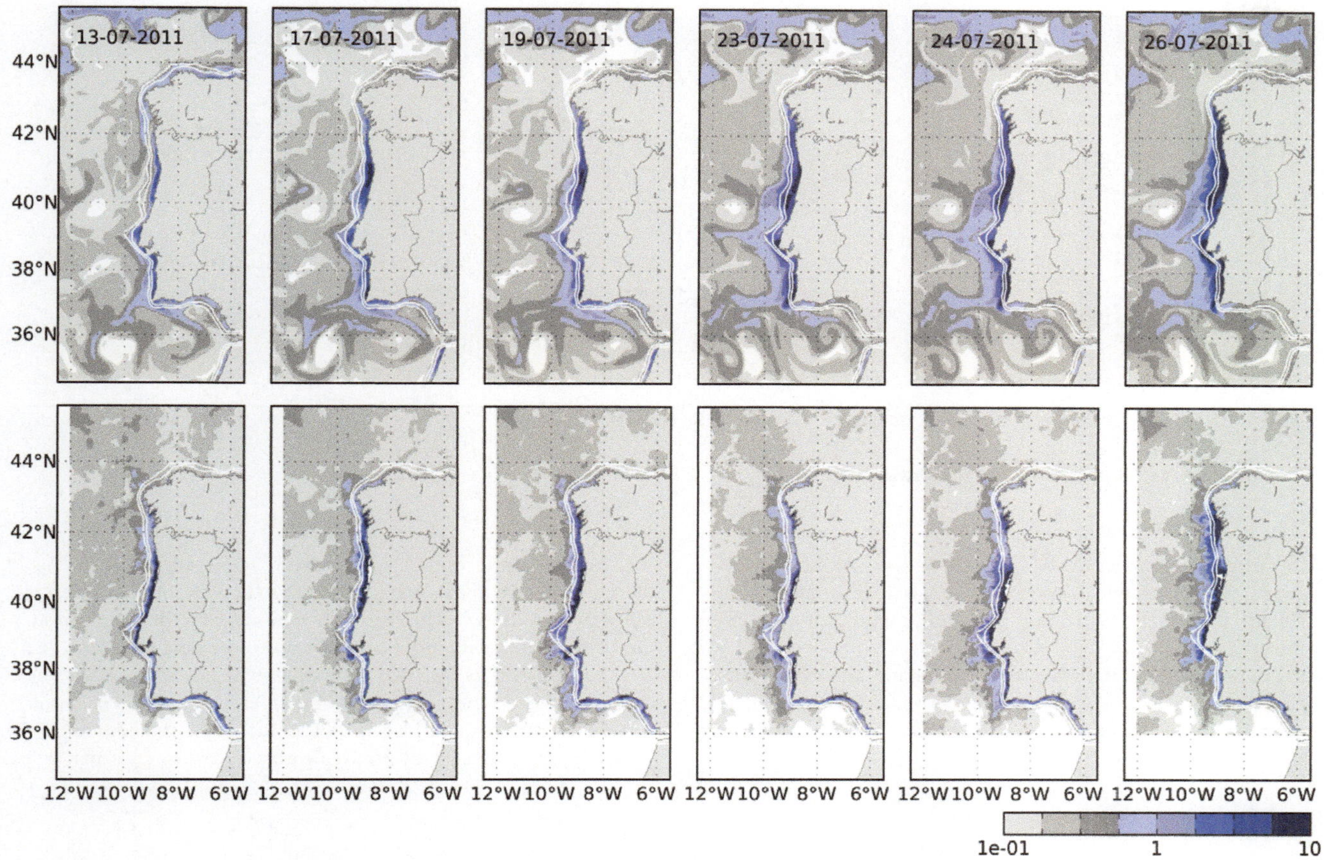

Figure 6. Same as previous figure but showing the chlorophyll coastal bloom. Chlorophyll observations are provided by IFREMER and obtained from OC5 optimised interpolation of MODIS, SeaWiFS and MERIS observations.

Discussion

A ROMS based operational hydrodynamic system was implemented covering the Portuguese and Galician coast. The ocean model has high horizontal and vertical resolution (about 3 km and 60 s-levels) and is running since November 2008. It uses monthly climatological lateral boundaries provided by a climatological run of a larger scale model with about 10 km of horizontal resolution (see Figure 1). The surface is forced with realistic momentum, temperature and salinity fluxes modeled by a local implementation

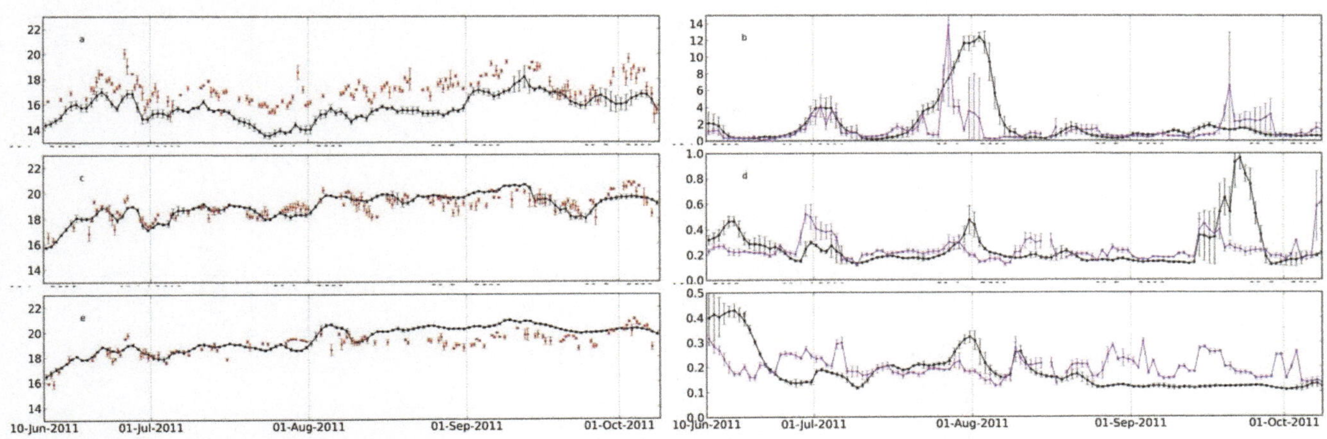

Figure 7. Observed and modeled (in black) sea surface temperature (a, c and e) and chlorophyll (b, d and f) at latitude 42°N and longitudes 9°W, 9.75°W and 10.5°W (locations indicated as triangles in Figure 2. The satellite observations are distributed by OSISAF and by IFREMER. Note the different vertical scales for chlorophyll. Each data points is the average of the data inside a circle of radius 7 km. Error bars are the corresponding standard deviations.

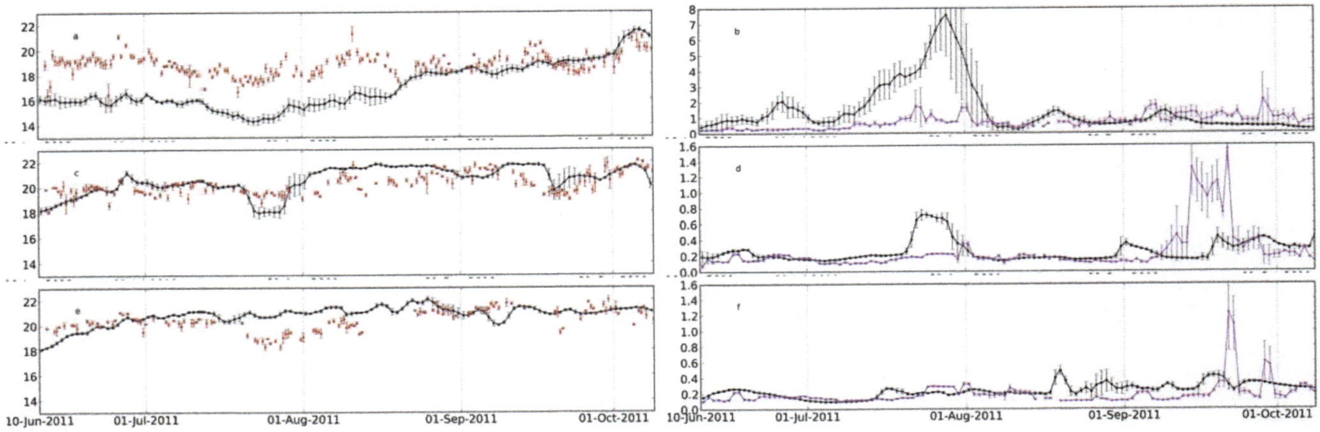

Figure 8. Same as Figure 7 but for the latitude 38°N.

of the atmospheric model WRF. The main rivers of the region are included in the model as monthly climatological means. Also included is the Peliz et al. [22] Gibraltar inflow/outflow parametrisation.

The operational system is totally automatic and runs systematically daily analysis and three days forecasts. This implementations relies on the OOFε (Operational Ocean Forecast Python Engine) described in [37].

More recently, on 10th June 2011, a 4-component nitrogen based biogeochemical model was initiated operational as a module of the ocean model. The rivers are considered a source of nitrates and chlorophyll with input concentrations constant along the year.

The modeled ocean and biological state variables of the first four months of the coupled hydrodynamic and biogeochemical model were compared with observations for validation. The observations consisted in radar measured surface currents, wind

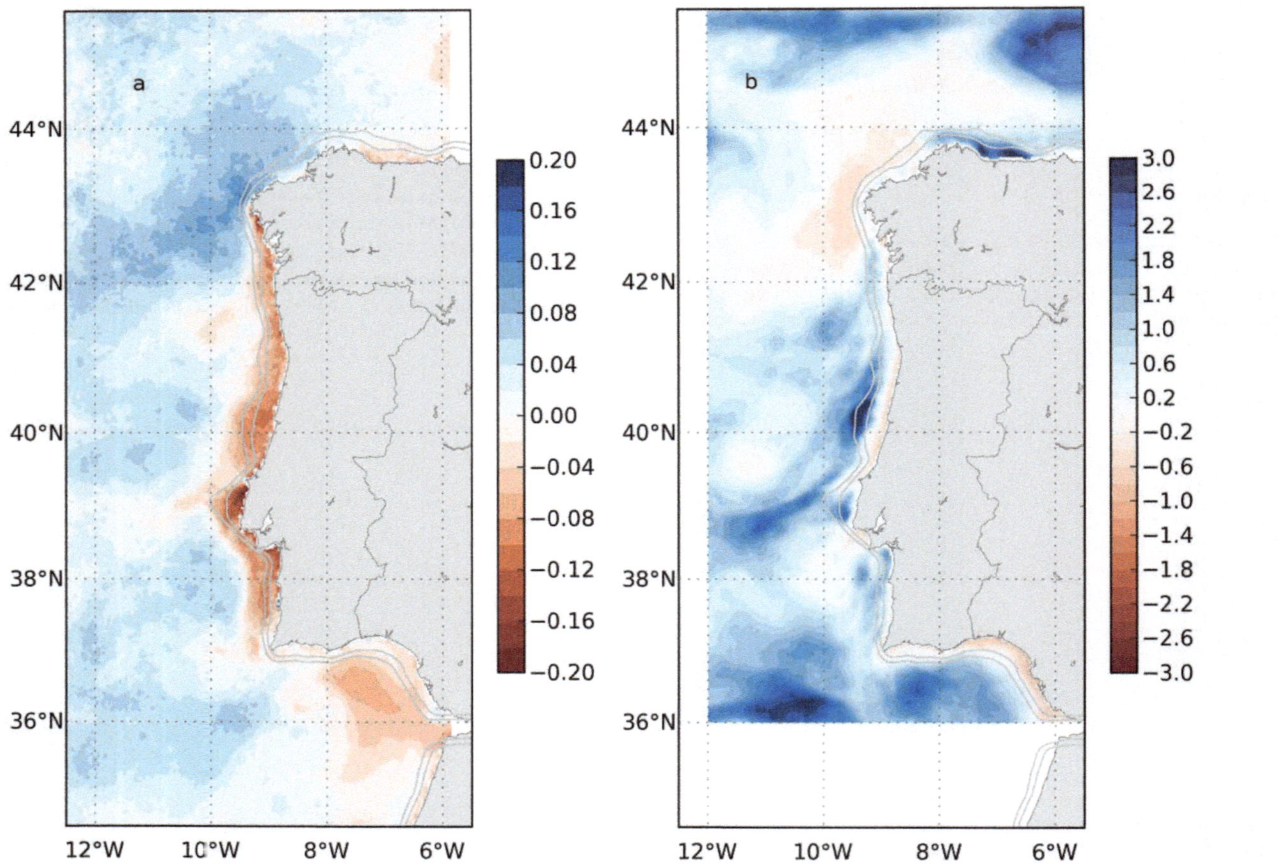

Figure 9. Model bias of sea surface temperature (a) and chlorophyll (b) for the period 10th June 2011 to 10th October 2011.

from a weather station and satellite sea surface temperature and chlorophyll.

The model reproduced adequately the wind-forced circulation over the shelf, typical of the western Iberia summer season. Namely the surface currents, the broad spatial and temporal variability of the upwelling circulation and consequent sea surface temperature. However, the consistent temperature deficit, reaching 2°C is an indication of overestimated upwelling. This arises from the difficulty of the atmospheric model to properly resolve the wind field near the coast. Such overestimation leads to artificial surplus of nutrients within the euphotic layer, and consequent overestimation of the intensity and duration of the phytoplankton bloom. It is worthy to note that the model has no relaxation of sea surface temperature, a numerical resource frequently used in long term simulations, to assure convergence to the correct seasonal pattern.

In the northern shelf the model was able to simulate the main blooms observed by the satellite along the upwelling season, and gave similar chlorophyll concentrations, although showing differences in the duration of the most intense coastal bloom. On the other hand, model results in the southern shelf tended to overestimate observed blooms. The main differences between model and satellite have hydrodynamic origin, namely the more intense upwelling and slower wind relaxation in the model. Nevertheless, in general, the base concentration and variability of surface chlorophyll is reasonably reproduced by the NPZD model. It is important to note that satellite observations can neither be taken as the "real value" in the ocean, since they have associated errors.

A more thorough validation of the model would require further comparisons of model data with high quality in situ measurements (buoys and sampling stations), which are not easily available. Comparisons for a longer time period are also needed. Though, work is underway to cope with the limitations of the current validation.

By providing information about the model skill, the comparison with observations gives clues about possible steps to improve the operational system. Among the future developments, the most important for the coastal and shelf regions, is to have a better representation of the wind field. The local WRF solutions should be improved, namely through optimisation of modelling parametrisations and data assimilation. Ensemble forecasts and assimilation of observations are currently in phase of test and implementation. The rivers have great impact on the shelf dynamics and the use of climatological monthly discharges is certainly an important limitation of the model. Another important limitation in the rivers is the use of constant concentration of nutrients (although the total amount of nutrients varies with river flux). Attempts to estimate at least seasonal patterns should be addressed. This can be done with ensemble Kalman filter optimisation techniques, for instance. The complexity of the biogeochemical model can also be increased to include more variables that may improve model results. Currently tests are underway adding a second subgroup of nitrates (ammonia) and subdividing the detritus in small and large size.

Because of the use of climatological lateral boundaries, the model lacks the ability to reproduce the eddy mesoscale features away from the continental shelf. Some operational eddy-resolving global models with data assimilation are available (like HYCOM, http://www.hycom.org, and MERCATOR http://www.mercator-ocean.fr) and our regional model can be offline nested in the nowcasts and forecasts of such models (e.g., [49]). Also modifying boundary information for the biogeochemical variables, by using data from a global of large-scale parent model, may improve the model results offshore. The results obtained so far with the operational oceanic and biogeochemical implementation are satisfactory and encourage the continuation of the work. The limitations detected incentive improvement efforts, which are currently being studied and tested. In the meantime, results of the oceanic and biogeochemical models are already available online at the web page of the operational system, http://neptuno.fis.ua.pt/oof. Also available are some satellite observations, for comparison. The full model input/output data sets are free to be accessed by any end user through OPenDAP, at the same web address.

Acknowledgments

The authors acknowledge OSISAF and OSTIA for satellite sea surface temperature, IFREMER for the satellite chlorophyll, INAG and EEA for rivers data and the Spanish Merchant Navy for the radar surface currents.

Author Contributions

Analyzed the data: MM-A RR CR NC JDLeS. Wrote the paper: MM-A RR CR RN TL AP MR-V. Conceived and designed modelling configurations: MM-A RR RN AR AP MR-V. Coordination: JD HQ.

References

1. Berx B, Dickey-Collas M, Skogen MD, Roeck YHD, Klein H, et al. (2011) Does operational oceanography address the needs of fisheries and applied environmental scientists? Oceanography 24: 166–171.
2. Borja A (2005) The European water framework directive: A challenge for nearshore, coastal and continental shelf research. Cont Shelf Res 25: 1768–1783.
3. Borja A, Elliott M, Carstensen J, Heiskanen AS, van de Bund W (2010) Marine management - Towards an integrated implementation of the European Marine Strategy Framework and the Water Framework Directives. Mar Pollut Bull 60: 2175–2186.
4. Brachet G (2004) From initial ideas to a European plan: GMES as an exemplar of European space strategy. Space Policy 20: 7–15.
5. Pinardi N, Woods J, eds (2002) Ocean forecasting: conceptual basis and applications. London: Springer-Verlag.
6. Siddorn JR, Allen JI, Blackford JC, Gilbert FJ, Holt JT, et al. (2007) Modelling the hydrodynamics and ecosystem of the North-West European continental shelf for operational oceanography. J Mar Syst 65: 417–429.
7. Neumann T (2000) Towards a 3D-ecosystem model of the Baltic Sea. J Mar Syst 25: 405–419.
8. Lazzari P, Teruzzi A, Salon S, Campagna S, Calonaci C, et al. (2010) Pre-operational short-term forecasts for the Mediterranean Sea biogeochemistry. Ocean Sci 6: 25–39.
9. Teruzzi A, Salon S, Bolzon G, Lazzari P, Campagna S, et al. (2011) Operational forecasts of the biogeochemical state of Mediterranean Sea. Mercator ocean Newsletter 40.
10. Shchepetkin AF, McWilliams JC (2005) The Regional Ocean Modeling System (ROMS): A split-explicit, free-surface, topography-following coordinates ocean model. Ocean Model 9: 347–404.
11. Penven P, Debreu L, Marchesiello P, McWilliams JC (2006) Evaluation and application of the ROMS 1-way embedding procedure to the central california upwelling system. Ocean Model 12: 157–187.
12. Haidvogel DB, Arango H, Budgell WP, Cornuelle BD, Curchitser E, et al. (2008) Ocean forecasting in terrain-following coordinates: Formulation and skill assessment of the Regional Ocean Modeling System. J Comput Phys 227: 3595–3624.
13. Marta-Almeida M, Dubert J, Peliz A, Queiroga H (2006) Influence of vertical migration pattern on retention of crab larvae in a seasonal upwelling system. Mar Ecol Prog Ser 307: 1–19.
14. Marta-Almeida M, Dubert J, Peliz A, dos Santos A, Queiroga H (2008) A modelling study of the Norway Lobster (*Nephrops norvegicus*) larval dispersal in southern Portugal: predictions of larval wastage and self-recruitment in the Algarve stock. Can J Fish Aquat Sci 65: 2253–2268.
15. Peliz A, Marchesiello P, Dubert J, Marta-Almeida M, Roy C, et al. (2007) A study of crab larvae dispersal on the Western Iberian Shelf: Physical processes. J Mar Syst 68: 215–236.

16. Silva PA, Ramos M, Marta-Almeida M, Dubert J (2007) Water Exchange Mechanisms Between Ria de Aveiro and the Atlantic Ocean. J Coastal Res 29: 1622–1626.
17. Otero P, Ruiz-Villarreal M, Peliz A (2008) Variability of river plumes off Northwest Iberia in response to wind events. J Mar Syst 72: 238–255.
18. Otero P, Ruiz-Villarreal M, Peliz A (2009) River plume fronts off NW Iberia from satellite observations and model data. ICES J Mar Sci 66: 1853–1864.
19. Otero P, García-García MRVL, Marta-Almeida M, Cobas M, González-Nuevo G, et al. (2011) Walking on the sea side: Modeling and observational efforts of the Iberian Margin Ocean Observatory (RAIA). In: OCEANS 2011. Santander: OCEANS-IEEE doi: 10.1109/Oceans-Spain.2011.6003564.
20. Ruiz-Villarreal M, González-Pola C, Otero P, Días del Río G, Lavín A, et al. (2006) Circulation in the Galicia-Southern Bay of Biscay: Reanalysis of the circulation influencing the Prestige oil spill. In: European Operational Oceanography: present and future. EuroGOOS, 342–348.
21. Sotillo MG, Fanjul EA, Castanedo S, Abascal AJ, Menendez J, et al. (2008) Towards an operational system for oil-spill forecast over Spanish waters: Initial developments and implementation test. Mar Pollut Bull 56: 686–703.
22. Peliz A, Dubert J, Marchesiello P, Teles-Machado A (2007) Surface circulation in the Gulf of Cadiz: Model and mean flow structure. J Geophys Res 107.
23. Peliz A, Marchesiello P, Santos AMP, Dubert J, Teles-Machado A, et al. (2009) Surface circulation in the Gulf of Cadiz: 2. Inflow-outflow coupling and the Gulf of Cadiz slope current. J Geophys Res 114.
24. Levitus S, Boyer TP (1994) World ocean atlas 1994 volume 4: Temperature. NOAA Atlas NESDIS 4, U.S. Department of Commerce, Washington, D. C p.
25. Levitus S, Burgett R, Boyer TP (1994) World ocean atlas 1994 volume 3: Salinity. NOAA Atlas NESDIS 3, U.S. Department of Commerce, Washington, D. C p.
26. da Silva A, Young AC, Levitus S (1994) Atlas of surface marine data 1994, volume 1: Algorithms and procedures. NOAA Atlas NESDIS 6, U.S. Department of Commerce, Washington, D. C p.
27. Otero P, Ruiz-Villarreal M, Peliz A, Cabanas JM (2010) Climatology and reconstruction of runoff time series in northwest Iberia: influence in the shelf buoyancy budget off Ria de Vigo. Sci Mar 74: 247–266.
28. Skamarock WC, Klemp JB, Dudhia J, Gill DO, Barker DM, et al. (2008) A Description of the Advanced Research WRF Version 3. Technical report, MMMD-NCAR, Boulder, Colorado, USA.
29. NOAA (2003) The GFS Atmospheric Model. Technical report, Environmental Modelling Center, NOAA-NCEP, Waashington D.C., USA.
30. Ferreira AP (2007) Sensibilidade às parametrizações fisicas do WRF nas previsões à superficie em Portugal Continental. Relatório de estágio - MOF. Universidade de Aveiro.
31. Fasham MJR, Ducklow HW, McKelvie SM (1990) A nitrogen-based model of plankton dynamics in the oceanic mixed layer. J Mar Res 48: 591–639.
32. Geider RJ, MacIntyre HL, Kana TM (1997) Dynamic model of phytoplankton growth and acclimation: responses of the balanced growth rate and the chlorophyll a:carbon ratio to light, nutrient-limitation and temperature. Mar Ecol Prog Ser 148: 187–200.
33. Gruber N, Frenzel H, Doney SC, Marchesiello P, McWilliams JC, et al. (2006) Eddy-resolving simulation of plankton ecosystem dynamics in the California Current System. Deep Sea Res Part I 53: 1483–1516.
34. Koné V, Machu E, Penven P, Andersen V, Garçon V, et al. (2005) Modeling the primary and secondary productions of the southern Benguela upwelling system: A comparative study through two biogeochemical models. Global Biogeochem Cy 19: GB4021.
35. Garcia HE, Locarnini RA, Boyer TP, Antonov JI (2006) World Ocean Atlas 2005, Volume 4: Nutrients (phosphate, nitrate, silicate). Washington, D.C.: NOAA Atlas NESDIS 64.
36. Ferreira JG, Simas T, Nobre A, Silva MC, Schifferegger K, et al. (2003) Identification of Sensitive Areas and Vulnerable Zones In Transitional and Coastal Portuguese Systems. Application of the United States National Estuarine Eutrophication Assessment to the Minho, Lima, Douro, Ria de Aveiro, Mondego, Tagus, Sado, Mira, Ria Formosa and Guadiana systems. Technical report, INAG/IMAR.
37. Marta-Almeida M, Ruiz-Villarreal M, Otero P, Cobas M, Peliz A, et al. (2011) OOFε: A Python engine for automating regional and coastal ocean forecasts. Environ Modell Softw 26: 680–682.
38. Cornillon P, Adams J, Blumenthal MB, Chassignet E, Davis E, et al. (2009) NVODS and the Development of OPeNDAP. Oceanography 22: 116–127.
39. Signell RP, Carniel S, Chiggiato J, Janekovic I, Pullen J, et al. (2008) Collaboration tools and techniques for large model datasets. J Mar Syst 69: 154–161.
40. Marta-Almeida M, Pereira J, Cirano M (2011) Development of a pilot Brazilian operational ocean forecast system, REMO-OOF. J Oper Ocean 4: 3–15.
41. Zhang X, Marta-Almeida M, Hetland RD (2012) A high-resolution pre-operational forecast model of circulation on the Texas-Louisiana continental shelf and slope. J Oper Ocean 5: 19–34.
42. Allen JI, Holt JT, Blackford J, Proctor R (2007) Error quantification of a high-resolution coupled hydrodynamic-ecosystem coastal-ocean model: Part2. Chlorophyll-a, nutrients and SPM. J Mar Syst 68: 381–404.
43. Jolliff JK, Kindle JC, Shulman I, Penta B, Friedrichs MAM, et al. (2009) Summary diagrams for coupled hydrodynamic-ecosystem modelskillassessment. J Mar Syst 76: 64–82.
44. Wooster WS, Bakun A, Mclain DR (1976) The seasonal upwelling cycle along the eastern boundary of the north atlantic. J Mar Res 34: 131–141.
45. Large WG, McWilliams JC, Doney SC (1994) Oceanic vertical mixing: a review and a model with a nonlocal boundary-layer parameterization. Rev Geophys 32: 363–403.
46. Capet X, Marchesiello PP, McWilliams JC (2004) Upwelling response to coastal wind profiles. Geophys Res Lett 31: L13311.
47. Otero P, Ruiz-Villarreal M (2008) Wind forcing of the coastal circulation off north and northwest Iberia: Comparison of atmospheric models. J Geophys Res 113: C10019.
48. Dufois F, Penven P, Whittle CP, Veitch J (2012) On the warm near shore bias in Pathfinder monthly SST products over Eastern Boundary Upwelling Systems. Ocean Model 47: 113–118.
49. Barth A, Alvera-Azcárate A, Weisberg RH (2008) Benefit of nesting a regional model into a large-scale ocean model instead of climatology: Application to the West Florida Shelf. Cont Shelf Res 28: 561–573.

3

Complex, Dynamic Combination of Physical, Chemical and Nutritional Variables Controls Spatio-Temporal Variation of Sandy Beach Community Structure

Kelly Ortega Cisneros[1,2]*, Albertus J. Smit[2], Jürgen Laudien[3], David S. Schoeman[2,4]

1 Leibnitz Center for Marine Tropical Ecology, Bremen, Germany, **2** School of Biological and Conservation Sciences, University of KwaZulu-Natal, Durban, South Africa, **3** Alfred Wegener Institute for Polar and Marine Research, Bremerhaven, Germany, **4** School of Environmental Science, University of Ulster, Coleraine, Northern Ireland

Abstract

Sandy beach ecological theory states that physical features of the beach control macrobenthic community structure on all but the most dissipative beaches. However, few studies have simultaneously evaluated the relative importance of physical, chemical and biological factors as potential explanatory variables for meso-scale spatio-temporal patterns of intertidal community structure in these systems. Here, we investigate macroinfaunal community structure of a micro-tidal sandy beach that is located on an oligotrophic subtropical coast and is influenced by seasonal estuarine input. We repeatedly sampled biological and environmental variables at a series of beach transects arranged at increasing distances from the estuary mouth. Sampling took place over a period of five months, corresponding with the transition between the dry and wet season. This allowed assessment of biological-physical relationships across chemical and nutritional gradients associated with a range of estuarine inputs. Physical, chemical, and biological response variables, as well as measures of community structure, showed significant spatio-temporal patterns. In general, bivariate relationships between biological and environmental variables were rare and weak. However, multivariate correlation approaches identified a variety of environmental variables (i.e., sampling session, the C:N ratio of particulate organic matter, dissolved inorganic nutrient concentrations, various size fractions of photopigment concentrations, salinity and, to a lesser extent, beach width and sediment kurtosis) that either alone or combined provided significant explanatory power for spatio-temporal patterns of macroinfaunal community structure. Overall, these results showed that the macrobenthic community on Mtunzini Beach was not structured primarily by physical factors, but instead by a complex and dynamic blend of nutritional, chemical and physical drivers. This emphasises the need to recognise ocean-exposed sandy beaches as functional ecosystems in their own right.

Editor: Simon Thrush, National Institute of Water & Atmospheric Research, New Zealand

Funding: This study was financed within the framework of the NRF-IRDP project (KZN Coastal Trophic Flows, ICD2006072700035). Additional financial support was provided by a DAAD (Deutscher Akademischer Austausch Dienst) scholarship to KOC, a Royal Society grant to DSS. The funders had no role in study design, data collection and analysis, decision to publish, or preparation of the manuscript.

Competing Interests: The authors have declared that no competing interests exist.

* E-mail: kortegac@gmail.com

Introduction

In most ecological systems, including those of the oceans, community structure emerges from a complex interplay between biotic interactions and abiotic environmental factors [1–6]. An exception to this apparent rule seems to be ocean-exposed sandy beaches. These systems are highly dynamic in space and time, and as a result, are traditionally referred to as "physically stressed" [7], in the sense that individual resident macrobenthic species appear to respond independently to physical features of the environment, with the influence of biotic interactions being negligible [8]. Although evidence to the contrary is slowly accumulating [9–12], the role of biotic interactions in structuring the macrobenthic communities of intertidal beaches is still considered trivial in all but the least disturbed and lowest energy systems [13].

Anthropogenic influence is nowadays ubiquitous in marine communities [14]. Of particular concern for beaches is the rate of coastal urbanisation [15], which brings with it extensive coastal armouring as well as additional elements of disturbance. These all have adverse effects on local ecology [16–18], and should further minimise the influence of ecological interactions [13]. Under such circumstances, greatest abundance and diversity of resident macroinfauna would be expected on fine-sand, dissipative beaches [19], while coarse-sand, reflective beaches should harbour small populations of only a few species [20].

On the other hand, much of the evidence for physical control comes from studies that measured only physical beach features [8,20,21]; characteristics such as primary productivity and food availability have been largely ignored, or considered as being subsumed within sampling designs. Among the few studies that have assessed chemical variables, salinity has been the most prominent and most influential in community structure [22]. Despite the importance of biological interactions (including food availability) as a factor controlling the structure and dynamics of benthic communities in general [6,23–27], as well as those of other soft-sediment shores [2,3,28,29], few studies of beach community ecology have taken this into account [30–32]. It is therefore evident that broader consideration of the roles of a range of potential structuring processes should enhance understanding of the ways in which sandy beach macrobenthic communities are structured.

Opportunities to investigate the relative roles of physical, chemical and nutritional gradients in structuring intertidal beach macrobenthic communities arise around estuary mouths. Because intertidal sandy beaches are devoid of biogenic structure and support little primary production [33], they depend heavily on allochthonous subsidies. In this sense, adjacent ecosystems, such as estuaries, can provide significant inputs of both inorganic nutrients to the surf zone and particulate organic matter to the intertidal zone [34]. Moreover, estuaries mediate strong salinity gradients, which have been demonstrated to have a detrimental influence on population and community attributes of sandy beach assemblages [22,35–37]. Unfortunately, however, few integrated attempts have been made to include a wide range of physical, chemical and nutritional variables as potential explanatory variables for patterns of intertidal community structure at the meso-scale. Here, we report on a study that takes advantage of the strong seasonal changes in estuarine flow of the Mlalazi Estuary on the KwaZulu-Natal (KZN) coast of South Africa to investigate macrobenthic community structure at Mtunzini Beach in the presence and absence of various gradients associated with estuarine inputs.

KZN, on the east coast of South Africa (Fig. 1), experiences marked seasonal cycles in rainfall, resulting in strong temporal patterns of river flow to the extent that many estuaries in this region are isolated from the sea for much of the dry winter season by sandbars that block their mouths [38–40]. Because KZN marine waters are reasonably oligotrophic [41], seasonal freshwater inflow represents the main source of nutrients for estuarine food webs [42–46]. During the rainy season, estuarine flow increases in volume; consequently, larger proportions of estuarine seston and nekton are exported to coastal waters and adjacent systems, including sandy beaches [47]. Due to this cycle, estuarine input to Mtunzini Beach should be minimal toward the end of a dry season, with any depression in salinity or elevation in nutrient concentrations and associated surf-zone plankton productivity being limited to the immediate vicinity of the estuary mouth. If the beach macroinfauna are sensitive to estuarine inputs, there might be aggregation at (or avoidance of) areas immediately adjacent to the estuary mouth, with the communities beyond the reach of these putative gradients distributed along the shore according to physical conditions, with higher macrofaunal abundance and diversity on sections of the beach with fine sand and wide intertidal zones. By contrast, as rainfall increases and the river starts to fill, estuarine input to the surf zone should increase and any resulting effect should spread along the shore. Any corresponding response by the macroinfaunal communities is likely to be detectable as consistent relationships with chemical or nutritional variables and would provide evidence that responses to other variables can override the effect of physical beach descriptors.

By sampling an along-shore grid of stations on Mtunzini Beach at three discrete times that correspond to the end of the dry season, the start of the rainy season and the late rainy season, this study aims to determine whether the beach community responds overwhelmingly to the physical environment or if other factors are influential. To address this aim, the objectives are: (1) to determine if alongshore patterns for physical, chemical and nutritional features of the beach vary through time; (2) to determine whether alongshore patterns in intertidal macrobenthic community structure vary through time; and (3) to determine, which physical, chemical and nutritional variables are most influential in describing changes in macrobenthic community structure.

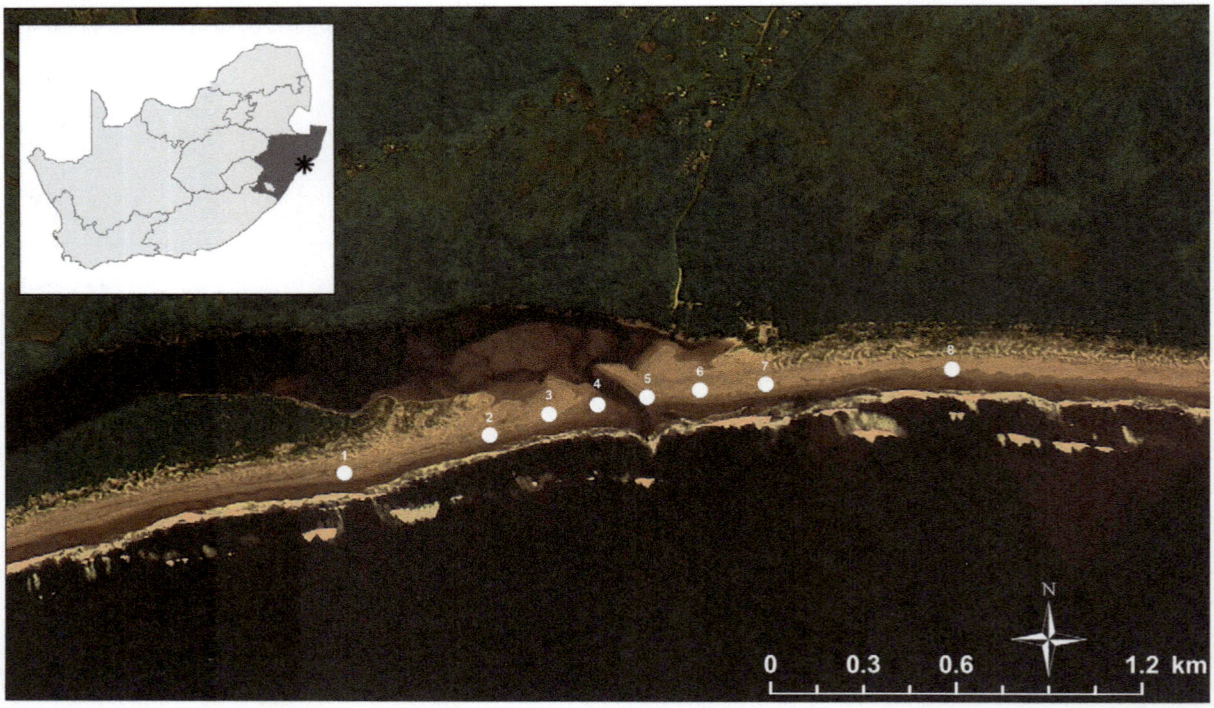

Figure 1. Map of the study area, indicating the position of the sample stations relative to the Mlalazi estuary. Numerals indicate station numbers (Station 1 is in the far south; Station 8 is in the far north).

Methods

The Mlalazi Estuary (28°58′S; 31°48′E, eastern coast of South Africa, Fig. 1) is a permanently open estuary with its mouth closing only during major droughts. The catchment of approximately 492 km^2 [48] is subject to high rainfall of about 1250 $mm.yr^{-1}$, peaking in austral summer (December–February). The estuary length extends from about 11 km in the dry season to 7 km during the rainy season [49]. The microtidal Mtunzini Beach (maximum tidal range is 2.13 m) borders the estuary and stretches about 5 km both north and south of the estuary mouth. Nel and Bezuidenhout [50] classified the grain size of Mtunzini Beach as medium-coarse and documented a well-sorted sediment and an intermediate-reflective beach morphodynamic state.

Biological samples and environmental data were collected bimonthly between October 2008 and February 2009 (three sessions). Sampling was conducted around low tide within four days of a spring tide. Sampling stations were arranged symmetrically around the mouth of the Mlalazi Estuary, with four stations to the north and four to the south (Fig. 1). In each direction, stations were positioned at 25 m, 150 m, 400 m, and 900 m from the mouth.

Ethics statement

A sampling permit for scientific investigation was issued to the School of Biological and Conservation Sciences, University of KwaZulu-Natal, by the Department of Environmental Affairs and Tourism, Republic of South Africa. Permit reference numbers were V1/1/5/1 and RES2008/14 for 2008 and 2009, respectively.

Sampling design for macrobenthos

At each station, biological sampling consisted of triplicate shore-normal transects at 10-m intervals alongshore. Each transect was sampled at ten across-shore sampling levels, arranged at uniform intervals between the drift line and the spring low-water mark. At each across-shore level, four sediment samples including macro-infauna were excavated using a stainless steel corer of 18-cm diameter (surface area of 0.0254 m^2) inserted to a depth of about 30 cm. These four samples were pooled and washed through a 1-mm nylon-mesh sieving bag in the swash zone to remove excess sand. All materials remaining in the sieve bag were stored in labelled, sealed polyethylene bags, frozen on return to the laboratory and stored at −20°C until analysis [51,52]. This field sampling design largely conforms to recommendations by Schlacher et al. [53] and represents a sampling effort of 3.05 m^2 per station and session. Although this sampling effort was smaller than that recommended by Schoeman et al. [54,55], our aim was to assess spatial and temporal trends in community structure on the basis of the most common species, rather than to accurately estimate macroinfaunal species richness or abundance. In this sense, the ability to sample more stations compensates for the slight loss in accuracy and precision of biotic measures at individual stations caused by smaller-than-ideal sample sizes.

Sampling design for environmental variables

For the measurement of physico-chemical variables (Table S1), triplicate water samples were collected below the water surface (ca. 5–10-cm depth) in the swash zone of the eight sandy beach stations, and in the mouth of the estuary during each session. Separate triplicate samples were taken for the determination of photopigment concentrations, organic matter content and dissolved inorganic nutrient concentrations.

For photopigment concentrations, 250-ml water samples were collected in acid-washed plastic bottles from each station. These samples were cooled on ice and stored in the dark for a maximum of two hours before further processing. For the determination of size-fractioned photopigment concentrations, the samples were serially filtered (vacuum, <50 kPa) through 20-µm Nitex mesh (for microplankton photopigments), 2-µm membrane filters (for nanoplankton photopigments), and 0.72-µm GF/F filters (for picoplankton photopigments), and then extracted in 90% acetone for 24 hrs in the dark at 4°C. Photopigment concentrations were determined using a Turner Trilogy fluorometer (Sunnyvale, California, US) before and after acidification with two drops of 1% HCl [56].

Corresponding 1-l water samples for the determination of Total Suspended Solids (TSS) and Particulate Organic Matter (POM) were collected in plastic bottles at each station, and were treated identically to the photopigment samples prior to processing. In the laboratory, the water was filtered (vacuum, <50 kPa) through pre-combusted and pre-weighed GF/F filters. The filters were then dried at 60°C for 24 hrs, reweighed and the amount of TSS determined by mass difference. Additionally, POM determinations were made from these filters after combusting them at 450°C for 12 hrs before reweighing. POM was estimated by mass difference before and after combustion.

For the determination of Particulate Organic Carbon (POC), Particulate Organic Nitrogen (PON) and C:N ratio, water samples were collected, stored and processed as described above for TSS determinations. GF/F filters, on which POM had been collected were acidified with 2N HCl and again dried at 60°C for 24 hrs. POC and PON were determined using a Flash EA 1112 series elemental analyzer (Thermo Finnigan, Milan, Italy). Results were expressed as percentages of carbon and nitrogen.

For the quantification of Dissolved Inorganic Nitrogen (DIN: nitrate+nitrite) and Dissolved Inorganic Phosphorus (DIP: ortho-phosphate) concentrations, 50 ml of water filtrate from the photopigment determination (after filtering through GF/F filters) were collected in acid-washed bottles and frozen at −20°C. Nutrient concentrations were determined using a Skalar San++ continuous-flow analyser (Skalar Analytica BV, The Netherlands), following the methods of Mostert [57]. Salinity measurements were determined from the resulting filtrate in the laboratory using a hand-held refractometer (ATAGO, Japan).

Sediment samples for the determination of the granulometric measures and Sediment Organic Matter (SOM) content were taken using a sediment core (internal diameter of 2 cm) to 10-cm depth at each shore level and station. To determine SOM, 5 g of oven-dried (60°C for 24 hrs) sediment was weighed and combusted at 450°C for six hours before reweighing. Organic matter content was determined as the difference in sediment masses. At each station, SOM content was determined as the mean value of the ten samples collected. For grain size analysis, the remaining dried sand was passed through nested sieves with a standard range of ϕ intervals [58], and conventional statistics derived [59].

Macrobenthic community analysis

After defrosting the material collected from the sieved macrofauna cores, organisms were separated from the remaining sediment by manual elutriation: the sample was placed in a 25-l plastic bucket containing 10 l of distilled water, and stirred vigorously; thereafter, the water was poured through a 1-mm mesh size sieve. This procedure was repeated five times. Govender [60] determined that three to four repetitions of elutriation are required to extract 95% of all macrofaunal organisms from sediment samples taken at nearby KZN beaches, and that five repetitions almost always extract all such organisms. Following elutriation, the remaining sediment was searched manually for large, heavy-bodied organisms, such as clams and whelks. Specimens were identified to the lowest taxonomic level possible and enumerated.

Data analysis

Granulometric parameters (mean grain size, sorting, kurtosis and skewness) were estimated with the software GRADISTAT Version 4.0 [61]. Beach morphodynamic state was classified according to Dean's dimensionless fall velocity (Ω) [62,63]. Sand-fall velocities were obtained from granulometric determinations; wave height and period data were obtained from the WindGuru website (http://www.windguru.cz, accessed on June 13th, 2009).

Species abundance was estimated as the number of individuals per strip transect (IST) [53,64,65]. Only beach-resident species were considered, and because we intended to conduct a range of multivariate analyses, all species that occurred in fewer than five transects (out of a total of 72 transects) were excluded from the analysis, to reduce the number of zeros in the data matrix. Data were fourth-root transformed (to further equalize the contributions to community structure by common and rare species) before the construction of a Bray-Curtis similarity matrix [66] to represent resemblances in community structure among transects from each station and session.

Inspection of correlograms of environmental variables revealed significant positive skew in many variables as well as several nonlinear relationships among variables. To ameliorate potential problems in subsequent analyses, all variables were transformed by their natural logarithm ($\ln(x+1)$) to linearise relationships and to provide more symmetrical distributions. All physical variables were then converted to z-scores (i.e., scaled to a mean of zero and unit variance) to avoid scale dependence in subsequent analyses. Potential collinearity among environmental variables was explored by inspecting the correlogram of the transformed data and by calculating condition indices of the principal components of the correlation matrix, as recommended by Quinn and Keough [67].

Hypotheses regarding spatio-temporal relationships were tested using Pearson correlation, while those concerning patterns were tested using PERMANOVA, as described by Anderson et al. [68]. In the latter case, session, side (north/south of the estuary mouth) and station (nested within side) were considered to be fixed factors. Following each PERMANOVA, *post-hoc* tests were used to further explore significant interactions or main effects, using Monte Carlo approximate p-values [68] where insufficient unique permutations existed for meaningful tests. Where multivariate hypotheses were tested, non-Metric Multidimensional Scaling (nMDS) was used to illustrate relationships among samples [66].

Because environmental data were collected on a per-station basis, rather than on a per-transect basis, all assessments of relationships between physico-chemical variables and community structure were made on per-station basis. To determine the set of physical variables that best described the observed spatio-temporal structure in the community data, the Envfit and Bioenv functions of the Vegan package in R were used [69]. The Envfit function fits individual environmental variables onto a given nMDS ordination of biotic variables in such a way as to maximize the correlation between the two data sources. By contrast, the Bioenv function investigates different combinations of environmental variables in an attempt to identify that subset for which the Euclidean distance resemblance matrix provides the maximum Spearman correlation with the Bray-Curtis community resemblance matrix. For the Envfit analyses, session and side were treated as factors (categorical variables) and the remaining variables as vectors (continuous variables). For the Bioenv analysis, we initially considered all environmental variables, including session and side (which had to be treated as continuous variables in this context). The procedure was then rerun without either session or side to assess whether some suite of other environmental variables might provide a solution to explain similar amounts of the variance.

In multivariate analysis (Bioenv), we are aware of no criteria that penalize improvement in fit relative to the number of parameters (as would be done in a more conventional model fitting by the Akaike Information Criterion, for example). We therefore set a simple *a priori* criterion for the "optimal" environmental matrix for the Bioenv analysis as that with the smallest number of variables, which improves the correlation by at least 10% relative to the strongest single-factor correlation. Although we acknowledge that this criterion is arbitrary, we feel that beyond this point, each additional environmental variable adds relatively little to the explanation of the ordination, and might potentially introduce spurious correlations.

Because we conducted many statistical tests, the probability of falsely rejecting at least one null hypothesis (i.e., the probability of a false positive result) would be substantially larger than the conventionally chosen α-level of 0.05 [70]. Although the sequential Bonferroni test is sometimes used to adjust inflated α-levels on a case-by-case basis, there are several mathematical, logical and practical objections to its use [71,72]. We therefore acknowledge that there is a strong possibility that some of our hypothesis tests will falsely reject true null hypotheses and thus set α to 0.01 to ameliorate this risk. While this concession appears arbitrary and will not eliminate false positives, it does provide a practical and consistent criterion for hypothesis tests and reduces the false positive rate by a factor of five.

Data manipulation and most analyses were carried out in R [73]; PERMANOVA and verification of multivariate analyses were conducted using PRIMER 6 [74].

Results

Physical environment

The environment along Mtunzini Beach, as described by beach morphodynamics and sediment characteristics, varied substantially in space and time (Fig. 2; Table 1, 2). The intertidal beach face fluctuated in width between 45 m and 90 m, being significantly ($p<0.01$) wider in October 2008 than in subsequent sessions (Fig. 2A; Table 1). Mean particle size showed that the sediment of Mtunzini Beach is coarse, moderately well sorted sand. Spatially, sediments were finer and better sorted south of the estuary than to the north, especially during February 2009 (Fig. 2F–I; Table 1, 2). Temporally, sediments during October 2008 were coarser and more coarse skewed than during other sessions, although skewness decreased with distance from the estuary mouth (Fig. 2F–I; Table 1, 2). According to Dean's dimensionless fall velocity (DFV), Mtunzini Beach was intermediate (mean $\Omega = 3.75$), with no spatial or temporal patterns in morphodynamic state, besides a tendency for the beach to be more reflective during October 2008 than during the other sessions (Fig. 2B; Table 1).

Chemical environment

Monthly rainfall varied from 45.4 mm in December 2008 to 184.5 mm in February 2009 (Fig. S1). Instead of increasing gradually through the study period, as one may predict, monthly rainfall was relatively uniform from October to December 2008, with an increase in January 2009. Salinity measurements matched this pattern (Fig. 2C; Table 1), being similar during October and December 2008, when they were uniform alongshore, but salinity fell significantly in February 2009 (Fig. 2C; Table 1). During this final session, salinity was significantly lower south of the estuary

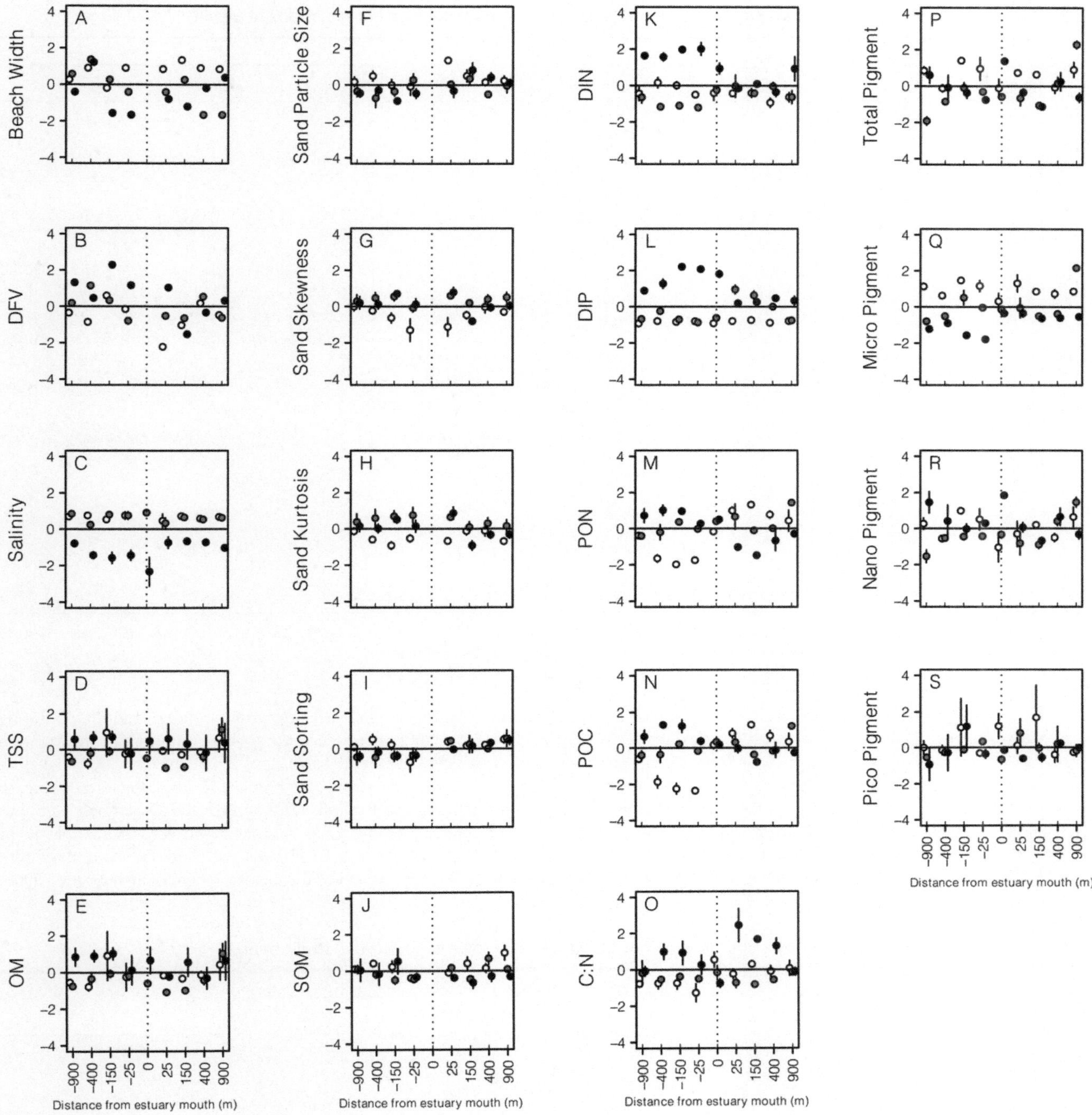

Figure 2. Plots of physical variables collected at the sample stations on Mtunzini Beach. All data are log-transformed and standardized to a mean of zero and SD of 1. Samples from October 2008 are indicated with white circles, those from December 2008 with grey circles and those from February 2009 with black circles. Error bars are standard deviations. For clarity, the *x*-axes are not plotted to scale. Negative distances are stations to the south of the estuarine mouth. Mean and standard deviation of untransformed data are: (A) Beach width (67.44±14.71 m), (B) DFV (3.69±0.52), (C) Salinity (33.25±5.89), (D) TSS (52.92±19.02 mg.l^{-1}), (E) OM (42.97±17.23 mg.l^{-1}), (F) Sand Particle Size (428.68±101.32 µm), (G) Sand Skewness (1.66±0.71), (H) Sand Kurtosis (8.29±4.31), (I) Sand Sorting (218.44±92.32), (J) SOM (1.29±1.05%), (K) DIN (0.62±0.65 µM), (L) DIP (0.9±1.21 µM), (M) PON (5.47±2.72%), (N) POC (45.79±24.11%), (O) C:N (8.44±2.27), (P) Total Photopigment concentrations (3.52±1.10 µg.l^{-1}), (Q) Microplankton Photopigment concentrations (1.08±0.62 µg.l^{-1}), (R) Nanoplankton Photopigment concentrations (1.95±0.87 µg.l^{-1}), (S) Picoplankton Photopigment concentrations (0.54±0.31 µg.l^{-1}).

than to the north, and was lowest in the estuary mouth (Fig. 2C). Spatio-temporal patterns in DIN and DIP concentrations reflected those in salinity, tending to be high where salinity was low (Fig. 2K, L; Fig. S1; Table 1). However, maximum concentrations were not recorded in the estuary mouth, and additional complexities were evident. Nevertheless, DIN and DIP concentrations tended to be higher during the latter sessions of the study, and this effect was more marked to the south of the estuary than to the north.

Table 1. Significant sources of variation in univariate descriptors of the environment on Mtunzini Beach.

Category	Type	Variable	Sources of variation
Physical	Morphodynamic	Beach width	Se
		DFV	Se
Physical	Sediment	Sand particle size	Se, Si
	characteristics	Sediment skewness	Se
		Sediment kurtosis	Se
		Sediment sorting	Si
Chemical	Water chemistry	Salinity	Se×Si, Se
Chemical	Nutrients	DIN concentration	Se×Si, Se, Si
		DIP concentration	Se×St (Si), Se×Si, St (Si), Se, Si
Nutritional	Photopigments concentrations	Micoplankton photopigment concentration	Se×St (Si), Se×Si, St(Si), Se, Si
		Nanoplankton photopigment concentration	Se×St (Si), Se×Si
		Picoplankton photopigment concentration	None
		Total photopigment concentration	Se×St(Si), Se×Si, St(Si), Se
Nutritional	Organic content	SOM	None
		POM	None
		POC	Se×St (Si), Se×Si, Se, Si
		PON	Se×St(Si), Se×Si, Se, Si
		C:N	Se, Si
		TSS	None

Significant ($p \leq 0.01$) results as identified by two-way fixed-effects PERMANOVA (Table S2). Abbreviations are as follows: Se = session; Si = side; St(Si) = station nested within side; and St = station (only for variables that were estimated without replication at the station level).

POC and PON concentrations exhibited very similar patterns and varied strongly between sides and sessions (Fig. 2M, N; Table 1). Concentrations of both differed significantly by session to the south of the estuary, with values increasing steadily through the sample period. To the north of the estuary the trend reversed, with both POC and PON concentrations significantly lower during February 2009 than during preceding sessions. This resulted in reversing spatial gradients along the beach, with increasing south-north concentration gradients in both POC and PON during December 2008, but a decreasing gradient in POC concentration during February 2009 (Table 2).

C:N ratios were slightly lower to the south of the estuary than to the north, and were similar in October and December 2008, but were significantly higher in February 2009 (Fig. 2O; Table 1). In October and December 2008, values measured in the estuary were among the highest along the beach, whereas in February 2009,

Table 2. Spatial trends in biotic and abiotic variables along Mtunzini beach during the study period.

Variable	Session	Trend	r	df	P
Sand particle size	February 2009	S-N	0.30	72	8.85×10^{-3}
Sand skewness	October 2008	Away from mouth	0.30	77	6.88×10^{-3}
Sand sorting	February 2009	S-N	0.36	72	1.56×10^{-3}
Sand sorting	All	S-N	0.22	222	8.38×10^{-4}
POC	December 2008	S-N	0.58	21	3.40×10^{-3}
POC	February 2009	S-N	−0.55	22	5.84×10^{-3}
PON	December 2008	S-N	0.61	21	1.93×10^{-3}
C:N	February 2009	Away from mouth	−0.52	22	9.15×10^{-3}
Micro pigment	December 2008	S-N	0.69	21	2.40×10^{-4}
Nano pigment	December 2008	S-N	0.78	21	5.05×10^{-6}
Total pigment	December 2008	S-N	0.85	21	2.00×10^{-7}
Species richness	February 2009	S-N	0.60	22	2.09×10^{-3}

Pearson correlations against distance from south-most station (Trend = S-N) or absolute distance from estuary mouth (Trend = Away from mouth). Tests were run on data from individual sessions, and for all sessions combined. Only significant ($p \leq 0.01$) results are tabulated. Descriptions of variables and their abbreviations are listed in Table S1.

estuarine values were lowest (Fig. 2O). Despite this, C:N ratios at beach stations decreased with distance from the estuary during February 2009 (Fig. 2O; Table 2).

The average concentrations of photopigments in the nanoplankton size range were almost twice those of the microplankton fraction, which in turn were twice those of in the picoplankton fraction (Fig. 2). Photopigment concentrations in the microplankton fraction decreased significantly through the study period on both sides of the estuary mouth (Fig. 2Q; Table 1, 2). Corresponding patterns for the nanoplankton fraction were less clear, and no trends were found for the picoplankton fraction; as a result, temporal patterns for total photopigment concentrations were inconsistent through time (Fig. 2R, S; Table 1). During December 2008, both microplankton and nanoplankton photopigment concentrations increased from south to north across the beach, a pattern reflected by total photopigment concentrations (Fig. 2P–R; Table 1, 2). In February 2009, this gradient broke down, but microplankton photopigment concentrations remained lower to the south of the estuary than to the north. Interestingly, the highest nanoplankton photopigment concentrations were recorded in the estuary mouth during February 2009.

Neither TSS nor POM concentrations exhibited any significant spatio-temporal variation (Fig. 2D, E; Table 1). Concentrations of SOM were similarly devoid of clear patterns (Fig. 2J; Table 1).

The correlogram of the scaled, transformed environmental variables (Fig. S2), indicated few potentially collinear environmental variables besides sand grain size and DFV ($r = -0.94$), PON and POC ($r = 0.94$), and OM and TSS ($r = 0.89$). This was confirmed by analysis of the condition indices of the principal components of the correlation matrix, of which only two were >30. Because the number of potentially collinear explanatory variables was small, we did not exclude any of them from subsequent analyses, but exercised appropriate care when interpreting results.

Macroinfaunal community structure

Trends in univariate measures of community structure were similar irrespective of whether they were viewed on a per-transect or per-station basis (Fig. 3). PERMANOVA and correlation analysis suggest that macroinfaunal abundance was lowest during October 2008, when both abundance and species richness were higher to the south of the estuary than to the north (Fig. 3; Table 2). By December 2008, these patterns had broken down, and by February 2009, very high numbers of recruiting mole crabs (*Emerita austroafricana*) and surf clams (*Donax madagascariensis*) immediately north of the estuary resulted in higher macroinfaunal abundance on the northern side of the beach than to the south.

The spatio-temporal patterns in univariate community data were reflected in the multivariate community structure (Bray-Curtis resemblance on fourth-root transformed abundance data). The nMDS biplot (Fig. 4A) illustrates changes in the arrangement of the community structure on Mtunzini Beach from October 2008 to February 2009, with samples from December 2008 showing an intermediate position on the ordination between those

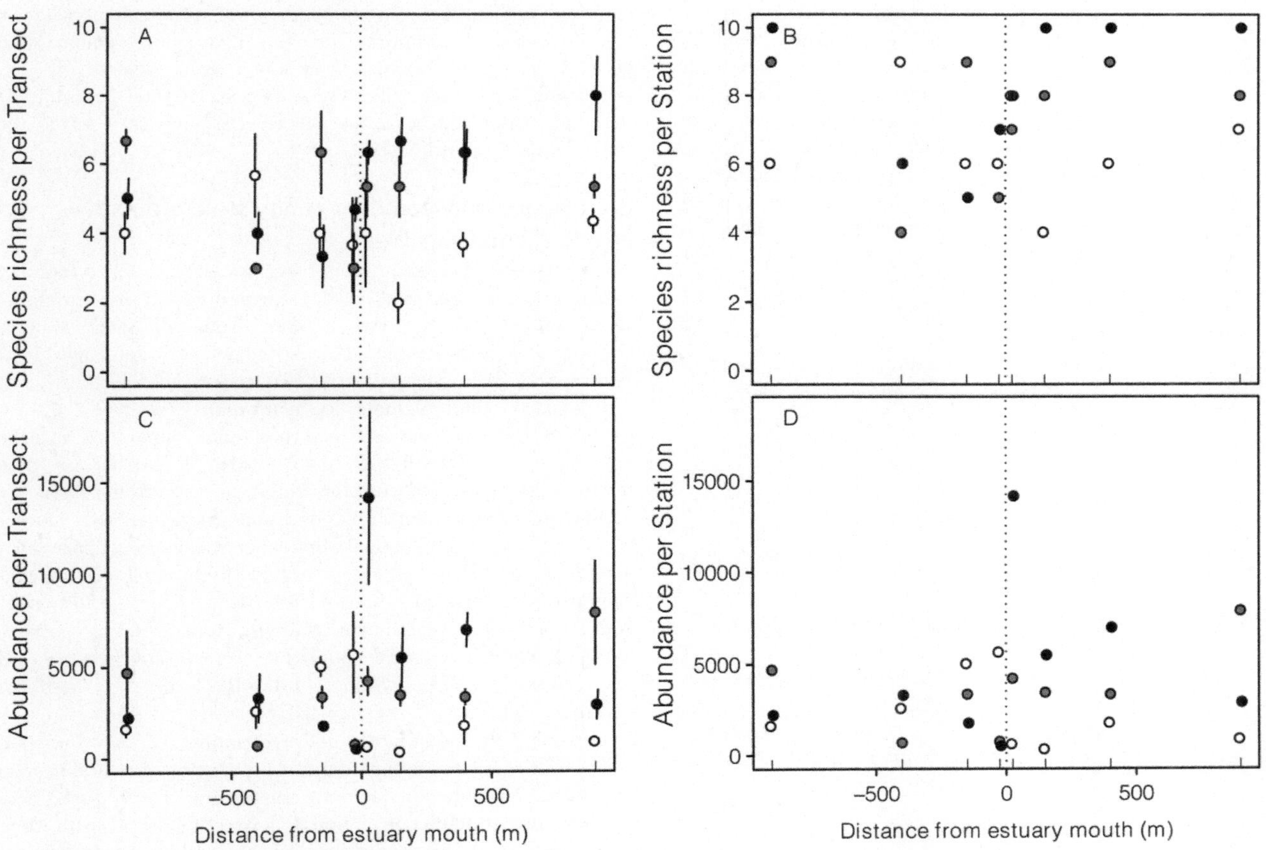

Figure 3. Plots of univariate measures of community structure per transect (A, C; mean ± SD) and aggregated by station (B, D). Samples from October 2008 are indicated with white circles, those from December 2008 with grey circles and those from February 2009 with black circles.

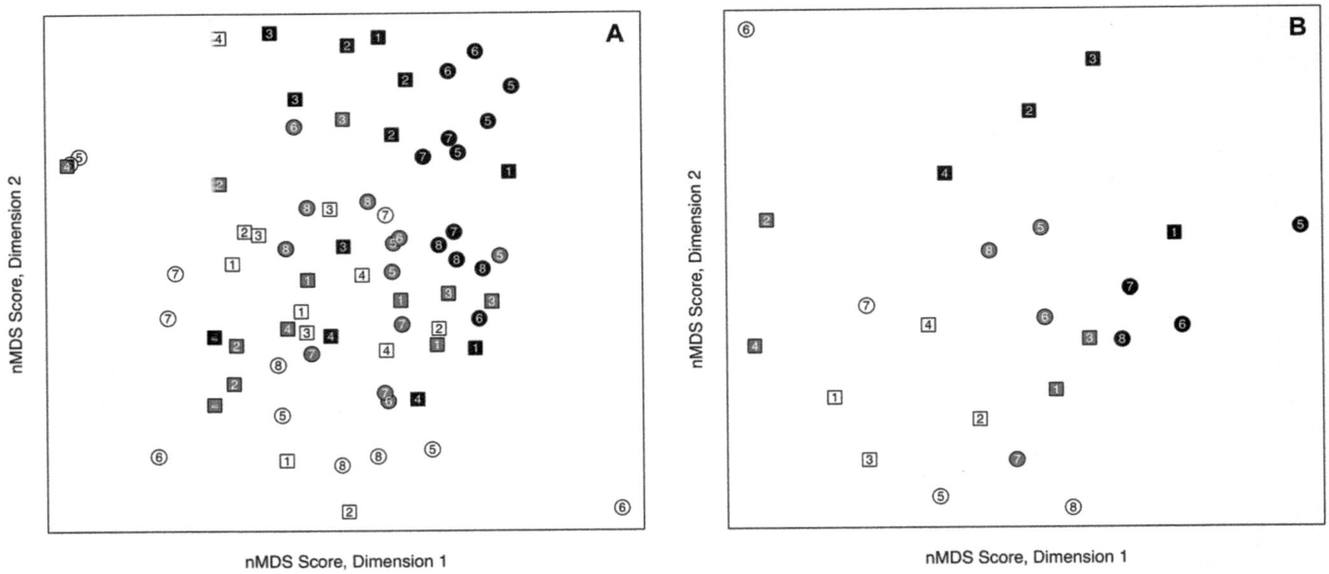

Figure 4. nMDS ordination of species abundance per transect (A) and aggregated by station (B), fourth-root-transformed using Bray-Curtis resemblance. Numerals indicate station numbers (Station 1 is in the far south; Station 8 is in the far north, as per Fig. 1); colors indicate session (white = October 2008; grey = December 2008; black = February 2009); and symbols indicate side of beach relative to the estuary mouth (square = south; circle = north). Stress = 0.25.

of the first and the last sessions. The stress value for the two-dimensional ordination was relatively high at 0.25, but the significant side × session interaction detected by PERMANOVA on these multivariate data (Table 3) nevertheless reveals significant spatio-temporal patterns. Specifically, although the community structure to the south of the estuary did not differ significantly between October and December 2008, all other pair-wise comparisons were significant. Similarly, the community structure to the north of the estuary differed significantly between all sessions. The strongest (and only significant) north-south difference in community structure was detected in February 2009. Together, these results mirror the univariate analysis in suggesting a graded change in community structure through time, with strongest spatial patterns in February 2009.

When pooling the samples by station (to accommodate comparison with environmental data, which were not measured at a per-transect resolution), replication and therefore power are lost from the PERMANOVA. Nevertheless, the nMDS ordination of this modified data set (Fig. 4B) broadly reflected that constructed on data from individual transects (Fig. 4A). The corresponding PERMANOVA (Table 3) detected a significant main effect of session, although due to reduced power, the

Table 3. Significant sources of variation in univariate and multivariate measures of community structure.

Type	Response variable	Sources of variation
Univariate	Species richness	Se×Si, Se
	Macrofaunal abundance	Se×St (Si), Se×Si, Se, Si
Multivariate	Bray-Curtis resemblance by transect	Se×St (Si), Se×Si, Se, Si
	Bray-Curtis resemblance by station	Se

Significant ($p \leq 0.01$) results as identified by two-way fixed-effects PERMANOVA (Table S3). Abbreviations as per Table 1.

community structure was found to differ significantly only between October 2008 and February 2009 (the p-value for the pair-wise test on differences in community structure between December 2008 and February 2009 was 0.019; although we considered this non significant due to our adjustment in α to account for multiple hypothesis tests, this value is nevertheless low, considering the reduced power of this test).

Correlations between community structure and environmental variables

Correlations between univariate measures of community structure and the physical beach environment did not reveal any significant relationships when data from all sessions were considered simultaneously. The strongest (non-significant) relationships in this context were between species richness and picoplankton photopigment concentrations ($r = 0.49$, df = 22, $p = 1.39 \times 10^{-2}$) and between macroinfaunal abundance and the C:N ratio ($r = 0.50$, df = 22, $p = 2.05 \times 10^{-2}$). Neither of these relationships were significant for individual sessions; nor were any of the observed relationships consistent across sessions. During October 2008, macroinfaunal abundance was inversely correlated with POC ($r = -0.90$, df = 6, $p = 2.05 \times 10^{-3}$) and PON concentrations ($r = -0.88$, df = 6, $p = 3.60 \times 10^{-3}$). During December 2008, there were no significant relationships and, during February 2009, species richness and abundance were positively correlated with salinity ($r = 0.87$, df = 6, $p = 4.83 \times 10^{-3}$) and C:N ($r = 0.85$, df = 6, $p = 8.28 \times 10^{-3}$), respectively.

Inspecting the explanatory power of individual environmental variables, the Envfit routine identified four environmental vectors and one environmental factor as significant (Table 4). Salinity provided the strongest correlation, followed by DIP concentration, C:N ratio and microplankton photopigment concentration. Session was the significant environmental factor.

When using the function Bioenv to seek matrices of environmental variables that best correlate with the biotic ordination, correlations tended to increase with increasing numbers of

Table 4. Fits of environmental vectors and factors to the biotic ordination of macroinfaunal community structure.

Variable	Correlation (*r*)	*p*
Vector		
Salinity	0.738	3.4×10^{-4}
DIP	0.737	3.6×10^{-4}
C:N	0.719	4.8×10^{-4}
Microplankton photopigments	0.600	9.7×10^{-3}
Factor		
Session	0.593	6.0×10^{-4}

Environmental vectors are continuous variables and factors are discrete variables. Correlations indicate the strength of the gradient in individual environmental variables across the biotic ordination. Only significant ($p\leq0.01$) results are tabulated.

environmental variables (Table 5). When session and side were included as explanatory variables, correlations peaked at seven variables ($r = 0.495$, this correlation before rounding was fractionally higher than that for the six-variable ordination). When excluding session and side, correlation peaked at six variables ($r = 0.475$, Table 5). When considering the relative increase in correlation with each additional variable (Table 5), the set of environmental variables identified as optimal by our *a priori* rule includes only three variables, irrespective of whether we include session and side as variables ($r = 0.443$), or not ($r = 0.440$). Permutation tests reveal that all of the selected correlations are significant ($p<0.01$).

To summarize, the most important variables explaining biotic community structure are: session (identified by PERMANOVA, Envfit and Bioenv); the C:N ratio (identified by Envfit and Bioenv, and it also had the strongest, though non-significant, correlation with macroinfaunal abundance over the study period), picoplankton photopigment concentrations (identified by Bioenv, and it also had the strongest, though non-significant, correlation with species richness over the study period), and DIP concentration (identified by Envfit and Bioenv). To these, we can tentatively add salinity, which was identified as influential only by Envfit, but did have strong correlations with macroinfaunal abundance during February 2009, when salinity values were lowest and spatial patterns in community structure were strongest.

Using 2-dimensional thin-plate splines fitted by General Additive Modeling, we overlaid the distribution of these selected environmental variables on the biotic ordinations (Fig. 5A–H). The results provide evidence that most of the influential environmental variables are non-linearly distributed across the biotic ordination. They also demonstrate that the variables selected first by the Bioenv routine explain outliers in the biotic ordination, and then more general patterns. For example, picoplankton photopigment concentration (the first variable selected by Bioenv) was unusually high at Station 6 in December 2008 (Fig. 2S) and the C:N ratio (the second variable selected) was especially high at Station 5 in February 2009 (Fig. 2O). Since corresponding community samples provided outliers in the biotic ordination, and the C:N ratio also provides some discrimination among sessions (especially separating samples from February 2009 from those of other sessions), it is not surprising that these variables together provide an ordination that is fairly strongly correlated ($r = 0.371$, Table 5) with the biotic ordination. DIP concentration adds to the abiotic ordination's ability to discriminate among all sessions; it also highlights the differences between the northern and southern sides of the beach during February 2009, when salinities were lowest and biotic patterns were strongest. The remaining variables add slowly to an increasingly complex pattern. Importantly, the nMDS biplots of both the optimal and the best explanatory environmental matrices (omitting session) showed a strong resemblance to that for the biotic variables (Fig. 5I, J). Although sessions and sides were grouped in a very similar way in both biotic and abiotic ordinations, individual stations were somewhat displaced relative to one another, reflecting the correlations of 0.440 and 0.475, respectively (Table 5).

Discussion

The concept of overriding physical control in harsh environments is one of three paradigms describing macro-scale community patterns on exposed sandy beaches [75]. Biological factors (including food availability, competition and predation), by contrast, have been considered to play a role in structuring macrofaunal communities at the meso-scale only in physically benign habitats, such as undisturbed dissipative beaches [13]. In this study, we aim to determine if the macroinfaunal community inhabiting a physically dynamic beach (intermediate morphodynamic type) really does

Table 5. Correlations between environmental and biotic ordinations for data from Mtunzini Beach.

	Analyses including Side and Session as environmental variables		Analyses excluding Side and Session as environmental variables	
No.	**Variables**	***r***	**Variables**	***r***
1	Session	0.307	Pico pigments	0.251
2	Session, Pico pigments	0.411	C:N, Pico pigments	0.371
3	*Session, C:N, Pico pigments*	*0.443*	*DIP, C:N, Pico pigments*	*0.440*
4	Session, DIN, C:N, Pico pigments	0.463	Kurtosis, DIP, C:N, Pico pigments	0.456
5	Session, Beach width, DIN, C:N, Pico pigments	0.483	Beach width, DIN, DIP, C:N, Pico pigments	0.464
6	Session, Beach width, Kurtosis, DIN, C:N, Pico pigments	0.495	**Beach width, Kurtosis, DIN, DIP, C:N, Pico pigments**	**0.475**
7	**Session, Beach width, Kurtosis, DIN, DIP, C:N, Pico pigments**	**0.495**	Beach width, Kurtosis, DIN, DIP, C:N, Pico pigment, Micro pigments	0.457

Correlations are listed with increasing numbers of variables. Left-hand panel includes factors session and side as variables, whereas the right-hand panel excludes them. Combinations of variables that maximize the correlation between biotic and abiotic ordinations are in bold, while optimal combinations of variables are italicized. Descriptions of variables and their abbreviations are listed in Table S1.

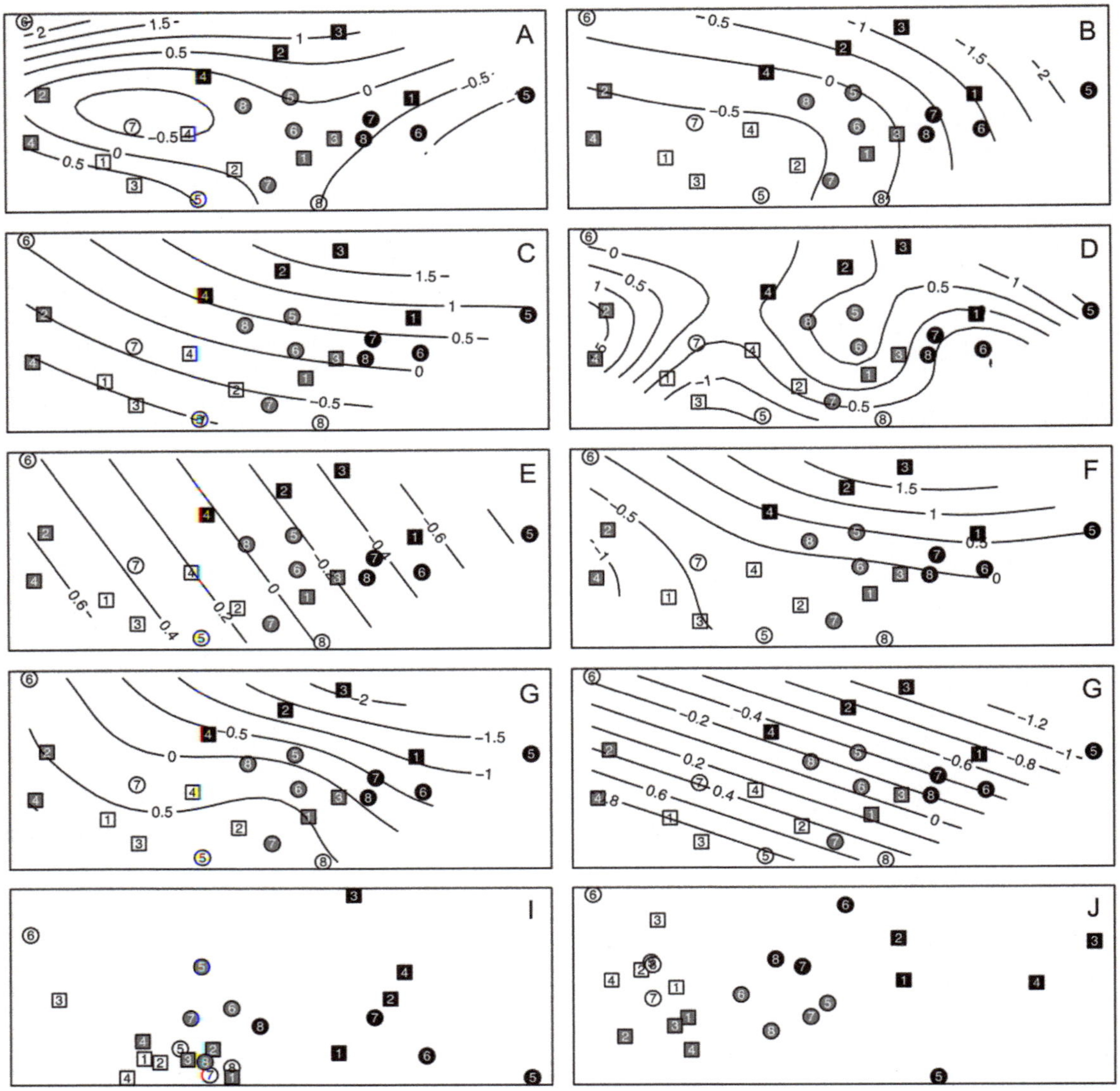

Figure 5. Bray-Curtis ordination of fourth-root transformed biotic variables overlaid with 2-dimensional thin-plate splines fitted by General Additive Modeling. A. Picoplankton photopigment concentrations. B. C:N ratio. C. DIP concentration. D. Sediment kurtosis. E. Beach width. F. DIN concentration. G. Salinity. H. Microplankton photopigment concentrations. I. nMDS ordination of "optimal" set of environmental variables. J. nMDS ordination of the set of environmental variables providing the best biotic-environmental correlation. A-F are the variables identified by the Bioenv routine in order of importance (i.e. the order in which they were selected). G & H are variables selected by Envfit, but that were not selected by Bioenv. I & J are Euclidean-distance nMDS ordinations of the matrices of environmental variables for the optimal and best solutions identified by Bioenv.

respond overwhelmingly to the physical environment at the meso-scale, as predicted by theory. The study site, Mtunzini Beach, is located on an oligotrophic, subtropical coast, and is bisected by the Mlalazi Estuary, which is considered the main source of nutrient inputs for the beach and surf zone. Our univariate and multivariate analyses demonstrate significant spatio-temporal variation in the physical and chemical features of the beach-surf-zone system, and also in its resident macroinfaunal community. If community structure on this beach were subject primarily to physical control, we would expect variables associated with the physical beach environment to consistently provide the best description of macrofaunal distribution through time and space.

In terms of the physical beach environment, macrobenthic abundance and species richness are expected to increase with beach width [21] and morphodynamic state (DFV) [13], and to decrease with sand particle size [20]. All of these physical variables differed significantly between October 2008 and subsequent sessions, but none exhibited significant spatial trends during either October or December 2008. Although univariate measures of community structure were similarly devoid of spatial pattern during December 2008, both macrofaunal abundance and species richness were lower to the south of the estuary than to the north during October 2008, when both biotic variables attained their minimum values and the beach was widest. This provides an initial indication that even when risks of competition and predation are most limited (i.e. when the densities of organisms are lowest) [11,13,76], factors other than physical characteristics of the beach can and do influence the spatial distribution of resident communities. Further evidence against the paradigm of physical control at the meso-scale is provided by the results from February 2009. During this session, the finest and best sorted sediments were found to the south of the estuary, but contrary to expectation, macrofaunal abundance was higher to the north of the estuary, and there was also a south-north increase in species richness. Our results therefore strongly suggest that physical factors are not adequate predictors of macroinfaunal abundance and species richness on Mtunzini Beach.

In terms of the chemical and nutritional composition of the surf-zone waters, significant spatio-temporal variations were observed for most variables and, as predicted, most of these patterns were associated with increased estuarine flow in February 2009. Estuarine input can directly influence macrobenthic community structure through two mechanisms. The first is a negative impact of low salinity on alongshore distribution patterns, abundance, biomass and life-history traits of sandy beach organisms [35–37]. The second is an increase in food resources, either through increased surf-zone primary production resulting from elevated concentrations of dissolved inorganic nutrients [34,77], or through increased inputs of particulate organic matter [34]. During our study, salinity showed no along-shore pattern during the dry season (October and December 2008), but was lower to the south of the estuary than to the north during the rainy season (February 2009), when salinity values were lowest overall. As anticipated, estuarine input to the surf zone strongly influenced patterns in the concentrations of dissolved inorganic nutrients (DIN and DIP). However, although surf-zone phytoplankton was dominated by nanoplankton (measured by photopigment concentrations), as is common in these waters [78–80], elevated nutrient concentrations in the surf-zone did not boost phytoplankton concentrations. This suggests either that the residence time of DIN and DIP in the surf-zone is too short to allow phytoplankton to fully respond, or that the reduced salinity inhibited surf-zone productivity [81,82]. Concentrations of particulate organic material were measured in our study by SOM and by swash-zone TSS, POM, POC and PON, with the C:N ratio providing an indication of the provenance of organic matter [83]. Of these variables, POM, TSS and SOM failed to show significant spatio-temporal patterns and were unrelated to salinity or inorganic nutrient concentrations. This suggests that estuarine inputs of these materials are unlikely to directly control community structure on Mtunzini Beach. PON and POC patterns provided more contrast, and peaked to the south of the estuary during February 2009, corresponding with the presence of a plume of low-salinity estuarine water (although to the north of the estuary, both variables were at their lowest during this session). C:N ratios also reflect a trend of increasing estuarine provenance of swash-zone organics throughout the study period, and in February 2009, when the C:N ratio was highest, there was a strong gradient of decreasing C:N ratio away from the estuary mouth reinforcing the notion that particulate organics were of estuarine rather than marine origin (although again, we could not account for the low C:N measured in the estuary mouth). Despite these anomalies, on a per-session basis, relationships between univariate measures of macroinfaunal community structure and descriptors of the chemical composition of the surf-zone waters were consistently stronger than those with physical beach variables. Further, when considering all data from the study simultaneously, the strongest (although not significant at $\alpha = 0.01$) biotic-abiotic relationships were trends of increasing species richness with picoplankton photopigment concentrations and of increasing macroinfaunal abundance with C:N ratio. Together, these results confirm that physical features of the beach face are not the most influential drivers of macroinfaunal community structure on Mtunzini Beach.

Using multivariate techniques to explore the covariation in ordinations of biotic and abiotic variables, the graded change in community structure between October 2008 and February 2009, and the increasing degree of spatial structure in February 2009 were best explained by sample session, the C:N ratio of organic matter, picoplankton photopigment concentration and DIP concentration. Sample session probably explains variability in community structure introduced by seasonal patterns such as recruitment dynamics. The other influential variables all describe features of the surf zone that are linked with food availability, either realized (particulate organics of estuarine origin, and picoplankton abundance) or potential (elevated inorganic nutrient concentrations of estuarine origin). Although we tentatively identified salinity as an additional influential variable, it is likely that most of its explanatory power derives from its strongly seasonal temporal pattern, which is already accounted for by sample session. Irrespective of the mechanisms by which the most influential variables structure the macrobenthic community of Mtunzini Beach, none of these variables describe physical aspects of the beach face. Even when increasing the number of explanatory variables beyond the "optimal" set to include all of those identified as contributing to the maximum correlation with the biotic ordination, beach width and sediment kurtosis are the only physical variables added. While macroinfaunal abundance and diversity are predicted to increase with beach width [21], the opposite appears to have occurred here, with lower abundance and fewer species present on the beach during October 2008, when the beach was widest. It is difficult to envisage a mechanism by which sediment kurtosis could influence community structure on beaches; it is therefore not surprising that kurtosis has not previously been identified as being important in this role. Results of multivariate analyses therefore support those of univariate analyses in contradicting the paradigm of physical control of community structure on Mtunzini Beach.

The present results do not contest the fact that patterns in physical processes explain significant amounts of variability in macroinfaunal species richness and abundance among geographic regions at macro-ecological scales [13,20,21]. They do, nevertheless, demonstrate that over smaller spatial scales, other processes are capable of driving the spatio-temporal distribution patterns of beach macroinfauna, even on physically dynamic beaches. This emphasises the fact that beaches are functional ecosystems [53,84] that have more in common with other coastal soft-sediment systems than is generally appreciated. This is important for many reasons, not least because beaches are increasingly threatened by accelerating coastal squeeze that is driven by climate change and coastal urbanisation [85,86]. The paradigm of physical control has encouraged coastal managers to consider the ecology of beaches subordinate to their sedimentary regime, but here we show that the ecology of individual beaches is considerably more complex. The challenge ahead lies in further exploring this complexity and in using resultant knowledge to ensure the persistence of beaches as ecosystems within the broader coastal zone.

Supporting Information

Figure S1 Plot of rainfall data in the catchment of the Mlalazi Estuary. Data corresponds to station 478, Empangeni - South African Sugarcane Research Institute (SASRI).

Figure S2 Correlogram illustrating the relationships among environmental variables. Variables are arranged according to the strength of relationships expressed in a corresponding PCA of all variables. Upper triangular panel contains pair-wise scatter plots for variables. Lower triangular panel contains concentration ellipses (1 SD, mean centered) for the bivariate data, with loess-smoothed curves of the relationships. Straight lines and narrow, diagonal ellipses indicate potential collinearities.

Table S1 List of the environmental variables analyzed in this study.

Table S2 Two-way fixed-effects PERMANOVA on univariate descriptors of the physical, chemical and nutritional environment on Mtunzini Beach.

Table S3 Two-way fixed-effects PERMANOVA on univariate and multivariate measures of community structure.

Acknowledgments

Thanks to Ezemvelo KwaZulu-Natal Wildlife for help with logistics during the fieldwork, in particular to Callum Beatie and Santosh Bachoo. Special thanks to Linda Harris, Natasha Govender, Troy Inman and other field assistants for their huge help during the fieldwork. Also thanks to Linda Harris for her assistance with Figure 1. We are grateful to two anonymous reviewers who made helpful comments to a previous version of the manuscript.

Author Contributions

Conceived and designed the experiments: KOC DSS AJS. Performed the experiments: KOC. Analyzed the data: KOC DSS AJS. Contributed reagents/materials/analysis tools: DSS AJS JL. Wrote the paper: KOC DSS AJS JL.

References

1. Hairston NG, Smith FE, Slobodkin LB (1960) Community structure, population control, and competition – a critique. Ecology 79: 1678–1687.
2. Hewitt JE, Pridmore RD, Thrush SF, Cummings VJ (1997) Assessing the short-term stability of spatial patterns of macrobenthos in a dynamic estuarine system. Limnol Oceanogr 42: 282–288.
3. Norkko A, Cummings VJ, Thrush S, Hewitt JE, Hume T (2001) Local dispersal of juvenile bivalves: implications for sand flat ecology. Mar Ecol Prog Ser 212: 131–144.
4. Richardson AJ, Schoeman DS (2004) Climate impact on plankton ecosystems in the Northeast Atlantic. Science 305: 1609–1612.
5. Cury PM, Shin YJ, Planque B, Durant JM, Fromentin JM, et al. (2008) Ecosystem Oceanography for global change in fisheries. Trends Ecol Evol 23: 338–346.
6. Blanchette CA, Schiel DR, Wieters EA, Broitman BR, Kinlan BP (2009) Trophic structure and diversity in rocky intertidal upwelling ecosystems: A comparison of community patterns across California, Chile, South Africa and New Zealand. Prog Oceanogr 83: 107–116.
7. McLachlan A (1983) Sandy beach ecology: a review. In: McLachlan A, Erasmus T, eds. Sandy beaches as ecosystems W. Junk, The Hague. pp 321–380.
8. McLachlan A (1990) Dissipative beaches and macrofauna communities on exposed intertidal sands. J Coast Res 6: 57–71.
9. Defeo O, Brazeiro A, de Alava A, Riestra G (1997) Is sandy beach macroinfauna only physically controlled? Role of substrate and competition in isopods. Est Coast Shelf Sci 45: 453–462.
10. Schoeman DS, Richardson AJ (2002) Investigating biotic and abiotic factors affecting recruitment of an intertidal clam on an exposed sandy beach using a generalized additive model. J Exp Mar Biol Ecol 276: 67–81.
11. Dugan J, Jaramillo E, Hubbard D, Contreras H, Duarte C (2004) Competitive interactions in macrofaunal animals of exposed sandy beaches. Oecologia 139: 630–640.
12. Duarte C, Jaramillo E, Contreras E, Acuña K, Navarro JM (2009) Importance of macroalgae subsidy on the abundance and population biology of the amphipod *Orchestoidea tuberculata* (Nicolet) in sandy beaches of south central Chile. Rev Biol Mar 44: 691–702.
13. Defeo O, McLachlan A (2005) Patterns, processes and regulatory mechanisms in sandy beach macrofauna: a multi-scale analysis. Mar Ecol Prog Ser 295: 1–20.
14. Halpern BS, Walbridge S, Selkoe KA, Kappel CV, Micheli F, et al. (2008) A Global Map of Human Impact on Marine Ecosystems. Science 319: 948–952.
15. Airoldi L, Beck MW (2007) Loss, status and trends for coastal marine habitats of Europe. Oceanogr Mar Biol Annu Rev 45: 345–405.
16. Lucrezi S, Schlacher TA, Robinson W (2009) Human disturbance as a cause of bias in ecological indicators for sandy beaches: Experimental evidence for the effects of human trampling on ghost crabs (*Ocypode* spp.). Ecol Indic 9: 913–921.
17. Lucrezi S, Schlacher TA, Robinson W (2010) Can storms and shore armouring exert additive effects on sandy-beach habitats and biota? Mar Freshw Res 61: 951–962.
18. Sheppard N, Pitt KA, Schlacher TA (2009) Sub-lethal effects of off-road vehicles (ORVs) on surf clams on sandy beaches. J Exp Mar Biol Ecol 380: 113–118.
19. Short AD (1996) The role of wave height, slope, tide range and embaymentisation in beach classification: a review. Rev Chil Hist Nat 69: 589–604.
20. McLachlan A, Dorvlo A (2005) Global patterns in sandy beach macrobenthic communities. J Coast Res 21: 674–687.
21. McLachlan A, Dorvlo A (2007) : Species area relationships for sandy beach macrobenthos in the context of intertidal width. Oceanologia 49: 1–15.
22. Lercari D, Defeo O (2006) Large-scale diversity and abundance trends in sandy beach macrofauna along full gradients of salinity and morphodynamics. Est Coast Shelf Sci 68: 27–35.
23. Pearson TH, Rosenberg R (1987) Feast and famine: structuring factors in marine benthic communities. In: Gee JHR, Giller PS, eds. The 27th Symposium of the British Ecological Society, Aberystwyth 1986 Blackwell Scientific Publications, Oxford. pp 373–395.
24. Grant J, Hargrave BT (1987) Benthic metabolism and the quality of sediment organic carbon. Biol Oceanogr 4: 243–264.
25. Thompson J K, Nichols FH (1988) Food availability controls seasonal cycle of growth in *Macoma balthica* (L.) in San Francisco Bay, California. J Exp Mar Biol Ecol 116: 43–61.
26. Graf G (1989) Benthic-pelagic coupling in a deep-sea benthic community. Nature 341: 437–439.
27. Basaguren A, Elosegul A, Pozo J (1996) Changes in the trophic structure of benthic macroinvertebrate communities associated with food availability and stream flow variations. Int Rev Gesamten Hydrobiol Hydrogr 81: 79–91.
28. Ambrose WG, Jr. (1984) Influence of predatory polychaetes on the structure of a soft-bottom community in a Maine estuary. J Exp Mar Biol Ecol 81: 115–45.
29. Ambrose WG, Jr. (1991) Are infaunal predators important in structuring marine soft-bottom communities? Am Zool 31: 849–60.
30. Dugan JE, Hubbard DM, McCrary MD, Pierson MO (2003) The response of macrofauna communities and shorebirds to macrophyte wrack subsidies on exposed sandy beaches of Southern California. Est Coast Shelf Sci 58 S: 25–40.
31. Lastra M, de la Huz R, Sánchez-Mata AG, Rodil IF, Aerts K, et al. (2006) Ecology of exposed sandy beaches in northern Spain: environmental factors controlling macrofauna communities. J Sea Res 55: 128–140.
32. Rodil IF, Lastra M, López J (2007) Macroinfauna community structure and biochemical composition of sedimentary organic matter along a gradient of wave exposure in sandy beaches (NW Spain). Hydrobiologia 579: 301–316.
33. McLachlan A, Brown AC (2006) The ecology of sandy shores. Burlington, MA, USA: Academic Press. 373 p.
34. Schlacher TA, Connolly RM (2009) Land-ocean coupling of carbon and nitrogen fluxes on sandy beaches. Ecosystems 12: 311–321.
35. Bergamino L, Muniz P, Defeo O (2008) Effects of a freshwater canal discharge on polychaete assemblages inhabiting an exposed sandy beach in Uruguay. Ecol Indic 9: 584–587.
36. Celentano E, Gutiérrez N, Defeo O (2010) Effects of morphodynamic and estuarine gradients on a sandy beach mole crab demography and distribution: implications for source sink habitat dynamics. Mar Ecol Prog Ser 398: 193–205.
37. Lozoya JP, Gomez J, Defeo O (2010) Modelling large-scale effects of estuarine and morphodynamic gradients on distribution and abundance of the sandy beach isopod *Excirolana armata*. Est Coast Shelf Sci 87: 472–478.
38. Day JH (1981) The nature, origin and classification of estuaries. In: Day JH, ed. Estuarine ecology with particular reference to southern Africa AA Balkema, Cape Town. pp 1–6.
39. Whitfield AK (1992) The characterization of southern African estuarine systems. S Afr J Aquat Sci 18: 89–103.
40. Cooper JAG, Wright CI, Mason TR (1999) Geomorphology and sedimentology. In: Allanson BR, Baird D, eds. Estuaries of South Africa Cambridge University Press, Cambridge. pp 5–25.
41. Pearce AF, Schumann EH, Lundie GSH (1978) Features of the shelf circulation off the Natal Coast. S Afr J Sci 74: 328–331.
42. Hodgson AN (1987) Distribution of macrobenthic fauna of the Kariega Estuary. S Afr J Zool 22: 153–62.

43. Wooldridge T (1999) Estuarine zooplankton community structure and dynamics. In: Allanson BR, Baird D, eds. Estuaries of South Africa Cambridge University Press, Cambridge. pp 141–166.
44. Vorwerk PD, Whitfield AK, Cowley PD, Paterson AW (2003) The influence of selected environmental variables on fish assemblage structure in range of southeast african estuaries. Environ Biol Fish 66: 237–247.
45. Froneman PW (2006) The importance of phytoplankton size in mediating trophic interactions within the plankton of a southern African estuary. Est Coast Shelf Sci 70: 693–700.
46. Skinner T, Adams JB, Gama PT (2006) The effect of mouth opening on the biomass and community structure of microphytobenthos in a small oligotrophic estuary. Est Coast Shelf Sci 70: 161–168.
47. Wooldridge TH, McGwynne L (1996) The estuarine environment, Institute for Coastal Research Report No C31, Port Elizabeth (South Africa). 91 p.
48. Department of Environmental Affairs and Tourism (DEAT) (2001) South Africa Estuaries: Catchment Land-Cover. National Summary Report. Pretoria. 27 p.
49. Hill BJ (1966) A contribution to the ecology of the Umlalazi estuary. Zoologica Africana 2: 1–24.
50. Nel R, Bezuidenhout K (2008) Ecological Effects of off-road vehicles (for boat launching) on macroinfauna of sandy beaches of northern KwaZulu-Natal. Report prepared for Ezemvelo KwaZulu-Natal Wildlife. 21 p.
51. Bosley KL, Wainright SC (1999) The effects of preservatives and acidification on the stable isotope ratios (15N:14N, 13C:12C) of two species of marine animals. Can J Fish Aquat Sci 56: 2181–2185.
52. Ponsard S, Amlou M (1999) Effects of several preservation methods on the isotopic content of *Drosophila* samples. C R Acad Sci, Sci Vie 322: 35–41.
53. Schlacher TA, Schoeman DS, Dugan J, Lastra M, Jones A, et al. (2008) Sandy beach ecosystems: key features, management challenges, climate change impacts, and sampling issues. Mar Ecol 29: 70–90.
54. Schoeman DS, Wheeler H, Wait M (2003) The relative accuracy of standard estimators for macrofaunal abundance and species richness derived from selected intertidal transect designs used to sample exposed sandy beaches. Est Coast Shelf Sci 58s: 5–16.
55. Schoeman DS, Ronel N, Soares AG (2008) Measuring species richness on sandy beach transects: extrapolative estimators and their implications for sampling effort. Mar Ecol 29: 134–149.
56. Holm-Hansen O, Riemann B (1978) Chlorophyll-a determination: Improvements in methodology. Oikos 30: 438–447.
57. Mostert SA (1983) Procedures used in South Africa for the automatic photometric determination of micronutrients in seawater. S Afr J Mar Sci 1: 189–198.
58. Gray JS (1981) The ecology of marine sediments: an introduction to the structure and function of benthic communities. Barnes RSK, Miller PL, Paul J, up Rees T, eds. Cambridge University Press, Cambridge and New York. 185 p.
59. Blott SJ, Pye K (2001) GRADISTAT: a grain size distribution and statistics package for the analysis of unconsolidated sediments. Earth Surf Proc Land 26: 1237–1248.
60. Govender N (2010) Durban's urbanised coastline: biodiversity, ecosystem services and implications for management. MSc. Dissertation. University of KwaZulu-Natal, Durban, South Africa. 97 p.
61. Blott S (2000) GRADISTAT Version 4.0: A grain size distribution and statistics package for the analysis of unconsolidated sediments by sieving or laser granulometer. http://onlinelibrary.wiley.com/doi/10.1002/esp.261/suppinfo.
62. Gourlay MR (1968) Beach and Dune Erosion Tests. Delft Hydraulics Laboratory, Report no. M935/M936.
63. Dean RG (1973) Heuristic models of sand transport in the surf zone. In: Proc. 1st Aust. Coastal Engr. Conf., Inst. Engrs Aust., Sydney. pp 208–214.
64. Brazeiro A, Defeo O (1996) Macroinfauna zonation in microtidal sandy beaches: is it possible to identify patterns in such variable environments. Estuar Coast Shelf Sci 42: 523–536.
65. Defeo O (1998) Testing hypotheses on recruitment, growth and mortality in exploited bivalves: an experimental perspective. Can Spec Publ Fish Aquat Sci 125: 257–264.
66. Clarke KR, Warwick RM (1994) Change in Marine Communities: An Approach to Statistical Analysis and Interpretation. 2nd edition PRIMER-E, Plymouth, UK. 172 p.
67. Quinn GP, Keough MJ (2002) Experimental Design and Data Analysis for Biologists Cambridge University Press. 537 p.
68. Anderson M, Gorley R, Clarke K (2008) PERMANOVA+ for PRIMER: Guide to Software and Statistical Methods PRIMER-E, Plymouth, UK.
69. Oksanen J, Guillaume Blanchet F, Kindt R, Legendre P, O'Hara RG, et al. (2010) vegan: Community Ecology Package. R package version 1.17-0. http://CRAN.R-project.org/package=vegan.
70. Peres-Neto PR (1999) How many statistical tests are too many? The problem of conducting multiple inferences revisited. Mar Ecol Prog Ser 176: 303–306.
71. Moran MD (2003) Arguments for rejecting the sequential Bonferroni in ecological studies. Oikos 100: 403–405.
72. García LV (2004) Escaping the Bonferroni iron claw in ecological studies. Oikos 105: 657–663.
73. R Development Core Team (2010) R: A Language and Environment for Statistical Computing. R Foundation for Statistical Computing, Vienna, Austria, http://www.R-project.org.
74. Clarke KR, Gorley RN (2006) PRIMER v6: User Manual/Tutorial PRIMER-E, Plymouth. UK. 192 p.
75. McLachlan A (2001) Coastal beach ecosystems. In: Lewin R, ed. Encyclopedia of Biodiversity Academic Press, New York. pp 741–751.
76. Begon M, Townsend CR, Harper JL (2006) Ecology: from individuals to ecosystems Blackwell Publishing. Oxford. UK. 738 p.
77. Schlacher TA, Skillington AJ, Connolly RM, Robinson W, Gaston TF (2008b) Coupling between marine plankton and freshwater flow in the plumes off a small estuary. Int Rev Hydrobiol 6: 641–658.
78. Perissinotto R, Nozais C, Kibirige I (2002) Spatio-temporal dynamics of phytoplankton and microphytobenthos in a South African temporarily open estuary. Est Coast Shelf Sci 55: 47–58.
79. Perissinotto R, Nozais C, Kibirige I, Anandraj A (2003) Planktonic food webs and benthic-pelagic coupling in South African temporarily-open estuaries. Acta Oecol 24: S307–S316.
80. Walker DR, Perissinotto R, Bally R (2002) Phytoplankton/protozoan dynamics in the Nyara Estuary, a small temporarily-open system in the Eastern Cape (South Africa). Afr J Aquat Sci 26: 31–38.
81. Defeo O (1993) The effect of spatial scales in population dynamics and modelling of sedentary fisheries: the yellow clam *Mesodesma mactroides* of an Uruguayan exposed sandy beach. PhD dissertation, CINVESTAVIPN, Merida. 308 p.
82. Lercari D, Defeo O (2003) Variation of a sandy beach macrobenthic community along a human-induced environmental gradient. Estuar Coast Shelf Sci 58: 17–24.
83. Stein R (1991) Accumulation of Organic Carbon in Marine Sediments Springer-Verlag. Berlin. 217 p.
84. Defeo O, McLachlan A, Schoeman DS, Schlacher TA, Dugan J, et al. (2009) Threats to sandy beach ecosystems: a review. Estuar Coast Shelf Sci 81: 1–12.
85. Schlacher TA, Dugan J, Schoeman DS, Lastra M, Jones A, et al. (2007b) Sandy beaches at the brink. Divers Distrib 13: 556–560.
86. Dugan JE, Defeo O, Jaramillo E, Jones AR, Lastra M, et al. (2010) Give beach ecosystems their day in the sun. Science 329: 1146.

4

Translocations as Experiments in the Ecological Resilience of an Asocial Mega-Herbivore

Wayne L. Linklater[1,2]*, Jay V. Gedir[1], Peter R. Law[3], Ron R. Swaisgood[4], Keryn Adcock[5], Pierre du Preez[6], Michael H. Knight[2,7], Graham I. H. Kerley[2]

1 Centre for Biodiversity and Restoration Ecology, School of Biological Sciences, Victoria University of Wellington, Wellington, New Zealand, 2 Department of Zoology, Centre for African Conservation Ecology, Nelson Mandela Metropolitan University, Port Elizabeth, South Africa, 3 PRLDB Modeling, Monroe, New York, United States of America, 4 San Diego Zoo Global Institute for Conservation Research, Escondido, California, United States of America, 5 IUCN African Rhino Specialist Group, Hilton, KwaZulu-Natal, South Africa, 6 Directorate of Scientific Services, Ministry of Environment and Tourism, Windhoek, Namibia, 7 Park Planning and Development, South African National Parks, Port Elizabeth, South Africa

Abstract

Species translocations are remarkable experiments in evolutionary ecology, and increasingly critical to biodiversity conservation. Elaborate socio-ecological hypotheses for translocation success, based on theoretical fitness relationships, are untested and lead to complex uncertainty rather than parsimonious solutions. We used an extraordinary 89 reintroduction and 102 restocking events releasing 682 black rhinoceros (*Diceros bicornis*) to 81 reserves in southern Africa (1981–2005) to test the influence of interacting socio-ecological and individual characters on post-release survival. We predicted that the socio-ecological context should feature more prominently after restocking than reintroduction because released rhinoceros interact with resident conspecifics. Instead, an interaction between release cohort size and habitat quality explained reintroduction success but only individuals' ages explained restocking outcomes. Achieving translocation success for many species may not be as complicated as theory suggests. Black rhino, and similarly asocial generalist herbivores without substantial predators, are likely to be resilient to ecological challenges and robust candidates for crisis management in a changing world.

Editor: Michael Somers, University of Pretoria, South Africa

Funding: U.S. Fish and Wildlife Service administered Rhinoceros and Tiger Conservation Act of 1994 (e.g., grant agreement numbers 98210-2-G363, 98210-4-G920, and 98210-6-G102); http://www.fws.gov/international/. The funders had no role in study design, data collection and analysis, decision to publish, or preparation of the manuscript.

Competing Interests: The authors have declared that no competing interests exist.

* E-mail: wayne.linklater@vuw.ac.nz

Introduction

Translocations, or movement of species between habitats, are remarkable experimental tests of the evolutionary capacity of species [1] and our ecological understanding [2]. Translocation success, or failure, at individual and population scales should be predicted by theoretical relationships between demographic and socio-ecological characteristics, and evolutionary fitness [3,4]. Translocations for reintroduction and restocking to restore and manage populations are also key to species rescue and recovery [5] and rapid progress demands that we find parsimonious guidelines for success [6]. The use of translocations as a conservation tool is expected to increase [7] due to the growing ranks of conservation-reliant species [8] and requirement for assisted migration [9] with climate change induced range shifts for many species [10]. With the need for this interventionist strategy on the rise, managers cannot afford to be unnecessarily timid or waste resources testing translocation strategies that bring only small, incremental improvements. General principles from evolutionary ecology that can be applied widely in the design of translocation programs are required.

Translocation success rates are generally poor [3,4,11,12]. The large number of elaborate hypotheses for translocation success and potential for important interactions amongst variables has led to complex uncertainty rather than simple solutions. The datasets required to test hypotheses for translocation success are also complex hierarchies of information because individuals may be released as groups to sites which may receive multiple releases over time. The datasets required to test such multi-level, nested hypotheses are necessarily large but rarely available and so most hypotheses for many species have not been tested. Further, hierarchical data and multivariate hypotheses cannot be treated using conventional correlation and regression [13,14] in the way that most hypotheses have been tested [3,4,11,12]. Consequently, current best-practice in the translocation of wildlife is based largely on anecdote or, at best, relationships that might be spurious, commensurate with under- and over-fitting of multivariate data [15,16]. The utility of general evolutionary ecological principles has not been tested.

The critically endangered black rhinoceros (*Diceros bicornis*) has an extraordinary documented history of translocations [17,18]. From our previous analyses of their post-release survival, we have drawn conclusions and made recommendations, particularly about the importance of an individuals' age over demographic or habitat influences (e.g., cohort size, population density and habitat quality) when restocking [19]. In comparison, analyses of reintroductions have not revealed strong influences on success. Outcomes were ambiguous, although the poorest habitats and small cohorts including only bulls or large cohorts including several mothers with dependent calves were weakly associated with greater mortality of bulls and calves, respectively, during the first year [19]. Thus, previous work has downplayed the role of socio-ecological influences on translocation success, although it does not address the potential complexity of influences on survival stemming from interactions amongst variables. Some variables, although not influential on their own, may nevertheless have a synergistic effect when they interact with other variables. In particular, the absence of interactions in previous analysis may explain why reintroduction success was so poorly explained and why ecological and demographic influences appeared unimportant to restocking success [19].

After restocking events we expect the socio-ecological context to be more complex and influential due to conflict when newly released individuals encounter established residents already occupying the better habitats. We expect residents to have a 'home advantage' during aggressive confrontations and post-release competition for habitat and mates. In reintroductions, where releases occur in areas that no longer support a resident population, social conflict and competition are likely to be less important. Thus, interactions amongst supported variables might improve the predictive power of models designed to improve post-release survival. Our objective here is to advance the understanding of establishment success after reintroduction and restocking by explicitly modeling interactions amongst socio-ecological, demographic and individual rhinoceros characters. To this end we applied an extraordinary record of 682 black rhinoceros released into 81 reserves in Namibia and South Africa over 25 years (1981–2005) to test hypotheses for establishment success (i.e., survival to one year post-release) after 89 reintroduction and 102 restocking events.

Results

The model describing the interaction between cohort size and habitat quality performed the best and improved substantially on previous leading models for reintroduction success (Table 1). Models including the interaction between cohort size and habitat quality contributed 92.5% of Akaike weights and were the only models to out-perform the base model without fixed-effects (Table 1). A simple, positive interaction between cohort size and habitat quality was supported (Fig. 1A) and substantially exceeded the explanatory power of two influences previously identified as potentially important for reintroduction success: the proportion of young and bulls in release cohorts which received no support ($\Delta AIC_c > 10$, $\omega < 0.004$, Table 1).

In contrast and unexpectedly, interactions of cohort and site-level variables representing population demography and habitat did not improve upon the restocking model including only age (Table 2) and so confirmed previous conclusions about the vulnerability of younger individuals when supplementing existing populations (Fig. 1B). Young are much less vulnerable after reintroduction (Age model: $\Delta AIC_c > 10$, $\omega < 0.003$, Table 1), a finding consistent with our prediction that the risks inherent in translocation to restock carry disproportionately higher risks for young rhinoceros. Importantly however, and contrary to predictions, the vulnerability of young after restocking is sufficient to account for variation in establishment success without recourse to complex interactions with demographic and ecological characters. Models including age class contribute 100.0% of Akaike weights and were the only models to out-perform a model without fixed-effects (Table 2).

Lastly, support for base models (including only random effects for introduction and reserve) would be evidence that important hypotheses and predictors were not represented by the current analysis. That the base models instead received such poor support (reintroduction: $\Delta AIC_c = 8.3$, $\omega = 0.010$; restocking: $\Delta AIC_c = 15.9$, $\omega = 0.000$) gives us confidence in the value of the leading models to explain and predict variation in the post-release survival of black rhinoceros amongst cohorts and sites.

Discussion

Information-Theoretic analyses deliver parsimony. An indication, therefore, of the extraordinary power of age to explain restocking success is that it contributed a large number of parameters to leading models (age is represented by four classes defining the first six years and an adult class) that performed better than smaller models. The risks posed to younger black rhinoceros when restocking are probably numerous and diverse (e.g., social asymmetries of competition and conflict, inter-specific conflict, resource unfamiliarity, disease, misadventure) such that no single risk dominates. While the addition of cohort and post-release adult sex ratio, sex, resident and post-release densities, cohort size, and habitat quality with age were also ranked highly, their explanatory power was not sufficient, at least for the dataset considered, to warrant any other action than avoiding the use of sub-adults, especially calves, when restocking black rhinoceros.

The interaction between cohort size and habitat quality resulted in extraordinarily high mortality rates after reintroductions of the smallest cohorts to habitat with the lowest carrying capacities. Large improvements in reintroduction success might be achieved by avoiding release of cohorts with fewer than four individuals, especially into poor quality habitat. Where reintroductions to poorer quality habitats are required, cohorts larger than six should be favored. The reason for the extraordinarily high mortality rates amongst individuals from small cohorts reintroduced to the poorest habitats is unclear. Perhaps normative social relationships amongst peers are important [20]? Even in the relatively asocial black rhinoceros, peers may help individuals refine habitat and food choices, especially in marginal habitat where resources are more heterogeneous in time and space. Normative behavior and conspecific attraction may facilitate habitat discovery and learning in novel environments. It is also possible that unsatisfied mate-choice or social behaviors encourage long-distance movements or displace maintenance behaviors after release such that individuals acclimate poorly in the absence of suitable mates or friends [21]. Small release cohorts might not provide the necessary peers or mates for successful post-release adjustment.

Mortality risks differ between reintroduction and restocking but model outcomes share similarities in those that are not supported, especially those describing interactions between individual characters (age, sex, experience) and metrics of population density and habitat quality. Thus, complex models representing themes of resource availability and competition and their interaction with individual characteristics continue to be unsupported [19]. Although authors have recommended large release cohorts [3,4,11,18] including individuals that are not predator-, competitor-, or translocation-naïve [22,23], and favoring large reserves with low conspecific density [24] and high-quality [3,4,18] or familiar [25] habitat, only two of these factors and their interaction were important for black rhinoceros, and only when reintroducing populations. The lack of support for the more complex models of translocation success, at least for black rhinoceros, indicates that the role of ecological and demographic influences is weaker than previously thought. Indeed, many relationships identified previ-

Table 1. Results summary of the 29 candidate models for reintroduction mortality risk among 89 cohorts and 414 reintroduced black rhino.

Model/Hypothesis	*K*	AIC_c	ΔAIC_c	ω_i
Cohort size * Habitat quality (Wolf et al. 1998)	5	222.8	0.0	0.635
Cohort size * Habitat quality * Proportion bulls in cohort	8	224.6	1.7	0.270
Cohort size * Habitat quality * Proportion mothers with calves in cohort	8	229.8	6.9	0.020
Base model (including only the random effect for reserve)	**2**	**231.1**	**8.3**	**0.010**
Cohort size * Proportion mothers with calves in cohort	5	231.2	8.4	0.010
Habitat quality	3	232.5	9.6	0.005
Post-release adult density (Linklater & Swaisgood 2008, Hitchins & Anderson 1983)	*3*	*232.6*	*9.7*	*0.005*
Cohort size (Fischer & Lindenmayer 2000)	*3*	*232.6*	*9.7*	*0.005*
Proportion bulls in cohort * Habitat quality	5	232.6	9.8	0.005
Proportion bulls in cohort	*3*	*233.0*	*10.1*	*0.004*
Cohort size+Habitat quality (Griffith et al. 1989)	*4*	*233.6*	*10.8*	*0.003*
Age	7	233.7	10.8	0.003
Proportion mothers with calves in cohort	*3*	*233.9*	*11.0*	*0.003*
Number of mothers with calves+Proportion mothers with calves in cohort	*4*	*233.9*	*11.1*	*0.003*
Cohort size * Post-release adult density	5	233.9	11.1	0.003
Cohort adult sex ratio * Sex	5	234.1	11.3	0.002
Reserve size+Post-release adult density (Linklater & Swaisgood 2008b)	*4*	*234.4*	*11.6*	*0.002*
Number of bulls+Proportion of bulls in cohort	*4*	*234.5*	*11.7*	*0.002*
Source-recipient reserve size difference+Previously translocated	*4*	*234.6*	*11.8*	*0.002*
No. sub-adults+Proportion of cohort that are sub-adults	*4*	*234.7*	*11.9*	*0.002*
Source-recipient habitat quality difference+Translocation distance	*4*	*235.0*	*12.2*	*0.002*
Post-release adult density * Proportion of carrying capacity occupied	5	235.5	12.7	0.001
Proportion of mothers with calves * Habitat quality	5	235.5	12.7	0.001
No. bulls * Proportion bulls	5	235.7	12.9	0.001
Reserve size * Post-release adult density	5	235.8	13.0	0.001
Cohort size * No. bulls	5	235.9	13.1	0.001
Cohort size * Proportion bulls	5	235.9	13.1	0.001
Age+Cohort size+Habitat quality (Brett 1998)	*9*	*236.6*	*13.8*	*0.001*
Age+Reserve size+Post-release adult density (Walker 1994)	*9*	*237.2*	*14.4*	*0.001*

Models are in descending order from most to least supported based on Akaike second-order Information Criteria (AIC_c). Leading models from previous analyses without interaction terms [*9,23*] are italicized. The model without fixed effects is indicated in bold type. A '*' indicates an interaction term in the regression between two variables and, by implication, predictors in interactions were also present additively in models (e.g., a*b refers to model including a+b+a*b as fixed effects).

ously from simple correlations or regressions are probably spurious. This analysis confirms that importance has been mistakenly attributed to complexes of ecological and demographic influences, albeit for strong theoretical reasons, that are instead more simply explained.

Why do black rhinoceros in southern Africa defy the expectations of adaptive theory for important relationships between socio-ecological characters and metrics of fitness like survival after translocation? Such a finding appears to contradict our knowledge about the intensity of intra- and inter-specific competition and conflict, and habitat preferences amongst rhinoceros. Certainly, the successful translocation of other species appears to be demographically and socio-ecologically complex. Such species, however, are either more selective herbivores requiring smaller amounts of higher quality browse or grass, cooperative or gregarious breeders, predators whose prey (cf. browse and grass) is elusive, or they are prey of sympatric predators [22,26,27,28]. Mega-herbivores appear to escape the constraints of predators that perhaps cause the failure of other ungulate introductions [29]. The simplicity of rules for black rhinoceros translocation, therefore, might be unusual, or at least confined to other similarly large, generalist herbivores with wide biogeographic ranges and largely asocial habits.

Importantly, the degree to which black rhinoceros are robust to a major life-history event like translocation, even into resident populations (i.e., restocking), raises important implications for the understanding of their ecology and conservation, especially in a changing climate. Species vary in their adaptive capacity for ecological change and their resiliency to drastic types of management like assisted migration [1]. Experience translocating black rhinoceros leads us to expect their populations to be resilient to ecological challenges like climate change compared to other species. Alternatively, they will be comparatively robust candidates for assisted migration, should it be required. The extraordinary success of white rhinoceros (*Ceratotherium simum var. simum*) reintroduction and recovery [30], might indicate the generality of our findings, at least amongst rhinoceros. So long as the anthropogenic causes of decline are treated (i.e., illegal hunting

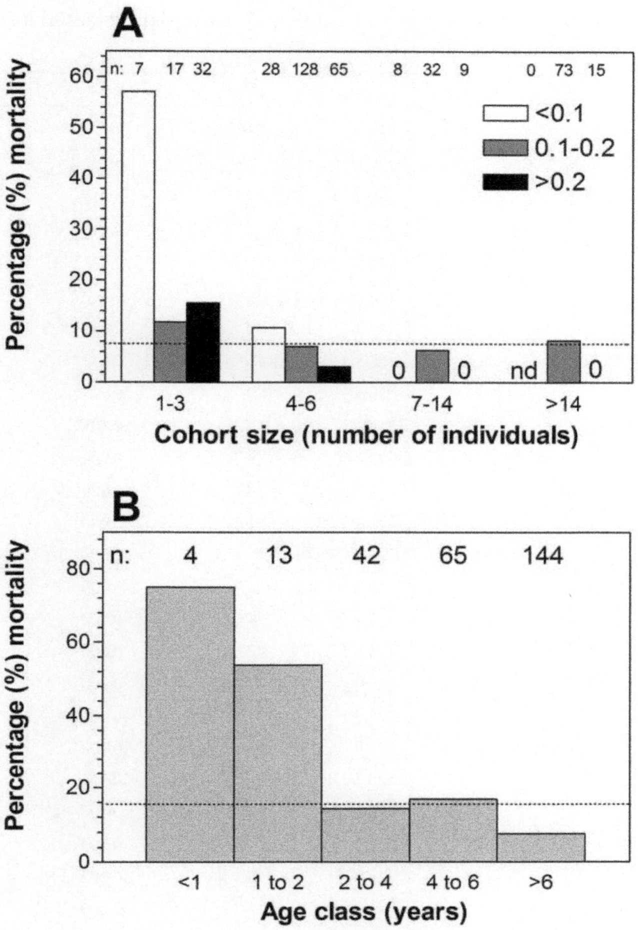

Figure 1. Post-release mortality in black rhinoceros (*Diceros bicornis*) after (A) reintroduction and (B) restocking. Cohort size and habitat quality (estimated carrying capacity <0.1, 0.1–0.2 or >0.2 rhino per km^2) explained reintroduction mortality while age class explained deaths after restocking. Age classes conform to Hitchins' A (calf) to F (adult) aging scheme [31]. Numbers of rhino (i.e., n) in each category are indicated above each bar. nd = no data. The dash line across each indicates mean mortality rate for all reintroduction (A) and restocking (B) events.

[31]), black rhinoceros recovery by reintroduction and restocking, and even assisted migration, should be comparatively easy. Groups of black rhinoceros of different size and composition can be moved successfully between different ecological contexts, and released into reserves that might already be stocked, and have poor habitat, so long as young are not used to restock populations and small cohorts are not reintroduced into the poorest habitats.

Our findings give confidence to the design of grand artificial meta-populations of similar conservation-reliant species that will require the translocation of individuals for assisted migration, reintroduction, and the genetic and demographic rescue of small populations by restocking. Achieving successful translocations of species like black rhinoceros, i.e., large asocial and biogeographically spread herbivores which are not predators and rarely prey, might not be nearly as socio-ecologically complicated as the literature leads us to believe. Such species will be robust to ecological challenge and resilient candidates for crisis management in a changing world.

Materials and Methods

Reports on properties with populations of black rhinoceros in Namibia and South Africa 1981–2005 [17] were consulted for translocations and post-release survival of individually identifiable rhinos of known sex- and age-class. Data from the three subspecies in the region (i.e., *D.b.* var. *micheali* n = 43, *minor* n = 338, *bicornis* n = 301) were pooled. Information from those reports was supplemented with estimates of each reserve's relative carrying capacity (i.e., 0.015 to 0.884 rhino.km^{-2}). Estimates of relative carrying capacity were derived from a regression model and sampling from 24 reserves in Kenya, Namibia and South Africa by Adcock *et al.* that are described in detail elsewhere [32,33]. Briefly, the regression model included indices representing each reserve's black rhinoceros browse standing crop (percentage volume of selected woody and forb plant leaves, twigs and small branches within the 0 to 2 m feeding height range of black rhinoceros), potential rainfall- and temperature-dependent browse growth (monthly rainfall and minimum mid-winter, July, temperatures), soil fertility and fire regimes.

Reintroduction events are attempts to establish a population in an area once part of the species range but from which it became extinct. Restocking events are attempts to add individuals to an existing population of conspecifics within the species range [5]. Subsequent releases into the same reserve were classified as restocking events if they occurred more than one month after the first release because black rhinoceros appear to have developed home ranges within 30 days post-release [24]. For each translocated individual we compiled 40 individual rhino, release cohort, or reserve characteristic predictors for survival to one year after release [19].

Over- and under-fitting are a problem in multivariate analyses for detecting important predictors or combinations of predictors, especially where variables interact. This is particularly problematic for regression when the variables are gleaned from pre-existing databases because some important variables may not have been measured or included. For this reason we adopted an Information-Theoretic approach to testing hypotheses about the causes of mortality after release [13,15] by constructing and comparing candidate models as hypotheses for translocation success.

To compile our candidate models we began with the leading models in our previous analysis [19] and appended a further suite of models describing interactions amongst predictors. The few leading models in our previous analysis for restocking success shared age class [34] in common and age was only found in the leading models. Age was also the only predictor with a credible interval (Bayesian measure of uncertainty, analogous to confidence intervals) that did not include zero and an effect size larger than any other predictor by a factor of two to three. Nevertheless, the possibility remains that the interaction of a number of other variables, particularly the quality of habitat, number or density of residents and post-release density, especially of bulls, exacerbates the vulnerability of young. So, to compile candidate models for our new analysis, we began with the leading models from our previous analysis and appended a further suite of models describing interactions amongst predictors.

The outcomes of previous analyses for reintroduction were ambiguous compared to those for restocking. Most models received similar support and no single hypothesis dominated [19]. Nevertheless, the coefficients of some variables and their effect sizes led us to speculate about their relative importance. In particular, the proportion of bulls and calves in the release cohort and quality of the habitat appeared to have greatest influence. We also speculated about the interaction between cohort size and the

Table 2. Results summary of the 23 candidate models for restocking mortality risk among 102 cohorts of black rhino released into 48 reserves and including 273 individuals.

Model/Hypothesis	K	AIC_c	ΔAIC_c	ω_i
Age	*7*	*203.5*	*0.0*	*0.321*
Age+Post-release adult sex ratio	8	204.4	0.8	0.212
Age+Cohort adult sex ratio	*8*	*205.2*	*1.6*	*0.142*
Age+Sex	*8*	*205.5*	*1.9*	*0.123*
Age+Resident bull density (Adcock et al. 1998)	*8*	*205.7*	*2.1*	*0.111*
Age+Reserve area+Post-release adult density (Walker 1994)	*9*	*207.8*	*4.3*	*0.038*
Age+Cohort size+Resident adult density+Habitat quality (Brett 1998)	*10*	*208.5*	*4.9*	*0.027*
Age+Sex+Cohort size+Post-release adult density+Habitat quality (Brett 1998)	*11*	*208.6*	*5.1*	*0.025*
Base model (including only the nested random effect for cohort and reserve)	**3**	**219.4**	**15.9**	**0.000**
Resident adult density * Habitat quality	6	223.2	19.7	0.000
Resident bull density * Habitat quality	6	224.6	21.1	0.000
Cohort size * Habitat quality	6	224.7	21.2	0.000
Habitat quality * Proportion of carrying capacity occupied	6	225.2	21.7	0.000
Post-release adult density * Habitat quality	6	225.2	21.7	0.000
Post-release adult sex ratio * Sex	6	221.9	18.4	0.000
Post-release adult sex ratio * Habitat quality	6	223.5	20.0	0.000
Cohort size * Final adult density	6	225.4	21.9	0.000
Cohort size * Resident adult male density	6	225.4	21.9	0.000
Cohort size * Resident adult density	6	225.2	21.7	0.000
Post-release adult sex ratio * Post-release adult density	6	223.9	20.4	0.000
Post-release adult density * Sex	6	222.8	19.3	0.000
Resident adult density * Sex	6	222.9	19.4	0.000
Resident adult male density * Sex	6	222.0	18.5	0.000
Resident adult density * Habitat quality	6	223.2	19.7	0.000
Resident adult male density * Habitat quality	6	224.6	21.1	0.000
Habitat quality * Sex	6	224.2	20.7	0.000

Models are in descending order from most to least supported based on Akaike second-order Information Criteria (AIC_c). Leading models from previous analyses without interaction terms [19,23] are italicized. The model without fixed effects is indicated in bold type. A '*' indicates an interaction term in the regression between the two variables and, by implication, predictors in interactions were also present additively in models (e.g., a*b refers to model including a+b+a*b as fixed effects).

contributions of bulls and calves to the release cohort since large cohorts with calves, and small cohorts (one to three individuals) consisting entirely of bulls, had disproportionately high mortality rates. Thus, it is possible that the performance of models testing interactions amongst cohort size, the contribution of bulls and calves to the cohort and habitat quality might show improvement over leading models. In particular, we modeled two- and three-way interactions between cohort size and metrics of cohort composition, especially the contribution of bulls and calves, metrics of habitat quality and post-release population density, and individual age (i.e., vulnerability).

We centered continuous predictors by subtracting the mean and dividing by two standard deviations [14] and left binary predictors unmodified. We conducted all logistic regressions using *lme4* package in R 2.11.1 (R version 2.11.1, 2010-05-31, Copyright (C) 2010 The R Foundation for Statistical Computing) and fitted general linear mixed-models using Laplace approximations of maximum likelihood to calculate Akaike Information Criterion (AIC) for each model [35]. We used AIC because we were interested primarily in the influence of fixed-effects and the random effects structure remained constant among models. We calculated a second-order Akaike Information Criterion (AIC_c) as our Information-Theoretic statistic because the number of structural parameters in models (*K*) was large relative to the number of contributing rhinoceros (n), particularly for the restocking dataset where $n/K<40$ [13]. We judged the relative power of candidate models by comparing their AIC_c and ratios of Akaike weights (w_i). Models with lowest AIC_c have most support from the data. Relative support between candidate models was the difference between each model's AIC_c and the minimum value ($AIC_{c,\ min}$) from all models (ΔAIC_c). We considered models with $\Delta AIC_c \leq 2$ to have compelling support from the data and models with $\Delta AIC_c > 10$ to have no support [13]. We included a base model including random effects, but without fixed effects, in the candidate set of models for comparison because we wanted to understand the amount of information in the data not explained by current theory. Models which performed worse than the base model could also be considered to be unsupported.

Acknowledgments

We acknowledge the contributions of the Ministry of Environment and Tourism, Namibia, and the Southern Africa Development Community, Rhino Management Group (RMG), and all private and state authorities who contributed data to the RMG.

Author Contributions

Conceived and designed the experiments: WL JG RS. Performed the experiments: WL KA PdP MK. Analyzed the data: WL JG PL. Contributed reagents/materials/analysis tools: WL JG PL. Wrote the paper: WL JG PL RS KA PdP MK GK. Collected, collated and evaluated the quality of multi-institutional data: WL JG KA PdP MK.

References

1. Dawson TP, Jackson ST, House JI, Prentice IC, Mace GM (2011) Beyond Predictions: Biodiversity Conservation in a Changing Climate. Science 332: 53–58.
2. Sarrazin F, Barbault R (1996) Reintroduction: challenges and lessons for basic ecology. Trends in Ecology & Evolution 11: 474–478.
3. Griffith B, Scott J, Carpenter J, Reed C (1989) Translocation as a species conservation tool: status and strategy. Science 245: 477–480.
4. Wolf CM, Garland T, Griffith B (1998) Predictors of avian and mammalian translocation success: reanalysis with phylogenetically independent contrasts. Biological Conservation 86: 243–255.
5. IUCN (1987) Position statement on the translocation of living organisms: Introductions, re-introductions, and re-stocking. Gland, Switzerland and Cambridge, UK: IUCN, The World Conservation Union.
6. Armstrong D, Seddon P (2007) Directions in reintroduction biology. Trends In Ecology & Evolution.
7. Morell V (2008) Conservation biology - Into the wild: Reintroduced animals face daunting odds. Science 320: 742–743.
8. Scott JM, Goble DD, Haines AM, Wiens JA, Neel MC (2011) Conservation-reliant species and the future of conservation. Conservation Letters 3: 91–97.
9. Hoegh-Guldberg O, Hughes L, McIntyre S, Lindenmayer DB, Parmesan C, et al. (2008) Assisted colonization and rapid climate change. Science 321: 345–346.
10. Parmesan C (2006) Ecological and evolutionary responses to recent climate change. Annual Review of Ecology and Systematics 37: 637–669.
11. Fischer J, Lindenmayer D (2000) An assessment of the published results of animal translocations. Biological Conservation 96: 1–11.
12. Dodd CK, Seigel RA (1991) Relocation, repatriation, and translocation of amphibians and reptiles - are they conservation strategies that work. Herpetologica 47: 336–350.
13. Burnham K, Anderson D (2002) Model Selection and Multimodel Inference: A Practical Information-Theoretic Approach. New York: Springer-Verlag New York, Inc.
14. Gelman A, Hill J (2007) Data Analysis using Regression and Multilevel/Heirarchical Models. Cambridge, U.K., 625 p.
15. Johnson JB, Omland KS (2004) Model selection in ecology and evolution. Trends in Ecology & Evolution 19: 101–108.
16. Whittingham MJ, Stephens PA, Bradbury RB, Freckleton RP (2006) Why do we still use stepwise modelling in ecology and behaviour? Journal of Animal Ecology 75: 1182–1189.
17. Adcock K (1995–2005) Sequence of reports on the status and management of black rhino in South Africa and Namibia: April 1989 to December 2003. Pietermaritzberg: Rhino Management Group. various p.
18. Brett RA (1998) Mortality factors and breeding performance of translocated black rhinos in Kenya: 1984–1995. Pachyderm 26: 69–82.
19. Linklater W, Adcock K, du Preez P, Swaisgood R, Law P, et al. (2011) Guidelines for large herbivore translocation simplified: black rhinoceros case study. Journal of Applied Ecology 48: 493–502.
20. Shrader A, Owen-Smith R (2002) The role of companionship in the dispersal of white rhinoceros (*Ceratotherium simum*). Behavioral Ecology and Sociobiology 52: 255–261.
21. Cameron EZ, Setsaas T, Linklater WL (2009) Social bonds between unrelated females increase reproductive success in feral horses. Proceedings of the National Academy of Sciences of the United States of America 106: 13850–13853.
22. Banks PB, Norrdahl K, Korpimaki E (2002) Mobility decisions and the predation risks of reintroduction. Biological Conservation 103: 133–138.
23. Frair JL, Merrill EH, Allen JR, Boyce MS (2007) Know thy enemy: Experience affects elk translocation success in risky landscapes. Journal of Wildlife Management 71: 541–554.
24. Linklater W, Swaisgood R (2008) Reserve size, release density and translocation success: Black rhinoceros movements, association, injury and death after release. Journal of Wildlife Management 72: 1059–1068.
25. Stamps J, Swaisgood R (2007) Someplace like home: experience, habitat selection and conservation biology. Applied Animal Behaviour Science 102: 392–409.
26. Somers MJ, Graf JA, Szykman M, Slotow R, Gusset M (2008) Dynamics of a small re-introduced population of wild dogs over 25 years: Allee effects and the implications of sociality for endangered species' recovery. Oecologia 158: 239–247.
27. Matson TK, Goldizen AW, Jarman PJ (2004) Factors affecting the success of translocations of the black-faced impala in Namibia. Biological Conservation 116: 359–365.
28. Nicoll MAC, Jones CG, Norris K (2003) Declining survival rates in a reintroduced population of the Mauritius kestrel: evidence for non-linear density dependence and environmental stochasticity. Journal Of Animal Ecology 72: 917–926.
29. Owen-Smith N (2003) Foraging behavior, habitat suitability, and translocation success, with special reference to large mammalian herbivores. In: Festa-Bianchet M, Apollonio M, eds. Animal Behaviour and Wildlife Conservation. Washington DC: Island Press.
30. Castley GH-M, A J (2003) The status of the southern white rhinoceros (Ceratotherium simum simum) on private land in South Africa in 2001. Pachyderm 34: 33–44.
31. Beech H, Perry A (2011) Killing fields: How Asia's growing appetite for traditional medicine is theatening Africa's rhinos. Time. New York: Time & Life. pp 28–35.
32. Adcock K (2001) Rhino Management Group Black Rhino Carrying Capacity Model Version 1.0: users guide (evaluation draft). Harare, Zimbabwe: IUCN-ROSA. Semester 3 task 4.2-2.1 Semester 3 task 4.2-2.1. 64 p.
33. Adcock K (2005) Visual assessment of black rhino browse availability version 3. Darwin Initiative and SADC Regional Programme for Rhino Conservation.
34. Hitchins PM (1978) Age determination of the black rhinoceros, *Diceros bicornis* Linn. in Zululand. South African Journal of Wildlife Research 8: 71–80.
35. Bolker BM, Brooks ME, Clark CJ, Geange SW, Poulsen JR, et al. (2009) Generalized linear mixed models: a practical guide for ecology and evolution. Trends in Ecology & Evolution 24: 127–135.

5

Don't Fall Off the Adaptation Cliff: When Asymmetrical Fitness Selects for Suboptimal Traits

Elodie Vercken[1]*, Maren Wellenreuther[2], Erik I. Svensson[2], Benjamin Mauroy[3]

1 Institut Sophia Agrobiotech, UMR 1355 ISA, Institut National de la Recherche Agronomique, Sophia-Antipolis, France, **2** Department of Biology, Lund University, Lund, Sweden, **3** Laboratoire J.A. Dieudonné, UMR CNRS 7351, Université de Nice-Sophia Antipolis, Nice, France

Abstract

The cliff-edge hypothesis introduces the counterintuitive idea that the trait value associated with the maximum of an asymmetrical fitness function is not necessarily the value that is selected for if the trait shows variability in its phenotypic expression. We develop a model of population dynamics to show that, in such a system, the evolutionary stable strategy depends on both the shape of the fitness function around its maximum and the amount of phenotypic variance. The model provides quantitative predictions of the expected trait value distribution and provides an alternative quantity that should be maximized ("genotype fitness") instead of the classical fitness function ("phenotype fitness"). We test the model's predictions on three examples: (1) litter size in guinea pigs, (2) sexual selection in damselflies, and (3) the geometry of the human lung. In all three cases, the model's predictions give a closer match to empirical data than traditional optimization theory models. Our model can be extended to most ecological situations, and the evolutionary conditions for its application are expected to be common in nature.

Editor: Martin Heil, Centro de Investigación y de Estudios Avanzados, Mexico

Funding: This work was supported by the Institut National de la Recherche Agronomique and the Centre National de la Recherche Scientifique. The funders had no role in study design, data collection and analysis, decision to publish, or preparation of the manuscript.

Competing Interests: The authors have declared that no competing interests exist.

* E-mail: elodie.vercken@sophia.inra.fr

Introduction

Evolutionary theory predicts that trait means in populations should evolve towards the value that maximizes fitness [1], which is also a central assumption in most optimality analyses [2]. However, in many cases the evolution of fitness-related traits might be constrained by genetic or physiological trade-offs that cause negative genetic correlations between traits [3]. Fitness in these situations will be maximized in a way that depends on the balance between the counteracting selective forces on traits, and the net fitness functions are expected to be bell-shaped [4]. For instance, increased annual reproductive effort is expected to negatively affect adult survival in long-lived species, leading to a fitness optimum where lifetime reproductive success will be maximized by intermediate reproductive effort [5]. Optimality theory predicts that natural selection will drive the population towards this 'optimal' trait value that maximizes fitness [6], which has a close connection to the concept of 'adaptive peaks' in population genetics [7]. This classical optimization approach has been successfully applied in evolutionary ecology to predict the population mean of many phenotypic traits.

In particular, the evolution of reproductive traits, such as offspring number, has received considerable interest and has provided evolutionary ecologists with a solid conceptual foundation for optimality theory in life-history evolution [8–10]. However, in many species of birds and mammals, the number of offspring most commonly observed is often less than the maximum [11–16]. Several alternative theories have been advanced to explain this pattern [15,17]: (1) costs of reproduction due to trade-offs with parental survival or future reproduction [18]; (2) inter-annual variation in juvenile survival related to variation in environmental quality [12]; (3) individual optimization in relation to individual condition and local resource availability [19]; and (4) the interaction between asymmetrical fitness costs and individual variance in brood size ('cliff-edge hypothesis') [12,20].

Among these hypotheses concerning the evolution of litter size, cliff-edge effects have the potential to provide a unifying framework for understanding the optimization of phenotypic traits. This theory predicts that when juvenile survival is asymmetrically low in large broods, moderate variance around the optimal brood size will result in large differences in survival between clutches slightly smaller or larger than the optimal. As a consequence of these asymmetric costs, females producing larger than the most productive broods will leave fewer descendants than females producing smaller than the most productive broods, and the evolutionary optimal should be smaller than the most productive brood size.

In what follows, the relationship between phenotypic value and reproductive value will be referred to as 'phenotype fitness' (which defines the single most productive brood size). In contrast, the relationship between genotypic value and the reproductive value averaged over the phenotype range for each genotype will be referred to as 'genotype fitness' (which defines the evolutionary optimum). The difference between these two definitions of fitness is illustrated on Figure 1. In the absence of any phenotypic variance, these two definitions merge.

On a more general note, cliff-edge effects are related to the properties of convex functions known as Jensen's inequality.

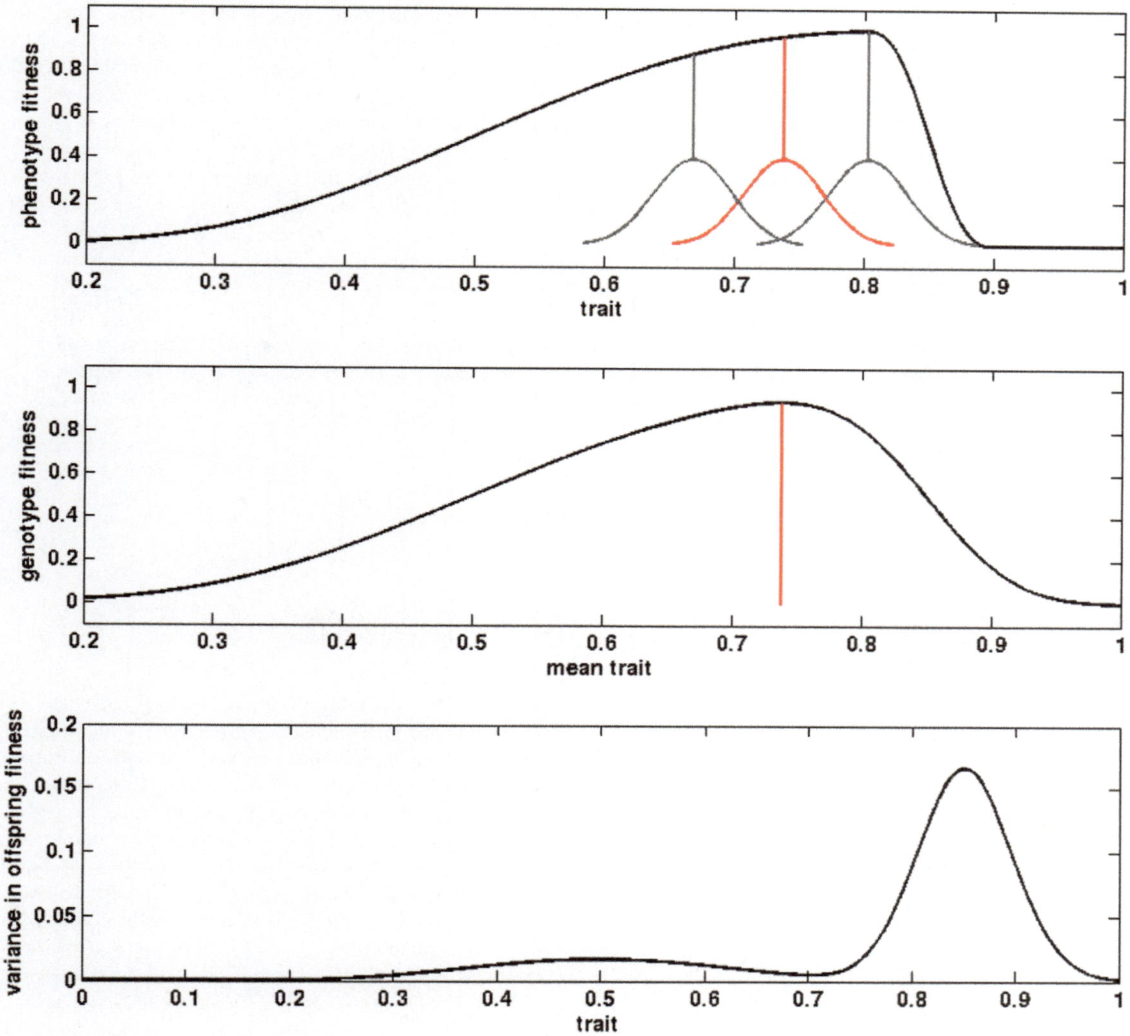

Figure 1. Comparison between (a) the phenotype fitness function, and (b) genotype fitness function relative to the trait variation (after Martin and Huey [27]). Because of the steep part (cliff) of function (a), the optimal trait value (red phenotypic distribution) is shifted downwards from the trait value that maximizes phenotype fitness. Function (c) represents the variance in offspring phenotype fitness for different mean values of the trait.

Jensen's inequality states that, for a convex function f and a set of values $x = (x_1,\ldots,x_n)$ with a mean of $\overline{x}$ and non-zero variance, the average image of x, $\overline{f(x)}$ does not equal the image of the average x, $f(\overline{x})$. In other words, the trait value that gives the highest genotype fitness in the presence of phenotypic variance is not the value that gives the highest phenotype fitness. These analytical principles have been successfully applied to a wide range of evolutionary questions, including the evolution of reproductive systems and life-history strategies, individual behaviour or population dynamics in variable environments [21–28]. In this article, we propose a general mathematical formalization of the cliff-edge problem that can be applied to any fitness-related trait that exhibits phenotypic variability and has asymmetric fitness costs. These conditions are expected to be quite common in nature; hence the generality of a model incorporating these effects is likely to be high.

We develop an analytical model that describes the evolution of a population with random variation in the expression of a fitness-related trait and an asymmetrical fitness function. We demonstrate that these conditions select for apparent sub-optimal genotypes with regard to phenotype fitness, and we show that the optimal genotypic value depends on both the amount of variance of the trait and on the skewness of the fitness function. The model provides quantitative predictions of the position of the optimum and the distribution of phenotypic variance. To illustrate our method, we apply the model to three different evolutionary systems for which we were able to estimate realistic fitness functions from empirical data. Two of these examples (evolution of litter size in guinea pigs, and the evolution of male sexual ornaments in a damselfly) are in line with the classical framework of life history evolution. The third example (evolution of human respiratory tract geometry) stems from evolutionary medicine and demonstrates that cliff-edge effects can act on any trait that is targeted by natural selection.

Results

1- Model of the evolution of a population in presence of phenotypic variation

We consider a population with a continuous trait (e.g. reproductive effort, physiological parameters). We assume that there is no genetic variation for this trait in the population: all genotypes take the value g, but the phenotypic expression can vary randomly from φ_{min} to φ_{max}. This phenotypic variance can arise from several different processes (e.g. developmental instability, environmental variability, maternal effects, epistasis) as long as it is a random process such that any individual of genotype g can experience any phenotype between φ_{min} and φ_{max}. The simple evolutionary processes involved in the population model were chosen to isolate the effect of phenotypic variability on trait evolution and to increase the generality of model predictions.

The function $a(\varphi,t)$ denotes the frequency of individuals in the population having the phenotype value φ at time t. We assume that instantaneous fitness depends on the phenotype, so these individuals have a reproductive rate $b(\varphi)\geq 0$ and a mortality rate of $m(\varphi)>0$. Therefore, at each instant t and per unit of time, $m(\varphi)a(\varphi,t)$ individuals of phenotype φ die, while $b(\varphi)a(\varphi,t)$ descendants are produced by individuals of phenotype φ.

For any individual with genotype g, the phenotype φ of its offspring is randomly distributed around the value g following the distribution function $G(\varphi,g,\sigma)$. We consider that $G(\varphi,g,\sigma)$ is a Gaussian function (default hypothesis for a quantitative characters, see [29]) centred around g with a variance σ^2.

Therefore, offspring of phenotype φ are produced by parents of genotype g in the proportion $G(\varphi,g,\sigma)$, regardless of the parents' own phenotype, i.e. there are no cross-generational effects. Therefore, at each time a total quantity $\int_{\vartheta_{min}}^{\vartheta_{max}} b(l)\,a(l,t)dl$ of new individuals are produced in the population among which only a fraction $G(\varphi,g,\sigma)$ will have the phenotype φ.

The variation of the distribution of individuals of parameter φ over time is then given by the differential equation:

$$\frac{\partial a}{\partial t}(\vartheta,t)=\underbrace{-m(\vartheta)\,a(\vartheta,t)}_{mortality}\ \%hbrace \; \underbrace{+\,G(\vartheta,g,\sigma)\times\int_{\vartheta_{min}}^{\vartheta_{max}} b(l)\,a(l,t)dl\ \%hbrace}_{reproduction} \tag{1}$$

We show in Supplementary Text S1 that such a population does not go extinct as long as

$$F(g)=\int_{\vartheta_{min}}^{\vartheta_{max}}\frac{G(l,g,\sigma)\,b(l)}{m(l)}dl\ >\ 1 \tag{2}$$

The term in the integral represents the per capita growth rate $w(l)=b(l)/m(l)$ (phenotype fitness) of individuals with phenotype l multiplied by their frequency in the population. Thus, the function $F(g)$ represents the sum of phenotype fitness of all phenotypes weighted by their respective frequency, i.e. the weighted mean of phenotype fitness in the population. The growth rate function $F(g)$ is therefore the genotype fitness of genotype g. This result holds for populations with limited resources (Supplementary Text S1).

Then, we consider two populations with limited growth, one with a genotype g_1 and the other with a genotype $g_2\neq g_1$. They are represented by their respective distribution $a_1(\varphi,t)$ and $a_2(\varphi,t)$. They are interacting with each other due to mutually shared resources that are limited. In order to have true competition, we assume that each population does not go extinct if it is alone, which is equivalent to $F(g_1)>1$ and $F(g_2)>1$.

The equations that describe the evolution of these populations and their distributions along time are:

$$\begin{cases}\dfrac{\partial a_1}{\partial t}(\vartheta,t)=-m(\vartheta)\,a_1(\vartheta,t)+\left(1-\displaystyle\int_{\vartheta_{min}}^{\vartheta_{max}}(a_1(l,t)+a_2(l,t))dl\right)\\ \quad\times\,G(\vartheta,g_1,\sigma)\times\displaystyle\int_{\vartheta_{min}}^{\vartheta_{max}} b(l)\,a_1(l,t)dl\\ \dfrac{\partial a_2}{\partial t}(\vartheta,t)=-m(\vartheta)\,a_2(\vartheta,t)+\left(1-\displaystyle\int_{\vartheta_{min}}^{\vartheta_{max}}(a_1(l,t)+a_2(l,t))dl\right)\\ \quad\times\,G(\vartheta,g_2,\sigma)\times\displaystyle\int_{\vartheta_{min}}^{\vartheta_{max}} b(l)\,a_2(l,t)dl\end{cases} \tag{4}$$

There are four equilibrium points: coexistence of both populations; extinction of both populations; only one population survives while the other goes extinct (two combinations). Coexistence is possible only if $g_1=g_2$, which is excluded by hypothesis. Moreover since $F(g_1)$ and $F(g_2)$ are assumed to be strictly greater than 1 for each population, it can be shown that the extinction of both populations is not possible. Thus, under these hypotheses one population must invade the other. Then, a successful invasion of population g_1 into population g_2 (i.e. equilibrium $a_1\neq 0$ and $a_2=0$ stable) is possible if and only if:

$$F(g_1)=\int_{\vartheta_{min}}^{\vartheta_{max}}\frac{G(l,g_1,\sigma)\,b(l)}{m(l)}dl\ >\ F(g_2)=\int_{\vartheta_{min}}^{\vartheta_{max}}\frac{G(l,g_2,\sigma)\,b(l)}{m(l)}dl \tag{5}$$

Hence, the evolutionary stable strategy (ESS) corresponds to the genotype g^* that maximises the growth rate function $F(g)$. The population of parameter g^*, also called the super-mutant population, will invade any population of parameter $g\neq g^*$, while it cannot be invaded by other populations with a parameter $g\neq g^*$. Thus, the genotype g^* is an ESS and should be observed in population at the equilibrium state, although the phenotype varies in the population.

To determine g^*, it is necessary to calculate the maximum of the growth rate function $F(g)$. Hence, the most efficient genotype is the one that maximises the success of the whole population by cumulating the relative success of each phenotypic trait weighted by their frequency.

When the fitness function is symmetric or if there is no phenotypic variance at all, the genotypic value g^* associated with the maximum of the function F is equal to the value that maximizes the phenotype fitness w. However, when the fitness function is asymmetric and the phenotypic variance is non-zero, these two values do not match anymore, as predicted by Jensen's inequality. In this case, the optimum genotype g^* value is systematically shifted from the maximum of the phenotype fitness in the direction of the least slope (Supplementary Text S2).

2- Application of the model using three biological examples

Most heritable traits are expected to include some degree of non-additive genetic, environmental or developmental variability that affects their optimal expression. Our first example focuses on optimal litter size in laboratory strains of Guinea pigs (*Cavia porcellus*). For this, we re-analyzed Mountford's original data set [20] that was used in his development of the cliff-edge hypothesis. Our second example deals with a secondary sexual trait in the damselfly *Calopteryx splendens*, using survival and mate choice data obtained in the field ([30], M. Wellenreuther, E. Vercken and E. Svensson, unpublished data). The third example is based on modelling work about the impact of lung geometry on respiratory performance in humans [31]. In all these examples, fitness functions are not symmetrically shaped around their maximum value, and we show consistent matching between empirical data and model predictions.

2-1 Example 1: Optimal litter size in Guinea pigs. In his seminal paper, Mountford [20] showed that the litter size that leads to the maximum number of surviving offspring in Guinea pigs was not the most frequent one. He suggested that the phenotypic variability associated with high asymmetric fitness costs for large litter sizes could explain this observation. To prove the validity of his theory, he produced a theoretical example based on simple numeric calculations. In what follows, we show that the frequency distribution of litter size in Guinea pigs could be predicted by the general model described above, and that this model can also provide information about both the optimal genotype as well as the variance in the expressed phenotype.

In our model, the litter size is called L and extents from $L_{min} = 0$ to $L_{max} = 9$. Reproductive rate $b(L)$ corresponds to offspring survival in relation to the litter size (data reproduced from Mountford [20]). Life-history theory predicts that, as a consequence of trade-offs between present and future reproduction, female survival should decrease when litter size increases ([32–36] but see [15,37,38] for counter-examples). In the absence of any data on the precise relationship between litter size and female mortality rate in Guinea pigs, we assumed a simple linear model $m(L) = m_r + \alpha \times L$, where m_r is the mortality rate for non-reproductive individuals and α is a constant estimated by least-squares method. The intercept value m_r does not affect the position of the genotype fitness maximum as it plays the role of a scaling factor once α is chosen. The phenotype fitness function is asymmetric around a maximum plateau for litter sizes between 2 and 3 (Figure 2a).

The general method to compare the predictions of our model to the measured data is similar in the three different examples used in this paper and is described in detail in Supplementary Text S3. First, we determine the optimal genotype $L_o(\sigma)$ for each of the possible values of phenotypic variance σ^2. Second, we find the value of $(\sigma, L_o(\sigma))$ that best fits empirical data using the mean-square method, here in the first example, the distribution of litter size in Guinea pigs.

Figure 2b represents the relationship between the standard deviation σ and optimal genotype $L_o(\sigma)$ (first step). Because the phenotype fitness function has a maximum plateau, this relationship is not monotonic. For small values of σ, most offspring will fall within the phenotype fitness plateau. However, because of the long tails of the Gaussian distribution, a small proportion will be outside of it. Therefore, the optimal genotype fitness is initially shifted away from the steepest slope (left side).

When σ increases, at first a significant proportion of offspring phenotypes reaches values of $L>3$, while fewer phenotypes reach values of $L<2$ (because the optimum is initially right-shifted). Therefore, the optimum genotype value decreases as it shifts away from the closest fitness slope, accounting for the initial negative relationship in Figure 2b.

For higher values of σ, a more significant proportion of offspring phenotype reaches values of $L<2$ and thus 'fall off the cliff'. Because the phenotype fitness loss is higher for phenotypes that reach values of $L<2$ than for those with values of $L>3$, the optimum genotype will shift away from the steepest slope and tends to obtain higher values of L as σ increases (positive relationship in Figure 2b).

In Guinea pigs, the optimal phenotypic trait predicted was *3.05* with a phenotypic variance of *1.30* ($\sigma = 1.14$). The difference between the distribution of the trait in the population predicted by

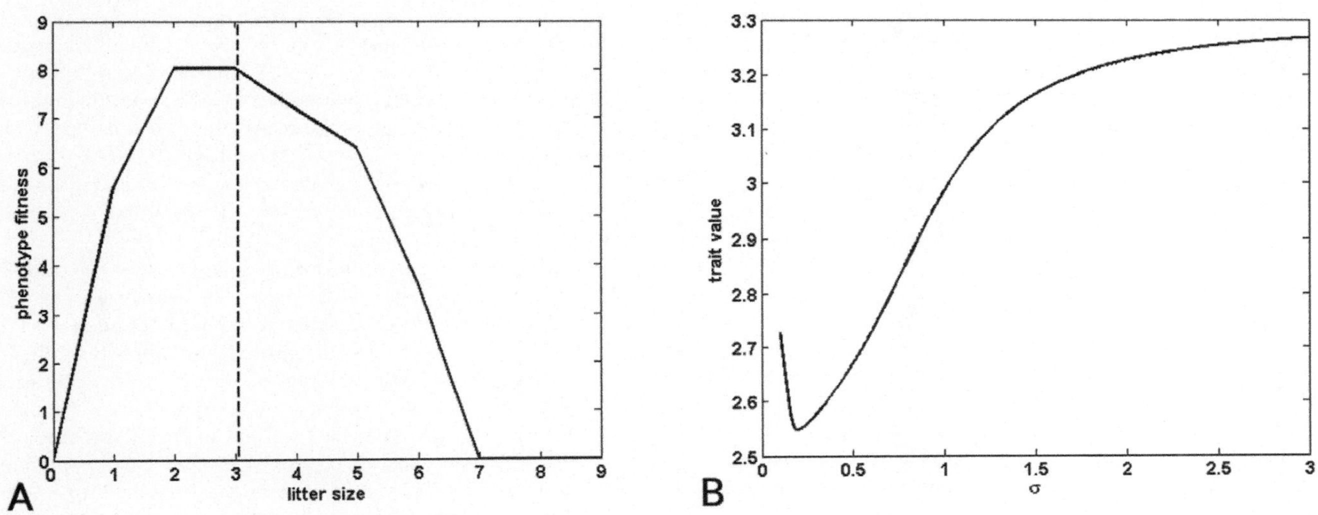

Figure 2. Effect of phenotypic variance on optimal litter size in Guinea pigs. (a) Asymmetric fitness function $b(L)/m(L)$. The curve reaches its maximum value on the plateau between $L = 2$ and $L = 3$. The dashed line corresponds to the optimal genotype $L = 3.05$ ($\sigma = 1.14$) that best fits the empirical data from Mountford [20]. (b) Position of the optimal genotype relative to the standard deviation σ.

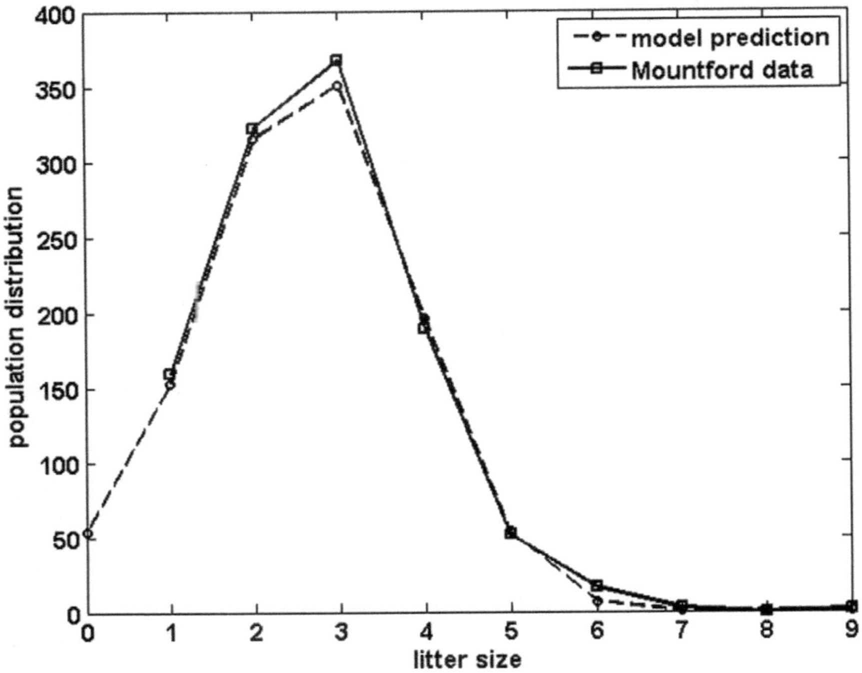

Figure 3. Comparison between model predictions (dashed line) and empirical data (Mountford [20], solid line) for the distribution of litter sizes in the population. The model predicts an optimal genotypic value for litter size at *3.05* with a variance σ^2 of *1.30*.

the model and the distribution observed by Mountford [20] was less than *3.5%* (Figure 3).

2-2 Example 2: Selection on male wing patch size in *Calopteryx splendens*. In the damselfly *Calopteryx splendens*, males have a dark melanized wing patch that covers approximately 50% of the wing [39]. Wing patches in this species function as secondary sexual traits and are only carried by males [40–42]. Wing melanization affects male predation risk, as males with larger wing patches suffer higher mortality from avian predators [30]. This trait is also under sexual selection in this population, where female mating response increases with male wing patch size (Supplementary Text S4).

The phenotype fitness function is asymmetric (Figure 4) around a maximum at wing patch length $x = 17.52\ mm$. The fitness decrease is steeper for larger patches; hence the optimal genotype is expected to be smaller than the phenotype fitness maximum.

The best fit was obtained with an optimal trait x_o of *16.93 mm* and a standard deviation σ of *2 mm*. The difference between the values predicted by the model and those measured in the field was less than *17%* (Figure 5).

2-3 Example 3: Estimating optimal lung geometry. Mauroy et al. [31] developed a model of the human bronchial tree to study the relationship between the geometry of the tree and its hydrodynamical resistance. They modelled the distal part of the lungs as a dichotomical tree branching in a homothetical way: at each bifurcation: each branch divides in two identical smaller branches, whose length and diameter are reduced by a constant factor h, the homothetical factor (Figure 6, for $h_l = h_d = h$).

They showed that the mean phenotypic parameter h, observed from empirical data [43], is around *0.8470*, while the optimum value predicted by their model was *0.7937*. Although small, this difference is expected to have major effects on the resistance and volume of the lung because of the multiplicative nature of the homothetical transformation (i.e. if the tree bifurcates 10 times, the deepest branches will be h^{10} smaller than the first generation branch). Large lung hydrodynamic resistance (small h) requires more energy for lung ventilation, while large lung volume (large h) results in a reduced exchange surface (less volume is available for alveoli). We consider that the fitness of an individual is a function of respiratory efficiency and thus depends on the value of the parameter h.

We further extended the model originally formulated by Mauroy et al. [31] to estimate a more realistic fitness function for h. Based on morphometric data [44] we assumed that lengths and diameters of the tree branches are not reduced by the same factor at each bifurcation (Figure 6, $h_l \neq h_d$). Under this hypothesis, the resulting phenotype fitness function (Figure 7) is asymmetrical around the optimum (Supplementary Text S4). The steepest decrease in fitness occurs for values below the optimum, thus we expect the optimal value of the genotype fitness to be higher than the most efficient phenotype.

Empirical data indicate a mean phenotypic value of $h = 0.8470$ [36]. The model predicts that this mean phenotype is reached for an optimal genotype $h = 0.8504$ and a standard deviation $\sigma = 0.2$. Compared with the value maximizing the phenotype fitness function $h_{min} \sim 0.7937$, the optimal genotype corresponds to a resistance that is *3.3* times smaller and a volume that is *2.4* times larger. This result quantitatively confirms the hypothesis stated by Mauroy et al. [31] that this shift from the phenotypic optimum acts as a security margin to protect the bronchial tree from phenotypic variations. This analysis is representative of an optimality problem related to the geometry of a transport tree, and as such it can be extended to many other well-known theoretical contexts, such as the Metabolic Theory of Ecology [44,45].

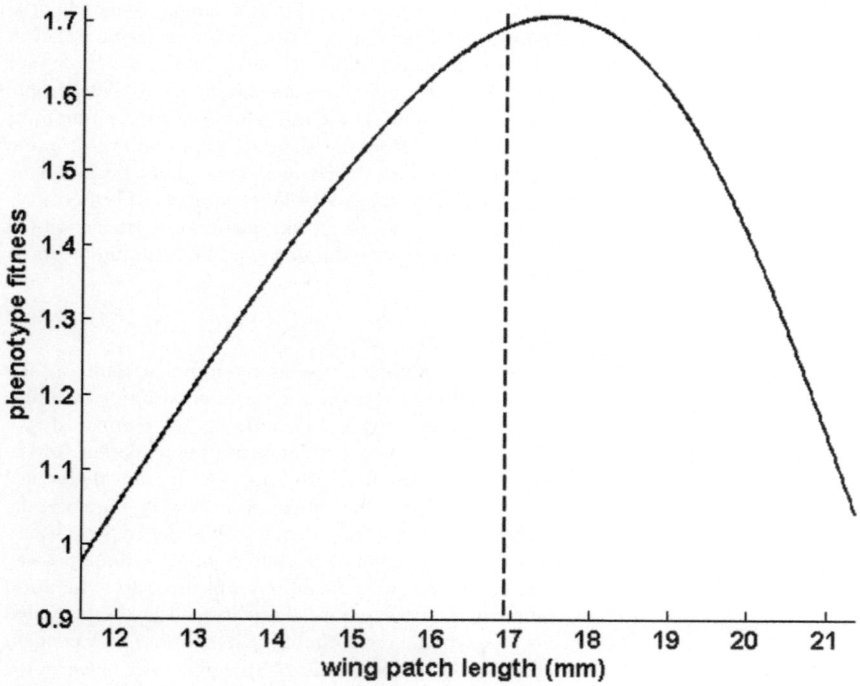

Figure 4. Estimate of the asymmetric phenotype fitness function ***z*** **for** ***Calopteryx splendens*** **in relation with wing patch length** ***x*****.** The maximum is reached for a wing patch length of *17.52 mm*. The dashed line represents the optimal genotype predicted by our model that best fits the observed population distribution, i.e. a patch length of *16.93 mm*.

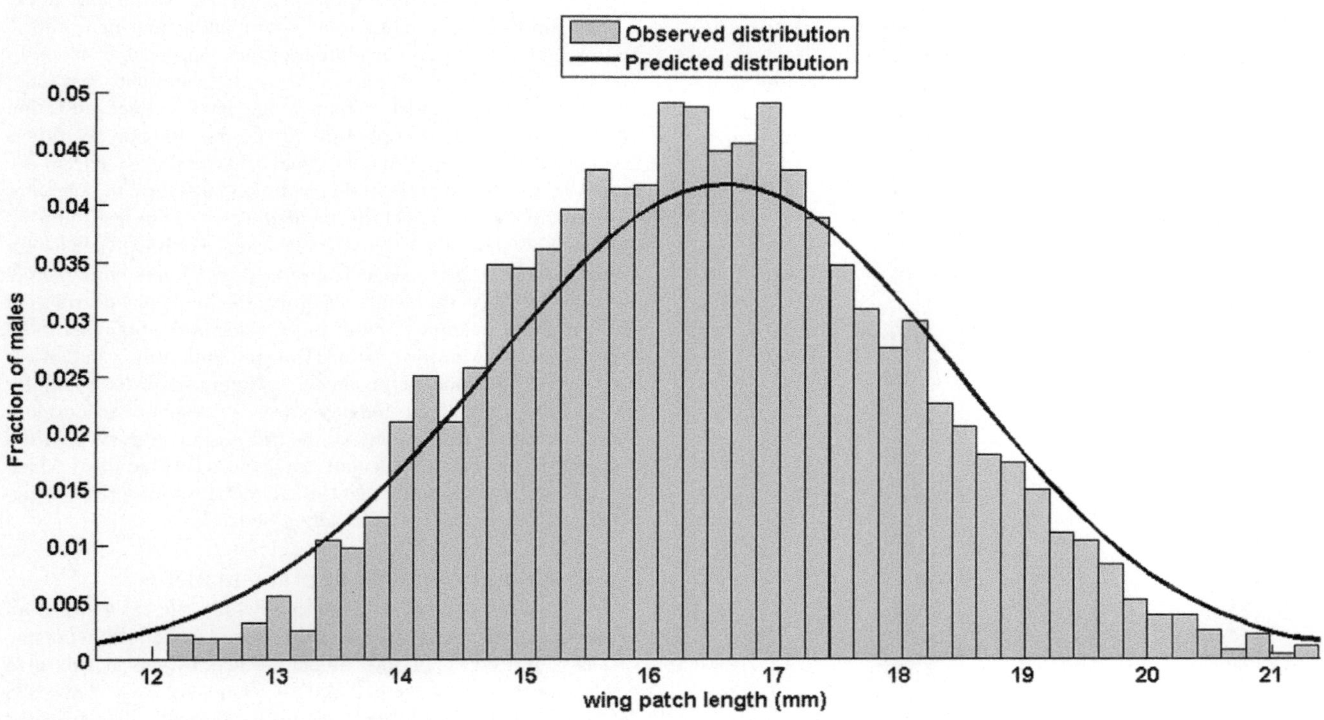

Figure 5. Distribution of wing patch length of ***Calopteryx splendens*** **at the population Naturreservat Klingavälsån (55.6384, 13.54142) in southern Sweden.** The bars correspond to males caught in the field. The line shows the distribution predicted by the model with an optimal patch length of *16.93* mm and a standard deviation of 2 mm.

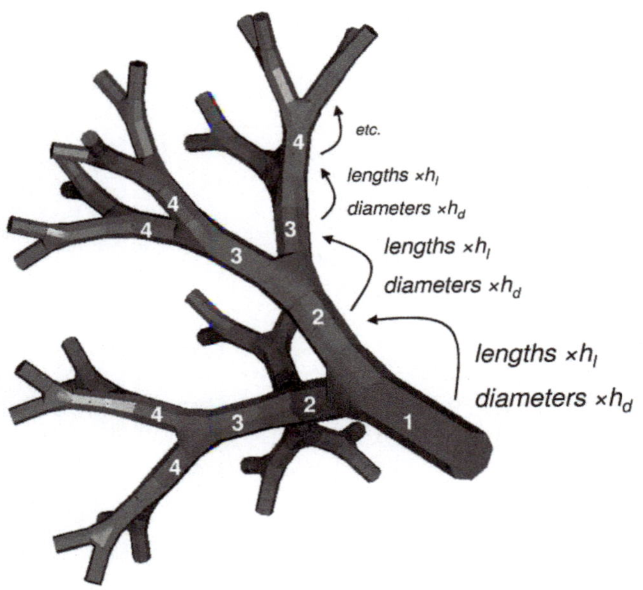

Figure 6. Model of the distal part of the bronchial tree used in this study. After each bifurcation the generation index is incremented by one (white numbers). The full model consists of *11* generations.

Discussion

Environmental variation is ubiquitous in nature and can generate substantial levels of phenotypic variation in fitness-related traits [20,23,46]. Such unpredictable variation (e. g. developmental plasticity) can have profound effects on optimal trait values [22,47]. Our model explicitly incorporates such unpredictable variation through its effects on phenotypic variance and shows that the position of the genotype fitness optimum will ultimately depend on both the amount of phenotypic variance and the shape of the fitness function. For symmetrical fitness and variance functions, the optimal value for the fitness-related trait matches the value that maximizes phenotype fitness. However, this classical optimization scenario does not hold when fitness functions are asymmetrical and when environmental variance leads to a variable expression of genetic traits. Then, the genotype fitness optimum is instead expected to shift from the phenotype optimum value in the direction of the least slope. Such qualitative predictions can be driven directly from the shape of the fitness function. Furthermore, as illustrated in our three examples, quantitative predictions of random phenotypic variance and genotype fitness optima can be derived from empirical data.

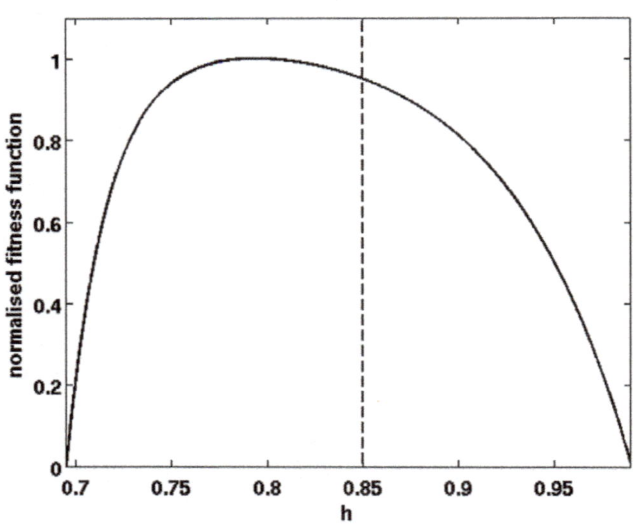

Figure 7. Relationship between homothetical factor *h* and fitness (trade-off between lung volume and hydrodynamical resistance). The vertical dashed line corresponds to the optimal genotype $h = 0.8504$ with $\sigma = 0.2$, which best fits empirical data.

Model applications

In the first two examples, we used empirical data on the frequency distribution of traits and their relationship with fitness components (reproductive rate and mortality) to estimate σ, the amount of random phenotypic variance in the population, and to validate the predictions from the model. If the theoretical distribution predicted by the value of σ closely matches the observed distribution of the trait, then the cliff-edge hypothesis is a sufficient condition to explain the shift in the distribution away from the phenotype fitness maximum. In this case, the most frequent phenotype in the population is located at the predicted value of the genotype fitness optimum. Furthermore, independent of the intrinsic quality of the prediction, the difference between the theoretical and observed descriptors provides information about the importance of processes other than cliff-edge effects in the evolution of the trait. In these two examples, we obtained less than 3.5% deviation for the Guinea pig data and less than 17% for the damselfly data. Data on litter size in Guinea pigs are more likely to meet the model's assumptions (e.g. laboratory strains with low amount of additive genetic variance, controlled environment with few external selective pressures). In contrast, natural damselfly populations should contain more genetic variability, and many selective pressures in addition to predation and female mate choice are expected to affect the evolution of male wing patch size [40]. These processes are partly responsible for the variability that is not incorporated in the model. Yet, the model gives a closer prediction of the actual fitness optimum than the phenotype fitness maximum, thus supporting the claim that cliff-edge effects are likely to play a strong role in the evolution of natural populations.

Alternatively, the model can be used as an *a priori* hypothesis to predict the value of key parameters when empirical data is not available, which is a classical approach in physics and biomechanics modelling. In the lung example, the frequency distribution of the trait *h* is unknown, but its mean value in the population has been estimated. Mauroy et al. [31] suggested that the value of *h* is shifted from the phenotype fitness optimum in order to confer higher robustness to the lung geometry in response to developmental variation. In this context, the cliff-edge hypothesis provides a formal framework to calculate the expected value of σ, which can then be used to implement other models and derive further predictions that can be empirically tested.

Generality of the conditions of the model

The keystone hypotheses of the model are the existence of an asymmetrical fitness function and a certain amount of phenotypic variance, and the qualitative model predictions appear robust to the precise shape of these functions (Supplementary Text S2). Although the exact geometry of trade-offs will differ between different ecological situations, asymmetrical trade-offs are likely to be the rule rather than the exception, especially for traits under stabilizing selection [48]. Indeed, if a trade-off results from the interaction between two unrelated traits, there is no reason why their respective effects on phenotype fitness should be exactly opposing each other, i.e. completely symmetric. Similarly, there are many processes that can generate random phenotypic variation. For instance, condition-dependence, phenotypic plasticity, developmental effects, environmental fluctuations and/or non-additive genetic effects (e. g. epistasis) are all common and well-known processes that are likely to increase phenotypic variance [49,50]. These processes are expected to have a significant influence on traits with low heritability, which is a common characteristic of traits strongly related to fitness [29]. Therefore, sources of stochastic phenotypic variation have been suggested to be key factors in the evolutionary ecology of populations [51–54].

Finally, we modelled the effects of phenotypic variance during development of individual phenotypes. These phenotypes are then assumed to be stable during life, i.e. we explicitly considered inter-generational phenotypic variance. However, the model can be generalized and the same predictions can also be made in the case of intra-individual variance, for example, when the value of a fitness-related trait changes during an individual's life. Such an example was recently documented by Martin and Huey [27] in the context of thermoregulation in reptiles. The authors showed that the optimal range of body temperatures for an individual should not be centred at the temperature for which the instantaneous fitness is maximised, but should be shifted towards a lower temperature and that the magnitude of the shift increased with the asymmetry of the fitness function.

Phenotypic variability and species adaptation

In the context of optimization problems, stochastic effects can influence the predictions of theoretical models (e.g. bet-hedging strategies, optimization of the geometric mean fitness, [23,55,56], this study). However, these effects depend quite strongly on the relationship between an individual's genotype and the variability of its phenotype.

First, if the variability in phenotype expression is independent from the individual's genotype (as is the case in our model), its association with an asymmetrical fitness functions can be a significant limit to adaptation. Our example of bronchial tree geometry illustrates this situation and provides an adaptive explanation relating to the fact that the human lung is probably not as efficient as it could be. Several other examples in evolutionary medicine appear to be consistent with the existence of cliff-edge effects [57–60]. Strong directional selection for traits that are globally advantageous would sometimes drive their mean too close to the 'fitness cliff', which could set the stage for counter-selection of extreme phenotypes. Such mechanisms would limit the long-term directional evolution of heightened physiological, mental and immune capacities in humans, and the average performance of individuals would be lower than their maximum potential. These examples tentatively suggest that similar processes might have operated in many different species to constrain the evolution of phenotypic traits within a smaller range than their full physiological potential.

Alternatively, when phenotype variability is related to the individual's genotype, the evolutionary consequences are likely to be quite different. In certain conditions, selection can favour the ability of phenotypes to resist random developmental or environmental perturbations, a process known as canalization [61]. Canalization is expected to be favoured in spatially or temporally variable environments, or in environments connected by high levels of gene flow, because it allows the persistence of high genetic variation and evolutionary potential [62]. In contrast, different genotypes might display different reaction norms in response to environmental variations, i.e. the amount of phenotypic variance and the shape of its distribution can differ between individuals. In this case, asymmetrical variance functions can be selected as a way to compensate for an asymmetrical fitness function by avoiding most detrimental phenotypes (Supplementary Text S2). However, such a strategy could be selected only if environmental variations do not affect the fitness function itself, and if they are restricted within a limited range so that extreme phenotypes will be rare.

Conclusion

In this study, we propose a simple formalization, validated by three empirical examples, of an evolutionary process known as the 'cliff-edge' effect. Our predictions stand for any trait (i) associated with an asymmetrical fitness function and (ii) when phenotypic expression is subjected to random variation, which are conditions expected to be common in nature [63,64]. In this framework, future studies should aim at analysing the optimization of genotype fitness instead of phenotype fitness. If only cliff-edge effects are shaping the evolution of the trait, then the most frequent value of the trait should match the genotype fitness optimum, i.e. the genotype fitness optimum is the null hypothesis for trait optimization. On the contrary, if the trait is non-optimal with regard to the genotype fitness, other evolutionary processes should be considered. For instance, unmeasured fitness components might cause undetected trade-offs that constrain the adaptation of the trait.

Supporting Information

Text S1 Analysis of the population model.

Text S2 Influence of the shapes of the phenotype fitness and variance on growth rate function.

Text S3 Methods for model fitting to empirical data.

Text S4 Detailed analysis of empirical examples 2 and 3.

Author Contributions

Conceived and designed the experiments: EV BM MW ES. Performed the experiments: EV BM MW ES. Analyzed the data: EV BM. Wrote the paper: EV BM MW ES. Wrote and analyzed the population model: EV BM.

References

1. Fisher RA (1930) The genetical theory of natural selection. Oxford: Clarendon Press. 272 p.
2. Maynard Smith J (1978) Optimization theory in evolution. Ann Rev Ecol Syst 9: 31–56.
3. Stearns SC (1992) The evolution of life histories. Oxford: Oxford University Press. 262 p.
4. Fairbairn DJ, Reeve JP (2001) Natural selection. In: Fox CW, Roff DA, Fairbairn DJ, eds. Evolutionary Ecology, Concepts and Case Studies. Oxford: Oxford University Press. pp 29–43.
5. Roff DA (1992) The evolution of life histories: theory and analysis. New York: Chapman and Hall. 535 p.
6. Parker GA, Maynard Smith J (1990) Optimality theory in evolutionary biology. Nature 348: 27–33.

7. Wright S (1932) The roles of mutation, inbreeding, crossbreeding and selection in evolution. Proceedings of the 6th International Congress on Genetics 1: 355–366.
8. Lack D (1947) The significance of clutch size. Ibis 89: 302–352.
9. Charnov EL, Skinner SW (1984) Evolution of host selection and clutch size in parasitoid wasps. Flor Entomol 67: 5–21.
10. Mock DW, Parker GA (1998) The evolution of sibling rivalry. Oxford: Oxford University Press. 480 p.
11. Lessells CM (1986) Brood size in canada geese - a manipulation experiment. J Anim Ecol 55: 669–689.
12. Boyce MS, Perrins CM (1987) Optimizing great tit clutch size in a fluctuating environment. Ecology 68: 142–153.
13. Moller AP (1991) Ectoparasite loads affect optimal clutch size in swallows. Funct Ecol 5: 351–359.
14. Boutin S, Moses RA, Caley MJ (1988) The relationship between juvenile survival and litter size in wild muskrats (*Ondatra zibethicus*). J Anim Ecol 57: 455–462.
15. Morris DW (1992) Optimum brood size: tests of alternative hypotheses. Evolution 46: 1848–1861.
16. Risch TS, Dobson FS, Murie JO (1995) Is mean litter size the most productive? A test in Columbian ground squirrels. Ecology 76: 1643–1654.
17. Risch TS, Michener GR, Dobson FS (2007) Variation in litter size: a test of hypotheses in Richardson's ground squirrels. Ecology 88: 306–314.
18. Williams GC (1966) Natural selection, the costs of reproduction, and a refinement of Lack's principle. Am Nat 100: 687–690.
19. Smith CC, Fretwell SD (1974) The optimal balance between size and number of offspring. Am Nat 108: 499–506.
20. Mountford MD (1968) The significance of litter size. J Anim Ecol 37: 363–367.
21. Smallwood PD (1996) An introduction to risk sensitivity: the use of Jensen's inequality to clarify evolutionary arguments of adaptation and constraint. Am Zool 36: 392–401.
22. Ruel JJ, Ayres MP (1999) Jensen's inequality predicts effects of environmental variation. Trends Ecol Evol 14: 361–366.
23. Pasztor L, Kisdi E, Meszéna G (2000) Jensen's inequality and optimal life history strategies in stochastic environments. Trends Ecol Evol 15: 117–118.
24. Wilson WG, Harder LD (2003) Reproductive uncertainty and the relative competitiveness of simultaneous hermaphroditism versus dioecy. Am Nat 162: 220–241.
25. Drake JM (2005) Population effects of increased climate variation. Proc Roy Soc B Biol Sci 272: 1823–1827.
26. Freedberg S, Taylor DR (2007) Sex ratio variance and the maintenance of environmental sex determination. J Evol Biol 20: 213–220.
27. Martin TL, Huey RB (2008) Why "suboptimal" is optimal: Jensen's inequality and ectotherm thermal preferences. Am Nat 171: 102–118.
28. Turnbull LA, Rees M, Purves DW (2008) Why equalising trade-offs aren't always neutral. Ecol Lett 11: 1037–1046.
29. Falconer DS (1981) Introduction to Quantitative Genetics, 2nd Edition. New-York: Longman. 340 p.
30. Svensson EI, Friberg M (2007) Selective predation on wing morphology in sympatric damselflies. Am Nat 170: 101–112.
31. Mauroy B, Filoche M, Weibel ER, Sapoval B (2004) An optimal bronchial tree may be dangerous. Nature 427: 633–636.
32. Millar JS (1977) Adaptive features of mammalian reproduction. Evolution 31: 370–386.
33. Stearns SC (1989) Trade-offs in Life History Evolution. Funct Ecol 3: 259–268.
34. Boyd IL, Croxall JP, Lunn NJ, Reid K (1995) Population demography of Antarctic fur seals - the costs of reproduction and implications for life-histories. J Anim Ecol 64: 505–518.
35. Ruusila V, Ermala A, Hyvarinen H (2000) Costs of reproduction in introduced female Canadian beavers (*Castor canadiensis*). J Zool 252: 79–82.
36. Koivula M, Koskela E, Mappes T, Oksanen TA (2003) Cost of reproduction in the wild: Manipulation of reproductive effort in the bank vole. Ecology 84: 398–405.
37. Hare JF, Murie JO (1992) Manipulation of Litter Size Reveals No Cost of Reproduction in Columbian Ground Squirrels. J Mammal 73: 449–454.
38. Humphries MM, Boutin S (2000) The determinants of optimal offspring number in free-ranging red squirrels. Ecology 81: 2867–2877.
39. Rüppel G, Hilfert-Rüppel D, Rehfeldt G, Schütte C (2005) Die Prachtlibellen Europas. Hoehnwarsleben: Westarp Wissenschaften. 255 p.
40. Siva-Jothy M (1999) Male wing pigmentation may affect reproductive success via female choice in a Calopterygid damselfly (Zygoptera). Behaviour 136: 1365–1377.
41. Svensson E I, Kristoffersen L, Oskarsson K, Bensch S (2004) Molecular population divergence and sexual selection on morphology in the banded demoiselle (*Calopteryx splendens*). Heredity 93: 423–433.
42. Svensson EI, Eroukhmanoff F, Friberg M (2006) Effects of natural and sexual selection on adaptative population divergence and premating isolation in a damselfly. Evolution 60: 1242–1253.
43. Weibel ER (1963) Morphometry of the human lung. Berlin: Springer. 151 p.
44. West GB, Brown JH, Enquist BJ (1997) A general model for the origin of allometric scaling laws in biology. Science 276: 122–126.
45. Enquist BJ, West GB, Charnov EL, Brown JH (1999) Allometric scaling of production and life-history variation in vascular plants. Nature 401: 907–911.
46. Morris DW (1996) State-dependent life-histories, Mountford's hypothesis, and the evolution of brood size. J Anim Ecol 65: 43–51.
47. Yoshimura J, Shields WM (1992) Components of uncertainty in clutch-size optimization. Bull Math Biol 54: 445–464.
48. De Witt TJ (1997) Optimizing clutch size in birds. Trends Ecol Evol 12: 443.
49. van Noordwijk AJ, de Jong G (1986) Acquisition and allocation of resources: their influence on variation in life-history tactics. Am Nat 128: 137–142.
50. Mazer SJ, Damuth J (2001) Evolutionary significance of variation. Fow CW, Roff DA, Fairbairn DJ, eds. Evolutionary Ecology, Concepts and Case Studies. Oxford: Oxford University Press. pp 16–28.
51. Real LA, Ellner S (1992) Life history evolution in stochastic environments: a graphical mean-variance approach. Ecology 73: 1227–1236.
52. Fox GA (1993) Life history evolution and demographic stochasticity. Evol Ecol 7: 1–14.
53. Whitlock MC (1995) Variance-induced peak shifts. Evolution 49: 252–259.
54. Wilson AJ, Pemberton JM, Pilkington JG, Coltman DW, Mifsud DV, Clutton-Brock TH, Kruuk LEB (2006) Environmental Coupling of Selection and Heritability Limits Evolution. PLoS Biol 4: e216.
55. Philippi T, Seger J (1989) Hedging one's evolutionary bets, revisited. Trends Ecol Evol 4: 41–44.
56. Yoshimura J, Jensen VAA (1996) Evolution and population dynamics in stochastic environments. Res Pop Ecol 38: 165–182.
57. Nesse RM, Williams GC (1996) Why we get sick: the new science of Darwinian medicine. New York: Vintage Books. 290 p.
58. Burns JK (2004) Elaborating the social brain hypothesis of schizophrenia. Behav Brain Sci 27: 868–885.
59. Nesse RM (2004) Cliff-edged fitness functions and the persistence of schizophrenia. Behav Brain Sci 27: 862–863.
60. Stearns SC, Koella JK (2008) Evolution in health and disease (2nd edition). Oxford: Oxford University Press. 400 p.
61. Flatt T (2005) The evolutionary genetics of canalization. Q Rev Biol 80: 287–316.
62. Stearns SC (2002) Progress on canalization. Proc Natl Acad Sci USA 99: 10229–10230.
63. Lajus DL, Graham JH, Kozhara AV (2003) Developmental instability and the stochastic component of total phenotypic variance. In: Polak M, ed. Developmental Instability: causes and consequences. Oxford: Oxford University Press. pp 343–366.
64. Hansen TF, Carter AJR, Pélabon C (2006) On adaptive accuracy and precision in natural populations. Am Nat 168: 168–181.

6

Patterns of Chemical Diversity in the Mediterranean Sponge *Spongia lamella*

Charlotte Noyer[1], Olivier P. Thomas[2], Mikel A. Becerro[1]*

1 Center for Advanced Studies of Blanes (CEAB, CSIC), Blanes, Spain, **2** Université de Nice - Sophia Antipolis, Laboratoire de Chimie des Molécules Bioactives et des Arômes, LCMBA-UMR 6001 CNRS, Nice, France

Abstract

The intra-specific diversity in secondary metabolites can provide crucial information for understanding species ecology and evolution but has received limited attention in marine chemical ecology. The complex nature of diversity is partially responsible for the lack of studies, which often target a narrow number of major compounds. Here, we investigated the intra-specific chemical diversity of the Mediterranean sponge *Spongia lamella*. The chemical profiles of seven populations spreading over 1200 km in the Western Mediterranean were obtained by a straightforward SPE-HPLC-DAD-ELSD process whereas the identity of compounds was assessed by comparison between HPLC-MS spectra and literature data. Chemical diversity calculated by richness and Shannon indexes differed significantly between sponge populations but not at a larger regional scale. We used factor analysis, analysis of variance, and regression analysis to examine the chemical variability of this sponge at local and regional scales, to establish general patterns of variation in chemical diversity. The abundance of some metabolites varied significantly between sponge populations. Despite these significant differences between populations, we found a clear pattern of increasing chemical dissimilarity with increasing geographic distance. Additional large spatial scale studies on the chemical diversity of marine organisms will validate the universality or exclusivity of this pattern.

Editor: Simon Thrush, National Institute of Water & Atmospheric Research, New Zealand

Funding: This project was supported by the French Agence Nationale de la Recherche on Biodiversity (ECIMAR program ANR-06-BDIV-001; http://www.ecimar.org/cms/), the Spanish Ministry of Science and Technology (SOLID program CTM2010-17755, and Benthomics program CTM2010-22218-C02-01), and the European Union (BIOCAPITAL UE grant MRTN-CT-2004-512301; http://www.biocapital.de/welcome.htm). The funders had no role in study design, data collection and analysis, decision to publish, or preparation of the manuscript.

Competing Interests: This work is part of Charlotte Noyer PhD thesis. Charlotte Noyer was funded by the European grant MRTN-CT-2004-512301 BIOCAPITAL. The authors have declared that no competing interests exist.

* E-mail: mikel@ceab.csic.es

Introduction

The current global loss of species has surpassed scientific interest, placing biodiversity conservation in political agendas and media around the globe [1,2]. But diversity is a complex and dynamic concept that applies below and beyond the species level [3]. Specifically, the intra-specific level of diversity is particularly relevant for species evolution [4]. It occurs at the within and between population level and provides the variance on which natural selection acts [4,5]. Many organisms produce secondary metabolites that play multiple functions in nature [6]. Secondary metabolites have evolved under the pressure of natural selection and represent ecological responses of organisms to their environment [5,6,7]. Intra-specific diversity in secondary metabolites is therefore crucial to understand species biology, ecology, and evolution, yet, it has received partial attention. This is particularly true in marine environments [5,8].

Marine benthic organisms provide unique opportunities to investigate chemical diversity. Sessile or slow moving organisms without effective escape mechanisms or structural protection are likely to be chemically defended [9]. Sessile organisms also are under intense pressures to acquire and maintain space on where to live [10], to gain protection from ultraviolet radiation [11] or to prevent from fouling [12]. Intra-specific information on their chemical diversity is fundamental to look for general patterns of secondary metabolite production and to understand the processes that produce chemical diversity in benthic communities. Available evidence suggests that intra-specific spatial and temporal variation in secondary metabolism is widely distributed in the benthic realm including algae [13,14,15], sponges [16,17,18], cnidarians [19], tunicates and bryozoans [19], and worms [20].

Sponges are major components of benthic communities and among the richest source of secondary metabolites isolated from marine organisms [21,22,23]. Unsurprisingly, chemical ecology has contributed significantly to advance in our understanding of sponge ecology [24]. Many sponge secondary metabolites are known to deter feeding, confer competitive advantages, and protect against fouling among other ecological roles [8,25]. But the demand for studies on sponge chemical ecology looms large [24] and the study of chemical diversity at a geographic scale is an area that has received little attention. The sponge *Aplysina aerophoba* shows significant differences in the concentration of bromotyrosine derivatives in multiple locations around the Canary Islands but the same secondary metabolites were present in all specimens [26]. In a distinct way the sponge *Acanthella cavernosa* shows remarkable variation in the sesquiterpene composition of specimens from multiple locations [27]. Variation at a geographic scale is fundamental to understand population genetics, demographic history, and evolution of benthic species [28,29]. Similarly, geographic patterns of chemical variation can be critical to understanding the evolution and function of secondary metabolites, as well as unraveling large scale patterns and a hidden chemical diversity that are poorly understood [8].

The secondary metabolites of the genus *Spongia* are well described [30,31,32,33]. The sponge *Spongia lamella* is an Atlanto-Mediterranean species well studied for its chemical diversity and a careful examination of the secondary metabolites composition has already been established [31,33]. Traditionally collected as a bath sponge [34], recent disease epidemics have decimated populations of *S. lamella* [34,35,36] and it is now considered an endangered species under Annex 3 of Bern and Barcelona Conventions [37]. Microsatellite markers previously described for this species revealed that populations were genetically differentiated and most of them must have suffered from recent demographic reductions [38,39]. Loss of genetic diversity has a major impact on population survival especially for threatened species. It plays a crucial role in increasing extinction risk, and thus in depleting biodiversity [40]. Information on the chemical variability of this species is lacking especially at a geographic level. Loss of intraspecific variation in the production of secondary metabolites can also jeopardize species survival and limit the potential for direct and indirect uses of diversity for the future, such as identification of compounds of pharmaceutical or biotechnological interests [2,41]. Given the potential applications of sponge secondary metabolites as new drugs or antifoulants [42,43,44], understanding the hidden chemical diversity of endangered species is a priority that could have strong conservation consequences.

In this paper we investigate the chemical diversity of *Spongia lamella* in seven Mediterranean populations distributed over 1200 km of coast. Diversity is a multifaceted concept based on i) the number of entities (richness) and ii) their abundances (evenness), iii) dissimilarities, iv) and functional roles [45]. Our study aimed at three out of the four pillars that support diversity as we quantified chemical diversity (richness/abundance), and dissimilarity of *S. lamella* at an intra-specific level and at two geographic scales (populations and biogeographic areas).

Results

Compound Identification and Quantification

The HPLC-ELSD calibration curve of Nitenin (**1**), we obtained by HPLC purification, showed good linearity within the range of concentrations investigated ($R^2 = 0.999$). The concentration of **1** in our samples was very close to that measured by the group of Salva (mean±se, 18.50±6.15 $mg.g^{-1}$, dry mass) and varied between 6.30 $mg.g^{-1}$ in Ceuta and 26.25 $mg.g^{-1}$ in Els Bullents [33]. Other major compounds observed by our standardized HPLC-DAD-ELSD protocol were identified by comparison of MS and NMR data with literature but also on the basis of the metabolic composition described by the group of Salva [33] (Figure 1).

The less retained compounds between 35 and 40 min were only visible at 210 nm in the UV spectrum. According to their retention times and mass spectra we assumed they could be assigned to the sterol derivatives previously described in *S. lamella* [31]. Dihydronitenin (**2**) and nitenin (**1**) were easily identified at RT 42.5 and 43 min after after HPLC purification and interpretation of NMR and MS analyses (Figures 1 and 2). The MS spectrum of the compound at RT 43.5 min exhibited m/z at 911.8 $[2M+Na]^+$, 467.5 $[M+Na]^+$, 445.7 $[M+H]^+$, 385.5 $[M-CH_3CO_2H+H]^+$ which strongly suggested the presence of 12-episcalarin (**3**) [46] also isolated as one of the major metabolites of

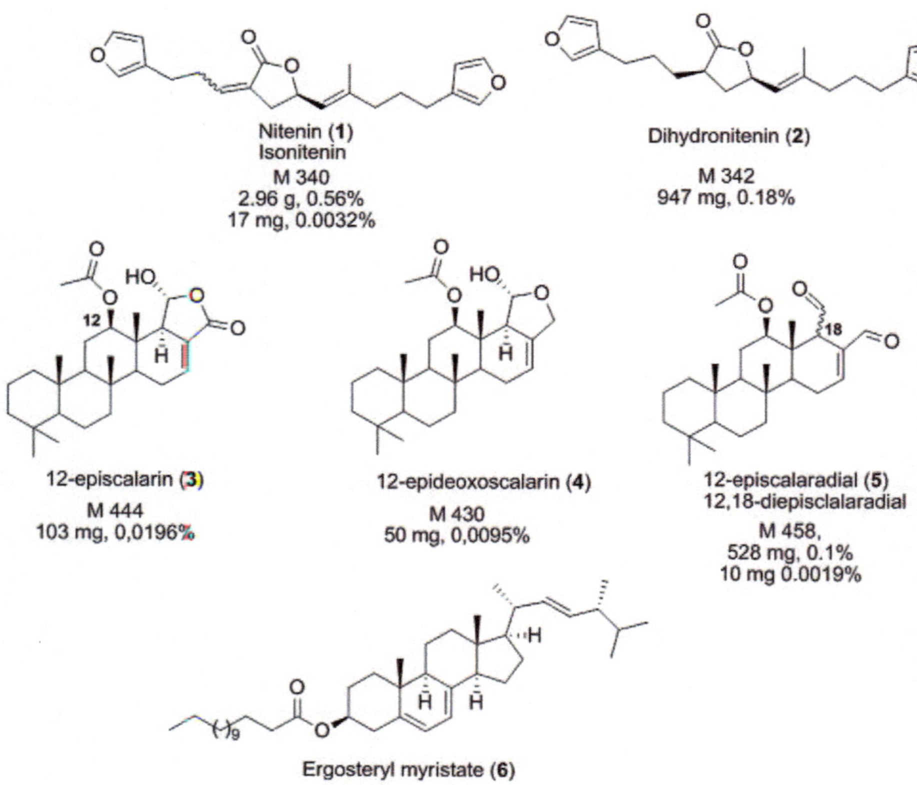

Figure 1. Secondary metabolites identified from *Spongia lamella*. Molecular mass and percentage in the extract (w/w) are included as described by Rueda and coworkers [33].

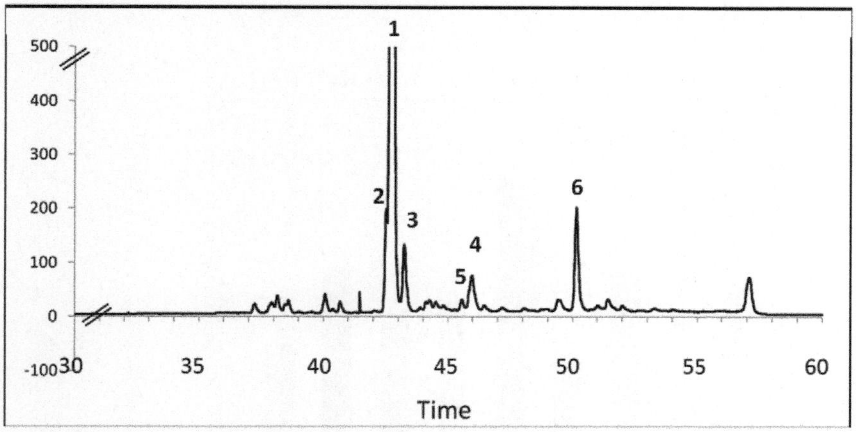

Figure 2. ELSD-HPLC chromatogram of *Spongia lamella* secondary metabolites. SPE-ELSD-HPLC chromatograms from desalting crude extract of *Spongia lamella.* The major compound identified as nitenin (1) was used as internal reference. Numbers are compounds described on Figure 2. (6) Second major compound described as an ergosterol ester.

this species [33]. Similarly, the MS spectrum of the compound at RT 46.0 min exhibited m/z 883.8 $[2M+Na]^+$, 453.5 $[M+Na]^+$, 353.4 $[M-CH_3CO_2H-H_2O+H]^+$ which suggested the presence of 12-epideoxoscalarin (**4**) [46]. Very close to this compound at RT 46.5 min the peak with m/z 879.8 $[2M+Na]^+$, 451.5 $[M+Na]^+$, 369.5 $[M-CH_3CO_2H+H]^+$ evidenced the absence of an hydroxyl group and then suggested the presence of 12-episcalaradial (**5**) (Figures 1 and 2). The presence of these biogenetically related metabolites was also interpreted as a confirmation of their structure. The major apolar compound did not show any peak in the mass spectrum and showed very distinct UV profile as it strongly absorbed at 254 and 280 nm. After HPLC purification the structure of this compound was elucidated on the basis of NMR spectra interpretation and comparison with literature data. The compound at RT 50.5 min exhibited signals at δ_H 1.02 (d, $J=6.5$ Hz, H-21), 0.94 (s, H-19), 0.92 (d, $J=6.7$ Hz, H-28), 0.86 (H-26), 0.83 (H-27) ppm in its ^{1}H NMR spectrum which were characteristic of an ergostane steroid. Furthermore, three unsaturations were evidenced in the steroid core by the signals at δ_H 5.57 (H-6), 5.38 (H-7), 5.20 (H-23) and 5.18 (H-22) ppm which were consistent with an ergosterol derivative. The signals at δ_H 2.35 (t, $J=7.1$ Hz, 2H, H-2′) and δ_C 33.7 ppm indicated the presence of a fatty acid ester at O-3. The length of the alkyl chain was finally determined by integration of the signal at 1.29 ppm. Comparing with the literature data this compound was identified as the previously described ergosteryl myristate [47].

Intra-specific Chemical Diversity

At the population level, the average number of compounds ranged from six in Plane to eleven in Ceuta and Pharillon. Compound richness was low and varied from 1.51 for Plane to over 3 for Pharillon and Ceuta. Shannon index varied from 0.69 in Els Bullents to 1.61 in Ceuta. The sponge populations of Els Bullents and Plane showed lower compound richness ($F=8.257$, $P<0.001$, Figure 3a, Table 1) and Shannon Index ($F=3.349$, $P=0.010$, Figure 3b, Table 1) than Ceuta, Pharillon, and Arenys. At the regional level, we found no significant differences in compound richness ($F=2.568$, $P=0.089$, Figure 3c) and Shannon index ($F=1.977$, $P=0.166$, Figure 3d).

Factor analysis (FA) of the chemical data led to six independent chemical factors that explained more than 84% of the total variance (Table 2). Both at local and regional scale, seasonal sampling did not influence chemical factors dissimilarities between populations and our data show the same degree of dissimilarity within and between seasons. In Marseille, winter chemical dissimilarities (10.45±1.29) were similar to winter-summer dissimilarities (13.01±2.07, *t-test*, $P=0.158$). In Catalonia, the summer dissimilarity (16.20) fell within winter-summer dissimilarities (15.84±1.05, *t-test*, $P=0.717$).

Multivariate analyses of variance showed significant chemical differences between sponge populations (*Wilks' Lambda* $F=4.168$, $P<0.001$; Table 3). Specific factors and compounds contributed differently to the observed chemical differences. Four out of the six factors and only ten compounds in these four chemical factors varied significantly between populations (Table 3).

We did not find significant chemical differences in the ten variable compounds between local population of *Spongia lamella* (Plane, Pharillon and La Ciotat, distant by less than 25 km, one way ANOSIM, global $R=0.091$, $p=0.107$, Figure 4a); However at larger geographical scale (Catalonia), we found significant chemical differences between populations (Cap de Creus, Els Bullents and Arenys distant by over 100 km, one way ANOSIM, global $R=0.449$, $p=0.002$, Figure 4b). We also detected a significant correlation between the chemical dissimilarities obtained with the ten variable compounds (Table 4) and the geographic distances between populations (*Mantel test* $P=0.017$, Figure 5a). Geographic distance explains 38.5% of the variance in chemical dissimilarity between sponge populations. Although we also found a relationship between bacterial dissimilarities and geographical distances between sponge populations [38], we failed to detect a correlation between chemical and bacterial dissimilarities (*Mantel test* $P=0.963$, Figure 5b).

Discussion

Understanding the intra-specific level of diversity in secondary metabolites is critical to advancing in chemical ecology [8]. However, studies addressing chemical diversity are underrepresented in marine systems [5,8]. Within-species diversity provides the variance on which evolution acts [4,5] and has important implications in population dynamics, conservation, and biotechnology [5,42,43,44]. To our knowledge, and beyond the variation in secondary metabolism across locations, our study is the first report of a geographic pattern in chemical diversity at an intra-

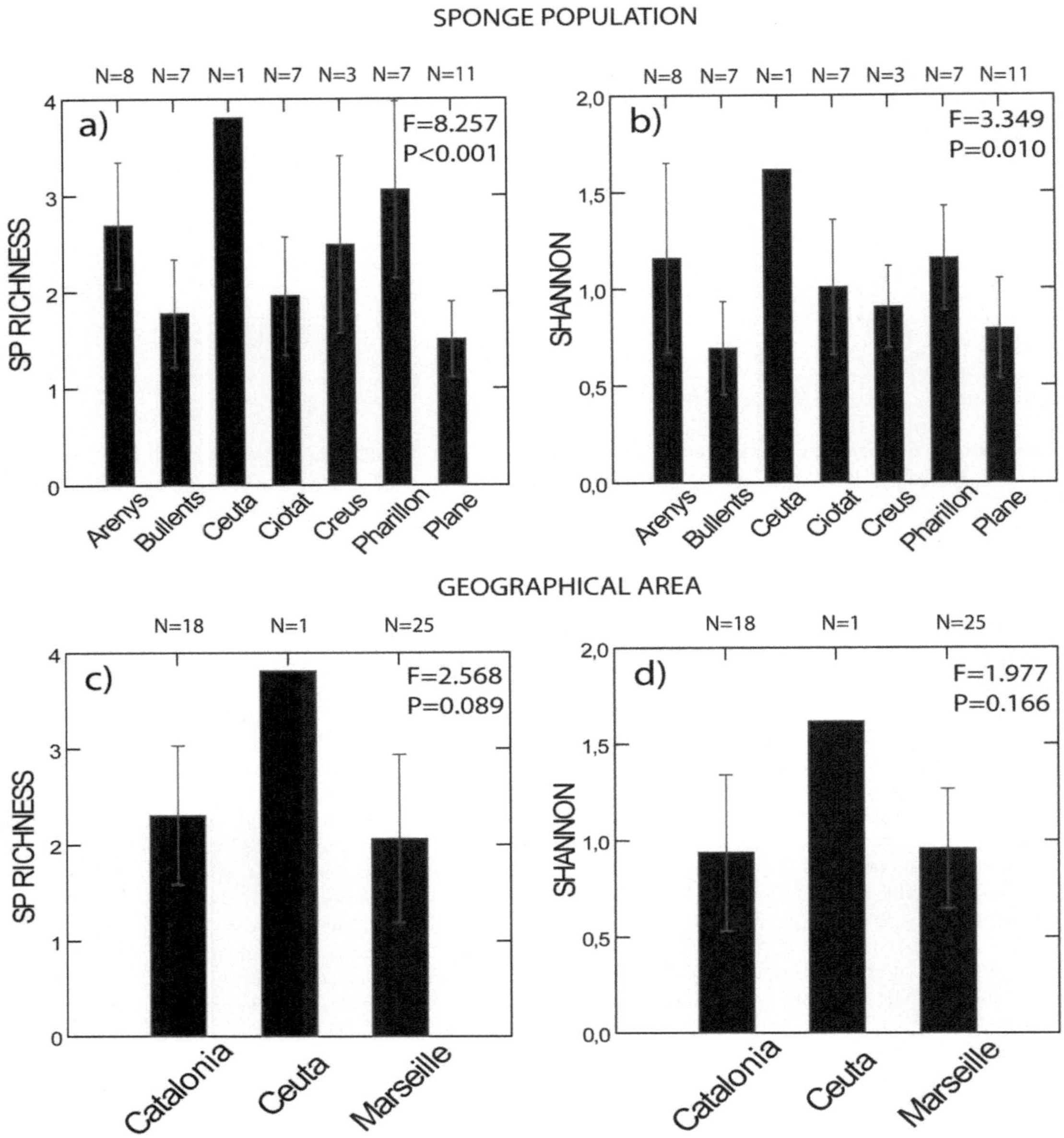

Figure 3. Chemical diversity of *Spongia lamella*. Chemical diversity was estimated in *Spongia lamella* at the population level a) and b), and at regional level c) and d) for peak richness and Shannon indices, respectively. Vertical bar is 1 standard deviation. Note lack of variation in Ceuta. A single specimen from this location was observed, sampled, and used in the analyses. One-way ANOVA: df = 6 at the population level, df = 2 at the regional level; F and P values included. See Table 1 for *Post hoc* Fisher's LSD tests.

specific level. We have documented the chemical diversity within the sponge *Spongia lamella* from nearby locations a few kilometers apart to populations spread over 1200 km. Chemical diversity in this species was relatively low, given the nature of the chemical profiles: the presence of the 2 major compounds led to the uneven repartition of diversity. We found significant differences in both Shannon and Margalef's indexes at the population level, sponge populations varied significantly in chemical richness and diversity. However, no differences in these diversity indexes were noticed at a regional geographic scale. This supported the capital role of the population level in analyses of diversity, especially when dealing with multiple components of biodiversity [4]. We also found that chemical dissimilarity in *Spongia lamella* increased with increasing geographical distances between populations.

We know little about geographic patterns of secondary metabolites at an intra-specific level in marine organisms. Data on this particular area are sparse, with contrasting results, and sometimes with methodological constrains. A study on bromotyrosine derivatives from *Aplysina fulva* evidenced chemical variability among specimens collected in multiple locations in Brazil and in a South Atlantic Bight reef in the United States); but multiple extraction and isolation protocols hampered ecological conclusions [48]. In contrast, the occurrence of brominated compounds in specimens of the sponge *Aplysina aerophoba* collected around the Canary Islands were identical, although their concentration varied [26]. Variation in the occurrence of secondary metabolites as a function of geographic location occurs in species from multiple animal and vegetal phyla [27,49,50,51]. The only geographic

Table 1. *Post hoc* Fisher's Least-Significant-Differences tests for *Spongia lamella* chemical diversity.

	Populations	SHANNON						
		Cio	Pla	Far	Cre	Bul	Are	Ceu
Species richness	Cio	-	NS	NS	NS	NS	NS	NS
	Pla	NS	-	**0.018**	NS	NS	**0.013**	**0.013**
	Far	**0.001**	**<0.001**	-	NS	**0.007**	NS	NS
	Cre	NS	**0.014**	NS	-	NS	NS	0.005
	Bul	NS	NS	**<0.001**	NS	-	**0.005**	**0.007**
	Are	**0.016**	**<0.001**	NS	NS	**0.004**	-	NS
	Ceu	**0.005**	**0.001**	NS	NS	**0.002**	NS	-

Species richness: below the diagonal; Shannon index: above the diagonal. Only values ≤5% are represented. NS: Non Significant.

pattern in the production of secondary metabolites that has received attention is the latitudinal gradient hypothesis [52]. Yet, most evidence for or against this hypothesis stems from the activities of secondary metabolites rather than from secondary metabolism itself [53,54,55]. Within-species variation of secondary metabolites from multiple locations can be so high as to cast doubt on the role of latitude [56]. Our study proved that pattern detection is possible despite significant chemical variation.

Chemical profiles are an excellent approach to quantify the relative abundance of secondary metabolites and assess their variations [57], and to address three out of the four concepts on

Table 2. Independent chemical factors (in columns) obtained after the Factor Analysis of chemical data.

Peaks	Chemical Factors					
	FC1	FC2	FC3	FC4	FC5	FC6
18	0.965	-	-	-	-	-
19	0.965	-	-	-	-	-
17	0.964	-	-	-	-	-
23	0.881	-	-	-	-	-
20	0.729	-	-	-	-	-
21	0.659	-	-	-	-0.691	-
14	0.653	-	-	-	-	-
(5)	-	-	-	-	-	0.746
(6)	-	0.916	-	-	-	-
25	-	0.913	-	-	-	-
24	-	0.877	-	-	-	-
(4)	-	0.719	-	-	-	-
(1)	-	-	-0.609	-	-	-
5	-	-	-0.852	-	-	-
4	-	-	-0.837	-	-	-
9	-	-	-0.808	-	-	-
7	-	-	-0.774	-	-	-
2	-	-	-	-0.891	-	-
3	-	-	-	-0.739	-	-
8	-	-	-	-0.638	-	-
6	-	-	-	-	-0.917	-
(3)	-	-	-	-	-	0.922
12	-	-	-	-	-	0.913
% of total variance explained	22.37	17.97	14.12	10.43	7.70	11.55

For a better interpretation, only peaks with absolute value greater than 0.600 are shown. Numbers in bracket are compounds from Figures 2 and 3.

Table 3. Multivariate and univariate analyses in chemical factors and chemical compounds between populations of *S. lamella*.

Chemical factors	F	P	Chemical compounds	F	P
FC1	**4.243**	**0.002**		**16.961**	**<0.001**
			18	Ns	
			19	Ns	
			17	Ns	
			23	7.269	0.010
			20	15.748	<0.001
			21	16.654	<0.001
			14	Ns	
FC2	Ns				
FC3	**2.947**	**0.014**		**44.223**	**<0.001**
			5	8.605	0.006
			4	4.295	0.045
			9	13.646	0.001
			7	30.78	<0.001
			(1) nitenin	172.782	<0.001
FC4	**2.775**	**0.020**		**6.267**	**0.002**
			2	Ns	
			3	18.611	<0.001
			8	Ns	
FC5	**21.830**	**<0.001**		**13.500**	**<0.001**
			6	12.892	0.001
			21	16.654	<0.001
FC6	Ns				

MANOVAs were run independently for each chemical factor. F-ratio and p-value of significant differences (P≤0.05) included. Ns: non-significant values.

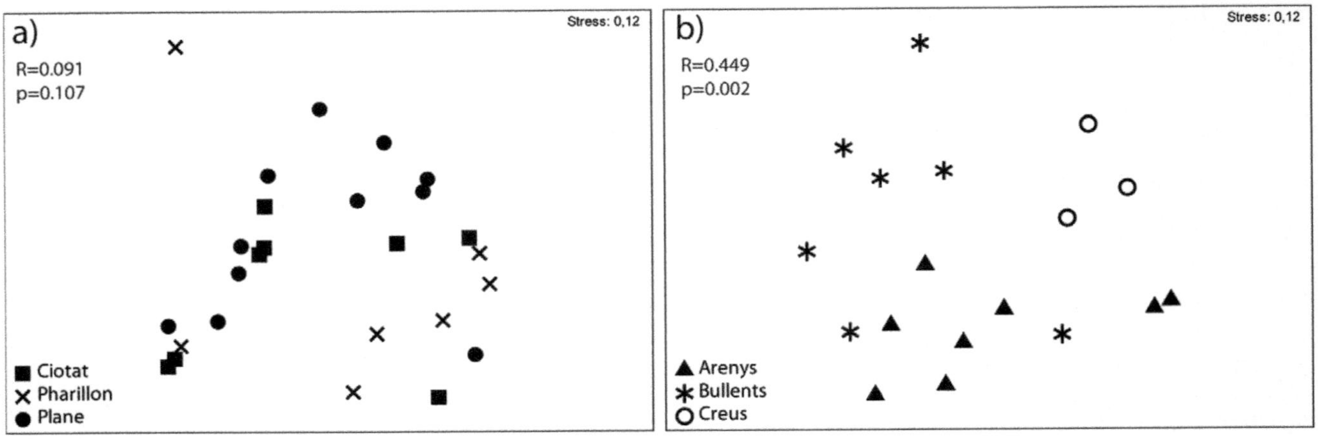

Figure 4. Chemical dissimilarities of *Spongia lamella*. Chemical dissimilarities were estimated by non-metric multidimensional scaling (MSD) and analysis of similarity (ANOSIM) from each the individual chromatograms using the 10 compounds responsible for variations, at local scale of Marseille a), and regional scale of Catalonia b).

which diversity is based. The universal ELSD detector was found to be the most appropriate detection mode for the visualization of the global secondary metabolism of a desalted marine extract. We used UV and MS to confirm the identity of the main quantified compound but these chromatograms were less representative of the real chemical composition of our fraction than our ELSD profile. Indeed responses in UV or MS of some of the compounds could strongly differ depending on their structure. However, the use of ELSD detector as well as the 280 nm chromatograms under our HPLC conditions allowed the visualization of the second major secondary metabolite which had remained undetected so far. Our study on the chemical diversity of *S. lamella* led us to identify five known terpenes. An ergosterol derivative is for the first time described for this species [47] and maybe a precursor of the previously described oxidized ergosterols in this species [33]. Ergosterols and their ester derivatives have already been reported from several fungi and are known to exhibit physiological roles, ergosteryl esters being a supply storage form [58]. The ergosterol ester identified in *S. lamella* (compound **6**) was included in the Factor 2 (FC2) of the Factor Analysis which did not show significant variation between sponge populations. However, the role of this compound in the primary or secondary metabolism is not known. Previous studies reported that nitenin and dihydronitenin are the major compounds of *S. lamella* [31,33,59]. In agreement with these studies, we found nitenin in high abundance but we failed to detect a high amount of dihydronitenin. Dihydronitenin was identified by HPLC-MS, however its low concentration prevented any quantification.

Given the distance covered in our surveys (1200 km), factors such as water temperature, light exposure, food availability, and others varied tremendously between locations and could affect the production of secondary metabolites in this sponge [60]. A particular constraint we faced was the difficulty to find *S. lamella* in some locations. For example, we observed and collected a single sponge in Ceuta, which limits ecological interpretations. However, given its important geographic location, we believe it was worth considering the information that this single specimen provides, so we used it for our diversity and dissimilarity analyses. Nevertheless, the use of replicates in the remaining locations in Catalonia and Marseille makes analyses robust. Populations were collected as we became aware of their existence, which provides additional limitations to our data set. Because of a few samples collected in two seasons in Catalonia and Marseille, we could check for the effect of season in the variations of chemical profiles in our samples. With our data, seasonal differences failed to increase chemical dissimilarities between populations in Catalonia or Marseille, supporting the role of space rather than time in the chemical differences quantified in our study. However, seasonal variation in the production of secondary

Table 4. Chemical (below the diagonal) and bacterial (above the diagonal) dissimilarities as estimated by the SIMPER procedure (PRIMER software) between pairwise sponge populations.

Area	Populations	Cio	Pla	Far	Cre	Bul	Are	Ceu
Marseille	Cio	-	28.55	23.31	28.61	-	27.32	27.77
	Pla	19.01	-	27.07	22.85	-	21.25	21.56
	Far	21.85	23.21	-	27.03	-	31.70	24.33
Catalonia	Cre	28.64	31.14	28.75	-	-	28.36	27.63
	Bul	24.14	22.56	25.84	32.24	-	-	-
	Are	23.06	23.54	28.07	27.77	29.25	-	25.95
Gibraltar	Ceu	32.97	33.50	28.97	39.84	38.77	31.53	-

Chemical dissimilarities were calculated from the 10 compounds that showed significant variations (Table 2). Bacterial dissimilarities were calculated from DGGE profiles of 4 specimens per populations (except for Els Bullents).

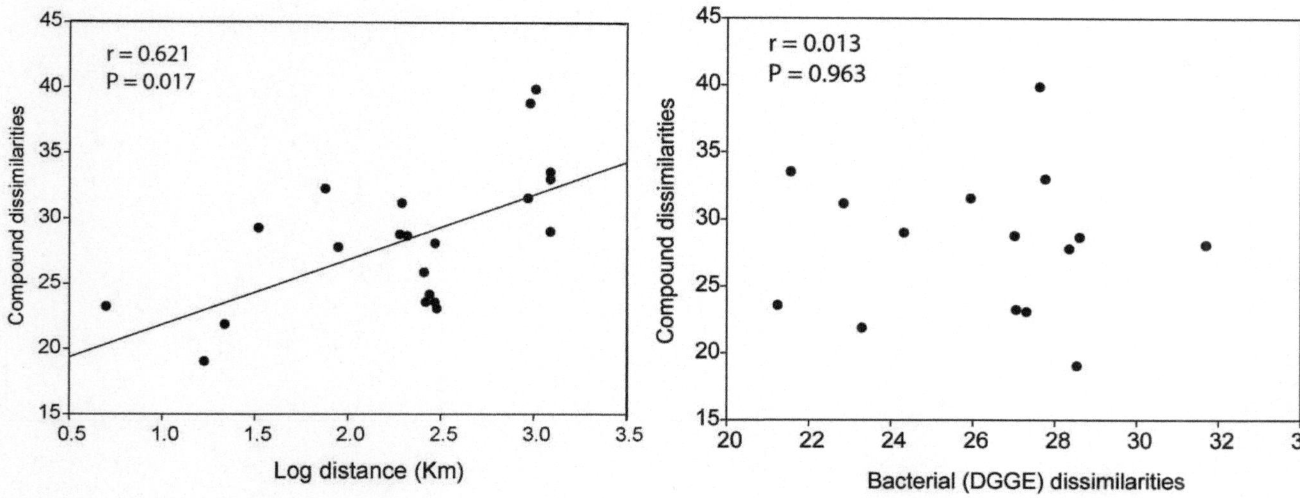

Figure 5. Relationship between chemical dissimilarities, geographical distances and bacterial dissimilarities. Chemical dissimilarities were estimated on the ten compounds responsible for variation between pair of sponge populations. Geographical distances were estimated as the logarithm of the geographical distances that separated pair of populations. Bacterial dissimilarities were estimated on the abundance/intensity of bacterial bands from DGGE gels using 4 sponge specimens per population. Chemical and bacterial dissimilarities values are included in table 4.

metabolites could occur in *S. lamella*, as it has been reported for other sponges [16,61,62,63]. Moreover, temporal variation can be site specific [18] and could act as a confounding factor for spatial analysis even if samples are collected within the same season. Ideally, pluriannual studies at multiple locations are crucial to understand chemical diversity at an intra-specific level. Qualitative and quantitative differences in the metabolites of *Spongia lamella* occurred between populations. The pattern of increasing chemical dissimilarity with geographic distance between sponge populations was remarkable. Geographic distance explained nearly 40% of the variance in chemical dissimilarity.

Whether variation in secondary metabolites is genetic or environmentally controlled is still an unresolved issue. Spatial and temporal variations could affect trophic and competitive interactions, increasing the chemical patchiness of the environment and promoting biodiversity at both genetic and species levels [41]. Increasing small-scale heterogeneity is crucial to create opportunities for new species to settle down. Alternatively, sponge genetic diversity could promote chemical diversity if secondary metabolites are produced by the sponge and the production has a heritable basis. Production of a cocktail of chemical compounds could confer advantages to an individual sponge and promote survival when facing specific predators, competitors, pathogens or environmental factors. If so, chemical diversity would increase sponge fitness. However, if just a limited number of chemical compounds are advantageous, increasing diversity of chemical compounds with ecologically irrelevant metabolites might dilute this benefit.

Distinct patterns of variation in secondary metabolites could also result from the dynamic nature of true sponge cells and microbial communities. There is little doubt that sponge associated microbes could account for many of the secondary metabolites assigned to the host [64], so changes in microbial populations that produce secondary metabolites will certainly lead to changes in sponge secondary chemistry. In fact, there is considerable debate as to whether sponges, their associated endobionts, or both parts are responsible for the production of secondary metabolites [65]. *Spongia lamella* hosts a very abundant and dynamic bacterial community in the western Mediterranean [38,66]. Our preliminary analyses showed independence between bacterial and chemical diversities [38]. Regardless bacteria do not influence or are influenced by metabolite production; secondary metabolites could protect sponges from pathogens by preventing them to overrun the sponge tissues. This would still have positive effects both on the sponge and associated bacterial communities, as it could limit competition between pathogens and symbiotic bacteria. Further analyses of these types of links between chemistry and bacteria would provide alternative ways to understand the production of secondary metabolites [67]. Clearly, this is a challenging area in sponge chemical ecology that will benefit tremendously from the new generation of genetic techniques [64].

Our study also highlights the complexity of the secondary metabolism within a single species. In *Spongia lamella*, we examined and quantified the relative abundance of 25 compounds that formed six coherent chemical factors. Ten compounds (41%) in four chemical factors varied significantly between populations while the remaining compounds showed no significant variation. Although uncertain at this time, these groups of compounds might represent inducible and constitutive secondary metabolites, although we cannot completely rule out the presence of some primary metabolites, particularly in the highly lipophilic section of the chemical profile. Constitutive defenses are always present in the organisms while inducible defenses can be synthesized or activated when needed [68,69]. Although both types of defenses can vary spatially, inducible defenses have the additional benefit of increased variability [70]. In our samples, the major compound (nitenin) varied significantly between populations while the second major compound (ergosteryl myristate) showed no significant variation. This could suggest distinct roles of diverse metabolites, and highlighted the importance of secondary metabolites as response of populations to their environment. Some of the metabolites isolated from this sponge showed cytotoxic activities against multiple tumor cell lines [33] but the ecological roles remain mainly unknown. We need further research to assess the functional role of these compounds in nature and how they respond to biotic or abiotic variables. This would unravel the last aspect of biodiversity: the functional role, which could give a

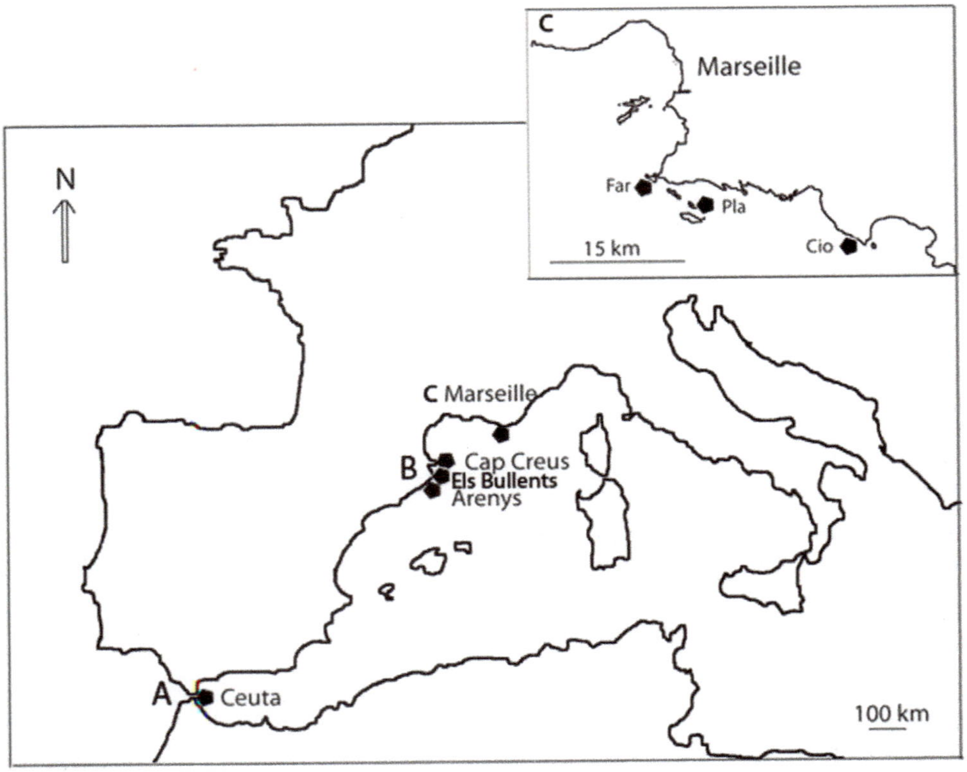

Figure 6. Map of sampling locations in the Western Mediterranean Sea. Samples were collected at various geographical scales in Ceuta (A), Catalonia (B) and in Marseille (C) areas. The area of Catalonia (regional scale) included populations at Cap Creus, Els Bullents, and Arenys; and the area of Marseille (local scale) included populations at Plane (Pla), Pharillon (Far) and La Ciotat (Cio).

complete insight of the chemical diversity of this endangered sponge [45,71].

Conclusion

Beyond an exhaustive structural characterization of all compounds, our study assessed chemical diversity by looking relatively at the number of compounds, their abundances, and dissimilarities in multiple populations of *S. lamella*. Our data showed qualitative and quantitative variation in the chemical profiles between populations and distance appeared as a major mechanism of chemical differentiation in *S lamella*. Environmental and genetic factors could be responsible for such variation [72,73]. In marine seaweeds, quantitative variation in furanones from *Delisea pulchra* showed significant heritability, indicating a potential for an evolutionary response [73,74]. In fact, variability per se could be target of selection [70] and chemical diversity could promote species evolution and biodiversity through the variety of biotic interactions that it mediates [6,41]. A better understanding of the patterns of chemical diversity is crucial to assess the mechanisms that generate and maintain chemical diversity and its consequences in species biology, ecology, and evolution.

Materials and Methods

Target Species

Spongia agaricina first description was performed on a now missing specimen from the Indian Sea [75]. Sharing similar morphology and shape with the Mediterranean sponge, the name *S. agaricina* was generalized and used for the Mediterranean specimens. Nowadays *S. agaricina* may be better known as a Mediterranean than Indian species. However, new evidence showed significant differences between the Philippine and Mediterranean specimens [75,76], which should be referred to as *S. agaricina* Pallas 1766 and *Spongia lamella* Schultze 1879 respectively [75].

Collection Sites

We sampled a total of 44 specimens from six populations of *Spongia lamella* in two geographic areas from the Northwestern Mediterranean Sea: Marseille (France), Catalonia (Spain), and an additional specimen from Ceuta (Strait of Gibraltar) (Figure 6). In December 2006, we collected 24 specimens by SCUBA diving in three locations near Marseille that spread over a distance of 22 km (Pharillon: 43°12′N, 5°20′E; Plane: 43°11′N, 5°23′E; La Ciotat: 43°9′N, 5°35′E). We also collected three additional samples around Plane in July 2007 and June 2008. In Spain, 18 more specimens were collected by SCUBA from three locations spread over a distance of 100 km (Arenys de Mar: 41°34′N, 2°33′E; Els Bullents: 41°42′N, 2°53′E; Cape Creus: 42°17′N, 3°18′E). Samples were collected as we became aware of populations with multiple specimens in November 2006 (Arenys de Mar), June 2006 (Cap de Creus), and Els Bullents (July 2008). In June 2007, a single specimen was observed and collected in Ceuta (35°53′N, 5°17′O).

All samples were individually placed underwater in plastic bags containing seawater, placed in a cooler for about 1 hr to reach the shore, and then manually cleaned to remove epibionts. They were then placed back in their plastic bags in the cooler to be transferred to the laboratory where they were placed in a freezer at −20°C. All specimens were then freeze-dried and stored in a −20°C freezer until chemical extraction.

Sample Preparation and Chemical Profiles

This analytical process has already been used for a chemotaxonomic approach among Homoscleromorpha sponges [77]. A piece of the sponge was freeze-dried and ground to yield a fine and homogenous powder. 20 ml of a 1:1 volumetric mixture of $MeOH/CH_2Cl_2$ were then added to a beaker containing two grams of this powder (dry mass). After 10 min in the ultrasonic bath, the organic phase was carefully recovered by pipeting and the process was repeated twice. The three organic phases were pooled and evaporated under reduced pressure after addition of 1 g of C_{18} silica powder. Desalting was then performed by SPE on a Phenomenex Stata™ C_{18} cartridge previously washed with 10 ml of $MeOH/CH_2Cl_2$ 1:1 and conditioned with 10 ml of water. The cartridge containing the solid extract was first washed with 50 ml of milliQ to remove salts elution with $MeOH/CH_2Cl_2$ 1:1 yielded the organic phase. Ten µl of this organic fraction were then analyzed by HPLC composed of a Waters™ 717plus autosampler, a Waters 600 controller, a Waters 996 photodiode array detector (DAD), and a SEDERE France evaporative light scattering detector (ELSD) to obtain quantitative profiles (Figure 3). High performance liquid chromatography (HPLC) experiments were run on a Phenomenex Gemini C_6-Phenyl (250×3 mm, 5 µm) column, using a linear gradient elution of A (H_2O+0.1% formic acid v/v) and B (acetonitrile+0.1% formic acid v/v) from 90% A at t 5 min to 100% A at t 35 min then isocratic until t 45 min (flow 0.5 $mL.min^{-1}$). Ultraviolet (UV) chromatograms were extracted at 210, 254, and 280 nm from the DAD detector. ELSD profiles were also used for chemical fingerprints. An Esquire 3000 Bruker mass spectrometer (MS) was also used with this HPLC system for m/z determination of selected compounds. HPLC-MS were run on a single randomly selected sample for each population. Data were visualized and processed by Bruker Daltonics Data Analysis 3.4.

Chromatographic profiles of all samples were processed by the Empower software with automatic integration of peak area on ELSD profiles. Integrations were individually checked to remove any artifacts. Nitenin was used as an internal standard. After background subtraction, the relative abundance of each compound was measured by integrating the area under the peak and was expressed as percentage of the total abundance in the chromatogram. Relative peak area allows an accurate and relative quantification of peaks between sponge samples. We calculated the absolute abundance of each compound proportionally to the nitenin concentration.

Isolation of Nitenin and Quantification

Three sponge samples from populations of Marseille were used for secondary metabolites isolation and identification. A pool of 25 g of sponges (dry mass) was extracted as described before, deposited on 2.5 g of C_{18} phase and subjected to Flash Chromatography on a C_{18} silica column (Armen Spot Flash, 10 $mL.min^{-1}$). The crude extract was fractionated with 250 mL of 1) H_2O 100%, 2) $H_2O/MeOH$ 1:1, 3) MeOH 100%, 4) $MeOH/CH_2Cl_2$ 3:1, and 5) in CH_2Cl_2 100%. HPLC-MS and nuclear magnetic resonance (NMR) analyses of the fractions showed that nitenin was a pure compound in fraction 3 (564 mg) [30], and was also present as a mixture in other fractions. A calibration curve of nitenin was obtained by HPLC-ELSD using seven $MeOH/CH_2Cl_2$ 1:1 solutions of nitenin at concentrations ranging from 0.08 mg to 10 mg/ml. Nitenin concentration in each sample was quantified by relating the peak area in the ELSD profile to the standard curve.

Isolation of the major lipophilic compound

For isolation of the second major compound, HPLC was conducted on the same equipment as previously described. The sample extract 4 ($MeOH/CH_2Cl_2$ 3:1) was separated and analyzed by using a Semi-preparative Diol Hypersil column in a normal phase at room temperature. We used an isocratic gradient consisting of solvent A (Ethyl acetate, 12%) and solvent B (hexane, 88%). Chromatographic peaks were identified by comparing the retention times, and UV spectra. After peak recollection, NMR spectra were recorded on a Bruker Avance 500 spectrometer in CDCl3 (Figure S1).

Bacterial profiles

We used the universal bacterial primers BAC358F (5′- CCT ACG GGA GGC AGC AG -3′) and BAC907RM (5′- CCG TCA ATT CMT TTG AGT TT -3′) to amplify fragments approximately 560 bp long in 4 specimens per sponge populations (except for Els Bullents population sampled afterward). We then performed Denaturing Gel Gradient Electrophoresis (DGGE) to obtain a first insight into the bacterial diversity inhabiting the sponges. DGGE analysis was performed using a Bio-Rad Dcode universal Mutation Detection System (BioRad) on 6% polyacrylamide gel in 1× TAE (40 mM Tris Base, 20 mM Sodium acetate trihydrate and 1 mM EDTA). We used a 40–75% vertical denaturant gradient (100% denaturant agent is 7 M urea and 40% deionized formamide). Comparable amounts of PCR products were loaded for each sample. Gels were run for 5 hours at 195 V at 60°C. Images of the gels were analyzed using the Gels plot lanes tool of ImageJ software 1.38× (Wayne Rasband, National Institutes of Health, USA). After background subtraction, the intensity of each band was measured by integrating the area under the peak and was expressed as percentage of the total intensity in the lane [38,66]. We measured the relative abundance of each band and compared bacterial profiles among and within sponge populations. We considered bacterial richness as the number of bacterial bands, and evenness as relative intensity of these bands [67].

Data Analysis

We tested for differences in chemical diversity between populations and geographical regions. For each specimen, we quantified the compound richness (Species Richness Margalef's index) and the Shannon's index using the DIVERSE procedure available in PRIMER 6 (Clarke and Warwick, 2001 Plymouth Marine Laboratory, UK [78]). Margalef's index is widely used in ecological studies to provide a measure of species richness (SR). SR is roughly normalized for sample size and is calculated as $SR = (S-1)/\log(N)$, where N is the number of individuals per population and S is the number of species. Using this index in our data, N was the abundance of sponge chemical compounds in a population and S the number of chemical compounds (e.g. number of peaks on the profiles). Chemical diversity was also calculated using Shannon's index, H ($H = -\sum pi \log(pi)$, where pi is the proportion i.e. abundance of the ith compound of an individual chromatogram). We used SYSTAT 12 software to perform analyses of variance (one-way *ANOVA*) on both SR and H.

Given the large number of secondary metabolites analyzed in this study, we used factor analysis (FA) to look for independent chemical factors formed by related chemical compounds [67,79]. Specifically, we used a principal component analysis extraction (PCA) with varimax rotation. The varimax rotation minimizes the number of variables that load highly on a factor and maximizes the loading variance across factors, which facilitate factor interpretation. The resulting independent factors were used as chemical variables in our analyses.

Because samples were collected over several seasons, we first tested whether seasonality could distort our spatial analysis. To do so, we used t-tests to compare the extent to which chemical dissimilarities within and between seasons differed in Marseille and Catalonia. We used the Multivariate Analysis of Variance (*MANOVA*) procedure in SYSTAT 12 to test whether chemical factors differed between sponge populations and to identify the secondary metabolites behind any significant factor.

We also performed statistical calculations available in the PRIMER v6.1.9 computer program [75] to analyze differences in the 10 compounds as function of sponge population. Square root data were used to calculate Bray-Curtis similarity. Chemical dissimilarities were analyzed graphically non-metric multidimensional scaling (MDS), and statistically by Analysis of Similarity (ANOSIM) at local (Marseille) and regional (Catalonia) scale. An exploratory similarity breakdown using the SIMPER procedure was used to the chemical dissimilarities between pair of sponge populations (percentages of dissimilarities between groups of samples). We then used a Mantel test to investigate whether 1) chemical dissimilarity and geographical distance, and 2) chemical and bacterial dissimilarities were related. Since the Mantel test can detect a significant association between two matrices we built a distance matrix with the geographical distances between populations, and matrixes containing the chemical dissimilarities, and bacterial dissimilarities between populations, respectively, as obtained from the SIMPER procedure.

Supporting Information

Figure S1 RMN spectrum of ergosteryl myristate −1[H+] in CDCl3.

Acknowledgments

We would like to thank Thierry Perez for the samples in the area of Marseille, Oscar Ocaña for the samples in Ceuta, and Josep Carreras Carbonel, Andrea Blanquer, and Xavier Turon for their support during field sampling in Spain. We are also grateful to Mohamed Mehiri and Julijana Ivanisevic for their HPLC-related help and advices.

Author Contributions

Conceived and designed the experiments: CN MAB OPT. Performed the experiments: CN OPT. Analyzed the data: CN MAB. Contributed reagents/materials/analysis tools: OPT. Wrote the paper: CN OPT MAB.

References

1. Wilson EO (1988) Biodiversity; Wilson EO, Peter FM, eds. National Academy Press. Washington, D.C.
2. Gaston KJ, Spicer JI (2004) Biodiversity: An Introduction. Oxford: Blackwell Science. 191 p.
3. Whittaker RH (1972) Evolution and Measurement of Species Diversity Taxon. pp 231–251.
4. Feral JP (2002) How useful are the genetic markers in attempts to understand and manage marine biodiversity? J Exp Mar Biol Ecol 268: 121–145.
5. Hay ME (1996) Marine chemical ecology: What's known and what's next? J Exp Mar Biol Ecol 200: 103–134.
6. Paul VJ, Vanalstyne KL (1992) Activation of chemical defenses in the tropical green-algae *Halimeda spp.* J Exp Mar Biol Ecol 160: 191–203.
7. Agrawal AA, Lau JA, Hamback PA (2006) Community heterogeneity and the evolution of interactions between plants and insect herbivores. Quarterly Rev Biol 81: 349–376.
8. Paul VJ, Puglisi MP, Ritson-Williams R (2006) Marine chemical ecology. Nat Prod Rep 23: 153–180.
9. Ianora A, Boersma M, Casotti R, Fontana A, Harder J, et al. (2006) New trends in marine chemical ecology. Estuar Coasts 29: 531–551.
10. Connolly SR, Roughgarden J (1999) Theory of marine communities: Competition, predation, and recruitment-dependent interaction strength. Ecol Monographs 69: 277–296.
11. Usher KM, Bergman B, Raven JA (2007) Exploring cyanobacterial mutualisms. Ann Rev Ecol Evol Syst 38: 255–273.
12. Wahl M (1989) Marine epibiosis .1. Fouling and antifouling - some basic aspects. Mar Ecol Prog Ser 58: 175–189.
13. Wright JT, de Nys R, Poore AGB, Steinberg PD (2004) Chemical defense in a marine alga: Heritability and the potential for selection by herbivores. Ecology 85: 2946–2959.
14. Matlock DB, Ginsburg DW, Paul VJ (1999) Spatial variability in secondary metabolite production by the tropical red alga *Portieria hornemannii*. Hydrobiologia 399: 267–273.
15. Marti R, Uriz MJ, Turon X (2004) Seasonal and spatial variation of species toxicity in Mediterranean seaweed communities: correlation to biotic and abiotic factors. Mar Ecol Prog Ser 282: 73–85.
16. Abdo DA, Motti CA, Battershill CN, Harvey ES (2007) Temperature and spatiotemporal variability of salicylihalamide A in the sponge *Haliclona sp.* J Chem Ecol 33: 1635–1645.
17. Becerro MA, Uriz MJ, Turon X (1997) Chemically-mediated interactions in benthic organisms: the chemical ecology of *Crambe crambe* (Porifera, Poecilosclerida). Hydrobiologia 355: 77–89.
18. Page M, West L, Northcote P, Battershill C, Kelly M (2005) Spatial and temporal variability of cytotoxic metabolites in populations of the New Zealand sponge *Mycale hentscheli.* J Chem Ecol 31: 1161–1174.
19. Marti R, Uriz MJ, Turon X (2005) Spatial and temporal variation of natural toxicity in cnidarians, bryozoans and tunicates in Mediterranean caves. Scientia Marina 69: 485–492.
20. Fielman KT, Targett NM (1995) Variation of 2,3,4-tribromopyrrole and its sodium sulfamate salt in the hemichordate saccoglossus-kowalevskii. Mar Ecol Prog Ser 116: 125–136.
21. Sara M, Vacelet J (1973) Ecologie des Demosponges. In: PP. G, ed. Traite de zoologie (anatomie, systematique, biologie). Paris: Masson. pp 462–576.
22. Harper MK, Bugni TS, Copp BR, James RD, Lindsay BS, et al. (2001) Introduction to the chemical ecology of marine natural products. Mar Chem Ecol. pp 3–69.
23. Blunt JW, Copp BR, Munro MHG, Northcote PT, Prinsep MR (2010) Marine natural products. Nat Prod Rep 27: 165–237.
24. Becerro MA (2008) Quantitative trends in sponge ecology research. Mar Ecol 9: 167–177.
25. Becerro MA, Turon X, Uriz MJ (1997) Multiple functions for secondary metabolites in encrusting marine invertebrates. J Chem Ecol 23: 1527–1547.
26. Teeyapant R, Woerdenbag HJ, Kreis P, Hacker J, Wray V, et al. (1993) Antibiotic and cytotoxic activity of brominated compounds from the marine sponge *Verongia-aerophoba*. Z Naturforsch 48: 939–945.
27. Jumaryatno P, Stapleton BL, Hooper JNA, Brecknell DJ, Blanchfield JT, et al. (2007) A comparison of sesquiterpene scaffolds across different populations of the tropical marine sponge *Acanthella cavernosa.* J Nat Prod 70: 1725–1730.
28. Avise JC, Arnold J, Ball RM, Bermingham E, Lamb T, et al. (1987) Intraspecific phylogeography - the mitochondrial-DNA bridge between population-genetics and systematics. Ann Rev Ecol Syst 18: 489–522.
29. Duran S, Giribet G, Turon X (2004) Phylogeographical history of the sponge *Crambe crambe* (Porifera, Poecilosclerida): range expansion and recent invasion of the Macaronesian islands from the Mediterranean Sea. Mol Ecol 13: 109–122.
30. Fattorusso E, Minale L, Sodano G, Trivellone E (1971) Isolation and structure of nitenin and dihydronitenin, new furanoterpenes from *Spongia nitens*. Tetrahedron 27: 3909–3917.
31. Aiello A, Ciminiello P, Fattorusso E, Magno S (1988) 3-Beta,5-Alpha-Dihydroxy-6-Beta-Methoxycholest-7-Enes from the Marine Sponge *Spongia-agaricina*. J Nat Prod 51: 999–1002.
32. Fontana A, Albarella L, Scognamiglio G, Uriz M, Cimino G (1996) Structural and stereochemical studies of C-21 terpenoids from Mediterranean Spongiidae sponges. J Nat Prod 59: 869–872.
33. Rueda A, Zubia E, Ortega MJ, Carballo JL, Salva J (1998) New metabolites from the sponge *Spongia agaricina*. J Nat Prod 61: 258–261.
34. Pronzato R (1999) Sponge-fishing, disease and farming in the Mediterranean Sea. Aquatic Conserv Mar Freshw Ecosyst 9: 485–493.
35. Perez T, Garrabou J, Sartoretto S, Harmelin JG, Francour P, et al. (2000) Mass mortality of marine invertebrates: an unprecedented event in the Northwestern Mediterranean. C.R. Acad Sci Paris/Life Science 323: 853–865.
36. Webster NS (2007) Sponge disease: a global threat? Environ Microbiol 9: 1363–1375.
37. Templado J, Calvo M, Garcia A, Luque AA, Maldonado M, et al. (2004) Guía de Invertebrados y Peces Marinos protegidos por la legislación nacional e internacional. In: técnica) NyPNs, ed. Guía de Invertebrados y Peces Marinos protegidos por la legislación nacional e internacional M.M.A. & C.S.I.C. pp 214.
38. Noyer C (2010) relationship between genetic, bacterial and chemical diversity of the Mediterranean sponge *Spongia agaricina* (*S. lamella* Schultze). University of Barcelona. PhD thesis. 278 p.
39. Noyer C, Agell G, Pascual M, Becerro MA (2009) Isolation and characterization of microsatellite loci from the endangered Mediterranean sponge *Spongia agaricina* (Demospongiae: Dictyoceratida). Conserv Genet 10: 1895–1898.
40. Frankham R (2005) Genetics and extinction. Biologic Conserv 126: 131–140.
41. Hay ME, Fenical W (1996) Chemical ecology and marine biodiversity: insights and products from the Sea. Oceanography 9: 10–20.

42. Sipkema D, Franssen MCR, Osinga R, Tramper J, Wijffels RH (2005) Marine sponges as pharmacy. Mar Biotechnol 7: 142–162.
43. Laport MS, Santos OCS, Muricy G (2009) Marine Sponges: Potential Sources of New Antimicrobial Drugs. Cur Pharmac Biotechnol 10: 86–105.
44. Raveendran TV, Mol VPL (2009) Natural product antifoulants. Cur Sci 97: 508–520.
45. Magurran AE (2004) Measuring Biological Diversity. Oxford: Blackwell Science Ltd.
46. Cimino G, De Stefano S, Minale L, Trivellone E (1977) 12-epi-scalarin and 12-epi-deoxoscalarin, Sesterterpenes from the sponge *Spongia nitens*. J Chem Soc Perkin Trans 1 13: 1587–1593.
47. Lee DI, Choi JS, Yang MR, Lee WK, Kim DS, et al. (1999) Ergosteryl mirystate, a new ergosterol derivative from unidentified marine algicolous fungus. Nat Prod Sci 5: 93–96.
48. Nunez CV, de Almelda EVR, Granato AC, Marques SO, Santos KO, et al. (2008) Chemical variability within the marine sponge *Aplysina fulva*. Biochem System Ecol 36: 283–296.
49. Puglisi MP, Paul VJ, Slattery M (2000) Biogeographic comparisons of chemical and structural defenses of the Pacific gorgonians *Annella mollis* and *A. reticulata*. Mar Ecol Prog Ser 207: 263–272.
50. Fahey SJ, Garson MJ (2002) Geographic variation of natural products of tropical nudibranch *Asteronotus cespitosus*. J Chem Ecol 28: 1773–1785.
51. Hemmi A, Jormalainen V (2004) Geographic covariation of chemical quality of the host alga *Fucus vesiculosus* with fitness of the herbivorous isopod *Idotea baltica*. Mar Biol 145: 759–768.
52. Bakus GJ, Green G (1974) Toxicity in sponges and holothurians: a geographical pattern. Science 185: 951–953.
53. Bolser RC, Hay ME (1996) Are tropical plants better defended? Palatability and defenses of temperate vs tropical seaweeds. Ecology 77: 2269–2286.
54. Siska EL, Pennings SC, Buck TL, Hanisak MD (2002) Latitudinal variation in palatability of salt-marsh plants: Which traits are responsible? Ecology 83: 3369–3381.
55. Becerro MA, Thacker RW, Turon X, Uriz MJ, Paul VJ (2003) Biogeography of sponge chemical ecology: comparisons of tropical and temperate defenses. Oecologia 135: 91–101.
56. Targett NM, Coen LD, Boettcher AA, Tanner CE (1992) Biogeographic comparisons of marine algal polyphenolics - evidence against a latitudinal trend. Oecologia 89: 464–470.
57. Fiehn O (2002) Metabolomics - the link between genotypes and phenotypes. Plant Mol Biol 48: 155–171.
58. Yuan JP, Wang JH, Liu X, Kuang HC, Zhao SY (2007) Simultaneous determination of free ergosterol and ergosterylesters in Cordyceps sinensis by HPLC. Food chemistry 105: 1755–1759.
59. Fontana A, Gavagnin M, Mollo E, Trivellone E, Ortea J, et al. (1995) Chemical Studies of *Cadlina* Mollusks from the Cantabrian Sea (Atlantic-Ocean). Biochem Mol Biol 111: 283–290.
60. Thompson JE, Murphy PT, Bergquist PR, Evans EA (1987) Environmentally induced variation in diterpene composition of the marine sponge *Rhopaloeides odorabile*. Biochem System Ecol15: 595–606.
61. Ferretti C, Vacca S, De Ciucis C, Marengo B, Duckworth AR, et al. (2009) Growth dynamics and bioactivity variation of the Mediterranean demosponges *Agelas oroides* (Agelasida, Agelasidae) and *Petrosia ficiformis* (Haplosclerida, Petrosiidae). Mar Ecol Evol Perspect 30: 327–336.
62. Turon X, Becerro MA, Uriz MJ (1996) Seasonal patterns of toxicity in benthic invertebrates: The encrusting sponge *Crambe crambe* (Poecilosclerida). Oikos 75: 33–40.
63. Turon X, Marti R, Uriz MJ (2009) Chemical bioactivity of sponges along an environmental gradient in a Mediterranean cave. Scientia Marina 73: 387–397.
64. Taylor MW, Radax R, Steger D, Wagner M (2007) Sponge-Associated Microorganisms: Evolution, Ecology, and Biotechnological Potential. Microbiol Mol Biol Rev 71: 295–347.
65. Becerro MA, Uriz MJ, Turon X (2001) Sponge chemical ecology: considering cytology into the equation. In: SG P, ed. Recent Research Developments in Ecology Transworld Research Network. pp 81–96.
66. Noyer C, Hamilton A, Sacristán-Soriano O, Becerro MA (2010) Quantitative comparison of bacterial communities in two Mediterranean sponges. Symbiosis 51: 239–243.
67. Sacristán-Soriano O, Banaigs B, Casamayor EO, Becerro MA (2010) Exploring the links between natural products and bacterial assemblages in the sponge *Aplysina aerophoba*. Appl Environ Microbiol.
68. Wittstock U, Gershenzon J (2002) Constitutive plant toxins and their role in defense against herbivores and pathogens. Cur Opin Plant Biol 5: 300–307.
69. Thoms C, Schupp PJ (2008) Activated chemical defense in marine sponges - a case study on Aplysinella rhax. J Chem Ecol 34: 1242–1252.
70. Karban R, Agrawal AA, Mangel M (1997) The benefits of induced defenses against herbivores. Ecology 78: 1351–1355.
71. Sala E, Knowlton N (2006) Global marine biodiversity trends. Ann Review Environ Res 31: 93–122.
72. Cronin G, Hay ME (1996) Susceptibility to herbivores depends on recent history of both the plant and animal. Ecology 77: 1531–1543.
73. Wright JT, Zuccarello GC, Steinberg PD (2000) Genetic structure of the subtidal red alga *Delisea pulchra*. Mar Biol 136: 439–448.
74. Wright JT, de Nys R, Steinberg PD (2000) Geographic variation in halogenated furanones from the red alga *Delisea pulchra* and associated herbivores and epiphytes. Mar Ecol Prog Ser 207: 227–241.
75. Pronzato R, Manconi R (2008) Mediterranean commercial sponges: over 5000 years of natural history and cultural heritage. Mar Ecol 29: 146–166.
76. Castritsi-Catharios J, Magli M, Vacelet J (2007) Evaluation of the quality of two commercial sponges by tensile strength measurement. J Mar Biol Ass U K 87: 1765–1771.
77. Ivanisevic J, Thomas OP, Lejeusne C, Chevaldonné P, Perez T (2010) Metabolic fingerprinting as an indicator of biodiversity: towards understanding inter-specific relationships among *Homoscleromorpha* sponges. Metabolomics;DOI: 10.1007/s11306-010-0239-2.
78. Clarke KR, Warwick RM (2001) Change in marine communities: an approach to statistical analysis and interpretation. 5-Primer-E Ltd, Plymouth Marine Laboratory. Plymouth.
79. McGarigal K, Cushman S, Stafford S (2000) Multivariate statistics for wildlife and ecology research. New York: Springer-Verlag.

7

Heritability of Asymmetry and Lateral Plate Number in the Threespine Stickleback

John Loehr[1,2], Tuomas Leinonen[1], Gabor Herczeg[1], Robert B. O'Hara[3¤], Juha Merilä[1]*

1 Ecological Genetic Research Unit, Department of Biosciences, University of Helsinki, Helsinki, Finland, 2 University of Helsinki, Lammi Biological Station, Lammi, Finland, 3 Department of Mathematics and Statistics, University of Helsinki, Helsinki, Finland

Abstract

The estimation of individual fitness and quality are important elements of evolutionary ecological research. Over the past six decades, there has been great interest in using fluctuating asymmetry (FA) to represent individual quality, yet, serious technical problems have hampered efforts to estimate the heritability of FA, which, in turn, has limited progress in the investigation of FA from an evolutionary perspective. Here we estimate the heritability of number of lateral plates, their FA and directional asymmetry (DA) in threespine stickleback, *Gasterosteus aculeatus*. By (i) using a meristic trait and (ii) basing our calculations on a large half-sib design experiment involving 2,079 offspring from 84 families, we overcame many of the difficulties faced by earlier FA studies. Both lateral plate number and FA in lateral plates were heritable ($h^2 = 0.46$ and 0.21, respectively), even after controlling for marker genotypes linked to *EDA* (the major locus influencing plate number). Likewise, DA in lateral plates was heritable $h^2 = 0.23$). The additive genetic component of FA in lateral plates makes it a prime candidate for further investigation into the evolutionary implications of FA and the genetic underpinnings of developmental instability. This discovery in an evolutionary model species holds the possibility to invigorate the study of FA from an evolutionary perspective.

Editor: Hans Ellegren, University of Uppsala, Sweden

Funding: This work was supported by the Academy of Finland (www.aka.fi; grant nos. 108601, 118673 and 213491), the LUOVA graduate school funded by the Ministry of Education (TL), and the LOEWE program of Hesse's Ministry of Higher Education, Research and the Arts. The funders had no role in study design, data collection and analysis, decision to publish, or preparation of the manuscript.

Competing Interests: The authors have declared that no competing interests exist.

* E-mail: juha.merila@helsinki.fi

¤ Current address: Biodiversity and Climate Research Centre, Frankfurt am Main, Germany

Introduction

The estimation of the quality or fitness of individuals is an important part of many biological research projects, especially in evolutionary biology. One relatively easy way that has been proposed to approximate individual quality/fitness is to measure the asymmetry of bilateral characters. While directional asymmetry (a consistent bias towards a given side) and antisymmetry (consistent asymmetry towards a random side) result from normal development, fluctuating asymmetry (FA; small non directional departures from perfect symmetry) is a result of disturbed development [1,2]. As one would expect that high quality individuals had a more stable development, they might express lower levels of FA ([3,4], but see [5]) as the random component left after other sources of variation in asymmetry (i.e. antisymmetry or directional asymmetry) is diminished.

The study of FA was initially received with enthusiasm because it appeared to be a useful measure of individual quality in various contexts in evolutionary ecology and conservation biology research [2,3,6,7]. For instance, assessment of FA can be a simple way to measure the amount of stress encountered by an individual during growth and development (greater stress can result in greater asymmetry), thus acting as a proxy measure of the underlying trait of Developmental Instability (DI) (e.g. [8]). During development individuals of higher quality may be more resistant to stress, which may be environmentally or genetically induced [9].

Despite the potential usefulness of FA, its overall value in evolutionary ecology has been questioned for a number of reasons. From an evolutionary perspective, the usefulness of FA resides in the possibility that it can predict lifetime reproductive success [10]. However, in the light of conflicting evidence, Lens *et al.* [11] cautioned that it should not be universally assumed that FA reflects fitness (also see [12,13]). Many of the conflicting results found for FA may actually be due to difficulties in measuring it accurately [8,14–16], and a lack of understanding of the underlying genetic architecture that affects development and FA [11,13]. Even the estimation of the heritability of FA (which is inevitable for evolutionary considerations) has turned out to be a particularly 'slippery fish', and the struggle towards this goal has been ongoing for over half a century (reviewed in [8]).

Artificial selection experiments in *Drosophila melanogaster* have shown responses to selection indicating that FA is heritable [17–20]. However, the vast majority of studies, on many species, have found little or no heritability of FA. Fuller and Houle [21] found that only a small fraction of studies (15 out of 179) assessing heritability of FA found evidence for it, and possible confounding factors were present in most of those cases. Meta-analyses have also been used to combine multiple estimates of heritability of FA. While initial meta-analysis estimated the average heritability of FA to be as high as 0.27 [22] this analysis proved to be controversial, and subsequent analyses suggested much lower estimates ranging from 0.026 to 0.08 [21,23]. The end result of all this work is the general acceptance that FA heritabilities, if present, are very small [13].

Using simulations, Fuller and Houle [21] and van Dongen [24] explored why significant heritability estimates of FA are so infrequent in the literature. These treatments point to the many pitfalls that have been unearthed over decades of research in the measurement of the heritability of FA (also see [8]). The experimental designs have frequently not been appropriate and sample sizes and sire to offspring ratios have not been optimal for producing meaningful estimates of heritability [21,24]. Measurement error appears to be particularly difficult to overcome and can obscure results (e.g., [8,14]). The conclusion from these studies was that the inability to find significant heritabilities for FA traces down to methodological difficulties.

The inability to determine whether there is a genetic basis for FA is unfortunate in that it has prevented progress on understanding the underlying evolutionary processes that affect development and asymmetry. In their review, Leamy and Klingenberg [13] pointed out that "... *a better understanding of the genetic architecture of FA should provide a much-needed perspective for sorting out the sometimes unexpected or contradictory patterns of differences in FA*." In the present manuscript our goal was to attempt to lay a foundation for future work on the investigation of the underlying genetic architecture of FA through the rigorous testing of the heritability of FA in a model trait of a model species.

An ideal trait to study the genetics and evolutionary implications of FA is the number of lateral plates of threespine sticklebacks (*Gasterosteus aculeatus*). Apart from being easy to count with little measurement error (see below), lateral plates are important structures in the defense against predatory attacks [25]. Lateral plate FA may be influenced by predators [26–28], and plate asymmetry is also an indicator of immunocompetence, with more asymmetric fish having higher rates of infection by endoparasites [29].

While the selective milieu influencing FA patterns in sticklebacks is reasonably well understood, results from research in heritability of lateral plate FA in threespine sticklebacks have produced conflicting estimates. Both Hagen [30] and Hermida *et al.* [31] estimated FA heritability from parent-offspring regressions arriving at estimates of 0.63 (SE $=0.16$) and -0.12 (SE $=0.14$), respectively. These estimates remain difficult to interpret because 1) the experimental designs could not properly separate environmental and genetic factors [21] and 2) the method used (total count of plates on left and right sides) to assess FA may not accurately reflect true FA ([26], and see Materials and Methods for a discussion of total count vs. homologous pairs assessment).

Here, our main aim was to estimate the heritability of FA for threespine stickleback lateral plates. We did this by conducting a large scale breeding experiment involving 42 sires and 84 dams, with a total number of 2,079 offspring, and subjecting the data to a rigorous Bayesian analysis. Hence, we were able to circumvent the two main problems faced by most previous analyses of heritability of FA: the small sample size and the large measurement error in the trait. The secondary aims of our research were to estimate the heritability of plate number and DA. The fish in our crosses varied greatly in plate number, and ranged from 30+ plates to a minimum of nine plates on one side of the fish, which correspond to full and partial plated morphs [32]. Here, we estimated the heritability of plate number and DA in sticklebacks with an approach that is statistically (cf. fitting an improved model) and genetically sound (cf. large sample size and appropriate design). Moreover, by accounting for the *EDA* (ectodysplasin A) gene in our analyses, we were able to address the question to what extent plate number is heritable after the variance due to this major gene influencing plate number [33,34] has been accounted for.

Results

Most fish (63%) had plates at all of the myomeres, and the variation in fish that had some plates missing appeared continuous (Fig. 1a). About 35% (732 of 2,079) of all fish were asymmetric, with most only having one or two asymmetries (Fig. 1b). When plates were missing they were frequently absent between myomeres 12 and 25 (Fig. 2).

Plates were directionally asymmetric with more plates on the right than left side (Fig. 3). The effects associated with the marker genotypes in the model were different (Fig. 3). Stn380 showed a large additive effect ($a=5.5$, 95% Highest Posterior Density Interval (HPDI): 4.6–6.5): if a Long homozygote myomere has a probability of 0.99 of being plated, substituting genes to a Short homozygote changes the probability of being plated to 0.28; 93% of the genetic variance in Stn380 being additive. In contrast, only 19% of the genetic variance in Stn381 was additive: most of the variance was dominance variance, due to the contrast between genotypes including the 173 allele: the combination with allele 186 gave less plates than either the homozygote or the 173/192 genotypes (Fig. 3).

The heritability of lateral plate FA on the latent scale was estimated as $h^2=0.24$ (95% HPDI: 0.14–0.36) (Fig. 4). Similarly, the heritability of lateral plate DA on the latent scale was estimated as $h^2=0.23$ (95% HPDI: 0.15–0.36; Fig. 4). The heritability of plate number (also latent scale) was estimated as 0.46 (95% HPDI: 0.27–0.65) (Fig. 4). In this model, the variance explained by the two marker loci linked to *EDA* was $h^2=0.19$ (95% HPDI: 0.14–0.25). When the heritability of plate number is estimated without taking *EDA*-linked loci into account a somewhat lower estimate is produced 0.36 (95% HPDI: 0.12–0.62) (Fig. 4). The above heritability estimates were barely affected by the choice of model (latent or observed; Fig. 4), with the largest change being a reduction in the FA heritability to 0.18 (95% HPDI: 0.12–0.26). The variance components underlying all estimated effects for plate number, DA and FA are shown in Fig. 5.

Discussion

The heritability estimate of $h^2=0.24$ for lateral plate FA is high compared to general estimates of FA heritability stemming from meta-analyses [8,21]. To our knowledge, only one previous study using a half-sib design has found evidence for a significant heritability of FA [35]. This massive effort involved over 10,000 *D. melanogaster* individuals and estimated h^2 bristle count FA to be <0.05. When our results are viewed in light of previous work showing that predation pressure selects for symmetric fish [27,36], it is evident that selection on lateral plate FA can have evolutionary consequences in populations subject to predation.

The detection of FA heritability over the last half century of research has been surprisingly difficult, yet it appears that we are now just beginning to understand what factors must be taken into account to estimate heritability of FA effectively. In this light, we examine the characteristics of our study that have aided in the detection of FA heritability. The sample size used in our study is large when compared to most other attempts to measure FA heritability and is only exceeded by work with fruit flies [35]. While our dam to sire ratio (2:1) is well below the recommended 25 to 1 [21], the sire to offspring ratio of 1-2 to 100 advised by van Dongen [24] was fulfilled.

The use of a meristic trait, such as presence of lateral plates, to measure FA has some distinct advantages that may aid the

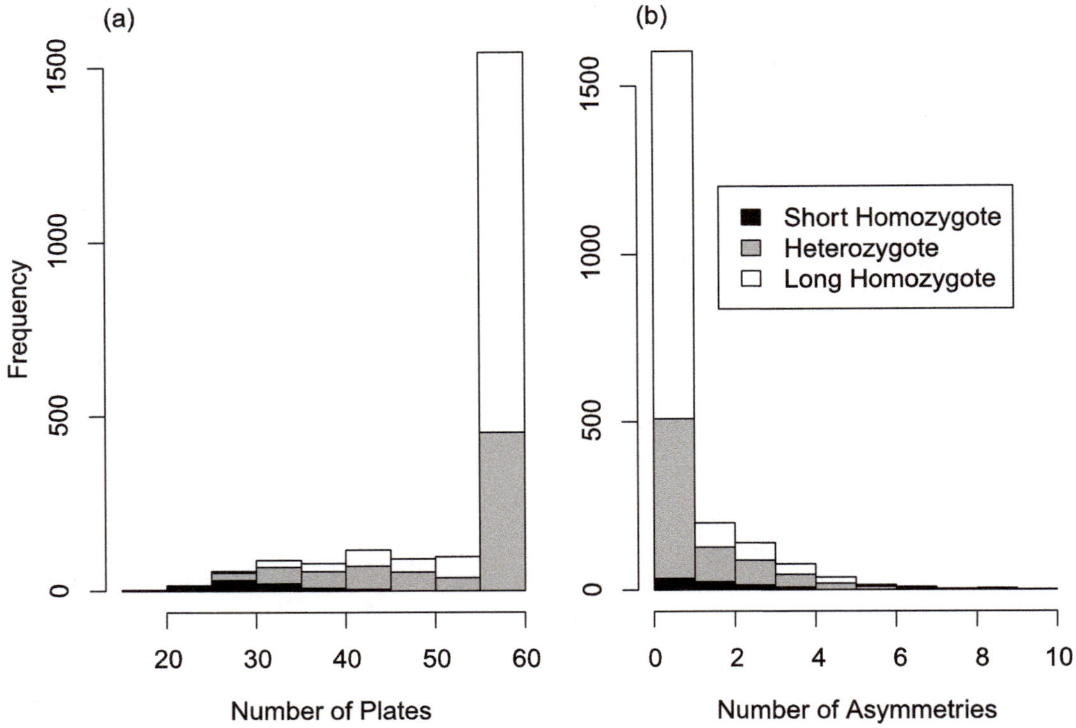

Figure 1. Histograms of (a) distribution of total plates on first 30 myomeres on each fish, (b) number of asymmetric myomeres, i.e. with only one plate. Shaded areas within bars show the proportion of *EDA* marker genotypes (see Table 1, Stn380) observed in the data.

accuracy to which heritability can be estimated. Meristic traits are less prone to measurement error than metric traits, because they can simply be counted as present or absent. Meristic traits also allow for a highly accurate assessment of asymmetry if the trait in question can be compared to its homologous pair on the other side of the organism (Fig. 6; [1,36,37]).

When FA is estimated using homologous pairs of a meristic trait, a few other important advantages emerge. Repeated

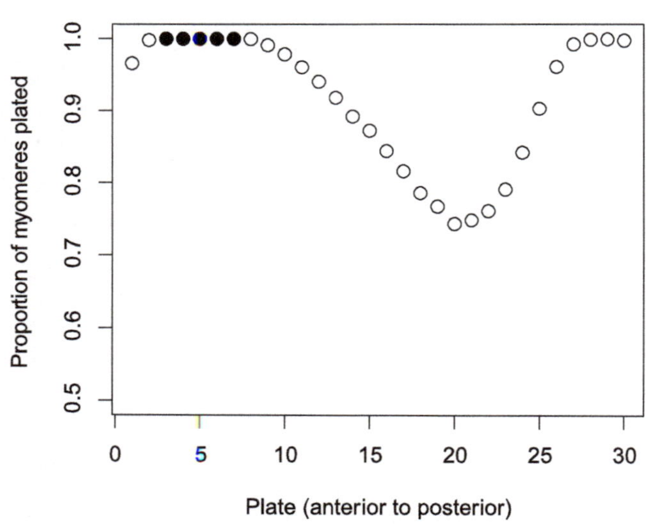

Figure 2. Proportion of myomeres (segments of fish) that have plates present on them in 2079 threespine sticklebacks. Filled circles: myomeres that are completely plated in all fish.

measurements of the same trait can be made (although not all plates are informative for FA, see Fig. 2), and more power can be gained by more accurately modeling the probability of being plated. In our particular case, individuals with all plates will not show FA. Basing FA estimates on the total numbers of plates underestimates any underlying FA because individuals with all plates appear in the data as having zero FA. In contrast, our analysis allows them to have an underlying FA in the propensity to have plates, but this will be masked when individuals have all or nearly all plates, i.e. there is little variation in platedness to be used to estimate heritability. The information about FA thus largely comes from the partially plated individuals.

Several heritabilities can be defined, depending on how exactly the trait is defined. In practice, there was little difference between following Swain [38] and defining the trait as the liability, or as the observed FA. This suggests that the "sampling" process only has a small effect on the variance in FA, which in turn suggests that it is a good indicator of the individual's DI because the actual FA and expected FA are similar [8]. This is, in part, because we averaged over all 30 myomeres: the FA for a single binary trait will have a larger contribution from sampling variance, and may therefore provide a less accurate measure of an individual's FA.

We found a right-side bias in plate number indicating that directional asymmetry (DA) prevails in our study population and is also heritable ($h^2 = 0.23$). Although it is commonly accepted that DA is genetically determined [39], and QTL's for DA have been found [40], additive genetic variance in DA has been shown to be lacking in a number of selection experiments [41,42,43]. However, two recent selection experiments have shown that DA in a population can respond to selection with changes occurring in the population mean [44,45]. Thus, our result provides further

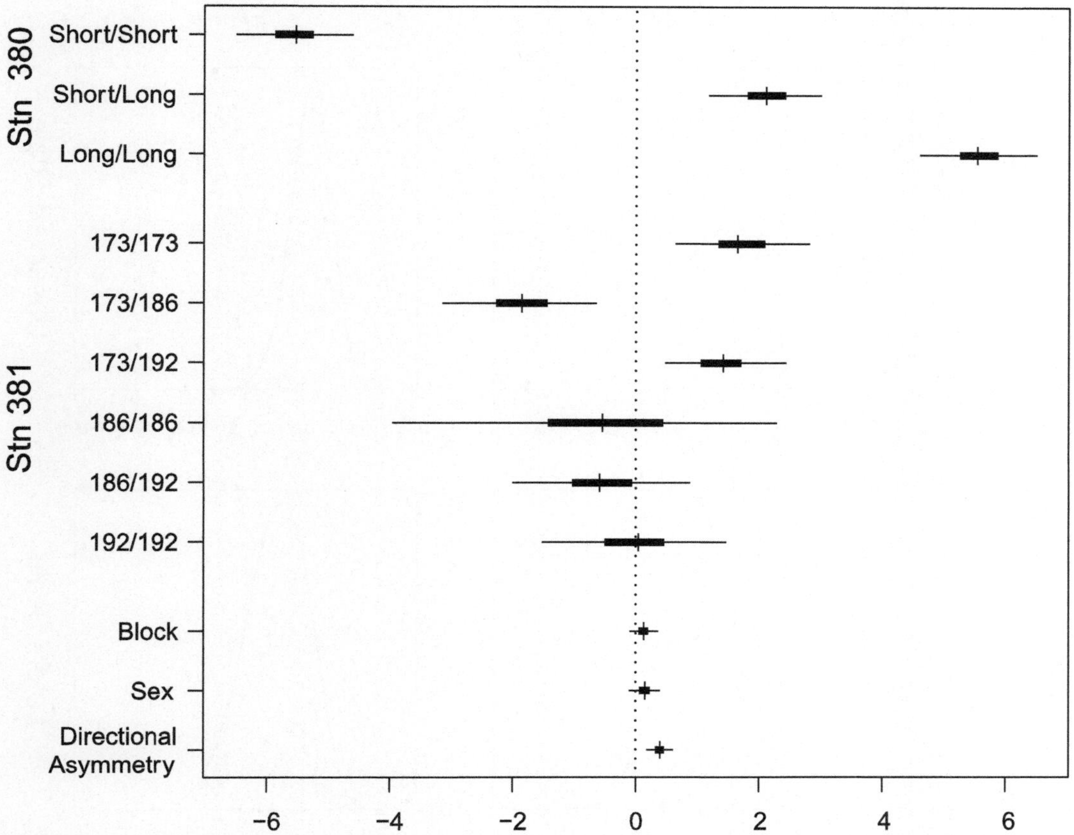

Figure 3. Log odds ratios of plate presence for fixed effects in the estimation plate number, directional asymmetry (DA) and fluctuating asymmetry (FA). A value of 0 means there is no effect of the factor. In 'Sex' refers to the effect of gender and 'Block' refers to block in the experimental set-up on plate number. The allelic effects associated with loci Stn 380 and Stn 381, which are closely linked to the EDA gene, also affect plate number. Posterior mode, and 50% (thick bar) and 95% (thin bar) highest posterior density intervals are shown.

evidence that DA can have a heritable component and has the possibility to respond to selection.

Reimchen and Nosil [29] measured asymmetry of over 10,000 adult *G. aculeatus* from a lake in western Canada. In their population there was a left side bias making the trend opposite to that found in the present study. DA in stickleback lateral plates could be due to interactions with predators or to basic biases in vertebrate developmental pathways [28]. Reimchen and Bergstrom [28] suggested that lateral plate DA may be a result of left/right biases in capture techniques of avian predators. If this is correct, our study would suggest that major predators of European populations use different capture techniques, or that the escape movement of fish is biased in opposite directions in Europe and North America.

We found that heritability of plate number was 0.46 and the variance explained by the two marker loci linked to the *EDA* gene was 0.19. This result contrasts with previous studies which have shown that the *EDA* gene can have a much stronger effect on plate number (e.g., [34,46,47]). In cases where *EDA* has been shown to have explained 80% of variation in plate number, other (unknown) modifier genes have been inferred to have a smaller effect [33,34]. The difference between our results and previous ones may be found in the pattern of plate morphs in study population. In our crosses, there was large variation in plate number, which varied between 9 and 35 on one side of a fish. Although fish in our crosses occasionally had very few plates, a few posterior plates were always present. When *EDA* has shown a large effect, the populations studied have had individuals with all plates present as well as fish where all posterior plates were missing [33,34].

An important caveat to consider is the possibility that the markers we used are not directly linked to *EDA* in our study population. However, this seems unlikely since previous research has shown that the same *EDA* alleles found by Colosimo *et al.* [33,34] for North American populations predict plate morph in European populations [46]. Yet, as the results stand, they suggest that there may be other genetic factors beyond *EDA* having significant additive effects on plate number in a Baltic Sea population of threespine sticklebacks.

We have shown that there is significant heritability in the FA of threespine stickleback lateral plates. By identifying a trait with significant FA heritability in a model species, our work should open avenues for more detailed evolutionary and genetic investigations in the role and indicator value of FA. For instance, future work on threespine sticklebacks can begin to identify the underlying genetic architecture of FA, as well as to elucidate the implications of spatially varying predation mediated natural selection on plate FA among different populations. Furthermore, the fact that FA has a genetic basis also has implications for the use of FA in conservation biology as a measure of stress [6,7]. While previous work has suggested that FA is purely influenced by the environment [21,23], our results suggest that genetic *vs* environmental causes of FA should be evaluated on a case by case basis, and our results add weight to the idea that the inability to find significant heritabilities for FA may be due to methodological difficulties inherent to the study of FA [21,24].

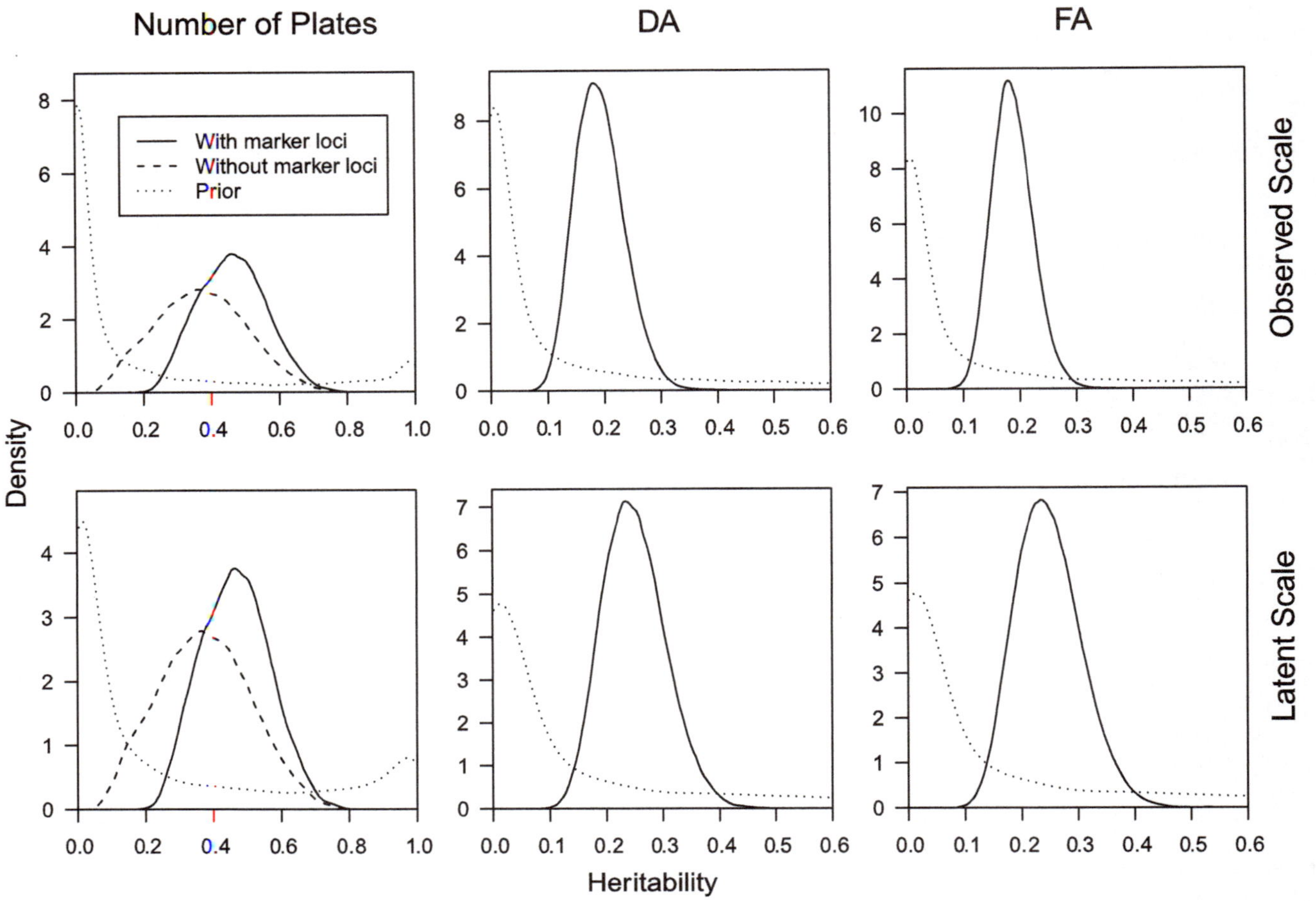

Figure 4. Posterior estimates of heritability for plate number, DA and FA in *G. aculeatus* lateral plates. Solid black: model including *EDA*-linked loci, dashed line: model excluding *EDA*-linked loci, dotted line: prior probability.

Materials and Methods

Ethic Statement

The experiment was approved and conducted under the license (HY 121-06) from the Helsinki University Animal Experimentation Committee.

Sampling and Fish Rearing

Mature male and gravid female threespine sticklebacks were collected during the breeding season in June 2006 from a Baltic Sea population (Vuosaari, Helsinki; 60°10′N, 25°00′E). A seine net with 6 mm mesh size was used for trapping. The fish were transported to the laboratory in 30 L tanks supplied with a battery-operated air pump, and the crosses were made immediately upon arrival. The crosses were performed using a nested paternal half-sib design, i.e. each male was crossed with two different females (North Carolina I design). In total, 42 males were crossed with 84 females. Males were anaesthetized and killed with an overdose of MS-222 (tricane methanesulphonate) before the testes were extracted and finely chopped in a few drops of water. The sperm solution was used to fertilise eggs *in vitro*, which were obtained by gently pressing the abdomen of the ripe females. The fertilised eggs were placed in cylindrical plastic containers with a plastic mesh bottom. The containers were submerged in 10 L plastic tanks with air supply to keep the water saturated with oxygen. Throughout the experiment the water temperature was set to 17 °C and the photoperiod to 12 L : 12 D. Once the eggs hatched, each clutch was divided into two replicate 10 L plastic tanks, and fed daily to excess with *Artemia* sp. nauplii. After three months, the fish were also fed daily with chopped chironomid larvae.

Initially 25 sticklebacks were kept in each replicate tank, two months after hatching, to ensure the fish did not suffer stress from abnormally high density, the number of fish was reduced to 15. To accomplish this, all fish were caught in a hand net and then 15 fish were randomly picked, so as to avoid any potential bias in catching fish. Families with less than 15 fish per replicate (11 out of 84 half sib families) were pooled with other families to make the density in each tank 15. The fish from different families were marked with fluorescent elastomers (Northwest Marine Technology, Inc.) prior to pooling. The fish were killed after six months, when they had reached ca. 4 cm in standard length and thus completed their lateral plate development [32,48]. The fish were fixed in 10% formalin and stored horizontally for a minimum of one month, and then stained with Alizarin Red S using the procedure described by Pritchard and Schluter [49].

Lateral Plate Asymmetry

The presence or absence of each lateral plate on both sides of each myomere for 2,079 fish was assessed from photographs (Fig. 6). We modeled presence/absence of plates at homologous

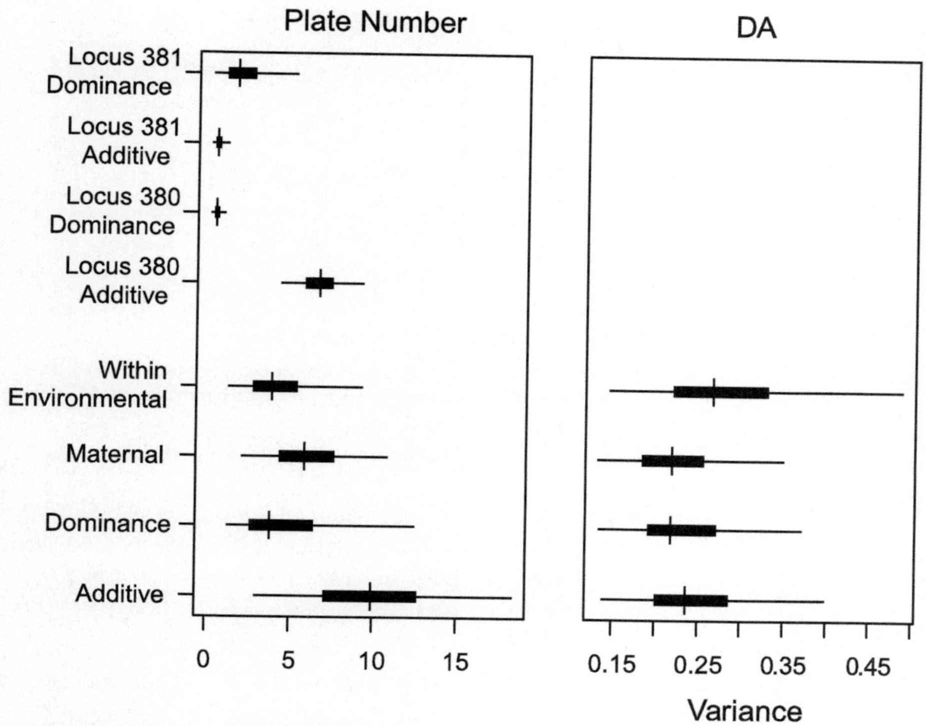

Figure 5. Posterior estimates of variance components for plate number, DA and FA in *G. aculeatus* lateral plates. Posterior mode, and 50% (thick bar) and 95% (thin bar) highest posterior density intervals are shown. Note that the maternal and dominance effects are partially confounded.

myomeres on both sides of the fish [26,36]. This leads to a more accurate measure of asymmetry than comparing total counts of plates on both sides, as it avoids asymmetries canceling out. However, error can occur if myomere location or plate presence/absence is incorrectly assessed. For the analyses, we only included the 30 first anterior myomeres (following Bergstrom and Reimchen [26]), as the small size of the posterior keel plates and myomeres can reduce accuracy of presence/absence observations. Bergstrom and Reimchen [26] assessed the amount of error versus the amount of asymmetries in their study and found that error was about five times less than the incidence of asymmetry. Following their example, we took a random sample of 40 individuals and reassessed plate presence/absence. We found that error occurred in 0.67% of myomeres, while asymmetries were found in 3.42% of myomere positions. Thus, error was about five times less than the incidence of asymmetry also in our study, suggesting that the results accurately reflect FA.

Sex Identification and *EDA* Genotyping

Individuals were sexed using DNA obtained from fins. DNA was extracted following Duan and Fuerst [50], and sex identification was done by amplifying a part of 3′UTR of Idh-gene [51]. The final PCR reaction volume was 10 µl and consisted of 1 µl template DNA diluted 1:10, 1x NH4 Reaction Buffer (Bioline), 1.5 mM $MgCl_2$ (Bioline), 20 µM of each dNTP (Finnzymes), 0.32 µM of Idh exon II 37F and Idh exon II 290R primers, as well as 0.25 U of BioTaq polymerase (Bioline). PCRs were conducted with an MBS thermal cycler (Thermo) according to following thermal profile: 94°C for 3 min, 38 cycles of 94°C for 30 s, 56°C for 30 s, 72°C for 1 min and the final extension at 72°C for 5 min. After PCRs, 5 µl of amplicon was run on a 2% Agarose LE (Cambrex) with 1x Loading dye (Fermentas). Fragment sizes were determined against a size standard (GeneRuler, Fermentas). Males had two fragments (~280 bp and 300 bp), females only one (300 bp). Each PCR plate contained certified male DNA as a positive control and a negative control (no sample). In some cases - either due to failed DNA-extraction or poor quality of DNA - the sex-specific region failed to amplify despite the further optimization of used PCR-protocol.

To investigate the effect of a known genetic determiner (*EDA* gene, [33,34]) on plate number and FA, individuals were typed for two *EDA*–linked microsatellite loci using primers for Stn380 and Stn381 [34]. PCRs were conducted in total reaction volume of 10 µl with 1xMultiplex Mastermix, 1 µl Q-solution (206145, Qiagen) and 5 pmol of each primer (forward primers labeled with fluorochromes FAM and TET) together with 1 µl of template. Temperature profile consisted of preliminary denaturation at 95°C for 15 min followed by 34 cycles in 95°C for 30 s, 56°C for 90 s, 72°C for 1 min and final extension at 60°C for 5 min. Amplicons were run diluted 1:100 together with ET-ROX 400 size standard in MegaBACE1000 (GE Healthcare) capillary electrophoresis instrument and scored in Fragment Profiler 1.2 software (GE Healthcare).

To avoid possible run-to-run and capillary-to-capillary allele size differences together with reader based errors, the Flexibin algorithm [52] was used to estimate the final allele sizes. For each lot of 94 samples, an individual from already published dataset [53] as well as a negative (no template) control were run and scored. Divergent alleles were verified by re-running PCR and capillary electrophoresis.

Statistical Analyses

Plate Model. Presence of a plate on a myomere (on one side of a fish) was modeled as a Bernoulli trial, so that.

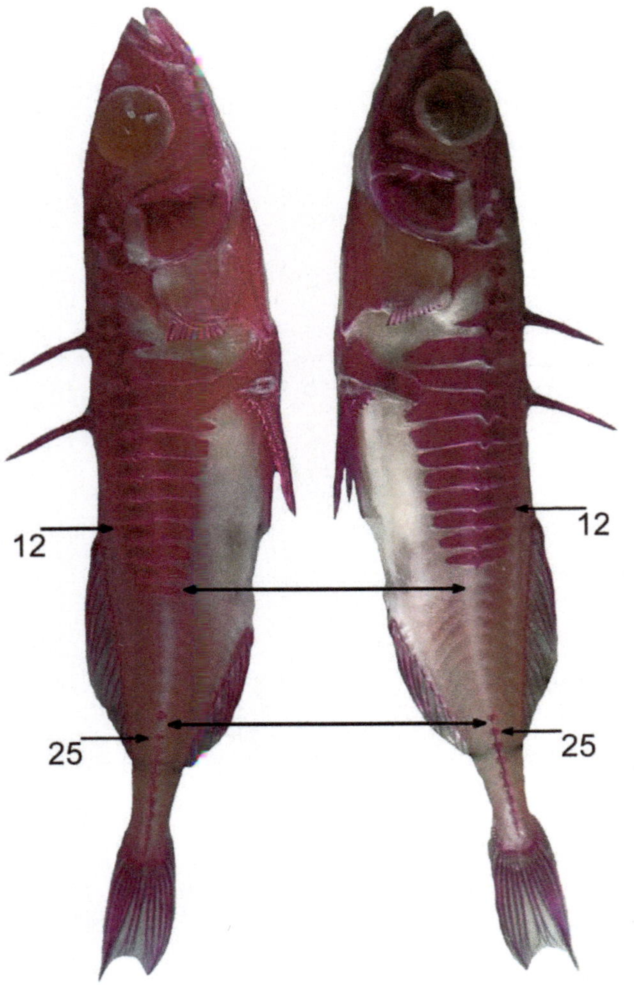

Figure 6. Example of a partially plated threespine stickleback stained for purposes of assessing plate morphology. Asymmetry in plates was assessed by comparing presence and absence of plates in homologous myomeres on the left and right sides of the fish (asymmetrical myomeres indicated by double headed arrows). Most variation in plate presence/absence was between the 12th and 25th myomeres as indicated in the figure. (Note: Background digitally removed.).

$$\Pr(P(i,m,s)=1)=\phi(i,m,s). \tag{1}$$

for individual i the plate on side s of myomere m. This was modeled as a logistic regression:

$$\operatorname{logit}(\phi(i,m,s))=\eta(i)+\lambda(i)(s-1)+\pi(m) \tag{2}$$

which separates the probability into individual-level effects of the number of plates ($\eta(i)$), the asymmetry ($\lambda(i)$), and a plate effect ($\pi(m)$): $s E\{1,2\}$, so the asymmetry effect is for $s=2$, or the right hand side of the fish. $\eta(i)$ and $\lambda(i)$ are both constant over myomeres, so we can interpret the model as a repeated measures model over the myomeres of the fish.

Individual Level Model. We model three individual-level traits related to platedness: the number of plates, directional asymmetry (DA - i.e. the overall mean asymmetry in platedness) and fluctuating asymmetry (FA - i.e. random asymmetry fluctuating around DA). For each of these we use a standard genetic model [54] to model the trait, $\mu_t(i)$ ($t=1,2,3$ for number of plates, DA and FA respectively), at the individual level:

$$\mu_t(i)=\gamma(i)+\frac{a(s(i))+a(d(i))}{2}+\tau(d(i))+\delta(d(i))+\sqrt{2}\varepsilon_a(i) + \sqrt{3/4}\varepsilon_d(i)+\beta(b(i))+\xi(x(i)) \tag{3}$$

where

- $\alpha(s(i))$ and $\alpha(d(i))$ are the parents' breeding values
- $\tau(d(i))$ and $\delta(d(i))$ are the maternal and dominance effects: these are partially confounded because of the experimental design, so are not interpreted further.
- $\varepsilon_a(i)$ and $\varepsilon_d(i)$ are the additive and dominance deviations, due to segregation.
- $\varepsilon_e(i)$ is the residual/within-individual environmental deviation
- $\beta(b(i))$ is the block effect (each block was one 10 liter tank), for an individual i grown in block $b(i)$. This is constrained
- $\xi(x(i))$ is the sex effect. This is constrained so that the sum of the effects is zero.

$\beta(b(i))$ and $\xi(x(i))$ were each constrained so that so that the sum of their effects were zero. The other parameters are modeled as random effects, with variances given by the standard quantitative genetic model for the additive (V_{Add}), maternal (V_{Mat}), dominance (V_{Dom}) and within-family environment (V_{EnvW}) respectively:

$$\alpha(s(i)),\ \alpha(d(i)),\ \varepsilon_a(i) \sim N(0,\ V_{Add}) \tag{4a}$$
$$\tau(d(i)) \sim N(0,\ V_{Mat}) \tag{4b}$$
$$\delta(d(i)),\ \varepsilon_d(i) \sim N(0,\ V_{Dom}) \tag{4c}$$
$$\varepsilon_e(i) \sim N(0,\ V_{EnvW}) \tag{4d}$$

Plate effect and EDA. The plate effect is.

$$\eta(i)=\mu_1(i)+\gamma_{380}(g_{380}(i))+\gamma_{381}(g_{381}(i)) \tag{5}$$

where $\gamma_{380}()$ and $\gamma_{381}()$ were the effects of the *EDA* locus, was modeled through the joint effects of the two genetic markers, with genotypes $g_{380}(i)$ and $g_{381}(i)$.

Stn380 had five alleles, of which three were rare (Table 1), so this was simplified to two genotypes: Long (>190 bps) and Short (<190 bps). We thus used the standard bi-allelic model (e.g. [54] chapter 4) with the homozygotes ($\gamma_{380}(1)$ and $\gamma_{380}(3)$) having effects $-a$ and $+a$ (for Short and Long respectively), and the heterozygote ($\gamma_{380}(2)$) having an effect d. The additive genetic variance due to this locus was then calculated as $V_{Add}(380) = 2p_{380}(1-p_{380})(a+(1-2p_{380})d)^2$, and the dominance variance is $V_{Dom}(380) = 4p_{380}^2(1-p_{380})^2(a+(1-2p_{380})d)^2$, where p_{380} is the frequency of the Long allele.

The Stn381 locus had three alleles (Table 1), and all three were used in the estimation of the genotypic effect. A separate effect was estimated for each genotype, so six genotypic effects ($\gamma_{381}(j)$ for $j=1,\ldots,6$) were estimated in total, each with frequency $p_{381}(i)$. The genotypic effects were assumed to be normally distributed, i.e. $\gamma_{381}(i) \sim \mathcal{N}(0,\ V_{381})$. From this the average deviation of each genotype from the mean was calculated, and hence the breeding value, as the average effect of that gene in its offspring. The additive variance is then the weighted (by the frequency of the genotype) mean squared breeding value. The dominance effect for each genotype was calculated as the difference between the genotypic value and the breeding value of the genotype, and the

Table 1. Allele frequencies in microsatellite loci Stn380 and Stn381 which are strongly linked to the *EDA* gene [34]. For analysis the alleles of Stn380 were simplified to two genotypes: Long (>190 bps) and Short (<190 bps).

Allele	Absolute Frequency	Relative Frequency (%)
Stn380		
181	14	0.4
183	5	0.1
185	2991	77.0
197	798	20.5
199	78	2.0
Stn381		
Allele	**Absolute Frequency**	**Relative Frequency (%)**
173	2798	72.2
186	95	2.5
192	983	25.4

dominance variance then calculated as the weighted average squared dominance effect (see Appendix for details). The additive genetic variance due to Stn381, $V_{Add}(381)$, is thus the mean squared breeding value, and the dominance genetic variance, $V_{Dom}(381)$, is the mean squared deviation from the breeding value [54].

Asymmetry. The model above defines the asymmetry effect as zero for the "left" side of the fish, and was estimated for the "right" side. Both a directional effect and fluctuating effect were modeled: the direction of the fluctuating effect had to be allowed to fluctuate between observations in an individual. The asymmetry effect was thus:

$$\lambda(i) = \psi + \mu_2(i) + S(i,s)\mu_3(i) \qquad (6)$$

where ψ is the mean directional asymmetry, $\mu_2(i)$ and $\mu_3(i)$ are the DA and FA terms (see above), and $S(i,s)$ denotes the random direction of the FA effect: it can take values -1 or 1, with probability 0.5.

Model Fitting

The fixed effect sex had to be modeled because 12% of the individuals could not be sexed. It was assumed to be a Bernoulli random variable with probability p_s. Similarly, 6.6% and 6.8% of the Stn380 and Stn381 genotype data, respectively. were missing. These missing values were imputed assuming they had been drawn from a Dirichlet distribution [55].

The model was fitted with a Bayesian approach. The priors were chosen to be vague. For the variance components, wrapped t-distribution priors were placed on the variances [56]:

$$V_{Add}, V_{Dom}, V_{Mat}, V_{EnvW} \sim \mathrm{wt}_1(0.1) \qquad (7)$$

where $\mathrm{wt}_v(\sigma)$ is a wrapped t distribution (i.e it is restricted to values >0) with v "degrees of freedom" and scale σ. This is only weakly informative for the variance components in the range of likely values. The other priors were:

$$\pi(m),\ a,\ d \sim \mathrm{N}(0,\ 100) \qquad (8a)$$
$$\psi(2),\ \beta(b),\ \beta_A(b),\ \xi(x),\ \xi_A(x) \sim \mathrm{N}(0,\ 10) \qquad (8b)$$
$$p_{380},\ p_s \sim \mathrm{U}(0,1) \qquad (8c)$$
$$V_{381} \sim \mathrm{U}(0,10) \qquad (8d)$$
$$p_{381}() \sim \mathrm{Dirch}(\boldsymbol{\alpha}) \qquad (8e)$$

(Note the constraint that the $\beta(b)$'s, $\beta_A(b)$'s, $\xi(x)$'s and $\xi_A(x)$'s each sum to zero). Where Dirch($\boldsymbol{\alpha}$) is a Dirichlet distribution with $\boldsymbol{\alpha}$ being a vector of the same length as p_{381}: here length 6, and we set every element of $\boldsymbol{\alpha}$ to 1.

The model was fitted in OpenBUGS. 4 chains were run. After a burn-in of 10,000 iterations, another 10^6 iterations were run, giving 4×10^6 draws from the posterior. Convergence was judged by eye. Most parameters mixed well, with the exception of $\pi()$, a, d, and $g_{381}()$. But even these poorly mixing variables had effective samples size [57] of at least 200.

Heritability

Depending on our exact focus, we can calculate heritability in different ways. Swain (1987) recommended that for models of meristic traits (such as plate number) our attention should focus on the liability level, i.e. the trait of interest is the 'propensity' to be plated (here this is the log odds of any myomere being plated on one side). If we take this approach, we can calculate heritability on the liability scale:

$$h^2 = \frac{v_{Add} + v_g}{v_{Add} + v_g + v_d + v_{Dom} + v_{mat} + v_{EnvW}} \qquad (9)$$

where V_g and V_d are respectively the additive and dominance variance components due to the plate genotype, i.e. $V_g = V_{Add}(380)+V_{Add}(381)$ and $V_d = V_{Dom}(380)+V_{Dom}(381)$. The definition of heritability used above avoids the problem that the means and variance of the trait may differ [38]. However it does this by redefining the trait to remove a component of the environmental variation (i.e. the "sampling" variation, due to the trait being a realization of a random process).

We can calculate an approximate heritability for the trait on the trait scale (i.e. probability) by calculating the binomial sampling variance at the trait mean. For lateral plates, the trait is the number of plates divided by the maximum number of plates ($=2M=60$). The variance of this, Var(Pl), is the sum of the variances in the individual plate effects. We can calculate $p(i,m,s)$ from the model, evaluating it at the mean over the population. This is $\exp(\pi(m))/1+\exp(\pi(m)))$ for the left hand side of the fish, and $\exp(\pi(m)+\psi(s))/1+\exp(\pi(m)+\psi(s)))$ for the right hand side.

$$\sum Var(p(i,m,s)) \qquad (10)$$

The other variance components then need to be transformed from the liability scale (i.e. the logit scale) onto the same scale. We can make this transformation (approximately) with the delta method, i.e.

$$V_p = \left(\frac{d\,p}{d\,\eta}\right)^2 V_\eta = V_\eta \frac{e^{2\eta}}{(1+e^\eta)^4} \qquad (11)$$

where p is the probability at the expected value of the trait, $p = e^\eta/(1+e^\eta)$, and V_p and V_η are the variances on the p and η scales.

For DA and FA, the trait of interest is the proportion of myomeres that are asymmetric. The probability that a myomere is asymmetric is $q(i,m) = p(i,m,1)(1-p(i,m,2))+p(i,m,2)(1-p(i,m,1))$. The variance in FA is thus:

$$\sum Var(q(m,s)) \qquad (12)$$

Note that we are summing over myomeres, as it is pairs of plates that are asymmetric.

Acknowledgments

We thank Jose Manuel Cano Arias, Simo Rintakoski and Chikako Matsuba for their help in measuring and rearing fish, and Marika Karjalainen for assistance with molecular analyses.

Author Contributions

Conceived and designed the experiments: JM. Performed the experiments: TL JM. Analyzed the data: JL RBO. Contributed reagents/materials/analysis tools: RBO. Wrote the paper: JL RBO TL GH JM.

References

1. Van Valen L (1962) A study of fluctuating asymmetry. Evolution 16: 125–142.
2. Palmer AR, Strobeck C (1986) Fluctuating asymmetry: measurement, analysis, patterns. Annu. Rev. Ecol. Syst. 17: 391–141.
3. Møller AP, Swaddle JP (1997) Asymmetry, developmental stability and evolution. Oxford: Oxford University Press. 291 p.
4. Thornhill R, Møller AP, Gangestad SW (1999) The biological significance of fluctuating asymmetry and sexual selection: a reply to Palmer. Am. Nat. 154: 234–241.
5. Palmer AR (1999) Detecting publication bias in metaanalyses: a case study of fluctuating asymmetry and sexual selection. Am. Nat. 154: 220–233.
6. Parsons PA (1992) Fluctuating asymmetry: a biological monitor of environmental and genomic stress. Heredity 68: 361–364.
7. Clarke GM (1995) Relationships between developmental stability and fitness: application for conservation biology. Cons. Biol. 9: 18–24.
8. Van Dongen S (2006) Fluctuating asymmetry and developmental instability in evolutionary biology: past, present and future. J. Evol. Biol. 19: 1727–1742.
9. Leary RF, Allendorf FW (1989) Fluctuating asymmetry as an indicator of stress: implications for conservation biology. Trends Ecol. Evol. 4: 214–217.
10. Badyaev AV, Foresman KR, Fernandes MV (2000). Stress and developmental stability: vegetation removal causes increased fluctuating asymmetry in shrews. Ecology 81: 336–345.
11. Lens L, Van Dongen S, Kark S, Matthysen E (2002) Fuctuating asymmetry as an indicator of fitness: can we bridge the gap between studies? Biol. Rev. 77: 27–38.
12. Leung B, Forbes MR (1996) Fluctuating asymmetry in relation to stress and fitness: effects of trait type as revealed by meta-analysis. Ecoscience 3: 400–413.
13. Leamy LJ, Klingenberg CP (2005) The genetics and evolution of fluctuating asymmetry. Annu. Rev. Ecol. Syst. 36: 1–21.
14. Merilä J, Björklund M (1995) Fluctuating asymmetry and measurement error. Syst. Biol. 44: 97–101.
15. Whitlock M (1996) The heritability of fluctuating asymmetry and the genetic control of developmental stability. Proc. R. Soc. Lond. B 263: 849–853.
16. Björklund M, Merilä J (1997) Why some measurements of fluctuating asymmetry are so sensitive to measurement error. Ann. Zool. Fennici 34: 133–137.
17. Mather K (1953) Genetical control of stability in development. Heredity 7: 297–336.
18. Reeve ECR (1960) Some genetic tests on asymmetry of sternopleural chaeta number in Drosophila. Genet. Res. I: 151–172.
19. Gibson JB, Bradley BP (1974) Stabilizing selection in constant and fluctuating environments. Heredity 33: 293–301.
20. Bradley BP (1980) Developmental stability of *Drosophila melanogaster* under artificial and natural selection in constant and fluctuating environments. Genetics 95: 1033–1042.
21. Fuller RC, Houle D (2003) Inheritance of developmental instability. In: Polak M, editor. Developmental Instability: Causes and Consequences. Oxford: Oxford University Press. 157–186.
22. Møller AP, Thornhill R (1997) A meta-analysis of the heritability of developmental stability. J. Evol. Biol. 10: 1–16.
23. Leamy LJ (1997) Is developmental stability heritable? J. Evol. Biol.10: 21–29.
24. Van Dongen S (2007) What do we know about the heritability of developmental instability? Answers from a Bayesian model. Evolution 61: 1033–1042.
25. Reimchen TE (1992) Injuries on stickleback from a toothed predator (*Oncorhynchus*) and some implications for the evolution of lateral plates. Evolution 46: 1224–1230.
26. Bergstrom CA, Reimchen TE (2000) Functional implications of fluctuating asymmetry among endemic populations of *Gasterosteus aculeatus*. Behaviour 137: 1097–1112.
27. Bergstrom CA, Reimchen TE (2003) Asymmetry in structural defenses: insights into selective predation in the wild. Evolution 57: 2128–2138.
28. Reimchen TE, Bergstrom CA (2009) The ecology of asymmetry in stickleback defense structures. Evolution 63: 115–126.
29. Reimchen TE, Nosil P (2001) Ecological causes of sex-biased parasitism in threespine stickleback. Biol. J. Linn. Soc. 75: 51–63.
30. Hagen DW (1973) Inheritance of number of lateral plates and gill rakers on *Gasterosteus aculeatus*. Heredity 30: 303–312.
31. Hermida M, Fernández C, Amaro R, San Miguel E (2002) Heritability and "evolvability" of meristic caracters in a natural population of *Gasterosteus aculeatus*. Can. J. Zool. 80: 532–541.
32. Hagen DW, Gilbertson LG (1973) The genetics of plate morphs in freshwater threespine sticklebacks. Heredity 31: 75–84.
33. Colosimo PF, Peichel CL, Nereng K, Blackman BK, Shapiro MD, et al. (2004) The genetic architecture of parallel armor plate reduction in threespine sticklebacks. PLoS Biol. 2: 1–7.
34. Colosimo PF, Hosemann KE, Balabhadra S, Villarreal G, Dickson M, et al. (2005) Widespread parallel evolution in sticklebacks by repeated fixation of ectodysplasin alleles. Science 307: 1928–1933.
35. Scheiner SS, Caplan RL, Lyman RF (1991) The genetics of phenotypic plasticity III. Genetic correlations and fluctuating asymmetries. J. Evol. Biol. 4: 51–68.
36. Moodie GEE, Reimchen TE (1976) Phenetic variation and habitat differences in *Gasterosteus* populations of the Queen Charlotte Islands. Syst. Zool. 25: 49–61.
37. Herczeg G, Szabó K, Korsós Z (2005) Asymmetry and population characteristics in dice snakes (*Natrix tessellata*): an interpopulation comparison. Amph. Rept. 26: 422–426.
38. Swain DP (1987) A Problem with the use of meristic characters to estimate developmental stability. Am. Nat. 129: 761–768.
39. Palmer AR (2004) Symmetry breaking and the evolution of development. Science 306: 828–833.
40. Leamy LJ, Routman EJ, Cheverud JM (1997) A search for quantitative trait loci affecting asymmetry of mandibular characters in mice. Evolution 51: 957–969.
41. Coyne JA (1987) Lack of response to selection for directional asymmetry in Drosophila melanogaster. J. Hered. 78: 119.
42. Tuinstra EJ, Dejong G, Scharloo W (1990) Lack of response to family selection for directional asymmetry in *Drosophila melanogaster*– left and right are not distinguished in development. Proc. R. Soc. Lond. B 241: 146–152.
43. Carter AJR, Osborne E, Houle D (2009) Heritability of directional asymmetry in *Drosophila melanogaster*. Int. J. Evol. Biol. 2009: 1–7.
44. Pélabon C, Hansen TF, Carter JR, Houle D (2006) Response of fluctuating and directional asymmetry to selection on wing shape in *Drosophila melanogaster*. J. Evol. Biol. 19: 764–776.
45. Pélabon C, Hansen TF, Carter JR, Houle D (2010) Evolution of variation and variability under fluctuating, stabilizing, and disruptive selection. Evolution 64: 1912–1925.
46. Cano JM, Matsuba C, Mäkinen H, Merilä J (2006) The utility of QTL-linked markers to detect selective sweeps in natural populations - a case study of the *EDA* gene and a linked marker in threespine stickleback. Mol. Ecol. 15: 4613–4621.
47. Bell MA, Gangavalli AK, Bewick A, Aguirre WE (2010) Frequency of *Ectodysplasin* alleles and limited introgression between sympatric threespine stickleback populations. Environ. Biol. Fish. 89: 189–198.
48. Bell MA (1981) Lateral plate polymorphism and ontogeny of the complete plate morph of threespine sticklebacks (*Gasterosteus aculeatus*). Evolution 35: 67–74.
49. Pritchard JR, Schluter D (2001) Declining interspecific competition during character displacement: summoning the ghost of competition past. Evol. Ecol. Res. 3: 209–220.
50. Duan W, Fuerst PA (2001) Isolation of a Sex-Linked DNA Sequence in Cranes. J. Hered. 92: 392–397.
51. Peichel CL, Ross JA, Matson CK, Dickson M, Grimwood J, et al. (2004) The master sex-determination locus in threespine sticklebacks is on a nascent Y chromosome. Curr. Biol. 14: 1416–1424.
52. Amos W, Hoffman JI, Frodsham A, Zhang L, Best S, et al. (2007) Automated binning of microsatellite alleles: problems and solutions. Molecular Ecology Notes 7: 10–14.
53. Mäkinen HS, Cano JM, Merilä J (2008). Identifying footprints of directional and balancing selection in marine and freshwater three-spined stickleback (*Gasterosteus aculeatus*) populations. Mol. Ecol. 17: 3565–3582.
54. Lynch M, Walsh JB (1998) Genetics and Analysis of Quantitative Traits. Sunderland, MA: Sinauer Assocs, Inc. 980p.
55. Nakagawa S, Freckleton RP (2008) Missing inaction: the dangers of ignoring missing data. Trends Ecol. Evol. 23: 592–596.
56. Gelman A (2006) Prior distributions for variance parameters in hierarchical models. Bayesian Analysis 1: 515–533.
57. Geweke J (1992) Evaluating the accuracy of sampling-based approaches to the calculation of posterior moments (with discussion). In: Bernardo J, Berger J, Dawid A, Smith A, editors. Bayesian Statistics 4, Oxford: Oxford University Press. 169–193.

8

Multifractal Spatial Patterns and Diversity in an Ecological Succession

Leonardo Ariel Saravia[1]*, Adonis Giorgi[2], Fernando Momo[1]

[1] Instituto de Ciencias, Universidad Nacional de General Sarmiento, Los Polvorines, Buenos Aires, Argentina, [2] CONICET, Departamento de Ciencias Básicas, Universidad Nacional de Luján, Luján, Buenos Aires, Argentina

Abstract

We analyzed the relationship between biodiversity and spatial biomass heterogeneity along an ecological succession developed in the laboratory. Periphyton (attached microalgae) biomass spatial patterns at several successional stages were obtained using digital image analysis and at the same time we estimated the species composition and abundance. We show that the spatial pattern was self-similar and as the community developed in an homogeneous environment the pattern is self-organized. To characterize it we estimated the multifractal spectrum of generalized dimensions D_q. Using D_q we analyze the existence of cycles of heterogeneity during succession and the use of the information dimension D_1 as an index of successional stage. We did not find cycles but the values of D_1 showed an increasing trend as the succession developed and the biomass was higher. D_1 was also negatively correlated with Shannon's diversity. Several studies have found this relationship in different ecosystems but here we prove that the community self-organizes and generates its own spatial heterogeneity influencing diversity. If this is confirmed with more experimental and theoretical evidence D_1 could be used as an index, easily calculated from remote sensing data, to detect high or low diversity areas.

Editor: Ricard V. Solé, Universitat Pompeu Fabra, Spain

Funding: This work was supported by the Universidad Nacional de Luján. The funders had no role in study design, data collection and analysis, decision to publish, or preparation of the manuscript.

Competing Interests: The authors have declared that no competing interests exist.

* E-mail: lsaravia@ungs.edu.ar

Introduction

We know that space can be an essential factor controlling species' coexistence and biodiversity in many communities [1]. Spatial pattern can be the result of environmental heterogeneity or can be produced by species' internal dynamics and interactions [2]. The latter are called self-organized and could be an important determinant of the structure and functioning of ecosystems [3]. Understanding the self-organization of ecological communities and the levels of biological diversity that emerge represents a fundamental theoretical challenge for ecologists [4].

Most studies focus on the effects of spatial patterns or spatial structure on diversity [5,6] but can we know something about biodiversity by studying spatial pattern?

Biodiversity is usually related to environmental spatial heterogeneity or habitat complexity [7,8] but self-organizing ecosystems generate its own spatial patchiness. This can be seen in the distribution of canopy height in tropical forests [9]. These distributions are fractals and are suspected to be produced by self-organization [10]. The auto-generated spatial heterogeneity could in principle have a strong influence on diversity, but this was never investigated. If the self-organizing process acts over an ecological succession it may influence both diversity and spatial pattern.

Early studies suggested that a measure of spatial pattern could be used as an index of persistence of populations and successional stage [11,12]. This measure was the fractal dimension of patches: early successional stages should have a higher fractal dimension because they have more intricate shapes and they are less persistent; late successional stages should have a lower fractal dimension because they have more regular shape and they are more persistent. This relationship between shape and persistence is derived from modified Brownian diffusion models, but it could be applied without committing to any particular Brownian model. Though its application was promising [13] no further studies of fractal dimension followed this line (but see [14]).

Spatial pattern and succession has been linked by a conceptual model called nucleation [15], where some species act as a nucleus to facilitate the establishment of other species; patches of the colonizing species first grow and then start to decrease as they are replaced by other, late successional, species. Eventually patches of some species coalesce and the spatial pattern is relatively fixed unless some disturbance occurs. This was later developed as a "model of heterogeneity cycles" [16] where cycles of heterogeneity alternate through succession: high heterogeneity represents periods of species invasion and establishment, and low heterogeneity represents periods of exclusion. One problem with this approach is that the measurement of heterogeneity is scale dependent so different sampling units will give different results [17].

Fractals methods can provide indices that measure the spatial heterogeneity of the community independent of the scale of observation over a range of scales [18,19], and have been extensively used in ecology [20,21]. A natural extension of these is multifractal analysis [22] that has not been applied so widely to date. Fractal analysis usually looks at the geometry of sets, that is, patterns arising from presence/absence data, whereas multifractal analysis looks at the arrangement of quantities, like densities or proportions.

In ecology, multifractals have been used in the analysis of temporal variability in plankton biomass [23,24], as a model of extinction and the origin of species [25], in the analysis of the spatial distribution of gaps caused by falling trees in the rainforest [26,27], and more recently in the analysis of species-area relationships [28,29]. The biomass spatial distribution of intertidal microphytobenthos has been analyzed using a modified multi-fractal technique [30]. These communities can be used as a model to study the relationship between species diversity, succession and spatial pattern. In a broader sense these communities are called periphyton.

Periphyton grows attached to submerged surfaces and is a complex community composed of microorganisms and other components including algae, bacteria, fungi, animals, and inorganic and organic detritus [31]. It can be responsible for most primary production in shallow aquatic environments such as streams, lakes, coastal waters and wetlands [32,33].

One of the descriptive characteristics of periphyton communities is biomass, which is defined as the amount of organic matter (or carbon) that has accrued from the production per unit area [34]. Biomass is a measure that integrates the interactions of individual characteristics of species, abiotic environmental controls and effects of herbivory [35].

In turn, the spatial distribution of biomass is the result of colonization processes, growth, competition and herbivory, to mention only the most important ones. Colonization occurs in random places that act as nuclei for development of algal biomass. As growth progresses algal communities develop a complex architecture similar to that of tropical forests, where competition for light and nutrients plays an important role [36].

Thus we use experimental microcosms with attached micro-algae communities (periphyton) to explore the relationship between spatial pattern and diversity throughout succession. Our hypothesis is that spatial self-organizing processes, characterized by multifractals, influence diversity over ecological succession. We also discuss the application of the conceptual model of cycles of heterogeneity during succession.

Methods

Experimental procedures

The periphyton colonization was carried out on squares of 5×5×0.1 cm high impact polystyrene. Initially we added 20 squares at the bottom of one 60×60 cm glass aquarium and then 10 squares were added every week for six weeks. The experimental device was located in a controlled environment with a light period of 12 hours and a temperature of 20°C. The aquarium was filled with water filtered through a 30 μM pore mesh. After that we added 50 cm3 of water with high density of algae obtained by scraping *Egeria densa* plants to remove the natural periphyton community. Twice a week we added water with periphytic algae to compensate for evaporation, to accelerate colonization and to simulate natural conditions. In the seventh week, sixty artificial substrata were removed and photographed to estimate species' composition and abundance. In one square for each week a Nikon Optiphot microscope with an underwater lens and phase contrast was used to identify algae. A minimum of ten random fields were observed, and observation continued on subsequent fields until no new species was found. Algal density was then calculated, together with species richness and Shannon diversity index [37].

To obtain the biomass spatial pattern, the photographs were digitized and the brightness values of each pixel were converted to biomass following the method of Saravia et al. [38]. Thus we obtained the spatial distribution of the periphyton biomass, once a week, along the succession. Additionally ten squares were left in the tanks for 3 additional weeks but only the biomass pattern was determined.

Multifractal analysis

To characterize the scaling behavior of the system we estimated the Fourier power spectrum of the biomass spatial pattern [39]. Power laws characteristics of scale invariance were observed. The exponents obtained from these can be used to determine stationarity of the dataset [40], here we refer to the broad sense definition of stationarity that means stationarity of spatial autocorrelation function.

Although fractal analysis can be used with nonstationary data [41], stationarity was a desirable property to obtain stable and accurate estimates [42,43].

The estimated exponents were in all cases greater than one so we used a gradient to transform the biomass data and obtain a stationary data set (Figure S1) [40].

Thus, we used the following transformation :

$$\nabla b(i,j)=\sqrt{(b(i+1,j)-b(i-1,j))^2+(b(i,j+1)-b(i,j-1))^2}$$

were $b(i,j)$ is the original biomass data, i and j have a range determined by the resolution of the image. We normalized the data using:

$$X(i,j)=\frac{\nabla b(i,j)}{\sum\sum \nabla b(i,j)}$$

After that we divide the spatial dataset with boxes of size ε, and compute:

$$X_q(\varepsilon)=\sum_i^{N(\varepsilon)} X_i^q$$

being $N(\varepsilon)$ the number of boxes needed to cover the image and q a real number called moment order.

We then estimated the exponents $K(q)$:

$$X_q(\varepsilon)\propto \varepsilon^{-K(q)}$$

as the slope of $log(X_q)$ versus $log(\varepsilon)$ using standard least squares.

A nonincreasing hierarchy known as generalized or Renyi dimensions [44,45] can be defined $D(q)=K(q)/(q\text{-}1)$. The most important exponent in this approach is quite possibly $D(1)$, the information dimension. When $q=1$ $D(q)$ is undefined, so it is estimated replacing $log(X_q)$ by the following:

$$Y=\sum_i^{N(\varepsilon)} X_i \log(X_i)$$

thus $D(1)$ is the slope of Y versus log(ε).

$D(1)$ provides us with a straightforward measure of the degree of intermittency or heterogeneity in the system [46]. Also the exponent $D(1)$ characterize the distribution and intensity of singularities, i.e. of high values of biomass with respect to the mean. If *D(1)* is smaller the distribution of singularities will be more sparse and data will be more spiky, if *D(1)* is greater (approaching two in our case) the singularities will have lower values of biomass and a more uniform distribution.

Other important values of *D(q)* are for $q=0$, *D(0)* is the dimension of the support of the distribution known as the standard fractal dimension. In our case, it is always two because the periphyton fills all the plate. For $q=2$, *D(2)* is a measure of scaling of spatial correlation: the correlation dimension. But the complete spectrum provides us with an enormous amount of additional information about the geometric structure of the underlying fractal distribution [27]. For a general interpretation positive *q* preferentially weight the denser regions of the distribution while negative *q* quantify the sparse regions of it [47].

A statistical permutation test was used to compare pairs of *D(q)* curves over the course of succession. The test statistic was the two-sample *t*-statistic to compare *D(q)* values between weeks, averaged over *q*. A *P*-value was obtained for the test statistic by simulation. Plates were randomly allocated to each of two groups (different weeks) and the mean *t* was recalculated for 10000 data sets permuted in this way [48]. Pairwise comparisons were performed between the weeks. The *P*-values were adjusted for multiple testing using a step-down Bonferroni procedure [49]. The statmod R package was used to perform these tests [50].

To determine if *D(q)* spectra are different from one produced by a random spatial distribution of biomass we performed a randomization test. We shuffle the pixel position and recalculate *D(q)* and obtained a confidence interval for a random pattern performing 1000 repetitions [48]. If the actual values of *D(q)* falls outside the interval the spatial pattern is not random.

Kendall's rank correlation and quantile regression were used to relate *D(1)* with Shannon's diversity across the weeks. Quantile regression allows the examination of the maximum response, rather than the mean response, of one variable to a predictor.

It can be used for analyzing data with non-constant variance, which is often meaningful for ecological processes in which many unmeasured variables may affect the response The ecological concept of limiting factors as constraints on organisms often focuses on rates of change in quantiles near the maximum response, when only a subset of limiting factors are measured [51].

Results

A data set can be called multifractal if the plots of log $X_q(\varepsilon)$ vs log(ε) are straight lines for wide range of log(ε) and several values of *q*. The biomass data from the photographs satisfy this assumption over the range of scales considered (Figure 1 and figures S2, S3, S4, S5, S6, S7, S8). The data used to calculate X_q at each scale was not independent because the squares used for smaller ε values are nested in the greater ones. This violates the assumptions implicit when performing statistical tests but does not invalidate the least square method to determine the exponents *K(q)*. The coefficient of determination R^2 can be used as a descriptive measure of goodness of fit [28]. The vast majority of R^2 were greater than 0.99, and never were less than 0.94.

The *D(q)* vs. *q* plot, show that the "singularities" in the spatial distribution of biomass decrease with time, i.e. the curves become flatter (Figure 2). This pattern can be observed looking at the biomass images of the community development (Figure 3). All weeks have significant differences (P<0.01) except weeks one and two. Looking at the images of biomass, weeks 1 and 2 appear to be different, but analysing the standard deviation of *D(1)* (Table 1) we realize that in these weeks the s.d. is much greater than in the rest of the weeks. This means that the succession development is more variable in its first stages, in fact, the s.d. tends to dimish as the succession advance suggesting a convergence of spatial pattern by the end of succession.

Each week, the values of *D(q)* are closer to two than the values of the previous week. This monotonic relationship is maintained

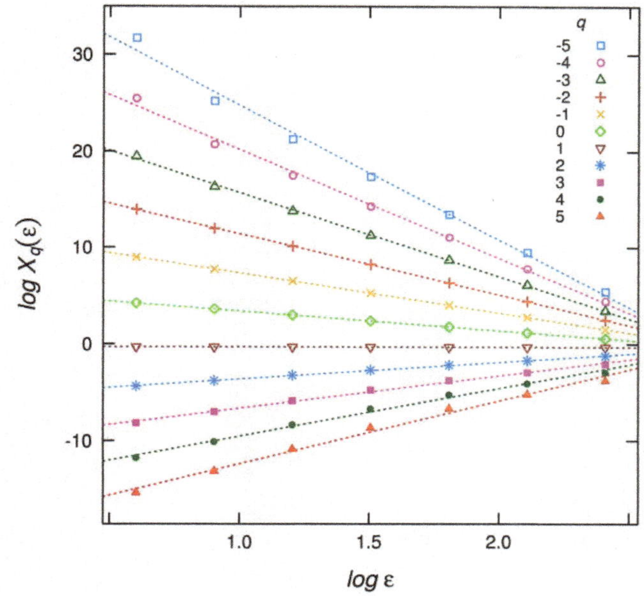

Figure 1. A typical graph in logarithmic scale corresponding to 3 weeks of development of the biomass spatial distributions of X_q versus ε for *q* varying from -5 to 5, showing a very good linear fit for all q considered, the R^2 values were always greater than 0.99. All other biomass distributions showed similar fits with R^2 values between 0.94 and 0.99.

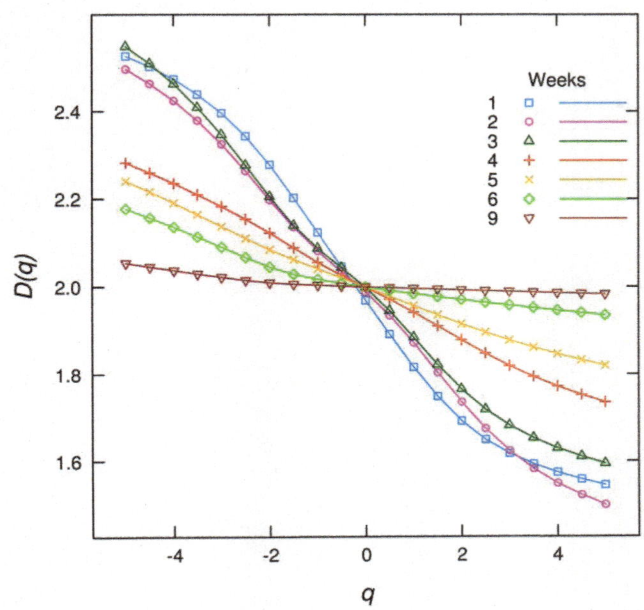

Figure 2. Spectrum of generalized dimensions *D(q)* for different times of the periphyton succession. The graphs are averages for all the plates with the same weeks of development.

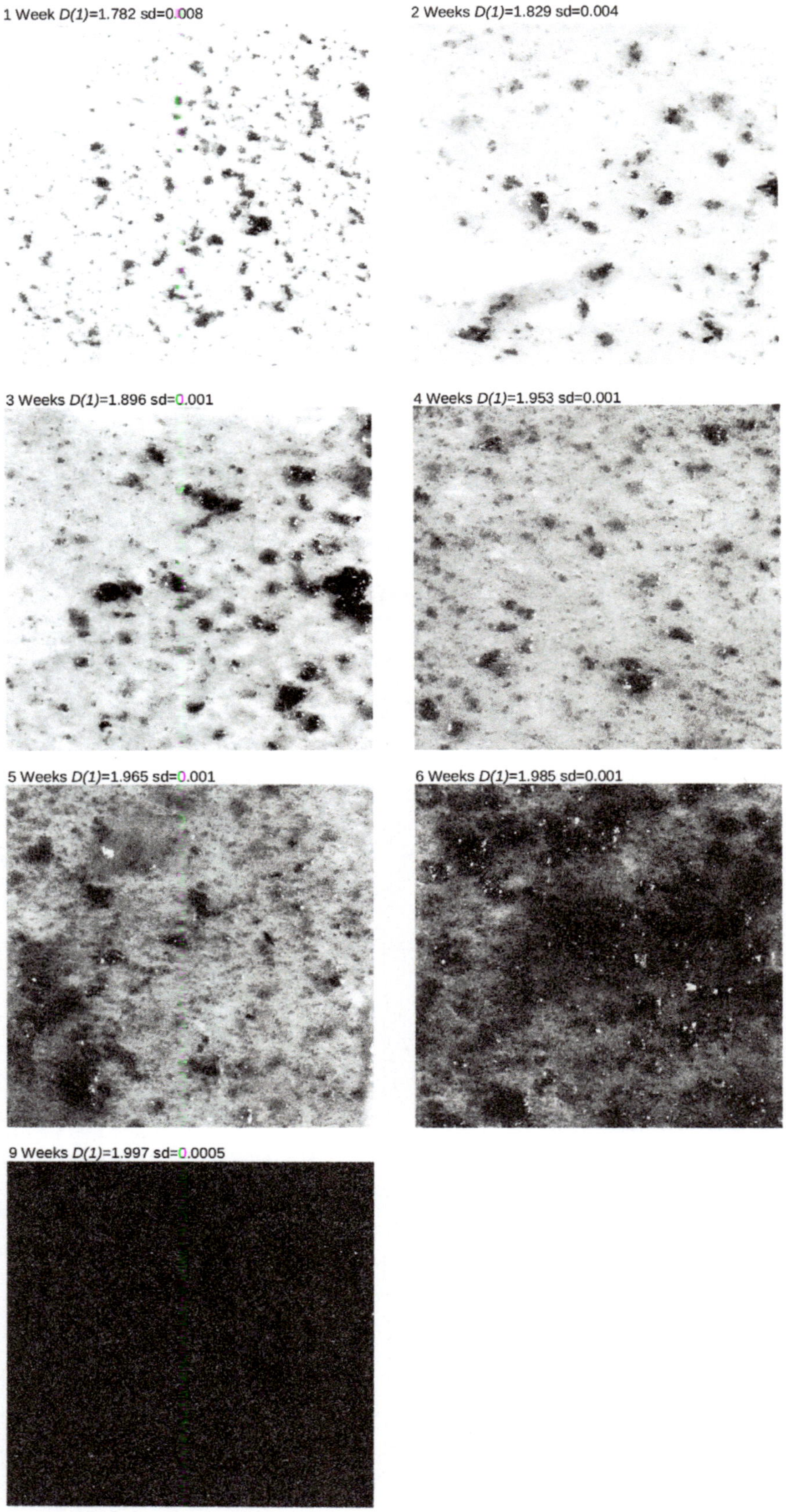

Figure 3. Images of periphyton biomass for different weeks of development. D(1) is the information dimension, and sd the standard deviation.

Table 1. Information dimension *D(1)* plus standard deviation, number of species and Shannon diversity (*H*) calculated for each week of the periphyton succession.

	Weeks						
	1	**2**	**3**	**4**	**5**	**6**	**9**
D(1)	1.780±0.038	1.881±0.039	1.897±0.013	1.943±0.017	1.957±0.006	1.985±0.002	1.996±0.001
No.Species	20	11	19	12	16	23	–
H	3.57	2.69	2.26	2.11	2.77	1.62	–

for all *D(q)* but the negative values of the second and third week, which are inverted. Regions of lower biomass of the third week have more irregularities than ones from the second week.

At the 9th week *D(q)* varies very little with *q* and is very close to two. A uniform surface with equal values of biomass will give a straight line at two, so we could say that the biomass is close to not being a multifractal anymore. There is very little structure on the image of 9 weeks, the dominant feature seems to be random noise produced by the method of biomass estimation. Indeed there is some structure because the test for random spatial distribution was rejected for all q, for all plates and weeks (Figure 4). This stage is rarely reached in natural scenarios because spates in streams and other natural disturbances remove most of the attached biomass and reset succession to its first phases.

In the passage from the third to fourth week there is a big separation between curves, and this coincides with a change in species dominance from *Oocystis sp.* to *Chlorella sp.* (Figure 5). This is the the only place where this kind of change occurs and can be related to a structural change in the community.

Chlorella sp is the only species occurring at high densities along the whole succession. This species has a cyclical pattern of dominance. In the second week it co-dominates with *Oocystis sp*, and in the third week the community is dominated by *Scenedesmus sp*, replacing *Chlorella sp* (Figure 5). The number of species is around 20 and has a cyclic behavior with peaks at one, three and six weeks and the Shannon diversity index *H* decreases as the succession advances, except for the 4th week where *H* is higher than the previous week (Table 1). The information dimension *D(1)*, used to characterize spatial heterogeneity, increases along the succession. This means that spatial heterogeneity decreases since a *D(1)* approaching two implies a flatter surface (Table 1).

The *D(1)* seems to be related to the Shannon diversity index *H*, Kendall's rank correlation is highly significant ($\tau = -0.52$, $P<0.05$) (Figure 6). All quantiles are different from zero ($P<0.05$) and all of them show a decreasing tendency, which means that the higher H is, the lower *D(1)*, and the more heterogeneous is the biomass spatial distribution. The 90% quantile can be used as a guide of the minimum diversity of a site given its *D(1)* (Figure 6).

In fact *H* and *D(1)* are both expressions of Shannon entropy, one estimated from the biomass distribution and another from species abundances [52]. Thus we found a way to relate the

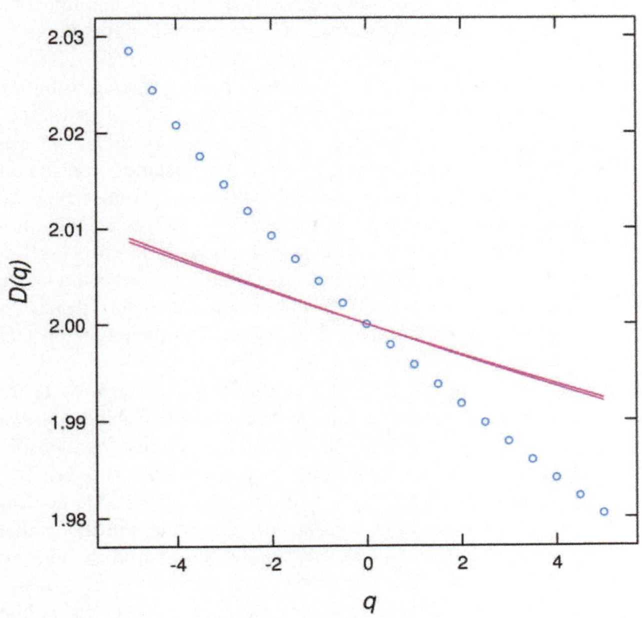

Figure 4. Example of confidence intervals (P = 0.01) for generalized dimensions *D(q)* for one plate of 9 weeks. These were calculated using a randomization of biomass spatial distribution and indicates that the spatial distribution is not random. All other plates give similar results.

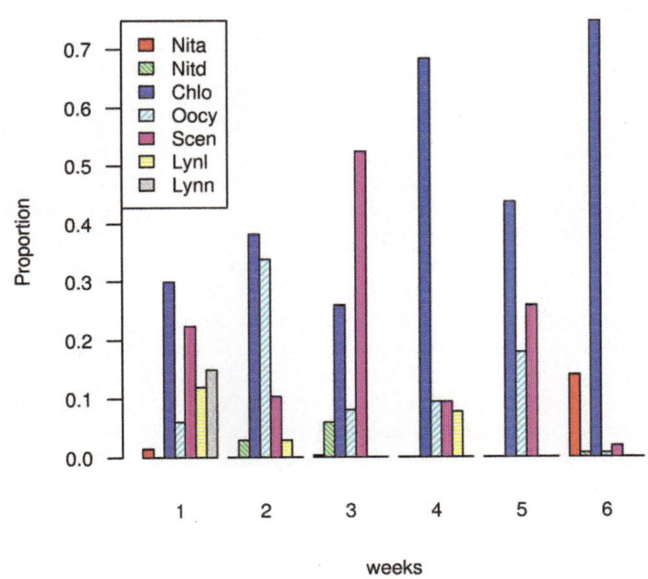

Figure 5. Proportion of the most abundant species for each week. The species that provide more than 5% of the total abundance for each week are included. The legend abbreviations are: Chlo: Chlorella sp, Scen: Scenedesmus sp, Oocy: Oocystis sp, Nita: Nitzschia amphibia, Nitd: Nitzschia dissipata, Lynl: Lyngbia limnetica, Lynn: Lyngbia nordgardii.

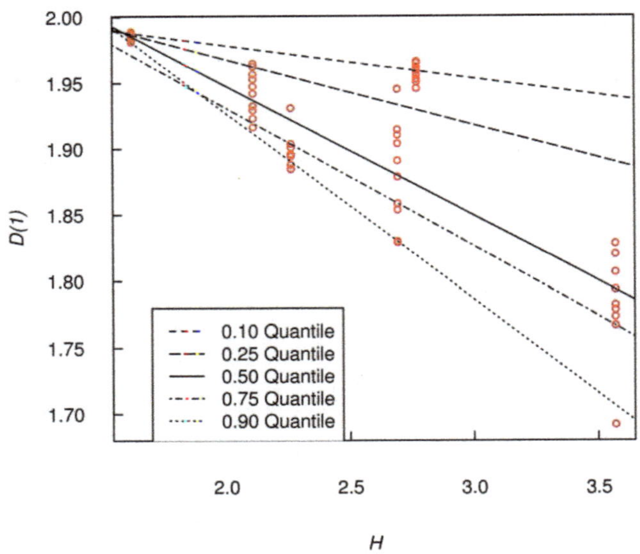

Figure 6. Information dimension *D(1)* by Shannon's diversity index *H*. The lines are estimated using quantile regression, all slopes are statistically different from zero ($P<0.05$), and all slopes are less than zero indicating a negative relationship with diversity which means a positive relationship with heterogeneity.

information contained in the spatial pattern to the information contained in the abundance and the number of species in this ecosystem.

Discussion

Abiotic systems maintained far from equilibrium can produce macroscopic ordered patterns in space emerging from what looks like microscopic chaos. This is especially evident when dealing with biotic ecosystems where similar patterns appear replicated at different scales [27,53].

Starting from homogeneous substrata the colonizing microalgae produces a self-similar spatial pattern. Thus the spatial pattern is not imposed by external conditions but is generated endogenously by the interactions and growth of algae and other components of the periphyton, therefore the conditions for self-organization are met [54]. This is the first part of our hypothesis: the spatial pattern of periphyton communities is self-organized.

The system produces a self-similar spatial pattern resembling those occurring in critical states, a description of the kind of criticality and the possible underlying mechanisms is outside of the scope of this paper and is reviewed elsewhere [55].

The model of cycles of heterogeneity assumes that when one species replaces another a high heterogeneity should be observed. We noted that in the third week of development, the two codominant species (*Chlorella sp* and *Oocystis sp*) were replaced by *Scenedesmus sp*, but we did not observe an increase in heterogeneity as measured by the information dimension *D(1)*. In fact we see that *D(1)* always decreases over time, but this could occur because the sampling interval is one week and the replacement of species has already been completed. A greater heterogeneity was only detected for negative *q*, that means that the lower biomass regions the plates are more heterogeneous. This evidence, which can only be observed using the multifractal approach, could lend partial support to the model of cycles of heterogeneity.

The link between persistence and fractal dimension has been studied in the context of patch dynamics. Both theory and experimentation have shown that when patch limits are more intricate and complex, these correspond to early succession stages, with less persistent patches and higher fractal dimension [13,56]. On the other hand when patches are more uniform, fractal dimension is lower corresponding to late successional stages. In our case *D(1)* has a different meaning: when it is lower it corresponds to more complex and heterogeneous spatial biomass distributions. When it is higher, closer to two, it corresponds to more uniform distributions. In fact when *D(1)* is equal to two the distribution is a completely flat surface. Our observations confirm previous studies, in that early successional stages correspond to more complex distributions and late successional to more uniform ones. This could be useful for applications in landscape management and classification of different areas by remote sensing [12]. The *D(q)* spectra can be used to characterize different successional stages, but more studies are needed to see if it can be used in systems with habitat heterogeneity.

The patch distribution of vegetation was extensively studied in arid regions [57]. It has been hypothesized that the loss of a self-organized spatial pattern in the vegetation patchiness can be used as signature of imminent catastrophic shifts between alternative states [54]. Evidence that patch distributions are self-similar (fractal) was found in Mediterranean ecosystems and it was also found that with increasing grazing pressure, vegetation patterns deviated from self-similarity [58]. Thus the loss of self-similar behavior can be used as a warning signal for the onset of desertification. In our case the loss of self-similiarity can be related to the loss of biodiversity. Which is not a catastrophic change but a gradual change produced by succession.

We observed a relationship between the information dimension estimated from the spatial pattern, that is from a snapshot of the community, and the Shannon's diversity estimated from the species abundances. To put it in another way, we could infer a minimum diversity of a site without either species identification nor estimation of abundances. As the spatial distribution of periphyton's biomass tends to lose the fractal behavior, diversity also tends to decrease. This is observed when the information dimension approaches an integer value of two.

The negative relationship between biodiversity and *D(1)* was also found in semi-arid shrubland and grasslands ecosystems [59,60]. In that case the relationship can be produced by an increased habitat heterogeneity that permits the coexistence of a greater number of species, or by the self-organization process, but is difficult to separate the two possible sources of heterogeneity. In our case we are confident that the spatial heterogeneity is produced by self-organization, i. e. by the interaction of species in the succession process.

This relationship may be due to different algal growth forms and thus estimating functional diversity could provide a more exact picture. Other studies have found a positive relationship between functional diversity and spatial biomass stability [6], which means less heterogeneity. This is opposite to our results, but the heterogeneity was measured at only one scale and the spatial extent used in that study was very limited compared to ours, so their results may not be comparable to ours.

Our approach could be applied to other ecosystems where the biomass distribution can be estimated, such as tropical forest [61], and used to identify zones with high and low diversity as a warning signal for conservation planning and management.

We are in the process of building a model that can reproduce the self-similar patterns and the link between Shannon's diversity and the information dimension. More detailed studies and the extension to other ecosystems are essential to evaluate the relevance and applicability of the results presented here.

Supporting Information

Figure S1 Bidimentional power spectrum for the periphyton's biomass distribution

Figure S2 Graph in logarithmic scale of X_q versus ε for 1 week of community development

Figure S3 Graph in logarithmic scale of X_q versus ε for 2 weeks of community development

Figure S4 Graph in logarithmic scale of X_q versus ε for 3 weeks of community development

Figure S5 Graph in logarithmic scale of X_q versus ε for 4 weeks of community development

Figure S6 Graph in logarithmic scale of X_q versus ε for 5 weeks of community development

Figure S7 Graph in logarithmic scale of X_q versus ε for 6 weeks of community development

Figure S8 Graph in logarithmic scale of X_q versus ε for 9 weeks of community development

Acknowledgments

We would like to thank Carlos Coviella and Carrie Kappel for kindly reading and improving the quality of the manuscript.

Author Contributions

Conceived and designed the experiments: LAS AG FM. Performed the experiments: LAS AG. Analyzed the data: LAS FM. Wrote the paper: LAS.

References

1. Tilman D (1994) Competition and biodiversity in spatially structured habitats. Ecology 75: 2–16.
2. Solé RV, Bascompte J (2006) Self-organization in complex ecosystems Princeton University Press. 373 p.
3. van de Koppel J, Gascoigne JC, Theraulaz G, Rietkerk M, Mooij WM, et al. (2008) Experimental Evidence for Spatial Self-Organization and Its Emergent Effects in Mussel Bed Ecosystems. Science 322: 739–742.
4. Buttel LA, Durrett R, Levin SA (2002) Competition and Species Packing in Patchy Environments. Theoretical Population Biology 61: 265–276. doi:10.1006/tpbi.2001.1569.
5. Economo EP, Keitt TH (2008) Species diversity in neutral metacommunities: a network approach. Ecology Letters 11: 52–62. doi:10.1111/j.1461-0248.2007.01126.x.
6. Weigelt A, Schumacher J, Roscher C, Schmid B (2008) Does biodiversity increase spatial stability in plant community biomass? Ecology Letters 11: 338–347. doi:10.1111/j.1461-0248.2007.01145.x.
7. Dufour A, Gadallah F, Wagner HH, Guisan A, Buttler A (2006) Plant species richness and environmental heterogeneity in a mountain landscape: effects of variability and spatial configuration. Ecography 29: 573–584.
8. Thomaz SM, Dibble ED, Evangelista LR, Higuti J, Bini LM (2007) Influence of aquatic macrophyte habitat complexity on invertebrate abundance and richness in tropical lagoons. Freshwater Biology 53: 358–367. doi:10.1111/j.1365-2427.2007.01898.x.
9. Kellner JR, Asner GP (2009) Convergent structural responses of tropical forests to diverse disturbance regimes. Ecology Letters 12: 887–897. doi:10.1111/j.1461-0248.2009.01345.x.
10. Manrubia SC, Solé RV (1996) Self-organized criticality in rainforest dynamics. Chaos, Solitons & Fractals 7: 523–541. doi:10.1016/0960-0779(95)00091-7.
11. Hastings HM, Pekelney R, Monticciolo R, Vun Kannon D (1982) Time scales, persistence and patchiness. BioSystems 15: 281–289.
12. Meltzer MI, Hastings HM (1992) The use of fractals to assess the ecological impact of increased cattle population: case study from the Runde Communal Land, Zimbabwe. Journal of Applied Ecology 29: 635–646.
13. Sugihara G, May RM (1990) Applications of fractals in ecology. Trends in Ecology & Evolution 5: 79–86. doi:10.1016/0169-5347(90)90235-6.
14. Alados CL, Pueyo Y, Giner ML, Navarro T, Escos J, et al. (2003) Quantitative characterization of the regressive ecological succession by fractal analysis of plant spatial patterns. Ecological Modelling 163: 1–17. doi:10.1016/S0304-3800(02)00294-6.
15. Yarranton GA, Morrison RG (1974) Spatial dynamics of a primary succession: nucleation. Journal of Ecology 62: 417–428.
16. Armesto JJ, Pickett TA, McDonnell MJ (1991) Spatial heterogeneity during succession: a cyclic model of invasion and exclusion. In: Kolasa J, Pickett STA, eds. Ecological Heterogeneity. New York: Springer-Verlag. pp 256–269.
17. Rejmánek M, Rosén E (1992) Cycles of Heterogeneity during Succession: A Premature Generalization? Ecology 73: 2329–2331.
18. Alados CL, Elaich A, Papanastasis VP, Ozbek H, Navarro T, et al. (2004) Change in plant spatial patterns and diversity along the successional gradient of Mediterranean grazing ecosystems. Ecological Modelling 180: 523–535. doi:10.1016/j.ecolmodel.2003.10.034.
19. Loehle C, Li BL (1996) Habitat destruction and the extinction debt revisited. Ecological Applications 6: 784–789.
20. Brown JH, Gupta VK, Li B-L, Milne BT, Restrepo C, et al. (2002) The fractal nature of nature: power laws, ecological complexity and biodiversity. Philosophical transactions of the Royal Society of London Series B, Biological sciences 357: 619–626. doi:10.1098/rstb.2001.0993.
21. Li B-L (2000) Fractal geometry applications in description and analysis of patch patterns and patch dynamics. Ecological Modelling 132: 33–50.
22. Evertzs CJG, Mandelbrot BB (1992) Multifractal measures. In: Peitgen H-O, Jurgens H, Saupe D, eds. Chaos and Fractals: New Frontiers of Science. New York: Springer. pp 921–953.
23. Pascual M, Ascioti FA, Caswell H (1995) Intermittency in the plankton: a multifractal analysis of zooplankton biomass variability. Journal of Plankton Research 17: 1209–1232.
24. Seuront L (1999) Universal multifractal analysis as a tool to characterize multiscale intermittent patterns: example of phytoplankton distribution in turbulent coastal waters. Journal of Plankton Research 21: 877–822. doi:10.1093/plankt/21.5.877.
25. Plotnick RE, Sepkoski Jr. JJ (2001) A multiplicative multifractal model for originations and extinctions. Paleobiology 27: 126–139. doi:10.1666/0094-8373(2001)027<0126:AMMMFO>2.0.CO;2.
26. Solé RV, Manrubia SC (1995) Self-similarity in rain forests: Evidence for a critical state. Physical review E 51: 6250–6253.
27. Solé RV, Manrubia SC (1995) Are rainforest self-organized in a critical state. Journal of Theoretical Biology 173: 31–40.
28. Borda-de-Água L, Hubbell SP, McAllister M (2002) Species-Area Curves, Diversity Indices, and Species Abundance Distributions: A Multifractal Analysis. American Naturalist 159: 138–155.
29. Yakimov BN, Bossuyt B, Iudin DI, Gelashviliy DB (2008) Multifractal diversity-area relationship at small scales in dune slack plant communities. Oikos-doi:10.1111/j.2007.0030-1299.16121.x.
30. Seuront L, Spilmont N (2002) Self-organized criticality in intertidal microphytobenthos patch patterns. Physica A 313: 513–539.
31. Wetzel RG (1983) Limnology. Philadelphia: W.B. Saunders Company. 744 p.
32. Dodds WK, Biggs BJF, Lowe RL (1999) Photosynthesis-irradiance patterns in benthic microalgae: variations as a function of assemblage thickness and community structure. Journal of Phycology 35: 42–53.
33. Vadeboncoeur Y, Lodge DM, Carpenter SR (2001) Whole-lake fertilization effects on distribution of primary production between benthic and pelagic habitats. Ecology 82: 1065–1077. doi:10.1890/0012-9658(2001)082[1065:WLFEOD]2.0.CO;2.
34. Biggs BJF, Smith RA (2002) Taxonomic richness of stream benthic algae: Effects of flood disturbance and nutrients. Limnology and Oceanography 47: 1175–1186.
35. Biggs BJF, Stevenson RJ, Lowe RL (1998) A habitat matrix conceptual model for stream periphyton. Archiv fur Hydrobiologie 143: 21–56.
36. Lowe RL (1996) Periphyton Patterns in Lakes. In: Stevenson RJ, Bothwell ML, Lowe RL, eds. Algal Ecology: freshwater benthic ecosystems. San Diego: Academic Press. pp 341–373.
37. Magurran AE (1988) Ecological diversity and its measurement Cambridge University Press. 198 p.
38. Saravia LA, Giorgi A, Momo FR (1999) A photographic method for estimating chlorophyll in periphyton on artificial substrata. Aquatic Ecology 33: 325–330.

39. Renshaw E (1997) Spectral techniques in spatial analysis. Forest Ecology and Management 94: 165–174.
40. Davis A, Marshak A, Wiscombe W, Cahalan R (1996) Scale invariance of liquid water distributions in marine stratocumulus. Part I: Spectral properties and stationarity issues. Journal of the atmospheric sciences 53: 1538–1558.
41. Loehle C, Li B-L (1996) Statistical properties of ecological and geologic fractals. Ecological Modelling 85: 271–284.
42. Marshak A, Davis A, Wiscombe W, Cahalan R (1997) Scale invariance of liquid water distributions in marine stratocumulus. Part II: Multifractal properties and intermitence issues. Journal of Atmospheric Sciences 54: 1423–1444.
43. Lavallée D, Lovejoy S, Schertzer D, Ladoy P (1993) Nonlinear variability of landscape topography: multifractal analysis and simulation. In: Cola ELD, Lam N, eds. Fractals in geography. pp 158–192.
44. Grassberger P (1983) Generalized dimensions of strange attractors. Physics Letters A 97: 227–230.
45. Renyi A (1970) Probability Theory. Amsterdam: North-Holland. 685 p.
46. Davis A, Marshak A, Wiscombe W, Cahalan R (1994) Multifractal characterization of nonstationarity and intermitency in geophysical fields: observed, retrieved or simulated. Journal of Geophysical Research 99: 8055–8072.
47. Chhabra AB, Jensen RV (1989) Direct determination of the f(alfa) singularity espectrum. Physical Review Letters 62: 1327–1330.
48. Crowley PH (1992) Resampling methods for computation-intensive data analysis in ecology and evolution. Annual Review of Ecology and Systematics 23: 405–447.
49. Holm S (1979) A Simple Sequentially Rejective Multiple Test Procedure. Scandinavian Journal of Statistics 6: 65–70.
50. Smyth G, Dunn P, Phipson B, Hu Y (2011) statmod: Statistical Modeling. Available:http://cran.r-project.org/package = statmod. Accessed 14 February 2012.
51. Cade BS, Noon BR (2003) A gentle introduction to quantile regression for ecologists. Frontiers in Ecology and the Environment 1: 412–420. doi:10.1890/1540-9295(2003)001[0412:AGITQR]2.0.CO;2.
52. Ricotta C (2000) From theoretical ecology to statistical physics and back: self-similar landscape metrics as a synthesis of ecological diversity and geometrical complexity. Ecological Modelling 125: 245–253.
53. Margalef R (1979) The Organization of Space. Oikos 33: 152–159.
54. Rietkerk M, Dekker SC, de Ruiter PC, van de Koppel J (2004) Self-Organized Patchiness and Catastrophic Shifts in Ecosystems. Science 305: 1926–1929. doi:10.1126/science.1101867.
55. Saravia LA, Giorgi A, Momo FR (2012) Multifractal growth in periphyton communities. Oikos 0: 0. doi:10.1111/j.1600-0706.2011.20423.x.
56. Hastings HM, Sugihara G (1993) Fractals: A User's Guide for the Natural Sciences. Oxford, New York: Oxford University Press. 235 p.
57. Klausmeier CA (1999) Regular and Irregular Patterns in Semiarid Vegetation. Science 284: 1826–1828.
58. Kéfi S, Rietkerk M, Alados CL, Pueyo Y, Papanastasis VP, et al. (2007) Spatial vegetation patterns and imminent desertification in Mediterranean arid ecosystems. Nature 449: 213–217. doi:10.1038/nature06111.
59. Alados CL, Aich AE, Komac B, Pueyo Y, Garcia-Gonzalez R (2007) Self-organized spatial patterns of vegetation in alpine grasslands. Ecological Modelling 201: 233–242. doi:10.1016/j.ecolmodel.2006.09.014.
60. Alados CL, Gotor P, Ballester P, Navas D, Escos JM, et al. (2006) Association between competition and facilitation processes and vegetation spatial patterns in alpha steppes. Biological Journal of the Linnean Society 87: 103–113. doi:10.1111/j.1095-8312.2006.00559.x.
61. Chave J, Condit R, Lao S, Caspersen JP, Foster RB, et al. (2003) Spatial and temporal variation of biomass in a tropical forest: results from a large census plot in Panama. Journal of Ecology 91: 240–252.

9

Nitrogen and Phosphorus Limitation over Long-Term Ecosystem Development in Terrestrial Ecosystems

Duncan N. L. Menge[1,2,3]*, Lars O. Hedin[1], Stephen W. Pacala[1]

1 Department of Ecology and Evolutionary Biology, Princeton University, Princeton, New Jersey, United States of America, 2 National Center for Ecological Analysis and Synthesis, Santa Barbara, California, United States of America, 3 Department of Biology, Stanford University, Stanford, California, United States of America

Abstract

Nutrient limitation to net primary production (NPP) displays a diversity of patterns as ecosystems develop over a range of timescales. For example, some ecosystems transition from N limitation on young soils to P limitation on geologically old soils, whereas others appear to remain N limited. Under what conditions should N limitation and P limitation prevail? When do transitions between N and P limitation occur? We analyzed transient dynamics of multiple timescales in an ecosystem model to investigate these questions. Post-disturbance dynamics in our model are controlled by a cascade of rates, from plant uptake (very fast) to litter turnover (fast) to plant mortality (intermediate) to plant-unavailable nutrient loss (slow) to weathering (very slow). Young ecosystems are N limited when symbiotic N fixation (SNF) is constrained and P weathering inputs are high relative to atmospheric N deposition and plant N:P demand, but P limited under opposite conditions. In the absence of SNF, N limitation is likely to worsen through succession (decades to centuries) because P is mineralized faster than N. Over long timescales (centuries and longer) this preferential P mineralization increases the N:P ratio of soil organic matter, leading to greater losses of plant-unavailable N versus P relative to plant N:P demand. These loss dynamics favor N limitation on older soils despite the rising organic matter N:P ratio. However, weathering depletion favors P limitation on older soils when continual P inputs (e.g., dust deposition) are low, so nutrient limitation at the terminal equilibrium depends on the balance of these input and loss effects. If NPP switches from N to P limitation over long time periods, the transition time depends most strongly on the P weathering rate. At all timescales SNF has the capacity to overcome N limitation, so nutrient limitation depends critically on limits to SNF.

Editor: Caroline P. Slomp, Utrecht University, The Netherlands

Funding: This work was conducted in part while D.N.L.M. was a Postdoctoral Associate at the National Center for Ecological Analysis and Synthesis, a Center funded by the National Science Foundation (NSF) (www.nsf.gov; Grant #EF-0553768), the University of California, Santa Barbara (www.ucsb.edu), and the State of California (www.ca.gov). D.N.L.M. was also supported by the Carbon Mitigation Initiative (cmi.princeton.edu). L.O.H. was supported by a grant from the NSF (DEB #0614116) and the A. W. Mellon Foundation (www.mellon.org). The funders had no role in study design, data collection and analysis, decision to publish, or preparation of the manuscript. No additional external funding was received for this study.

Competing Interests: The authors have declared that no competing interests exist.

* E-mail: dmenge@princeton.edu

Introduction

Over the past few decades an elegant conceptual model of long-term ecosystem development has emerged that focuses on the roles of nitrogen (N) and phosphorus (P) in shaping terrestrial ecosystem dynamics. This conceptual model concerns the development of ecosystems during primary succession–i.e., after a catastrophic disturbance such as a volcanic eruption or glacial retreat–over time periods extending to millions of years in the absence of another catastrophic disturbance. A basic component of the model is the contrast between abiotic N versus P inputs as soils develop. P inputs decline over time because P weathers out of rocks, whereas abiotic N inputs do not change much with soil age because they come primarily from the atmosphere (although some rocks contain substantial amounts of N [1]). Declining P inputs have been proposed to produce a "terminal steady state" of P deficiency [2,3].

These abiotic P versus N input dynamics have been associated with a transition in nutrient limitation to net primary production (NPP), with NPP being N limited on young soils and P limited on old soils. For example, fertilization studies in Hawaiian montane rainforests show that forest growth is N limited on young (300 year old) soils, co-limited by N and P on intermediate (20,000 year old) soils, and P limited on old (4.1 million year old) soils [4]. While direct fertilization tests are lacking, data on nutrient distributions in plants and soils from long-term chronosequences in New Zealand [5], Arizona [6,7], and elsewhere [8] have been interpreted to be consistent with a transition to P limitation as soils age.

However, a number of issues complicate the input side of the nutrient limitation picture. The first concerns biotic N inputs via biological N fixation, which is at least as important as abiotic N deposition in most ecosystems. At the young end of the spectrum, many researchers [3,9–12] have pointed out that the scarcity of N in young soils often leads to the dominance of woody symbiotic N fixers (hereafter, "N fixers"). N fixers can often fix more than enough N to meet their own demand [13] (but see [14]) and their litter adds substantial N to the soil, so N limitation to NPP might actually be uncommon when N fixers are sufficiently common. Indeed, P has been shown to limit symbiotic N fixation (hereafter, SNF) and presumably NPP in some young Alaskan forests [15]. Thus, when N fixers are abundant and actively fixing N, NPP might be more likely to be P limited or co-limited than N limited.

At the old end of the age spectrum, P inputs via atmospheric deposition of dust and aerosols become at least as important as P inputs from rock weathering. Atmospheric P deposition likely depends on climate patterns and the properties of upwind ecosystems [16–18] rather than soil age. Furthermore, tectonic uplift [19] or erosion could replenish parent material P such that weathering P inputs remain significant even at the terminal steady state. If there is a continual P input in addition to a continual N input, is a transition to P limitation inevitable? Apart from some very old soils [3,5,6], unpolluted old-growth forests in temperate and boreal regions are typically thought to be N limited [20–22]. Are the soils on which these forests grow simply not old enough to have reached the terminal P-limited steady state? Or is their steady state N limited?

So far this discussion has focused on N and P inputs, but losses can play an equally vital role in determining nutrient limitation. In particular, losses of plant-unavailable nutrients are essential to the maintenance of nutrient limitation [23–27]. One such loss vector–leaching of dissolved organic forms of N or P (DON or DOP)–has been implicated in maintaining N limitation in old temperate forests [22,23,28] and P limitation in old tropical forests [24]. Other losses of plant-unavailable nutrients such as physical erosion or hydrologic losses of particulate N or P might also be important, but for brevity we often refer to losses of plant-unavailable nutrient as DON or DOP losses.

These examples illustrate that different states–N limitation versus P limitation–might be possible at different times during ecosystem development. Young forests might be N limited because rocks have little N, or P limited because N fixers dominate the communities and fix as much N as they need. Old forests might be P limited because the majority of P has weathered out of rocks and DOP losses are high enough, N limited because they are not old enough to reach the P-limited steady state, or N limited because sufficiently large dust P inputs or DON losses produce an N-limited terminal steady state.

These lines of evidence imply a diverse natural history of ecosystem development with several possible trajectories, and thus identify the need for a broader model with the quantitative rigor to develop alternative testable hypotheses. Under what conditions should N limitation and P limitation prevail? If there are transitions between these different ecosystem states, when do they occur? Which ecosystem properties and processes influence why and when these transitions occur, and which are less relevant? What effect do symbiotic N fixers have on these states and transitions between them? These questions about ecosystem development concern fundamental ecosystem properties that, in addition to being interesting in their own right, determine the extent to which ecosystems themselves will help mitigate anthropogenic environmental impacts. As atmospheric reactive N levels are increasing [29,30], the ability of ecosystems to sequester additional N and carbon (C) depends on whether N, P, or both limit production. N-limited ecosystems would respond to N deposition by absorbing additional N and CO_2 [31] and mitigating global warming, whereas P-limited ecosystems might not soak up additional N or CO_2 [32], and would likely exhibit negative effects of N saturation such as decreases in stream and lake water quality, downstream eutrophication, and increased emissions of NO and the greenhouse gas N_2O [33–36].

In the present work we develop and analyze a dynamical model to investigate the questions posed in the previous paragraph. In particular, we take advantage of the natural timescale separation inherent to biogeochemical processes [26,37] to solve for the transient and equilibrium dynamics of plant biomass, litter N and P, soil organic matter N and P, and plant-available N and P. Additionally, we derive the conditions under which N limitation and P limitation to NPP are expected and the times at which transitions between N and P limitation occur. We focus on three periods in ecosystem development: the initial stages of primary succession, the late stages of succession on relatively young soils, and the terminal steady state on geologically old soils. Our purposes in this work are three-fold. First, given a set of known biogeochemical mechanisms, we investigate possible nutrient limitation trajectories, looking specifically for counterintuitive dynamics that have received little attention in the literature. Second, we provide a quantitative framework in which to examine where these different nutrient limitation trajectories might occur. Third, we present an analytical multiple timescale framework that could render multitudes of other ecosystem-level questions mathematically tractable.

Methods

Model Description

The basic model we use is an extension of previous work [25,26,38–40], which draws on earlier models [27,41–43]. The present version tracks changes through time in plant biomass carbon (B); organically bound, plant-unavailable N and P in litter (L_N, L_P); organically bound, plant-unavailable N and P in soil organic matter (SOM, which is distinct from litter in that it is sufficiently decomposed to the point where its original source is uncertain [44]; D_N, D_P); and plant-available N and P in the soil

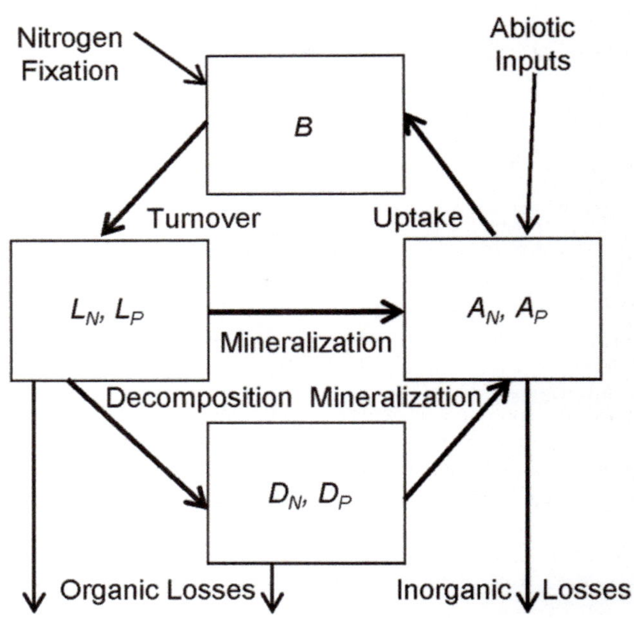

Figure 1. Model ecosystem described in equations 1–7. Living plant biomass (B) contains carbon, nitrogen (N), and phosphorus (P) in a fixed stoichiometric ratio. When plant biomass turns over it becomes litter. The model keeps track of litter N (L_N) and P (L_P) separately. Some litter is decomposed into plant-unavailable soil organic matter (SOM; D_N, D_P), some is mineralized to plant-available soil nutrients (A_N, A_P), and some is lost from the system. Some SOM is mineralized into plant-available nutrients, and some is lost. Plant-available nutrients come from mineralization and external inputs, are taken up by plants, and are lost from the ecosystem. Symbiotic N fixation enters directly into the plant pool. We use "mineralization" as shorthand for "mineralization and depolymerization" to indicate the conversion of nutrient to plant-available form, which can include some organic forms.

(A_N, A_P). Units of each variable are element (C, N, or P) mass per area [kg ha^{-1}]. We track N and P independently in litter and SOM because marked variation in soil N:P occurs throughout ecosystem development [5,45] (also see Discussion). In addition to inorganic N (e.g., nitrate and ammonium), plants can access N in some organic molecules such as amino acids [46,47]. We know of no evidence that plants directly take up organic P, but small P-containing molecules like nucleotides are likely candidates, and in effect plants can access some organic P via root phosphatases [47]. Therefore, we distinguish A_N and A_P from the other soil N and P pools by whether or not plants can acquire them rather than by their chemical form. The model, cartooned in Fig. 1, is defined by the equations:

$$\frac{dB}{dt}=B(g(A_N,A_P)-(\mu+\theta_\mu F)) \tag{1}$$

$$\frac{dL_i}{dt}=\frac{(\mu+\theta_\mu F)B}{\omega_i}-\delta_i L_i-h_i L_i \tag{2}$$

$$\frac{dD_i}{dt}=(1-\varepsilon_i)\delta_i L_i-m_i D_i-\phi_i D_i \tag{3}$$

$$\frac{dA_N}{dt}=I_N-k_N A_N+\varepsilon_N\delta_N L_N+m_N D_N - \frac{B}{\omega_N}(g(A_N,A_P)-\omega_N F) \tag{4}$$

$$\frac{dA_P}{dt}=I_P(t)-k_P A_P+\varepsilon_P\delta_P L_P+m_P D_P-\frac{B}{\omega_P}g(A_N,A_P) \tag{5}$$

$$g(A_N,A_P)=MIN[\omega_N(v_N A_N+F),\omega_P v_P A_P]-\theta_g F \tag{6}$$

$$I_P(t)=\alpha+\gamma e^{-\psi t}, \tag{7}$$

where the subscript i indicates N or P. Due to the nature of our questions we assume in this model that NPP is limited by N, P, or both, although other factors such as water, light, and top-down control also limit NPP in many real ecosystems. This assumption appears in the plant growth function $g(A_N, A_P)$, which is a Liebig's law function specifying that NPP increases with plant-available N, P, or at the exact ratio of plant N:P demand, both [41,48]. Although environmental factors such as moisture are known to influence most of the processes in our model [49,50], we do not include them explicitly here for the sake of simplicity. They appear implicitly in this model in that different sites would be represented by different parameter values, such as faster weathering [51] in wetter ecosystems compared to drier ecosystems.

Growth via N or P limitation depends on nutrient acquisition via uptake (v_i) and SNF (F) and the creation of new biomass C as mediated by nutrient use efficiencies (ω_i). Plant mortality and tissue turnover (μ + θ_μ F) transfer N and P from plants to litter. SNF carries costs associated with both growth (θ_g) and turnover (θ_μ) relative to uptake of soil N [21,39,40,52,53]. Litter N and P decompose at rates δ_i, with fractions ε_i for each being mineralized and the remaining fractions (1− ε_i) becoming SOM. SOM N and P are mineralized at rates m_i. There are losses from all soil pools–h_i from litter, φ_i from SOM, and k_i from plant-available pools–which can include leaching of dissolved and particulate nutrients (for all), erosion (for all), and gaseous losses (for plant-available N, e.g., from denitrification or nitrification). In this work we do not distinguish between different types of loss vectors from a given pool–such as leaching versus erosion of SOM or leaching versus gas loss of plant-available N–but include them all together. Abiotic inputs of N and P (I_N and $I_P(t)$) enter the plant-available soil pools. The definitions, units, and base values of all parameters are given in Table 1. Throughout this work we are careful to distinguish the term "rate"–a constant with units time^{-1} such as δ_i, φ_i, or ψ–from "flux," which has units of mass area^{-1} time^{-1} and may be a constant such as I_N or a variable such as $I_P(t)$.

The structure of equations 1–7 differs from those in [40] in two key ways. First, plant-unavailable N and P in the soil are now divided into two pools each–litter and SOM–whereas previous models included a single plant-unavailable pool per nutrient. There is ample evidence that organic matter is heterogeneous and decomposes at multiple timescales [54,55]. Second, to capture P input dynamics at long timescales, $I_P(t)$ includes both a constant flux (e.g., dust deposition, uplift, and erosion [19], α) and a flux that decreases with time (e.g., weathering, $\gamma e^{-\psi t}$, with the weathering rate ψ), as in [56]. True weathering dynamics might involve multiple timescales corresponding to P-containing minerals with different weathering rates, but we included only a single rate for simplicity. See Appendix S1 for a derivation of the P input term. Inclusion of inputs from weathering of rock N, which is important in some areas [1], would be a simple matter of altering I_N to be analogous to $I_P(t)$.

Because our primary focus in this paper is on timescales longer than a few months, we do not consider a number of short-term processes. For example, immobilization of N onto litter occurs in fresh litter with a high C:N ratio, but because there must be net mineralization over the timescales we consider [57], we ignore immobilization, in effect averaging over short timescales. As another example, rapid fluctuations in soil moisture are known to affect soil N [49,50] and P [58] dynamics, but our approach here is to treat our parameters as averages over these shorter timescales. To test whether this averaging of rapid fluctuations might unduly influence our longer timescale results, we conducted simulations where some of our parameter values (those likely to be influenced by soil moisture) fluctuated rapidly. For the variety of conditions we investigated, the effects on longer timescales were minimal (see Fig. S2).

Model Analysis

Although the model structure in the present paper differs from [40] in only two ways, our questions and analyses are entirely different. Menge et al. [40] examined how different degrees of regulation of SNF affected ecosystem properties. In the present paper we are concerned with how myriad ecosystem processes (including SNF, but also including many others) affect the development of N versus P limitation over long timescales. The analyses we use in the present paper build on [26], which examined how different loss types affect N limitation over multiple timescales. However, because [26] did not include P, it did not examine limitation by N versus P, which is the main focus of the present paper.

Our full model (equations 1–7) is sufficiently complicated that many of our analytical goals, such as solving for the transition times between N and P limitation, appear intractable. However, the natural diversity of timescales within biogeochemical systems allows solutions for the transient dynamics via the techniques of

Table 1. Variables, functions, and parameters.

Symbol	Definition	Units	Restrictions	Value or N	value P
B	Plant biomass C	kg C ha^{-1}	≥0		
L_i	Litter soil nutrients	kg i ha^{-1}	≥0		
D_i	SOM nutrients	kg i ha^{-1}	≥0		
A_i	Plant-available soil nutrients	kg i ha^{-1}	≥0		
$g(A_N,A_P)$	Plant growth function	yr^{-1}	≥0		
ω_i	Nutrient use efficiencies	kg C kg i^{-1}	>0	50	600
ν_i	Nutrient uptake parameters	ha kg C^{-1} yr^{-1}	>0	2	20
F	N fixation parameter	kg N kg C^{-1} yr^{-1}	≥0	0	…
μ	Biomass turnover rate	yr^{-1}	>0	0.5	…
θ_g	Growth cost of N fixation	kg C kg N^{-1}	$0-(\omega_N-\theta_\mu)$	5	…
θ_μ	Turnover cost of N fixation	kg C kg N^{-1}	$0-(\omega_N-\theta_g)$	5	…
δ_i	Litter decomposition rates	yr^{-1}	>0	4	4
h_i	Litter loss rates	yr^{-1}	>0	0.001	0.001
ε_i	Proportions of litter decomposition mineralized	unitless	0–1	0.4	0.5
m_i	SOM decomposition rates	yr^{-1}	>0	0.03	0.04
ϕ_i	SOM loss rates	yr^{-1}	>0	0.001	0.001
I_i	Plant-available nutrient input fluxes	kg i ha^{-1} yr^{-1}	>0	2	$\alpha + \gamma e^{-\psi t}$
A	P input flux from dust	kg P ha^{-1} yr^{-1}	>0	…	0.01
γ	P weathering flux from virgin rock	kg P ha^{-1} yr^{-1}	>0	…	0.4
ψ	P weathering rate	yr^{-1}	>0	…	0.00025
k_i	Plant-available nutrient loss rates	yr^{-1}	>0	6	4

Table notes: i refers to nitrogen (N) and phosphorus (P). C: carbon. SOM: soil organic matter. Parameters are from [40], except for δ_i and m_i (adapted from m in [40] to account for two soil organic pools per nutrient here rather than one), h_i (chosen to equal φ_i), ε_i (reasonable guesses), k_i (adjusted to reflect greater N than P loss due to greater soil mobility), and P input parameters (chosen to equal I_P in the previous work on young soils but for weathering to proceed at a relatively rapid rate).

timescale separation [26,59–61]. The basic idea of timescale separation is that when some ecosystem components (e.g., litter) change much more rapidly than others (e.g., weatherable P in rock), one can solve for the fast dynamics (e.g., litter turnover) assuming that the slow components (e.g., the P weathering input) are effectively constant, and solve for the slow dynamics assuming that the fast components are at their quasi equilibria. We distinguish quasi equilibria, which are the equilibria of the fast components on their timescales, from equilibria (of the slowest timescale, i.e., the terminal steady state) because the quasi equilibria depend on slower variables that are dynamic at the longer timescales. With timescale separation, despite the overall complexity of equations 1–7, we can solve approximately for how plant biomass, plant-available soil nutrients, litter, and SOM N and P change over time, which nutrient is limiting at any given time, and when limitation switches from N to P.

The first step in timescale separation analysis is to determine the different timescales based on empirical knowledge of the processes. We analyzed equations 1–7 with three distinct timescales (short, intermediate, and long), although our results indicate that more than three exist (i.e., there are two distinct timescales within the "short timescale" of our analyses). Our short timescale tracks changes in plant-available and litter N and P, assuming that plant biomass, SOM N and P, and weathering P inputs are constant. Nitrate, ammonium, and plant-available organic N (e.g., amino acids) turn over within hours to days in grasslands and forests [46,62–66]. Litter N and P turn over on the order of months to a few years [55,67]. Our intermediate timescale tracks changes in plant biomass over years to decades (averaging foliar, root, and wood turnover rates for different plant types), assuming that plant-available and litter N and P remain at their quasi equilibria and that SOM N and P and weathering P inputs are constant. Assuming that plant-available and litter N and P remain at their quasi equilibria does not mean they remain constant in time because the quasi equilibria depend on plant biomass, which changes over the intermediate timescale. Our long timescale tracks changes in SOM N and P and P inputs over centuries to millennia and beyond, assuming that plant-available and litter N and P and plant biomass are all at their quasi-equilibria.

After breaking the system into three timescales, we analyzed the short timescale assuming that plant biomass, SOM N and P, and weathering P inputs are constant. This tells us how litter N and P and plant-available N and P change in the first hours to years following a disturbance. The transient dynamics depend on whether N or P limits plant growth, so we solved both the N-limited and P-limited cases. To determine whether N or P limits plant growth at the beginning of the short timescale, we simply evaluated the growth equation (equation 6) with the starting conditions (i.e., the post-disturbance amounts of litter and plant-available N and P). If the limiting nutrient was different at the beginning versus the end of the short timescale, we solved for the transition time $t_{N\rightarrow P}$ or $t_{P\rightarrow N}$, which is defined as the time at which N and P co-limit growth. For the case of a transition from N limitation to P limitation, we use the expressions from when plants are N limited ($g(A_{N,N}(t_{N\rightarrow P})) = g(A_{P,N}(t_{N\rightarrow P}))$), and vice versa ($g(A_{P,P}(t_{P\rightarrow N})) = g(A_{N,P}(t_{P\rightarrow N}))$) for a transition from P limitation to N limitation. The first subscript in $A_{P,N}$ indicates the variable (in this case plant-available P) and the second subscript indicates the nutrient limitation case (in this case N limitation).

Switching from the short to the intermediate timescale (or the intermediate to the long) introduces a small discontinuity, or break, in the transient dynamics. Litter N and P and plant-available N and P approach their quasi equilibria asymptotically, meaning they never actually reach it, whereas at the intermediate timescale we assume they are exactly at their quasi equilibria. Deciding when to transition to the intermediate timescale involves balancing the jump to the quasi equilibrium–a better approximation the longer we wait–against the assumption that plant biomass remains constant–a worse approximation the longer we wait. We chose to transition to the intermediate timescale when litter and the limiting nutrient both came within 1% of their quasi equilibria. The intermediate timescale dynamics are then approximately linear (see Results section), so they can be solved assuming that litter and plant-available N and P are at their quasi equilibrium at that SOM N and P and weathering P inputs are constant. As above, we solve for N-limited and P-limited plant biomass, determine limitation at the beginning and end of the timescale, and solve for the time at which limitation switches.

Transitioning to the long timescale also involves a discontinuity, and we made the same assumption that the transition occurs when plant biomass comes within 1% of its quasi equilibrium. The long timescale dynamics and switches in limitation can then be determined as above. The transient solutions for each timescale will demonstrate the global stability at that timescale–e.g., stable if they approach an asymptote and unstable if they do not.

Timescale separation yields approximate solutions to the full system, and thus the results are only as good as the approximation. To check timescale approximations we numerically integrated the full system using the ode45 function in MATLAB 7.60. R2008a. For timescale approximations and numerical integrations we did not allow negative variable values because they are biologically meaningless. If timescale separation yields a decent approximation of the true model dynamics given by the numerical simulations, our analysis lends insight into the controls on transient dynamics at each timescale.

"Time" in the timescale approximations is the time since the last event that altered the variables in any way other than the system description. For example, a landslide that results in a rapid loss of soil and plants from a given site would reset all the variables; "time" would begin again at zero and the full set of timescales would be evaluated. Setting the initial condition for the time-dependent P input (γ, which depends on the initial amount of weatherable material; see Appendix S1) would be the trickiest initial conditions to parameterize in this case. A nitrogen fertilization event (say, with calcium nitrate) would only reset the plant-available N pool, but again, time would restart at zero and the full set of timescales would be evaluated.

Parameter Values and Assumptions

Because our primary results are analytical, they do not depend on the exact parameter values beyond the general bounds of timescale separation. That is, our analysis would yield strange results if rock weathering were faster than litter decomposition, but this is an unrealistic situation [51,67] that is beyond the scope of real ecosystems. For simulations we chose to use a reasonable parameter set for forests (Table 1). We used the same set of parameters as in [40], which explains the sources of each value, with exceptions noted in the Table 1 notes. Our explicit results are often too bulky to be informative, but simplifications based on assumptions about relationships between the parameters are quite informative. For instance, we assume that litter decomposition

Table 2. Parameter assumptions for analytical simplifications.

Assumption	Reasons/references	Exceptions
δ_i, $\delta_i(1-\varepsilon_i) >> h_i$	Recycling exceeds losses[1]	Heavy soil disturbance
$Bv_i >> k_i$	Recycling exceeds losses[1]	Low *B* (after disturbance)
$m_i >> \phi_i$	Recycling exceeds losses[1]	Heavy soil disturbance
$\phi_i >> \psi$	Weathering is very slow[2]	Rapid weathering rate
$\omega_P >> \omega_N$	Biological stoichiometry[3]	
$\varepsilon_P > \varepsilon_N$	Litter biochemistry[4]	Common ground fires
$m_P > m_N$	SOM biochemistry[4]	Common ground fires
$\delta_P \approx \delta_N$	Parameter definition[5]	
$h_P \approx h_N$	Same material[6]	Common ground fires
$\phi_P \approx \phi_N$	Same material[6]	Common intense fires
$m_iD_i >> \lvert I_i - k_iA_{i,i} \rvert$	Recycling exceeds inputs/losses[1]	Heavy pollution, Low *D*

Table notes and references: The subscript *i* refers to nitrogen (N) and phosphorus (P).

[1]In non-agricultural systems, internal recycling of both N and P (between plants and soils) typically exceeds total inputs and losses (to or from the atmosphere or waterways) many fold, both globally [44,101] and at individual sites [102]. Consequently, litter decomposition is much greater than litter loss ($\delta_iL_i >> h_iL_i$), plant uptake is much greater than losses of plant-available nutrients ($Bv_iA_i >> k_iA_i$), SOM mineralization is much greater than SOM loss ($m_iD_i >> \phi_iD_i$), and SOM mineralization is much greater than the balance of abiotic inputs and plant-available losses ($m_iD_i >> \lvert I_i - k_iA_{i,i} \rvert$). Exceptions to this pattern will occur where losses are very high (e.g., when ground fires or heavy erosion frequently remove substantial amounts of litter, it is unlikely that $\delta_i >> h_i$), where plant or SOM pools are very small (e.g., at the beginning of primary succession, it is unlikely that $m_iD_i >> \lvert I_i - k_iA_{i,i} \rvert$; see text and analysis), or in heavily polluted regions.

[2]See ref [51].

[3]Plants typically have an order of magnitude higher N content than P content [44,103].

[4]P is more readily cleaved from organic matter than N (see text).

[5]For decomposition, we have defined the parameters such that ε controls the relative mineralization of N versus P and δ is the overall litter decomposition rate.

[6]N and P loss rates via leaching and erosion should be similar because they come from the same organic material. In ecosystems where fire is important, organic N loss rates may be relatively higher than P.

rates are much higher than litter loss rates ($\delta_i >> h_i$). These simplifying assumptions are listed in Table 2 along with reasons for the assumptions and likely scenarios in which they do not hold.

Results

Goodness of Fit for Timescale Approximation

Our timescale approximations fit the full numerical integration extremely well for biomass (Fig. 2A), determining which nutrient is limiting (Fig. 2B), litter N and P (Fig. 2C–D), and the dynamics of the limiting nutrient in SOM and the plant-available pool (Fig. 2E,G when N is limiting, Fig. 2F,H when P is limiting). They fit less well for the non-limiting nutrient in SOM and the plant-available pool (Fig. 2E–H), so we do not put much emphasis on results for the non-limiting nutrient. Explanations for why the deviations occur are in Appendix S1.

Short Timescale

Our primary questions in this work concern successional timescales (years-centuries) and ecosystem development timescales (millennia and longer), so the short timescale dynamics (hours-years) are less relevant than the intermediate and long timescale

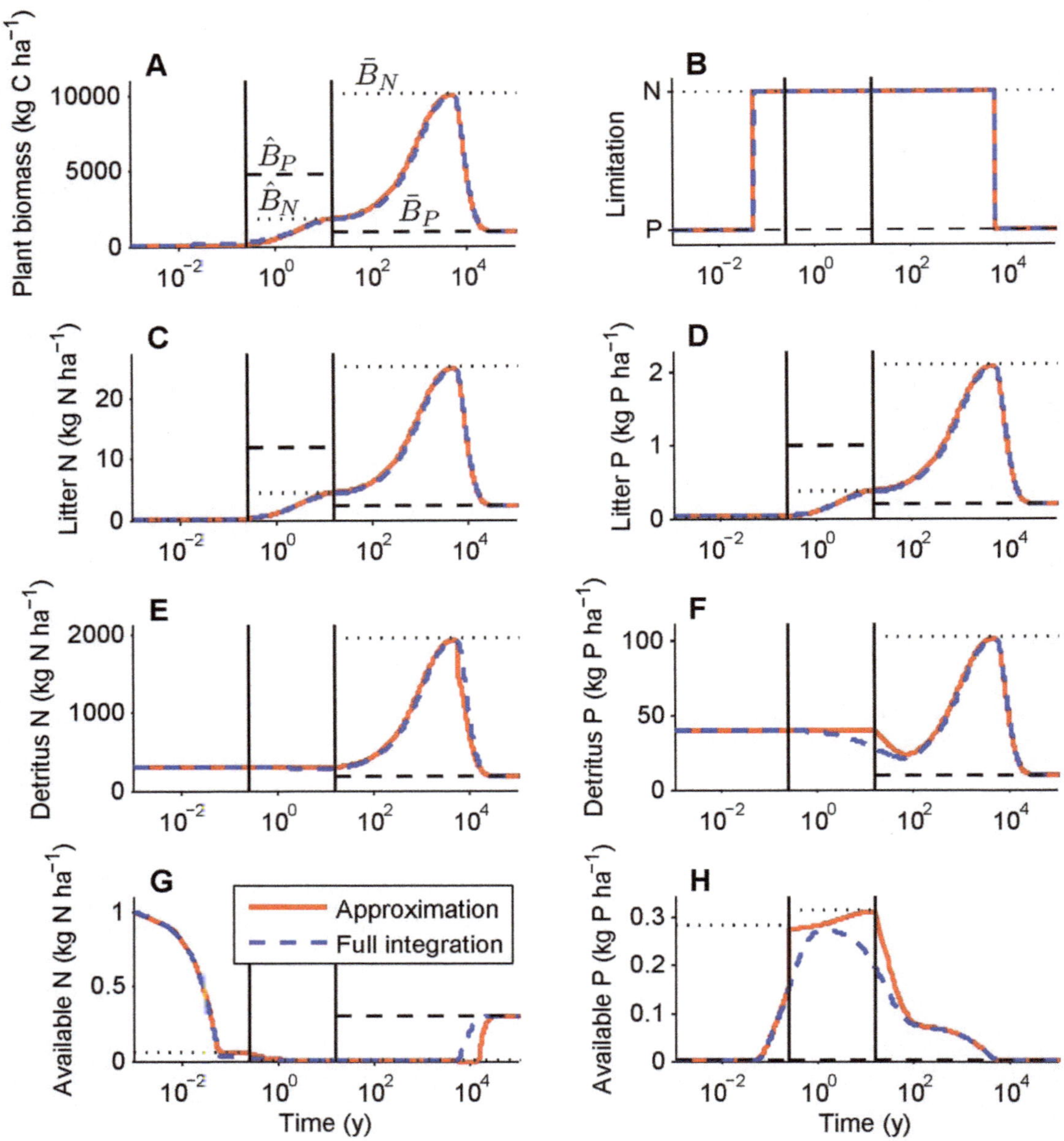

Figure 2. Fit of timescale approximation to full numerical integration. Parameters are as in Table 1, with no symbiotic N fixation. Both the approximation (solid red line) and the numerical integration (dashed blue line) began with biomass = 300 kg C ha^{-1}, litter N and P = 0.1 kg N ha^{-1} and 0.05 kg P ha^{-1}, soil organic matter (SOM) N and P = 300 kg N ha^{-1} and 40 kg P ha^{-1}, and plant-available N and P = 1 kg N ha^{-1} and 0.003 kg P ha^{-1}. Black dotted (N limited) and dashed (P limited) lines are quasi equilibrium and equilibrium values, displayed only for the relevant timescales. Vertical black lines indicate the timescale breaks: we used short timescale approximations to the left of the first black vertical line, intermediate timescale approximations between the two lines, and long timescale approximations to the right of the second line. Panels show (A) plant biomass, (B) which nutrient is limiting, (C) litter N, (D) litter P, (E) SOM N, (F) SOM P, (G) plant-available N, and (H) plant-available P. In this case net primary production begins P limited because there is an abundance of plant-available N, as from a small N fertilization, but becomes N limited within a couple weeks due to the large P:N ratio of inputs relative to plant demand and preferential P recycling. Plants remain N limited through succession and for thousands of years due to the preferential recycling of P and concomitant high loss ratio of dissolved organic N:P. At the long timescale plants become P limited because weathering inputs of P are negligible and dust P inputs are small. The time at which P limitation appears is controlled by the rock P weathering rate. See Fig. 4 for a different development trajectory. Note the logarithmic time axis.

results. In the interest of space, we present the short timescale results in Appendix S1 except for the following essential points. All short timescale dynamics converge to globally stable quasi equilibria. For everything but the non-limiting plant-available nutrient, all variables approach their quasi equilibria at the timescale of litter decomposition (a few months for our parameterization). Within the short timescale of our analysis, the limiting plant-available nutrient has two distinct timescales, with the shorter timescale controlled by plant uptake and the longer by litter decomposition. Therefore, the limiting plant-available nutrient quasi-equilibrates with litter before they both quasi-equilibrate at the end of the short timescale (Fig. S1B).

Intermediate Timescale

To solve for the plant dynamics on the intermediate timescale, we substitute the short timescale quasi equilibrium expressions for litter and plant-available N and P (Appendix S1) for the variables L_i and A_i and assume that SOM (D_i) and weathering P inputs (I_P) are constant. Importantly, these assumptions do not mean that the total amount of organic N and P in soil is constant on the intermediate timescale because litter N and P (L_i) change with plant biomass. This leaves a single differential equation (dB/dt) and a single variable (plant biomass, B) that can be N limited or P limited. Solving for both limitation cases will reveal which nutrient is limiting. If N is limiting, the quasi equilibrium of plant biomass (from setting equation 1 to zero) is

$$\hat{B}_N = \frac{\omega_N}{\mu'_N}\left(I_N + m_N D_N - k_N \hat{\hat{A}}_{N,N}\right) \tag{8}$$

$$\mu'_N = \frac{(\mu + \theta_\mu F)(\delta_N(1-\varepsilon_N) + h_N)}{\delta_N + h_N} - \omega_N F \tag{9}$$

The double hat indicates the intermediate timescale quasi equilibrium of A_N at the point where B_N quasi equilibrates. Equation 8 can be interpreted approximately as the late successional biomass for a given stage in long-term soil development. Equation 9, which is roughly the turnover rate of plants, defines the controlling rate of the intermediate timescale (see below). A few simplifications will aid the interpretation of these results. First, N limitation is unlikely if N fixers are present in late succession, and they are absent from many late successional forests [68], so we examine the case of no SNF ($F=0$). Second, mineralization fluxes from SOM far exceed the net balance of inorganic inputs and losses in unpolluted, late successional forests (see Table 2 notes), so we assume that $m_N D_N >> I_N - k_N \hat{\hat{A}}_{N,N}$. Using these and the assumption that litter loss is small relative to litter decomposition and transfer to SOM (δ_N, $\delta_N(1-\varepsilon_N) >> h_N$) (see Table 2),

$$\hat{B}_N \approx \frac{\omega_N m_N D_N}{\mu(1-\varepsilon_N)} \tag{10}$$

This means that the main controls on N-limited plant biomass late in succession are net N mineralization ($m_N D_N$ is net N mineralization from SOM, ε_N is the fraction of litter that is mineralized), the plant's N use efficiency (ω_N), and the amount of time N is retained within the plant ($1/\mu$).

After solving for the plant biomass quasi equilibrium, the next step is to solve for the transient dynamics by integrating equation 1. Assuming that N-limited plants take up far more plant-available N than is lost ($B v_N >> k_N$), the transient dynamics of plant biomass are

$$B_N(t) \approx \hat{B}_N + \left(B_N(0) - \hat{B}_N\right)e^{-\mu'_N t} \tag{11}$$

If SNF is below a threshold described in the following paragraph, μ_N' (equation 9) is positive, so N-limited plant biomass begins at $B_N(0)$ and saturates at its quasi equilibrium (Fig. 3A). The saturation rate–how quickly plant biomass approaches its quasi equilibrium–is given by μ_N'; equivalently, the intermediate timescale is $1/\mu_N'$. With no SNF, the intermediate timescale is approximately $1/\mu(1-\varepsilon_N)$, the turnover time of plant biomass divided by the proportion of litter decomposition that enters the SOM pool (see above and Table 2). Alternatively, when SNF exceeds the threshold, plants can escape N limitation (i.e., stability is controlled by the sign of μ_N'). This is visible in equation 11 in that μ_N' (equation 9) becomes negative if SNF is sufficiently large, leading to exponential growth of N-limited plant biomass and an unstable quasi equilibrium. In reality this would mean plant growth rapidly becomes P limited, so the N-limited case would become irrelevant, as illustrated in the short and intermediate timescale dynamics in Fig. 4. Fig. 4 differs from Fig. 2 in that obligate N fixers dominate the plant community until the end of the intermediate timescale and dust P inputs are higher.

The threshold that controls stability at the intermediate timescale is when the SNF flux equals the "loss" flux from the litter pool (leaving the system as well as entering the slow SOM pool, which is constant on this timescale). SNF can exceed 100 kg N ha^{-1} yr^{-1} when N fixers are dominant [13,69,70], whereas losses from the litter pool–which are less than litterfall inputs–rarely exceed 50 kg N ha^{-1} yr^{-1}. Therefore, abundant N fixers have the capacity to overcome N limitation at the intermediate timescale, although they would need to account for the majority of community biomass.

The quasi equilibria and transient dynamics of litter and plant-available N and P can be solved at the intermediate timescale by substituting the plant dynamics (equations 8, 9, and 11) for the plant constant (B) in the short timescale quasi equilibria. These are in Appendix S1, with the primary result being that they have the same timescale (μ_N') and stability as plant biomass.

If P limits NPP at the intermediate timescale, the dynamics of plant biomass, litter N and P, and plant-available P are always stable because there is no plant-controlled P input as there is for N,

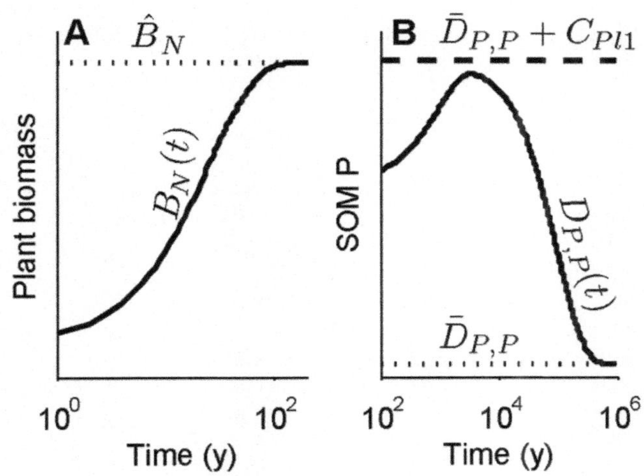

Figure 3. Examples of transient dynamics with single or double saturation rates. (A) On the intermediate timescale, N-limited plant biomass (solid curve) approaches its quasi equilibrium (dotted line) at the rate μ_N' (equation 9, which is the exponent in equation 11). With a logarithmic time axis (as here) this appears sigmoidal, but with a linear time axis it would be a saturating curve (similar to the Michaelis-Menten, or Type II, curve). P-limited plant biomass would have the same shape. (B) When NPP is P limited on the long timescale, SOM P has two controlling rates, the SOM P loss rate ($m_P(1-\kappa_{\delta P}) + \varphi_P \approx \varphi_P$) and the P weathering rate (ψ, equation 17). When these are sufficiently different, SOM P (solid line) first approaches an intermediate quasi equilibrium (dashed line) at the SOM P loss rate, then proceeds to its equilibrium (dotted line) at the P weathering rate. Here the rates are different enough to yield an overshoot, but not sufficiently different to yield full quasi equilibration at the intermediate point. Other details of timescale dynamics are given in Figure S1.

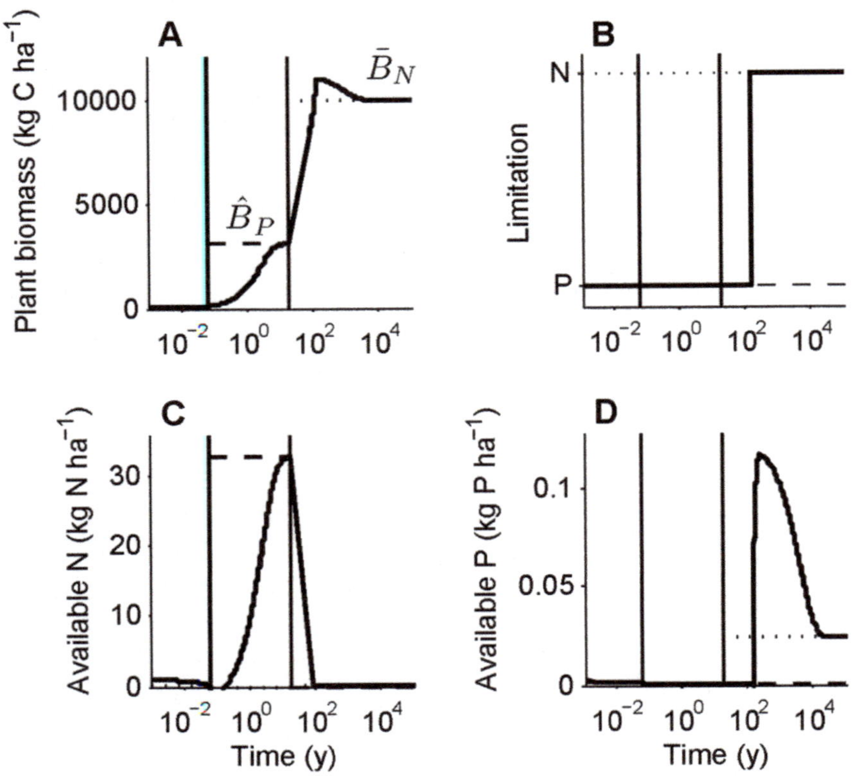

Figure 4. Timescale approximation illustrating the effects of obligate N fixation and high continual P inputs. Parameters and starting conditions are as in Fig. 2 except that $F = 0.07$ kg N kg C^{-1} y^{-1} for the short and intermediate timescales and $\alpha = 0.2$ kg P ha^{-1} y^{-1} throughout. At the end of the intermediate timescale F is set to 0 to simulate the exclusion of N fixers. (A) Plant biomass, (B) which nutrient limits net primary production (NPP), (C) plant-available N, and (D) plant-available P are shown. Breaks in timescales, quasi equilibria, and equilibria are shown as in Fig. 2. In this case obligate N fixers fix enough to overcome N limitation, hence P limits NPP through the short and intermediate timescales. Although P weathering depletes rock P as in Fig. 2, relatively high dust deposition and high losses of plant-unavailable N relative to P combine to make N limit NPP at the terminal steady state.

and the intermediate timescale is approximately $1/\mu(1-\varepsilon_P)$. The expressions for P-limited quasi equilibria and transient dynamics are identical to the N-limited cases (equations 8–11 and equations in Appendix S1) with the subscript P substituted for N and vice versa, except for slight differences in the equation that describes plant-available N (see Appendix S1).

At the beginning of the intermediate timescale, the plant is N limited if N demand relative to N supply is greater than P demand relative to P supply. Mathematically, this is given by

$$\omega_N\left(I_N + m_N D_N - k_N \hat{\hat{A}}_{N,N}\right) - \omega_P\left(I_P + m_P D_P - k_P \hat{\hat{A}}_{P,P}\right) <$$
$$B(0)\left(\mu + \theta_\mu F\right)\left(\frac{\delta_N(1-\varepsilon_N) + h_N}{\delta_N + h_N} - \frac{\delta_P(1-\varepsilon_P) + h_P}{\delta_P + h_P} - \omega_N F\right) \quad (12)$$

This is a fairly complicated expression, but the assumptions in Table 2 simplify it substantially, and specific cases of interest are even more easily interpreted. First, consider the case without N fixers (which we examine hereafter in all simplifications about limitation status) during early primary succession. In this case there is virtually no SOM ($D_i \approx 0$), and because ecosystem N and P stocks grow rapidly during early succession it is likely that plant-available N and P inputs (I_i) far exceed losses ($k_i(A_i)$). Using these assumptions, condition 12 simplifies to

$$\omega_N I_N - \omega_P I_P < B(0)\mu(\varepsilon_P - \varepsilon_N) \quad (13)$$

This shows that the main control on N limitation early in primary succession is the balance of N and P inputs (the dominant controls on supply at this timescale) relative to plant N:P demand. The right-hand side of condition 13 represents the balance of litterfall N versus P that is mineralized before becoming SOM. This term is likely to be positive ($\varepsilon_P > \varepsilon_N$) because organic P–generally attached with ester bonds that are easily broken with phosphatase enzymes, which are produced by plants, fungi, and bacteria–is more easily mineralized than organic N [21]. However, this term is likely to be near zero because the overall quantity of litter is low early in succession. Therefore, N versus P limitation in early primary succession depends chiefly on whether the left-hand side of condition 13 is positive or negative. If P inputs (I_P) relative to plant P demand (ω_P) exceed atmospheric N inputs (I_N) relative to plant N demand (ω_N), the left-hand side is negative and N limits NPP.

Which nutrient is limiting during early primary succession (condition 13) is plotted on Fig. 5A as a function of input fluxes (I_P and I_N). The lines indicate the divide between N and P limitation for a range of plant N:P demand and relative mineralization rates (see Fig. 5 caption). Although our parameter set (open circle in Fig. 5A) yields N limitation, P limitation is plausible in ecosystems with low rock P or slow weathering release of rock P relative to atmospheric N deposition.

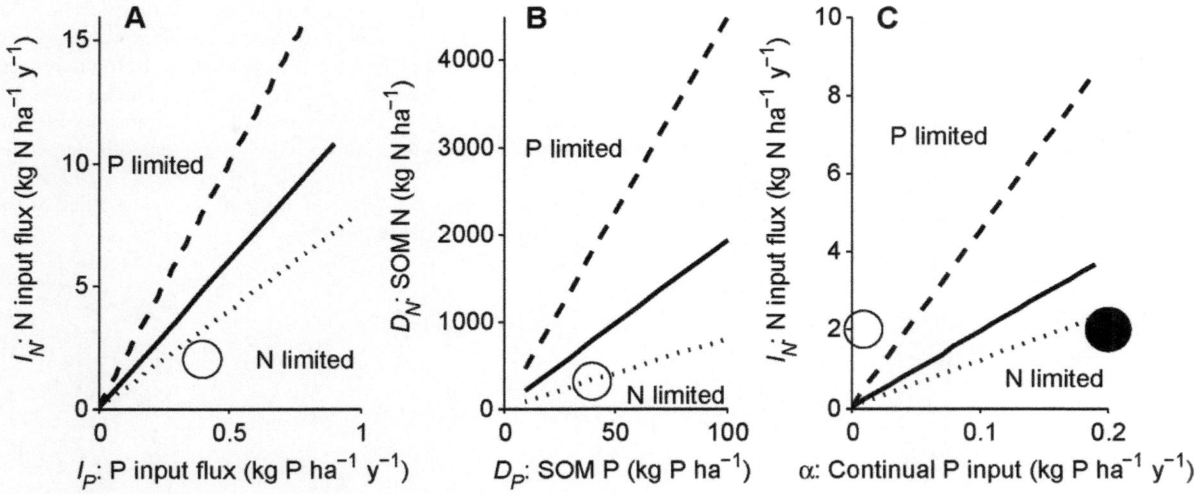

Figure 5. Approximate determination of N versus P limitation at different timescales. (A) The beginning of the intermediate timescale (beginning of succession), (B) end of the intermediate timescale (end of succession), and (C) terminal steady state (end of long-term ecosystem development) are shown with simplifying assumptions from Table 2 and no symbiotic N fixation. (A) At the beginning of succession there is negligible soil organic matter (SOM), so limitation is determined by input fluxes (the axes), plant demand, and litter N versus P mineralization. The dividing line between N and P limitation (condition 13) is plotted with parameters in Table 1 and 10 kg C ha^{-1} initial plant biomass (solid line) as well as for high (litter N:P = 20, P:N mineralization rates = 1.5; dashed line) and low (litter N:P = 8, P:N mineralization rates = 1; dotted line) parameter values. The circle represents parameters in Table 1. Ecosystems with high rock P inputs would be at the right end of the panel. Ecosystems with high N deposition would be near the top of the panel. (B) At the end of succession, mineralization from SOM far exceeds the balance of abiotic inputs and losses in unpolluted ecosystems, so N versus P limitation is determined by SOM N and P (the axes), plant N:P demand, and the mineralization of litter and SOM N versus P (condition 15). The lines are as in (A), and the circle is the initial SOM N and P in Fig. 2. (C) At the terminal steady state, N versus P limitation is determined by the balance of continual inputs and losses, plant N:P demand, and the mineralization of litter and SOM N versus P (condition 18). Axes are N and P input fluxes. Lines and open circle as in (A), with the closed circle indicating the input fluxes for Fig. 4.

The second case we consider is late in primary or secondary succession when plant biomass has quasi equilibrated with soil organic matter, i.e., the end of the intermediate timescale. In this case NPP is N limited at the intermediate quasi equilibrium when

$$\frac{\omega_P\left(I_P - k_P\left(\frac{\mu+\theta F}{\omega_P v_P}\right) + m_P D_P\right)}{\frac{(\mu+\theta_\mu F)(\delta_P(1-\varepsilon_P)+h_P)}{\delta_P+h_P}} > \frac{\omega_N\left(I_N - k_N\left(\frac{\mu-(\omega_N-\theta)F}{\omega_N v_N}\right) + m_N D_N\right)}{\frac{(\mu+\theta_\mu F)(\delta_N(1-\varepsilon_N)+h_N)}{\delta_N+h_N} - \omega_N F}, \qquad (14)$$

where $\theta = \theta_\mu + \theta_g$. As above, this is equivalent to when N demand relative to supply exceeds P demand relative to supply, but we are interested in going beyond this general statement to determine how each parameter affects the likelihood of N limitation and which are the most important. Decreasing N use efficiency (ω_N), N inputs (I_N), N uptake (v_N), SNF (F), or N mineralization (δ_N, ε_N, m_N, or D_N), or increasing N losses (k_N or h_N) pushes the system toward N limitation, and vice versa with the analogous P parameters. However, a few of these properties are much more relevant than others. Using the simplifying assumptions in Table 2, and again considering the case of no symbiotic N fixers, condition 14 can be simplified to

$$\left(\frac{1-\varepsilon_N}{1-\varepsilon_P}\right)\left(\frac{m_P}{m_N}\right) > \frac{\omega_N D_N}{\omega_P D_P} \qquad (15)$$

This shows that limitation at the end of the intermediate timescale is controlled primarily by preferential mineralization of P versus N, the N:P ratio of SOM, and plant N:P demand. The first term in parentheses on the left-hand side of condition 15 is the ratio of litter N to P that remains plant-unavailable during decomposition to SOM. As explained above, the biochemistry of organic matter suggests that P is easier to mineralize than N [21], so the first term is likely greater than 1. The second set of parentheses is also likely greater than 1 for the same reason since it is the ratio of the SOM P:N mineralization rates. The right-hand side of condition 15 is the ratio of SOM N:P relative to plant N:P demand. Plant N:P demand in this model is equivalent to the N:P ratio of litterfall, NUE:PUE, and if δN = δP and hN = hP (litter decomposition and loss rates are the same for N and P), the N:P ratio of litter. Therefore, the intermediate timescale without SNF ends N limited if P is preferentially mineralized over N by a greater factor than the ratio of SOM N:P to plant N:P demand.

Which nutrient is limiting at the end of the intermediate timescale (condition 15) is plotted on Fig. 5B with the amounts of SOM N and P on the axes, for the parameters in Table 1 (solid line) and a range of litter N:P and N:P mineralization values. The SOM N:P ratio (the x, y coordinate in Fig. 5B) is a strong determinant of N versus P limitation, although the degree of preferential mineralization and the plant N versus P demand (which control the dividing line in Fig. 5B) also play strong roles. The initial conditions for Fig. 2 yield N limitation (open circle in Fig. 5B), although many realistic scenarios would yield P limitation.

The effect of preferential mineralization on ecosystem N:P stoichiometry at the end of the intermediate timescale is shown on Fig. 6A, using parameters from Table 1 and initial SOM conditions from Fig. 2. At the intermediate timescale SOM N:P is constant and DON:DOP losses are nearly constant, regardless of preferential mineralization, so increasing P mineralization relative

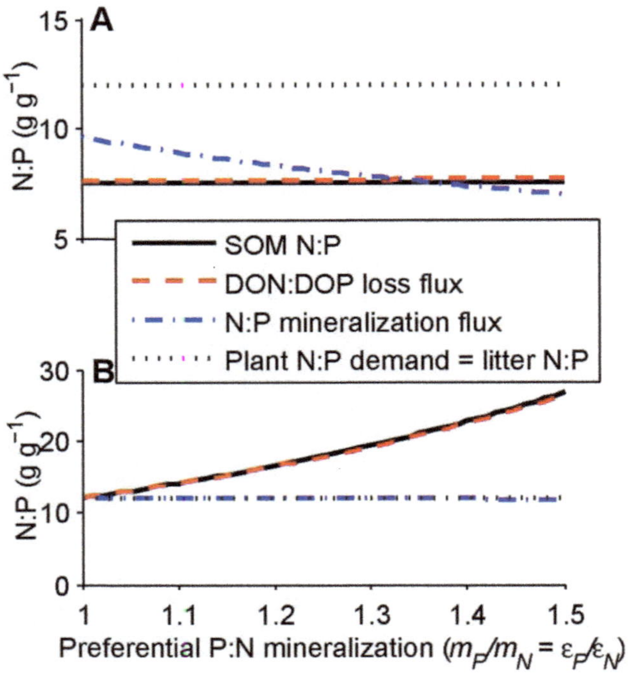

Figure 6. Effects of preferential P versus N mineralization. Effects are different for the (A) intermediate and (B) long timescales. Parameters are from Table 1 except for the mineralization rates m_N and m_P and the fractions of litter N and P decomposition mineralized, ε_N and ε_P, which vary to give the ratios along the horizontal axis. Timescale approximations of soil organic matter (SOM) N:P (solid black line), N:P losses of plant-unavailable nutrients from SOM and litter combined (termed DON:DOP in the legend for brevity, but in reality incorporating dissolved and particulate hydrologic losses and erosion losses; red dashed line), N:P mineralization fluxes from SOM and litter combined (blue dashed-dotted line), and plant N:P demand (black dotted line; which here is equivalent to P use efficiency:N use efficiency, litter N:P, and plant uptake) are plotted. The near-1:1 correspondence between SOM N:P and DON:DOP stems primarily from the equivalence of φ_N and φ_P, which is not always true (see Table 2 and [37]). However, any monotonic relationship would yield the same timescale dynamics exhibited here.

to N causes a decline in the N:P mineralization flux, exacerbating N limitation. If limitation switches during the intermediate timescale, the transition times can be calculated but the expressions are cumbersome, so they are given in Appendix S1.

Long Timescale

Our long timescale ends in the terminal steady state. The full equilibrium expressions, denoted by bars over the variables, are given in Appendix S1. Even more so than at the intermediate timescale, N limitation at the terminal equilibrium is unlikely if N fixers are present and actively fixing. At the long timescale SNF need only exceed losses of plant-unavailable N to overcome N limitation, as is generally the case in this class of ecosystem models [25]. Except where there are frequent, large fires or erosion, losses of plant-unavailable N are typically well under 10 kg N ha^{-1} yr^{-1} [22–24,71–73], whereas fixation fluxes can exceed that by an order of magnitude or more [13,69].

The long timescale transient dynamics of SOM (D_i) are solved by substituting the intermediate timescale equilibrium expressions into equation 3 and integrating equation 3. Plugging these equations for D_i back into the intermediate timescale quasi equilibrium expressions for plant biomass (equation 11 or its P limitation counterpart), litter, and plant-available nutrients (given in Appendix S1) completes the transient analysis. In the main text we focus on the dynamics of SOM N in the N-limited case and SOM P in the P-limited case because they illustrate the relevant dynamics, but solutions for all variables are presented in Appendix S1.

Transient dynamics of SOM N in the N-limited case and SOM P in the P-limited case are given by

$$D_{N,N}(t)=\bar{D}_{N,N}+(D_{N,N}(0)-\bar{D}_{N,N})e^{-\left(m_N\left(1-\kappa_{\delta_N}\right)+\phi_N\right)t} \quad (16)$$

$$D_{P,P}(t)=\bar{D}_{P,P}+C_{Pl1}e^{-\psi t}+ (D_{P,P}(0)-\bar{D}_{P,P}-C_{Pl1})e^{-\left(m_P\left(1-\kappa_{\delta_P}\right)+\vartheta_P\right)t} \quad (17)$$

The new terms $\kappa_{\delta N}$, $\kappa_{\delta P}$, and C_{Pl1} are combinations of parameters, as defined mathematically in Appendix S1. Biologically, at least in the case of no SNF, $\kappa_{\delta N}$ and $\kappa_{\delta P}$ describe the fractions of litter N and P output (decomposition and loss) that remain in the ecosystem. On the long timescale when N limits NPP, the equation describing SOM N dynamics (equation 16) has the same qualitative shape as the equation describing plant biomass dynamics at the intermediate timescale (Fig. 3A): a saturating curve with a single controlling rate ($m_N(1-\kappa_{\delta N}) + \varphi_N$), provided that the rate is positive. Some algebra shows that the rate is positive when losses of plant-unavailable N exceed SNF (see Appendix S1). With no SNF, $\kappa_{\delta N}$ is near 1, so the controlling rate is approximately the SOM N loss rate, φ_N. This is similar to a model that did not explicitly consider litter or SNF, in which the controlling rate of D_N was the SOM N loss rate [26]. On this long timescale plant biomass, litter N and P, and SOM P also have the same ultimate controlling rate as $D_{N,N}(t)$.

When P limits NPP on the long timescale, SOM P (equation 17) has two controlling rates, $m_P(1-\kappa_{\delta P}) + \varphi_P$ and ψ, giving it a qualitatively different shape than SOM N or plant biomass (Fig. 3B). The shape of SOM P depends on the relative values of the controlling rates. The fraction $(1-\kappa_{\delta P})$ is likely to be so small that the rate $m_P(1-\kappa_{\delta P})$ is small compared to the loss rate of plant-unavailable P from SOM, φ_P. Therefore, $m_P(1-\kappa_{\delta P}) + \varphi_P$ is approximately the SOM P loss rate, as was the case for N limitation above.

The second controlling rate, ψ, is the rock weathering rate, which is likely to be very slow relative to all other processes discussed in this paper. Assuming $m_P(1-\kappa_{\delta P}) + \varphi_P >> \psi$, SOM P begins at $D_P(0)$, approaches an intermediate saturation point–$\bar{D}_{P,P}+C_{Pl1}$–at the rate $m_P(1-\kappa_{\delta P}) + \varphi_P$, then proceeds to its terminal equilibrium at the rate ψ (Fig. 3B). Plant biomass, litter N and P, and SOM N share these two slow rates with SOM P. The SOM P buildup over millennia followed by a decline over longer timescales pictured in Fig. 3B is a typical trajectory, but a monotonic decline is also possible if SOM P begins at a high level.

With no or little SNF, and making the assumptions in Table 2, the terminal steady state is N limited when

$$\left(\frac{1-\varepsilon_N}{1-\varepsilon_P}\right)\left(\frac{m_P}{m_N}\right)>\frac{\omega_N\left(I_N-k_N\bar{A}_{N,N}\right)}{\omega_P\left(\alpha-k_P\bar{A}_{P,P}\right)} \quad (18)$$

and P limited if the inequality is reversed.

Condition 18 shows that limitation at the long timescale is controlled primarily by the balance of plant-unavailable N versus P losses and continual inputs of N versus P. The effect of plant-unavailable N versus P losses shows up on the left-hand side of condition 18. The left-hand side of condition 18 is identical to condition 15–the condition for N limitation at the end of the intermediate timescale–and is likely to exceed 1 because of the preferential mineralization of P over N. However, the mechanism behind this effect is different at the long timescale, owing to the feedback between preferential mineralization, soil N and P pools (which are now dynamic), and losses of plant-unavailable N and P. At the intermediate timescale, preferential P mineralization increases the amount of actively cycling P relative to N but does not affect the N:P ratio of SOM because SOM changes so slowly (Fig. 6A). However, at the terminal steady state preferential mineralization of P has no effect on the mineralization fluxes of P versus N since the greater ease of P mineralization is balanced by the increase in SOM N:P that results from preferential P mineralization.

This increase in the N:P ratio in SOM explains the role of plant-unavailable losses in determining N versus P limitation. Because the fluxes of plant-unavailable N and P losses are proportional to the amounts of N and P in SOM, the increase in the N:P ratio of SOM causes an increase in the N:P ratio of DON:DOP losses. This relatively higher loss of plant-unavailable N than P is what can maintain N poor conditions. Fig. 6B shows these long timescale effects using the parameters in Table 1. This increase in DON loss relative to DOP over long timescales suggests a tendency toward N limitation in old soils based on the loss side of the calculation.

The right-hand side of condition 18 illustrates the effects of abiotic N and P inputs on whether N versus P limits NPP at the terminal steady state. When continual P inputs (α, e.g., dust deposition and terminal weathering P inputs) are low, the right-hand side of condition 18 is large, suggesting that N limitation is unlikely even if P is preferentially mineralized over N. Alternatively, if dust deposition is high relative to atmospheric N deposition, the long-term equilibrium would more likely be N limited. Which nutrient is limiting at the long timescale (condition 18) is plotted on Fig. 5C with continual P and N input fluxes on the axes, for the parameters in Table 1 (solid line) and a range of litter N:P and mineralization N:P values (dashed, dotted dividing lines). Areas with low dust deposition and low terminal weathering rates are likely to be P limited (e.g., Fig. 2; open circle in Fig. 5C), but ecosystems with larger continual P inputs would more likely be N limited at the terminal steady state (e.g., Fig. 4; closed circle in Fig. 5C) particularly given the greater losses of SOM N than P.

The contrasting effects of inputs and plant-unavailable losses on nutrient limitation over the long timescale are summarized in Fig. 7. The N:P ratio of plant-unavailable losses increases at the beginning of the long timescale due to preferential P mineralization, which increases the N:P ratio of SOM and thus the ratio of plant-unavailable DON:DOP losses. This pushes the ecosystem toward or exacerbates N limitation. Over geological timescales the N:P ratio of inputs increases due to declining P inputs from rock weathering, which pushes the ecosystem toward or exacerbates P limitation. Ultimately the balance of input and plant-unavailable loss N:P ratios determine limitation at the terminal steady state (condition 18; Fig. 5). However, the contrast in the timescales of dissolved organic matter leaching (slow but not glacial) and rock weathering (glacial) means ecosystems that are N limited at the end of intermediate timescales will likely remain so at least until well into the geological timescale.

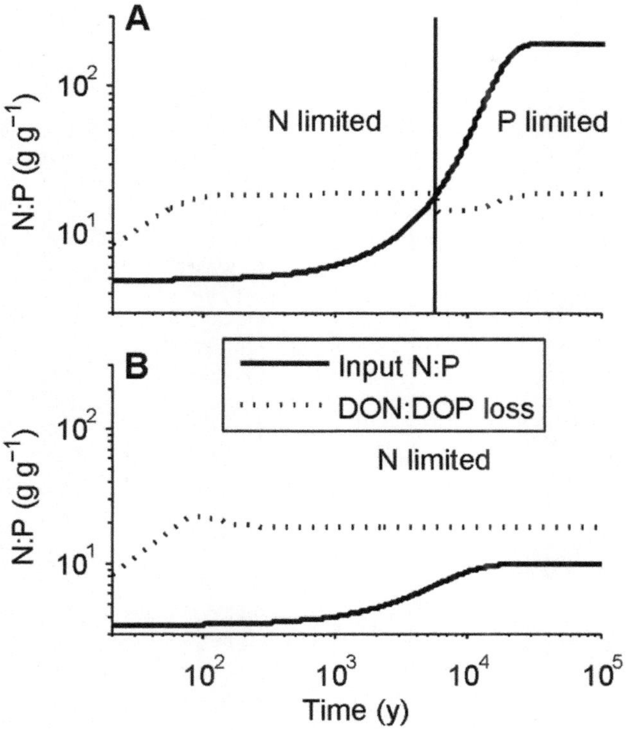

Figure 7. Contrasting effects of inputs versus losses in determining N versus P limitation over long timescales. Effects are shown for simulations in (A) Fig. 2 and (B) Fig. 4. The N:P ratio of plant-unavailable losses (termed DON:DOP in the legend for brevity, but in reality incorporating dissolved and particulate hydrologic losses and erosion losses; dotted lines) increases over decades due to preferential P mineralization (see Fig. 6), increasing the likelihood of N limitation over the intermediate and the beginning of the long timescales. On millennial and longer timescales the input N:P ratio increases due to the decline in P weathering, increasing the likelihood of P limitation over very long timescales. The transition time from N to P limitation in Fig. 2 corresponds to the crossing of the input and loss N:P ratio lines in (A) (indicated by the vertical line), which is controlled primarily by the weathering rate (see Fig. 8). In Fig. 4 and (B) there is no transition to P limitation over long timescales because the input N:P remains lower than the DON:DOP loss ratio, despite increases in both. The condition for whether limitation switches from one nutrient to the other is given in conditions 14, 15 and 18 (also see Fig. 5). The dip in the DON:DOP loss ratio when the input and loss line cross (when limitation switches) in (A) is an artifact of the timescale separation approximation–it remains constant at its saturation level in the full numerical integration (results not shown)–but all other trends reflect the true dynamics of the system. Note that both axes are logarithmic.

As with the intermediate timescale, there may be a transition from N to P limitation or vice versa at the long timescale. A transition from P to N limitation would be driven by the increase in the ratio of losses of plant-unavailable N:P over the SOM timescale. Conversely, a transition from N to P limitation would be driven by the decrease in P input over rock weathering timescales (Fig. 7). Because the SOM loss timescale is likely to be shorter than the rock weathering timescale, the DON:DOP loss ratio rises faster than the input N:P ratio (Fig. 7), so the transition from N to P limitation is much more likely to occur on the rock weathering timescale (see Appendix S1 for details). If so, the transition time from N to P limitation is

$$t_{N\to P} = \frac{-1}{\psi}\ln\left(\frac{-a_1}{\gamma}\right) \quad (19)$$

Faster weathering rates (larger ψ), lower initial parent material P content (smaller γ), and slower dust deposition fluxes (smaller α, which means smaller a_1; a combination of constants defined in Appendix S1) make the transition occur sooner (smaller $t_{N\to P}$). The primary control over the transition time is the weathering rate since it is the only term outside the logarithm and it varies substantially across ecosystems [51], so the transition time is plotted as a function of the weathering rate in Fig. 8. The solid line represents the parameters in Table 1, whereas the dashed and dotted lines represent a range of parent material P and continual P input fluxes that still yield long-term P limitation. The switch in limitation is highly dependent on the weathering rate, and for these parameters the transition occurs at approximately $1/\psi$ (i.e., at 10^3 years for a weathering rate of 10^{-3}/y, 10^6 years for $\psi = 10^{-6}$/y, and so on).

Discussion

Our model allows for a diversity of ecosystem development trajectories. One of these is a transition from N to P limitation over long timescales, as observed across the Hawaiian archipelago [4,37] and inferred in New Zealand [3,5] and Arizona [6,7]. In this case, depicted in Fig. 2, very young soils with negligible soil organic matter tend toward N limitation primarily because P weathering inputs relative to plant demand exceed atmospheric N deposition relative to plant demand. As succession proceeds and SOM accumulates over decades to centuries, this input imbalance is compounded by the higher fraction of organic P returned to the plant-available pool than organic N. This recycling imbalance increases the N:P ratio of SOM, and the resulting high N:P ratio of plant-unavailable nutrient losses exacerbates N limitation. However, when P inputs decline to a sufficiently low level over geological timescales the input effect trumps the plant-unavailable loss effect, yielding a terminal P-limited steady state. This case shows that declining P inputs–which are the primary cause of the transition to P limitation–are sufficient to explain declines in biomass, productivity, soil P, and soil N observed during ecosystem retrogression [8]. Other factors could also contribute to these retrogression patterns, as discussed in [8], and must if retrogression occurs when P is not limiting.

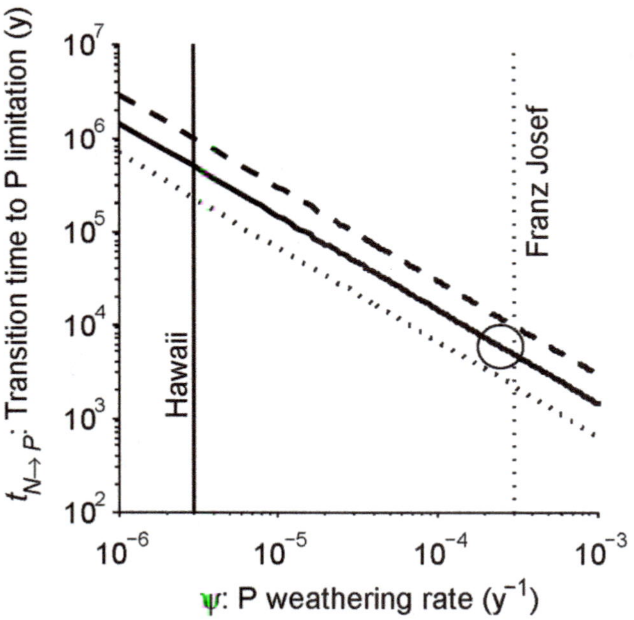

Figure 8. Transition time to P limitation on the weathering timescale. The main determinant of the transition time, if it occurs on the weathering timescale, is the weathering rate. Equation 19 is plotted as a function of the weathering rate, with the rest of the parameters as in Table 1 (solid diagonal line), high dust deposition and parent material P ($\alpha = 0.05$ kg P ha^{-1} y^{-1}, $\gamma = 1$ kg P ha^{-1}; dashed diagonal line) and low dust deposition and parent material ($\alpha = 0.001$ kg P ha^{-1} y^{-1}, $\gamma = 0.2$ kg P ha^{-1}; dotted diagonal line). The weathering rate and transition time in Table 1 and Fig. 2 are plotted as an open circle. All parameter sets plotted here yield a transition to P limitation, although many possible parameters sets do not (see Figs. 4, 5). Weathering rates from the Hawai'i (solid vertical line) and Franz Josef (dotted vertical line) chronosequences from [51] are also plotted. Note that both axes are logarithmic.

Our model also allows for P limitation on young soils and the persistence of N limitation on very old soils. Ecosystems with abundant and active N fixers are unlikely to remain N limited, although co-limitation might be more common than P limitation if SNF is regulated based on demand [40,74,75] or if community-level processes are fast enough to balance the abundance of N fixers versus non-fixers. Even without SNF, young ecosystems can be P limited if they have P-poor parent material, low weathering rates (termed "transactional" P limitation by [76]), or high N inputs. In these cases, however, a transition from P to N limitation during succession is possible if N fixers are excluded during succession or if preferential recycling of P over N is sufficiently large. On old soils with low or no SNF, ecosystems with high dust P deposition relative to atmospheric N inputs are likely to remain N limited indefinitely, particularly given that losses of plant-unavailable N are high relative to losses of plant-unavailable P. Fig. 4 illustrates one of these cases, where obligate N fixers overcome N limitation early in succession, but high dust deposition and the exclusion of symbiotic N fixers yields an N-limited terminal steady state.

A common feature of these trajectories is that regardless of their initial limitation status, ecosystems should experience increasing N deficiency relative to P over decades to millennia but increased P deficiency relative to N over geologic timescales. At decadal timescales the shift toward N limitation occurs because of increased P mineralization, which directly alleviates P limitation, and possibly because of a loss of N fixers from the community. At centurial to millennial timescales the shift toward N limitation occurs because of greater losses of SOM N relative to SOM P–which itself results from preferential P mineralization–and possibly due to a loss of N fixers as well. The shift toward P limitation over geologic timescales derives from declining P inputs, as in the classic argument of Walker & Syers [3]. These contrasting effects at different timescales stem from well-known biogeochemical mechanisms, yet are typically not discussed together in the context of ecosystem development.

Symbiotic N Fixation Overcomes N Limitation

As noted elsewhere [21,25,39,53,75,77], SNF has the capacity to overcome N limitation if N fixers are abundant and active. Our model agrees with this notion, and as with the models in [25,39], specifies that SNF need only exceed plant-unavailable "losses" to overcome N limitation if SNF is proportional to plant biomass. Unlike previous work, however, the model in the present work shows that the exact definition of plant-unavailable losses changes at different timescales. At the terminal equilibrium plant-unavailable "losses" are losses from the entire ecosystem such as leaching of DON. Alternatively, on the plant timescale plant-unavailable N "losses" include the flux of litter to SOM because SOM turns over so slowly that litter decomposed to recalcitrant SOM is effectively lost for decades. Strong plant-unavailable N loss vectors, such as frequent fires, may be able to maintain N limitation in the face of SNF, but except for these cases SNF can likely overcome N limitation at each timescale if N fixer abundance and activity are not constrained.

Because of this capacity to overcome N limitation, understanding N fixer abundance and activity is paramount to understanding N versus P limitation at all timescales, and has been the subject of much attention. Symbiotic N fixers are abundant in some early successional temperate forests [5,9,10,68], and there is evidence that these species–such as *Alnus rubra* [13,71,73] and *Coriaria arborea* [78]–are highly active N fixers regardless of soil N availability (i.e., they are functionally obligate). In such cases N would likely not limit these N fixers, so limitation by P or some other factor to SNF might be common [15,21,53,77]. In addition to limitation to SNF, P limitation to NPP might be common in early successional temperate forests dominated by active N fixers. This P limitation would seem paradoxical both because N fixers likely would not have established without N limitation and because N fixation beyond the point of co-limitation would appear energetically wasteful [24,75].

Despite these seeming paradoxes, observations of high plant-available N losses from forests co-dominated by alder in the Pacific Northwest of the United States [13,71,73] and high N availability after 60 years of *Coriaria arborea* dominance along the Franz Josef chronosequence in New Zealand [5] are consistent with–though not direct evidence of–P limitation, as our Fig. 4B,C shows. Obligate SNF might be due to high costs of being facultative or time lags in regulating SNF [40], although empirical tests of these hypotheses are currently lacking. As succession proceeds and non-fixers outcompete fixers, there can be a shift toward co-limitation and ultimately to N limitation on the intermediate timescale because much plant-available N is lost and preferential P recycling takes its toll. Long-term fertilization studies that track SNF and NPP simultaneously would help bolster these interpretations.

Although many young temperate forests are dominated by symbiotic N fixers, many have none [9,68], and symbiotic N fixers are virtually absent from old temperate forests [21,68]. Understanding N fixer rarity is equally important in the search to understand N versus P limitation in temperate forests. P limitation to SNF–which has been shown experimentally in Alaskan forests [15]–might explain N fixer rarity [21,53], as might high energetic costs of SNF [21,79], limitation of SNF by another nutrient such as molybdenum [21], preferential herbivory on symbiotic N fixers [21,39,53,80–82], low nitrogen use efficiency in N fixers [39], or a tradeoff between SNF and soil N uptake [39,79]. It is unlikely, however, that a phylogenetic constraint explains N fixer rarity in late successional temperate forests, at least in the United States [68].

Unlike temperate forests, many tropical forests seem to have abundant N fixers [69,75,83], although there are notable exceptions [37,75]. N limitation is therefore unlikely in tropical forests with abundant N fixers, regardless of their soil age, although because tropical N fixers seem to adjust SNF based on soil N [74] and/or P [84] content, co-limitation by N and P might be common [40,75].

Successional and Ecosystem Development Trajectories without Symbiotic N Fixation

Even without SNF, it is conceivable that many young ecosystems could be P limited due to high N deposition or low P inputs, particularly as atmospheric N deposition rises with anthropogenic activity [32,76]. In light of these input effects, we examined how well our model predictions matched observations of young forests in Hawai'i and Franz Josef. The available data suggest that the 300 year old Hawaiian site has relatively high abiotic N inputs (owing in part to volcanic N fixation) and moderate abiotic P inputs (owing mostly to weathering) [37]. Plotting these N and P inputs on our early succession figure suggests that NPP in young Hawaiian forests is N limited, but not far from co-limitation (Fig. 9; solid line and closed circle). This prediction of N-limited plant growth is consistent with fertilization experiments in the 300 year old site and a nearby 26 year old site [85]. The 26 year old site is a much better match for the assumptions in Fig. 9 (chiefly, negligible SOM). However, our model suggests that N limitation should become stronger through 300 years due to preferential P mineralization, so the prediction of N-limited NPP early in primary succession is consistent with N-limited NPP in the 300 year old site. The Franz Josef sites have relatively low abiotic N inputs and high estimated abiotic P inputs (see Fig. 9 caption), so our model indicates that NPP in the youngest Franz Josef site (5 years old, which has negligible SOM) would be strongly N limited without SNF (Fig. 9; dotted line and open circle). Although there have been no fertilization experiments

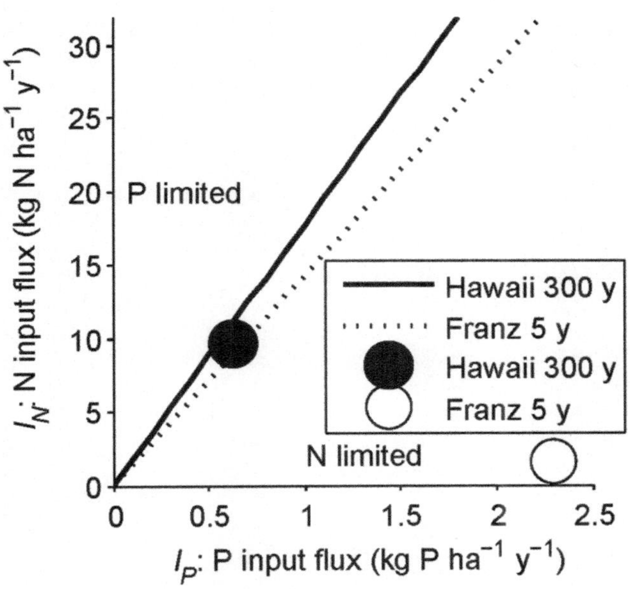

Figure 9. Model predictions for N vs. P limitation in the youngest Hawai'i and Franz Josef sites. Axes and equations are as in Fig. 5, but with Hawai'i in a solid line and closed circle and Franz Josef in a dotted line and open circle. At the beginning of primary succession the Hawai'i site would be N limited but fairly close to co-limitation, whereas the Franz Josef site would be strongly N limited without any symbiotic N fixation. Input and nutrient use efficiency parameters come from the 300 year old Hawaiian site ($I_N = 9.6$ kg N ha^{-1} y^{-1}, $I_P = 0.63$ kg P ha^{-1} y^{-1}, $\omega_N = 382$ g C g N^{-1}, $\omega_P = 6780$ g C g P^{-1} [37]) and the 5–7 year old New Zealand site ($I_N = 1.5$ kg N ha^{-1} y^{-1} [78], $I_P = 2.2$ kg P ha^{-1} y^{-1}–calculated from the decrease in soil inorganic P from the 5 to the 60 year old site–$\omega_N = 45$ g C g N^{-1}, $\omega_P = 643$ g C g P^{-1} [5]), with the other parameters as in Fig. 5a.

at Franz Josef, this prediction matches inferences based on soil nutrients and the dominance of symbiotic N fixers [5].

Without SNF, preferential recycling of P [21] is the dominant factor that favors N limitation at the intermediate timescale (i.e., the end of primary succession on young soils), although substantial N inputs from SNF or anthropogenic emissions could tip the scale toward P limitation. At the beginning of the long timescale, preferential P mineralization increases the SOM N:P ratio and consequently the N:P ratio of plant-unavailable losses, which exacerbates N limitation long before weathering P inputs decline. This highlights one of the counterintuitive results from our analysis: an increase in the soil N:P ratio over time should not be taken as an indication of N sufficiency or P limitation. Increasing soil N:P over time can stem from increased return of P to plants (over decadal to timescales) and larger losses of plant-uncontrollable N relative to P (over centurial timescales), both of which indicate increasing N deficiency.

These soil dynamics are observed at Franz Josef, where the ratio of N to organic P in the top 10 cm of mineral soil rises from less than 1 (5 year old site) to 3 (60 years) to 7 (130 years) to 12 (280 years) to 17 (500 years), then remains at 15–17 until 60,000 years before jumping to 34 at 120,000 years [5,45]. Mineral soil N to organic P does not rise substantially over time in Hawaii (27 at 0.3 ky, then 7–13 from 2.1–4100 ky in the top 50 cm [86]) or Arizona (25 at 1 ky, 22 at 55 ky, 30 at 750 ky, 17 at 3000 ky in the top 15 cm [6,7]), although both of these sequences start at an age when soil N:P may already have equilibrated (centuries to a millennium).

At the end of ecosystem development, nutrient limitation is determined by the balance of continual input (such as atmospheric N and dust P deposition) and loss fluxes. Although the Hawaiian chronosequence experiences little dust P deposition, even that small amount is biologically relevant [17], and a large fraction of the globe receives much greater dust P deposits. Large parts of Africa, Eurasia, and Australia and some parts of South America and the Caribbean receive between 0.1 and 1 kg dust P ha^{-1} y^{-1} [18], which according to our model could maintain P sufficiency indefinitely (Figs. 4, 5C). A careful analysis of global projections of N and P deposition [18,30] combined with plant-unavailable N versus P loss data could reveal which areas of the globe are likely to be N limited versus P limited at steady state. For example, parts of southern Chile and Argentina have substantial dust P deposition [18], low N deposition [30], and substantial DON losses [22,23], suggesting N limitation. Consistent with this, some old soils in Chile and Argentina display low plant-available N losses [22,23] and substantial retention of added N [65,87]. Replenishment of parent material P by tectonic uplift can also be substantial across much of the globe [19], strengthening the argument that a terminal N-limited steady state is plausible in many ecosystems.

If limitation transitions from N to P at the long timescale, our model suggests that the P weathering rate (with units of [time^{-1}]) is the dominant control over when the transition occurs, although properties such as the initial parent material P and dust P deposition also play roles. Weathering rates from Franz Josef (dotted vertical line) and Hawai'i (solid vertical line) estimated by [51] are plotted on Fig. 8 to illustrate these transition times. Based on these rate estimates and our model, the transition occurs sooner in Franz Josef than in Hawai'i–matching predictions from Richardson et al. [5]–somewhere between 1000 and 10,000 years. In the case of Hawai'i, our model predicts that the transition occurs between 200,000 and 1 million years, which is later than observed given the results showing co-limitation [4] at 20,000 years. However, the weathering rate is the only parameter fit to these sites on Fig. 8–all other parameters to make the transition time curve come from our generic set in Table 1–and uncertainty in the weathering rates would substantially alter these predictions.

Model Omissions and Opportunities for Future Development

As in any model, we have omitted many features of reality. Perhaps the most interesting and important of these concerns biotic responses to nutrient limitation. Theory [88–90] and data [91] suggest that organisms often respond to imbalances in nutrient supply versus demand. These responses can occur over physiological (plastic) or evolutionary (genetic) timescales, and tend to make organisms approach co-limitation in many models [40,92–97]. However, plasticity constraints [98,99], time lags in plastic and/or evolutionary change [100], or other constraints [39] can prevent nutrient co-limitation in models [25,40,95]. Therefore, unlike co-limitation between a nutrient and light (which is absent about half of each day) or water (which is highly variable over short timescales), co-limitation between multiple nutrients might not be ubiquitous.

For simplicity, our model does not include biotic adjustments to counteract limitation by a single nutrient; it stipulates that NPP is limited by either N or P except at the exact ratio that yields co-limitation. Previous theoretical work [38,40,79,95–97] shows that such biotic adjustments can yield co-limitation under some or all environmental conditions, depending on how they are incorporated into the model. This past work also shows that the approach to co-limitation may be very rapid, intermediate, or slow. Because such adjustments are likely to occur in nature, our statements about N versus P limitation should be taken as tendencies toward certain limitations, with the magnitude of the tendencies proportional to the distance from co-limitation.

Conclusion

In the present work we offer a simple, quantitative set of predictions for whether N or P is more likely to limit NPP at different stages in ecosystem development. Most of the qualitative predictions are in line with intuition. For example, increasing P inputs and preferential mineralization and retention of P decrease the likelihood of P limitation. However, our theoretical analysis goes beyond this intuition, specifying how strong these effects need to be to push the ecosystem from one state to another and the exact timescales at which they are important. From a theoretical standpoint, our analytical solutions of transient dynamics offer a framework to address myriad questions–such as how environmental factors influence N versus P limitation–in a more interpretable and rapid way than relying on simulations alone.

One counterintuitive result revealed by this analysis concerns the implications of changing N:P ratios in soil organic matter. Although rising SOM N:P seems like it might indicate increasing N sufficiency, our analysis shows that the opposite could be true. Preferential P mineralization acts to transfer P from SOM to the biota, and the resulting increase in SOM N:P yields greater loss of organic N relative to P, which in turn promotes N deficiency. This prediction could be supported empirically with data showing increasing N relative to P limitation (from fertilization experiments) as SOM N:P rises. Additionally, data showing a positive, monotonic relationship between SOM N:P and the N:P ratio of losses of plant-unavailable nutrients (the combination of erosion and dissolved and particulate hydrologic losses) would support the mechanism behind this result. A negative, monotonic relationship would refute this long timescale mechanism of increasing N limitation, whereas a non-monotonic relationship would suggest that the mechanism holds under some conditions but not others.

One of the most empirically useful contributions of this work might be that surprisingly few quantities are needed to answer certain questions. For example, knowledge of continual inputs, losses of plant-available nutrients, nutrient use efficiencies, and the degree of preferential mineralization of P versus N are sufficient to determine whether N or P is more likely to limit NPP at the terminal steady state (condition 18). Because most of our analytical results depend heavily on timescale separation, time series (for short or intermediate timescales) or chronosequence (for intermediate or long timescales) data exhibiting multiple saturations of the variables included here (as in Figs. 2, 3, 4 and Fig. S1, with time on a log scale) would provide strong support for these assumptions. More generally, the greatest empirical needs for evaluating this model are more quantitative data on P weathering and dust inputs and preferential P mineralization (for parameterization), and crucially, more fertilization studies testing for N versus P limitation in forest ecosystems on old soils around the world.

Supporting Information

Figure S1 Examples of transient dynamics at the short timescale. (A) Litter N approaches its quasi equilibrium in a saturating manner with a rate controlled by the exponent in equation S2 in Appendix S1. Litter P would be qualitatively similar. Depending on whether the perturbation increases (e.g., a storm that blows leaves down) or decreases (e.g., a ground fire) litter stocks, the saturation will approach from above or below. Short timescale litter dynamics do not depend on which nutrient limits plant growth. (B) Limiting plant-available nutrient dynamics if limitation does not switch on the short timescale. If plant uptake is faster than litter decomposition there are two separate saturations, which can take a variety of shapes (solid lines give two examples; see Appendix S1 for details).

Figure S2 Simulation to evaluate effects of forced rapid fluctuations. Pools and fluxes of plant-available nutrients fluctuate rapidly due to changes in soil moisture, among other things, which would mean that they would fluctuate around the quasi equilibria we present in this ms. We conducted additional simulations to evaluate whether these rapid fluctuations would propagate up to longer timescales. These simulations used the same conditions as in Figure 2 except that each of the soil (δ_i, h_i, m_i, φ_i, and k_i) and/or plant (v_i, F) rate parameters vary as sine functions of time, with 10–100 fluctuations per year. The parameters varied ±50–90% of their base values. Specifics for the run shown here were that soil parameters only (not plant parameters) varied 20 fluctuations per year and ±90% variation for each parameter, and the simulation was run for long enough to evaluate the fit at the long timescale (8000 y). This combination exhibited among the largest discrepancies from the results in Fig. 2 of any of the values we tried, which show up in the very short timescale for plant-available N and P. However, it is still very close to the original results, particularly for the longer timescale variables, leading us to conclude that rapid fluctuations such as these would not strongly affect our results.

Appendix S1 Derivations and additional results.

Author Contributions

Conceived and designed the experiments: DNLM LOH SWP. Performed the experiments: DNLM SWP. Analyzed the data: DNLM LOH SWP. Wrote the paper: DNLM LOH SWP.

References

1. Holloway JM, Dahlgren RA (2002) Nitrogen in rock: occurrences and biogeochemical implications. Global Biogeochemical Cycles 16: 65-1–65-17.
2. Stevens PR (1968) A chronosequence of soils near the Franz Josef Glacier. Christchurch: University of Canterbury. 350 p.
3. Walker TW, Syers JK (1976) The fate of phosphorus during pedogenesis. Geoderma 15: 1–19.
4. Vitousek PM, Farrington H (1997) Nutrient limitation and soil development: Experimental test of a biogeochemical theory. Biogeochemistry 37: 63–75.
5. Richardson SJ, Peltzer DA, Allen RB, McGlone MS, Parfitt RL (2004) Rapid development of phosphorus limitation in temperate rainforest along the Franz Josef soil chronosequence. Oecologia 139: 267–276.
6. Selmants PC, Hart SC (2008) Substrate age and tree islands influence carbon and nitrogen dynamics across a retrogressive semiarid chronosequence. Global Biogeochemical Cycles 22: 13.
7. Selmants PC, Hart SC (2010) Phosphorus and soil development: Does the Walker and Syers model apply to semiarid ecosystems? Ecology 91: 474–484.
8. Peltzer DA, Wardle DA, Allison VJ, Baisden WT, Bardgett RD, et al. (2010) Understanding ecosystem retrogression. Ecological Monographs 80: 509–529.
9. Walker LR (1993) Nitrogen fixers and species replacements in primary succession. In: Miles J, Walton DWH, editors. Primary Succession on Land. Oxford, UK: Blackwell Scientific. 249–272.
10. Chapin FS, Walker LR, Fastie CL, Sharman LC (1994) Mechanisms of primary succession following deglaciation at Glacier Bay, Alaska. Ecological Monographs 64: 149–175.
11. Crocker RL, Major J (1955) Soil development in relation to vegetation and surface age at Glacier Bay, Alaska. Journal of Ecology 43: 427–448.
12. Walker LR, del Moral R (2003) Primary Succession and Ecosystem Rehabilitation. New York, NY: Cambridge University Press. 442 p.
13. Binkley D, Cromack Jr K, Baker DD (1994) Nitrogen Fixation by Red Alder: Biology, Rates, and Controls. In: Hibbs D, DeBell D, Tarrant R, editors. The Biology and Management of Red Alder. Corvallis, OR: Oregon State University Press. 57–72.
14. Arnone JAI, Gordon JC (1990) Effect of nodulation, nitrogen fixation and CO_2 enrichment on the physiology, growth and dry mass allocation of seedlings of *Alnus rubra* bong. New Phytologist 116: 55–66.
15. Uliassi DD, Ruess RW (2002) Limitations to symbiotic nitrogen fixation in primary succession on the Tanana River floodplain. Ecology 83: 88–103.
16. Swap R, Garstang M, Greco S, Talbot R, Kallberg P (1992) Saharan dust in the Amazon basin. Tellus Series B-Chemical and Physical Meteorology 44: 133–149.
17. Chadwick OA, Derry LA, Vitousek PM, Huebert BJ, Hedin LO (1999) Changing sources of nutrients during four million years of ecosystem development. Nature 397: 491–497.
18. Okin GS, Mahowald N, Chadwick OA, Artaxo P (2004) Impact of desert dust on the biogeochemistry of phosphorus in terrestrial ecosystems. Global Biogeochemical Cycles 18: 10.
19. Porder S, Vitousek PM, Chadwick OA, Chamberlain CP, Hilley GE (2007) Uplift, erosion, and phosphorus limitation in terrestrial ecosystems. Ecosystems 10: 158–170.
20. Bormann BT, Sidle RC (1990) Changes in productivity and distribution of nutrients in a chronosequence at Glacier Bay National Park, Alaska. Journal of Ecology 78: 561–578.
21. Vitousek PM, Howarth RW (1991) Nitrogen limitation on land and in the sea: how can it occur? Biogeochemistry 13: 87–115.
22. Perakis SS, Hedin LO (2002) Nitrogen loss from unpolluted South American forests mainly via dissolved organic compounds. Nature 415: 416–419.
23. Hedin LO, Armesto JJ, Johnson AH (1995) Patterns of nutrient loss from unpolluted, old-growth temperate forests: evaluation of biogeochemical theory. Ecology 76: 493–509.
24. Hedin LO, Vitousek PM, Matson PA (2003) Nutrient losses over four million years of tropical forest development. Ecology 84: 2231–2255.
25. Menge DNL (2011) Conditions under which nitrogen can limit steady-state net primary production in a general class of ecosystem models. Ecosystems 14: 519–532.
26. Menge DNL, Pacala SW, Hedin LO (2009) Emergence and maintenance of nutrient limitation over multiple timescales in terrestrial ecosystems. American Naturalist 173: 164–175.
27. Vitousek PM, Hedin LO, Matson PA, Fownes JH, Neff J (1998) Within-system element cycles, input-output budgets, and nutrient limitation. In: Pace ML, Groffman PM, editors. Successes, Limitations, and Frontiers in Ecosystem Science: Springer. 432–451.
28. McGroddy ME, Baisden WT, Hedin LO (2008) Stoichiometry of hydrological C, N, and P losses across climate and geology: An environmental matrix

approach across New Zealand primary forests. Global Biogeochemical Cycles 22: GB1026.
29. Vitousek PM, Aber JD, Howarth RW, Likens GE, Matson PA, et al. (1997) Human alteration of the global nitrogen cycle: Sources and consequences. Ecological Applications 7: 737–750.
30. Galloway JN, Dentener FJ, Capone DG, Boyer EW, Howarth RW, et al. (2004) Nitrogen cycles: past, present, and future. Biogeochemistry 70: 153–226.
31. Reich PB, Hobbie SE, Lee T, Ellsworth DS, West JB, et al. (2006) Nitrogen limitation constrains sustainability of ecosystem response to CO_2. Nature 440: 922–925.
32. Menge DNL, Field CB (2007) Simulated global changes alter phosphorus demand in annual grassland. Global Change Biology 13: 2582–2591.
33. Aber JA, Nadelhoffer KJ, Steudler P, Melillo JM (1989) Nitrogen saturation in Northern forest ecosystems. Bioscience 39: 378–386.
34. Aber J, McDowell W, Nadelhoffer K, Magill A, Berntson G, et al. (1998) Nitrogen saturation in temperate forest ecosystems - Hypotheses revisited. Bioscience 48: 921–934.
35. Hall SJ, Matson PA (1999) Nitrogen oxide emissions after nitrogen additions in tropical forests. Nature 400: 152–155.
36. Hall SJ, Matson PA (2003) Nutrient status of tropical rain forests influences soil N dynamics after N additions. Ecological Monographs 73: 107–129.
37. Vitousek PM (2004) Nutrient Cycling and Limitation: Hawai'i as a Model System; Molina MJ, Morel FMM, Morgan JJ, Tilman D, editors. Princeton, NJ: Princeton University Press. 221 p.
38. Menge DNL, Ballantyne F, Weitz JS (2011) Dynamics of nutrient uptake strategies: lessons from the tortoise and the hare. Theoretical Ecology 4: 163–177.
39. Menge DNL, Levin SA, Hedin LO (2008) Evolutionary tradeoffs can select against nitrogen fixation and thereby maintain nitrogen limitation. Proceedings of the National Academy of Sciences of the United States of America 105: 1573–1578.
40. Menge DNL, Levin SA, Hedin LO (2009) Facultative versus obligate nitrogen fixation strategies and their ecosystem consequences. American Naturalist 174: 465–477.
41. Tilman D (1982) Resource Competition and Community Structure; May R, editor. Princeton, NJ: Princeton University Press.
42. DeAngelis DL (1992) Dynamics of nutrient cycling and food webs; Usher MB, editor. London, UK: Chapman and Hall. 270 p.
43. Daufresne T, Hedin LO (2005) Plant coexistence depends on ecosystem nutrient cycles: Extension of the resource-ratio theory. Proceedings Of The National Academy Of Sciences Of The United States Of America 102: 9212–9217.
44. Chapin FS, Matson PA, Mooney HA (2002) Principles of Terrestrial Ecosystem Ecology. New York: Springer.
45. Turner BL, Condron LM, Richardson SJ, Peltzer DA, Allison VJ (2007) Soil organic phosphorus transformations during pedogenesis. Ecosystems 10: 1166–1181.
46. Näsholm T, Ekblad A, Nordin A, Giesler R, Högberg M, et al. (1998) Boreal forest plants take up organic nitrogen. Nature 392: 914–916.
47. Lambers H, Chapin FSI, Pons TL (2008) Plant Physiological Ecology. New York, NY: Springer. 604 p.
48. Von Liebig J (1840) Organic chemistry in its application to agriculture and physiology: Friedrich Vieweg.
49. Porporato A, D'Odorico P, Laio F, Rodriguez-Iturbe I (2003) Hydrologic controls on soil carbon and nitrogen cycles. I. Modeling scheme. Advances in Water Resources 26: 45–58.
50. D'Odorico P, Laio F, Porporato A, Rodriguez-Iturbe I (2003) Hydrologic controls on soil carbon and nitrogen cycles. II. A case study. Advances in Water Resources 26: 59–70.
51. Buendía C, Kleidon A, Porporato A (2010) The role of tectonic uplift, climate, and vegetation in the long-term terrestrial phosphorous cycle. Biogeosciences 7: 2025–2038.
52. Gutschick VP (1981) Evolved strategies in nitrogen acquisition by plants. American Naturalist 118: 607–637.
53. Vitousek PM, Field CB (1999) Ecosystem constraints to symbiotic nitrogen fixers: a simple model and its implications. Biogeochemistry 46: 179–202.
54. Baisden WT, Amundson R, Cook AC, Brenner DL (2002) Turnover and storage of C and N in five density fractions from California annual grassland surface soils. Global Biogeochemical Cycles 16: 1117.
55. Adair EC, Parton WJ, Del Grosso SJ, Silver WL, Harmon ME, et al. (2008) Simple three-pool model accurately describes patterns of long-term litter decomposition in diverse climates. Global Change Biology 14: 2636–2660.
56. Porder S, Hilley GE (2011) Linking chronosequences with the rest of the world: predicting soil phosphorus content in denuding landscapes. Biogeochemistry 102: 153–166.
57. Manzoni S, Porporato A, D'Odorico P, Laio F, Rodriguez-Iturbe I (2004) Soil nutrient cycles as a nonlinear dynamical system. Nonlinear Processes in Geophysics 11: 589–598.
58. Runyan CW, D'Odorico P (2012) Hydrologic controls on phosphorus dynamics: A modeling framework. Advances in Water Resources 35: 94–109.
59. Ludwig D, Jones DD, Holling CS (1978) Qualitative analysis of insect outbreak systems: The spruce budworm and forest. Journal of Animal Ecology 47: 315–332.
60. McMurtrie RE, Comins HN (1996) The temporal response of forest ecosystems to doubled atmospheric CO_2 concentration. Global Change Biology 2: 49–57.
61. Ågren GI, Bosatta E (1996) Theoretical Ecosystem Ecology. Cambridge: Cambridge University Press. 234 p.
62. Jackson LE, Schimel JP, Firestone MK (1989) Short-term partitioning of ammonium and nitrate between plants and microbes in an annual grassland. Soil Biology & Biochemistry 21: 409–415.
63. Schimel JP, Jackson LE, Firestone MK (1989) Spatial and temporal effects on plant microbial competition for inorganic nitrogen in a California annual grassland. Soil Biology & Biochemistry 21: 1059–1066.
64. Zak DR, Groffman PM, Pregitzer KS, Christensen S, Tiedje JM (1990) The vernal dam - plant microbe competition for nitrogen in northern hardwood forests. Ecology 71: 651–656.
65. Perakis SS, Hedin LO (2001) Fluxes and fates of nitrogen in soil of an unpolluted old-growth temperate forest, Southern Chile. Ecology 82: 2245–2260.
66. Providoli I, Bugmann H, Siegwolf R, Buchmann N, Schleppi P (2006) Pathways and dynamics of $^{15}NO_3^-$ and $^{15}NH_4^+$ applied in a mountain *Picea abies* forest and in a nearby meadow in central Switzerland. Soil Biology & Biochemistry 38: 1645–1657.
67. Parton W, Silver WL, Burke IC, Grassens L, Harmon ME, et al. (2007) Global-scale similarities in nitrogen release patterns during long-term decomposition. Science 315: 361–364.
68. Menge DNL, DeNoyer JL, Lichstein JW (2010) Phylogenetic constraints do not explain the rarity of nitrogen-fixing trees in late-successional temperate forests. PLoS ONE 5: e12056.
69. Cleveland CC, Townsend AR, Schimel DS, Fisher H, Howarth RW, et al. (1999) Global patterns of terrestrial biological nitrogen (N_2) fixation in natural ecosystems. Global Biogeochemical Cycles 13: 623–645.
70. Ruess RW, McFarland JM, Trummer LM, Rohrs-Richey JK (2009) Disease-mediated declines in N-fixation inputs by *Alnus tenuifolia* to early-successional floodplains in interior and South-Central Alaska. Ecosystems 12: 489–502.
71. Binkley D, Sollins P, Bell R, Sachs D, Myrold D (1992) Biogeochemistry of adjacent conifer and alder-conifer stands. Ecology 73: 2022–2033.
72. Goodale CL, Aber JD, McDowell WH (2000) The long-term effects of disturbance on organic and inorganic nitrogen export in the White Mountains, New Hampshire. Ecosystems 3: 433–450.
73. Compton JE, Church MR, Larned ST, Hogsett WE (2003) Nitrogen export from forested watersheds in the Oregon Coast Range: The role of N_2-fixing red alder. Ecosystems 6: 773–785.
74. Barron AR, Purves DW, Hedin LO (2011) Facultative nitrogen fixation by canopy legumes in a lowland tropical forest. Oecologia 165: 511–520.
75. Hedin LO, Brookshire ENJ, Menge DNL, Barron AR (2009) The nitrogen paradox in tropical forest ecosystems. Annual Review of Ecology Evolution and Systematics 40: 613–635.
76. Vitousek PM, Porder S, Houlton BZ, Chadwick OA (2010) Terrestrial phosphorus limitation: mechanisms, implications, and nitrogen-phosphorus interactions. Ecological Applications 20: 5–15.
77. Vitousek PM, Cassman K, Cleveland C, Crews TE, Field CB, et al. (2002) Towards an ecological understanding of nitrogen fixation. Biogeochemistry 57/58: 1–45.
78. Menge DNL, Hedin LO (2009) Nitrogen fixation in different biogeochemical niches along a 120 000-year chronosequence in New Zealand. Ecology 90: 2190–2201.
79. Rastetter EB, Vitousek PM, Field CB, Shaver GR, Herbert DA, et al. (2001) Resource optimization and symbiotic nitrogen fixation. Ecosystems 4: 369–388.
80. Ritchie ME, Tilman D (1995) Responses of legumes to herbivores and nutrients during succession on a nitrogen-poor soil. Ecology 76: 2648–2655.
81. Hulme PE (1996) Herbivores and the performance of grassland plants: a comparison of arthropod, mollusc and rodent herbivory. Journal of Ecology 84: 43–51.
82. Knops JMH, Ritchie ME, Tilman D (2000) Selective herbivory on a nitrogen fixing legume (*Lathyrus venosus*) influences productivity and ecosystem nitrogen pools in an oak savanna. Ecoscience 7: 166–174.
83. Pearson HL, Vitousek PM (2002) Soil phosphorus fractions and symbiotic nitrogen fixation across a substrate-age gradient in Hawaii. Ecosystems 5: 587–596.
84. Pearson HL, Vitousek PM (2001) Stand dynamics, nitrogen accumulation, and symbiotic nitrogen fixation in regenerating stands of *Acacia koa*. Ecological Applications 11: 1381–1394.
85. Vitousek PM, Walker LR, Whiteaker LD, Matson PA (1993) Nutrient limitation to plant growth during primary succession in Hawaii Volcanoes National Park. Biogeochemistry 23: 197–215.
86. Crews TE, Kitayama K, Fownes JH, Riley RH, Herbert DA, et al. (1995) Changes in soil phosphorus fractions and ecosystem dynamics across a long chronosequence in Hawaii. Ecology 76: 1407–1424.
87. Perakis SS, Compton JE, Hedin LO (2005) Nitrogen retention across a gradient of ^{15}N additions to an unpolluted temperate forest soil in Chile. Ecology 86: 96–105.
88. Bloom AJ, Chapin FSI, Mooney HA (1985) Resource limitation in plants: an economic analogy. Annual Review of Ecology and Systematics 21: 363–392.
89. Chapin FS, Bloom AJ, Field CB, Waring RH (1987) Plant-responses to multiple environmental-factors. Bioscience 37: 49–57.

90. Field CB, Chapin FSI, Matson PA, Mooney HA (1992) Responses of terrestrial ecosystems to the changing atmosphere: a resource-based approach. Annual Review of Ecology and Systematics 23: 201–235.
91. Treseder KK, Vitousek PM (2001) Effects of soil nutrient availability on investment in acquisition of N and P in Hawaiian rain forests. Ecology 82: 946–954.
92. Abrams P (1987) The functional-responses of adaptive consumers of 2 resources. Theoretical Population Biology 32: 262–288.
93. Reynolds HL, Pacala SW (1993) An analytical treatment of root-to-shoot ratio and plant competition for soil nutrient and light. American Naturalist 141: 51–70.
94. van den Berg HA, Kiselev YN, Orlov MV (2002) Optimal allocation of building blocks between nutrient uptake systems in a microbe. Journal of Mathematical Biology 44: 276–296.
95. Klausmeier CA, Litchman E, Levin SA (2007) A model of flexible uptake of two essential resources. Journal Of Theoretical Biology 246: 278–289.
96. Menge DNL, Weitz JS (2009) Dangerous nutrients: Evolution of phytoplankton resource uptake subject to virus attack. Journal of Theoretical Biology 257: 104–115.
97. Rastetter EB, Shaver GR (1992) A model of multiple-element limitation for acclimating vegetation. Ecology 73: 1157–1174.
98. van Kleunen M, Fischer M (2005) Constraints on the evolution of adaptive phenotypic plasticity in plants. New Phytologist 166: 49–60.
99. Auld JR, Agrawal AA, Relyea RA (2010) Re-evaluating the costs and limits of adaptive phenotypic plasticity. Proceedings of the Royal Society B-Biological Sciences 277: 503–511.
100. Padilla DK, Adolph SC (1996) Plastic inducible morphologies are not always adaptive: The importance of time delays in a stochastic environment. Evolutionary Ecology 10: 105–117.
101. Smil V (2000) Phosphorus in the environment: Natural flows and human interferences. Annual Reviews of Energy in the Environment 25: 53–88.
102. Likens G, Bormann F (1995) Biogeochemistry of a Forested Ecosystem. New York: Springer-Verlag. 159 p.
103. Wright IJ, Reich PB, Westoby M, Ackerly DD, Baruch Z, et al. (2004) The worldwide leaf economics spectrum. Nature 428: 821–827.

10

Variation in Community Structure across Vertical Intertidal Stress Gradients: How Does It Compare with Horizontal Variation at Different Scales?

Nelson Valdivia[1,2]*, Ricardo A. Scrosati[3], Markus Molis[4], Amanda S. Knox[3]

1 Instituto de Ciencias Marinas y Limnológicas, Facultad de Ciencias, Universidad Austral de Chile, Valdivia, Chile, **2** Centro de Estudios Avanzados de Zonas Áridas (CEAZA), Facultad de Ciencias del Mar, Universidad Católica del Norte, Coquimbo, Chile, **3** Saint Francis Xavier University, Department of Biology, Antigonish, Nova Scotia, Canada, **4** Section Functional Ecology, Biologische Anstalt Helgoland, Alfred Wegener Institute for Polar and Marine Research, Helgoland, Germany

Abstract

In rocky intertidal habitats, the pronounced increase in environmental stress from low to high elevations greatly affects community structure, that is, the combined measure of species identity and their relative abundance. Recent studies have shown that ecological variation also occurs along the coastline at a variety of spatial scales. Little is known, however, on how vertical variation compares with horizontal variation measured at increasing spatial scales (in terms of sampling interval). Because broad-scale processes can generate geographical patterns in community structure, we tested the hypothesis that vertical ecological variation is higher than fine-scale horizontal variation but lower than broad-scale horizontal variation. To test this prediction, we compared the variation in community structure across intertidal elevations on rocky shores of Helgoland Island with independent estimates of horizontal variation measured at the scale of patches (quadrats separated by 10s of cm), sites (quadrats separated by a few m), and shores (quadrats separated by 100s to 1000s of m). The multivariate analyses done on community structure supported our prediction. Specifically, vertical variation was significantly higher than patch- and site-scale horizontal variation but lower than shore-scale horizontal variation. Similar patterns were found for the variation in abundance of foundation taxa such as *Fucus* spp. and *Mastocarpus stellatus*, suggesting that the effects of these canopy-forming algae, known to function as ecosystem engineers, may explain part of the observed variability in community structure. Our findings suggest that broad-scale processes affecting species performance increase ecological variability relative to the pervasive fine-scale patchiness already described for marine coasts and the well known variation caused by vertical stress gradients. Our results also indicate that experimental research aiming to understand community structure on marine shores should benefit from applying a multi-scale approach.

Editor: Simon Thrush, National Institute of Water & Atmospheric Research, New Zealand

Funding: Research was supported by grants from the Canada Research Chairs (CRC) program, the Canada Foundation for Innovation (CFI), and the Natural Sciences and Engineering Research Council of Canada (NSERC Discovery Grant) to RAS, by funds from Biologische Anstalt Helgoland to NV and MM, and by a grant from the Fondo Nacional de Desarrollo Científico y Tecnológico of Chile (FONDECYT, grant #3100014) to NV. The funders had no role in study design, data collection and analysis, decision to publish, or preparation of the manuscript.

Competing Interests: The authors have declared that no competing interests exist.

* E-mail: nelson.valdivia@uach.cl

Introduction

During the last several years, it has become evident that the different processes that affect species distribution operate and generate variability at different spatial scales [1,2]. Thus, ecological research has evolved from considering spatial variation in species assemblages as "noise" to appreciating that, in fact, identifying spatial scales of ecological variation is important to understanding how species assemblages are structured. Studies focused on determining spatial scales of variation have been especially prolific for rocky intertidal systems, for which patterns in community structure have indeed been shown to depend on the scale of sampling [3–5].

The rocky intertidal habitat is a unique system because a strong environmental stress gradient occurs within a few metres from low to high elevations [6]. The regular alternation of high and low tides determines that the duration of exposure to the air increases with elevation. Because of that, the abiotic stresses related to temperature, desiccation, irradiance, and osmotic potential increase towards the high intertidal zone, where they exhibit their most extreme values [7,8]. Ubiquitous patterns of species distribution along this vertical stress gradient (vertical variation) have inspired an extensive literature that accounts for competition [9], grazing [10], predation [11], facilitation [12], and the interplay between biotic and abiotic factors in structuring intertidal communities [13]. These processes are considered in environmental stress models that predict patterns of variation in biotic interactions and species diversity across environmental stress gradients [14–17].

In recent years, variability in intertidal community structure has been increasingly evaluated at different horizontal scales in terms of spatial resolution [4,17]. Resolution may be defined as the size of the individual areas among which ecological properties (e.g. community structure) are compared. Examples are patch scale, often meaning in the literature areas between 10s of cm and a few m in width, and site scale, often meaning areas 10s of m in width. At fine resolutions (differences within nearby patches), causes of horizontal variability in community structure include physical disturbance [18], frond sweeping by canopy-forming seaweeds [19], substrate pre-emption [20], availability of refugia [3], and variations in grazing activity [21]. At broader resolutions (differences among shores, each one commonly meant to be between 100s and 1000s of m in width), variation in community structure has been related to changes in wave exposure and ice scour [22], and recruitment [3,23]. Correlative evidence suggests that coastal geomorphology affects the distribution of primary producers at broad spatial scales [24], which is linked to broad-scale patterns in benthic community structure [25]. For regions harbouring the same basic biota, the highest variation in species abundance and community structure often occurs at fine resolutions [4].

Another approach to investigating scale effects in ecology is to focus on sampling interval, that is, the average distance between sampling units from which variation in ecological properties is calculated. Using this meaning of scale, it is thought that broad-scale processes may add extra variability to the patchiness commonly observed at fine scales on rocky shores [5]. Since many intertidal species have a planktonic stage, broad-scale processes are relevant to the structure of intertidal communities [15,26,27] and may cause a larger amount of variation than that resulting from the strong vertical stress gradient on rocky shores. However, so far only one study has contrasted vertical variation across intertidal elevations with horizontal variation measured along the shoreline at different sampling intervals [28]. Using intertidal systems from the NW Mediterranean coast, that study found that vertical variation in community structure was larger than horizontal variation measured at small scales but lower than horizontal variation measured at large scales. However, Benedetti-Cecchi [28] cautioned against considering such findings as universal for marine rocky shores, therefore calling for equivalent tests to be done on coasts representing different environments and biotas.

In this paper, we report the results of a study done using rocky intertidal systems from the cold-temperate shores of Helgoland Island, NE Atlantic. We tested the hypothesis that vertical variation in community structure is higher than fine-scale horizontal variation but lower than broad-scale horizontal variation. To test these predictions, we used a nested sampling design to quantify species abundances across the desired horizontal spatial scales in terms of differences in sampling interval. To help explain the observed community patterns, we also analysed the variability in the abundance of the species with the highest contribution to the spatial variation in community structure.

Methods

Study places

We evaluated vertical and horizontal patterns in rocky intertidal community structure in July–August 2008 on 3 shores on Helgoland Island, which is located off the coast of mainland Germany, NE Atlantic (Fig. 1): Bunker (54° 11′ 32″ N, 7° 52′ 35″ E), Kringel (54°10′ 60″ N, 7° 53′ 15″ E), and Nord-Ost-Hafen (54° 11′ 00″ N, 7° 53′ 34″ E). Our research adheres to the legal requirements of the Schleswig-Holstein state act of 24 April 1981 (classification number 791-4-37) that declared Helgoland Island a nature reserve and allows ecologists to access sites to accomplish field research.

A nested sampling design was employed. On each of the 3 study shores, we randomly established 3 sites that were spaced 5–10 m from one another following the shoreline. Within each site, we randomly established 2 patches that were spaced a few m from one another. At each patch, we randomly deployed three 25 cm×25 cm quadrats that were spaced a few 10s of cm from one another at each of 3 intertidal elevations (low, middle, and high intertidal zones).

On each shore, we surveyed the full vertical range of elevation between 0 m and an upper intertidal boundary determined using ecological indicators. On marine shores, waves during high tides wash the substrate at higher elevations than the high-water level indicated by tide tables (which provide predictions for still-water conditions). Thus, the regular occurrence of wave action expands the vertical range where sessile organisms can live, so different elevation zones (high, middle, and low) are wider and higher as wave exposure increases [30]. Because of that, to compare community structure from areas with a given degree of emersion-related stresses occurring on shores differing in wave exposure, community measurements must be done on each shore at elevations that account for the local effects of waves as described above. The sessile, perennial species occurring highest on the shore under all wave exposures in a given region are excellent indicators of the upper intertidal boundary, because their upper distribution limit on different shores represents a summary of the local wave regime. As the upper intertidal boundary for our study shores, we used the upper distribution limit of the brown seaweed *Fucus spiralis*, because this is the sessile, perennial species occurring highest on the 3 shores. Once the upper boundary was determined on each shore, we divided the intertidal range in three zones of equal vertical extent (high, middle, and low zones). We took all

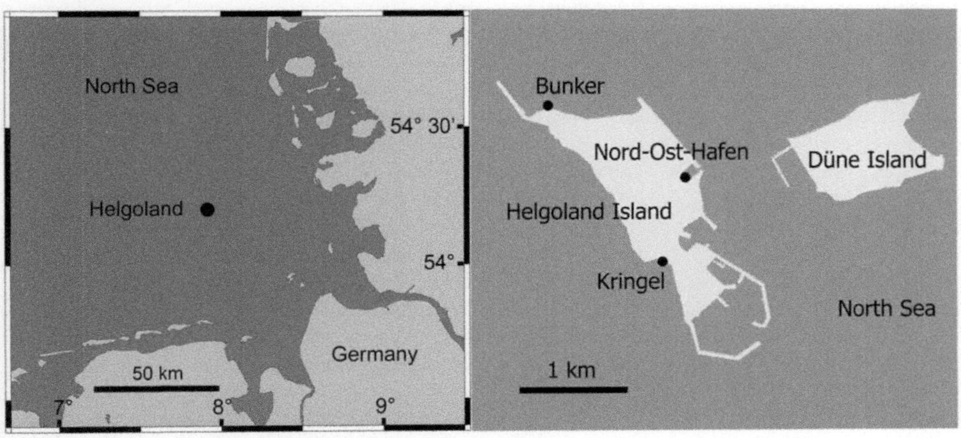

Figure 1. Location of the 3 studied shores on Helgoland Island, NE Atlantic.

elevation measurements to the nearest cm using a theodolite and a tide pole. All boundaries were permanently marked with screws bolted to the rocky substrate. The upper intertidal boundary was 2.3 m at Bunker, 1.8 m at Kringel, and 2.0 m at Nord-Ost-Hafen.

Sampling

For each quadrat, we identified all seaweeds and invertebrates (>1 mm) using field guides [29–31] and taxonomic keys [32,33]. Because of small size or morphological overlap with similar species, a few taxa were difficult to identify to the species level. Such organisms were classified to the lowest possible taxonomic level (usually genus), as normally done in field studies that identify all producers and consumers simultaneously in communities [34,35].

For each quadrat, we measured the percent cover of each species using a 25 cm×25 cm frame divided in 100 equal sections with monofilament line. For data analyses, it was important to have abundance data expressed in the same measurement unit for every species. We chose percent cover to quantify abundance because alternative measures (e.g. density of individuals) cannot always be determined reliably for clonal organisms [36] or (e.g. biomass) would have implied destructive sampling of numerous shore areas. We first measured the cover of canopy species before carefully moving the canopy aside to sample the understorey species. Organisms were not removed from the quadrats. Cover values were obtained from the projection of three-dimensional structures to the plane of the sampling frame. Total percent cover for a quadrat could exceed 100% because of the multilayered structure of assemblages. When the cover was <1% for a given species, we recorded it as 0.5% [37].

Data analysis

To reveal spatial patterns in community structure on the studied shores, we used canonical analysis on principal coordinates (CAP), [38]. Constrained multivariate methods such as CAP use an *a priori* hypothesis to produce an ordination plot, so they allow one to detect patterns that could be masked by overall dispersion in unconstrained methods such as multidimensional scaling. We generated CAP ordination plots for each shore on the basis of a matrix of factors (elevation, patch, and site) fitted to a matrix of Bray-Curtis dissimilarities. We assessed the significance of fits with 1000 permutations.

In nested designs, sampling effort (and thus statistical power) increases lower in the hierarchy of scales, which prevents a direct comparison of the magnitude of variation among the different scales though hierarchical analysis of variance [28]. Thus, to test our main hypothesis, we used an alternative type of analysis of variance described by Underwood and Chapman [39] and Benedetti-Cecchi [28], for which we obtained independent estimates of vertical and horizontal variation from different subsets of quadrats pooled at the appropriate spatial scales to keep a constant intensity of sampling across scales.

To measure vertical and horizontal variation in community structure, we used PERMANOVA based on Bray-Curtis distances between pairs of quadrats [40]. This multivariate technique tested for the effects of elevation (3 levels: high, middle, and low) on community structure using the abundance data for all species from a selected set of quadrats depending on the scale of interest. In PERMANOVA, the total multivariate variance (i.e. pseudo-variance) is partitioned into mean squares due to the factor of interest (elevation) and the residual or error mean squares [41]. We estimated the multivariate pseudo-variance component for elevation by equating the observed mean squares of the elevation factor to the expected mean squares derived from the linear model: $(MS_{elevations} - MS_{residual})/n$ [42], where n was the number of replicate quadrats within each elevation. The $MS_{residual}$ thus corresponded to the pseudo-variance component for horizontal variation. We applied PERMANOVA to measure horizontal variation at the different desired scales (different sampling intervals) in the way explained below.

To measure vertical variation in community structure across the three elevation zones and horizontal variation at the patch scale (variation among quadrats 10s of cm apart within patches), we randomly selected one site from each shore. We used one-way PERMANOVAs separately for each available patch to measure variation. Six patches were available for this analysis (3 shores ×1 site selected per shore ×2 available patches per site), yielding 6 independent estimates of vertical variation and 6 estimates of horizontal variation at the patch scale.

To measure horizontal variation at the site scale (variation among quadrats a few m apart within sites), we randomly selected another site from each shore. For each elevation zone within each site, we pooled the quadrats from both patches and, then, randomly divided the resulting 6 quadrats into 2 groups of 3 quadrats each. For each site, we then generated 2 data sets, each one including species abundance data from 3 quadrats from each elevation zone. We then applied one-way PERMANOVAs separately to each data set to generate 2 measures of horizontal variability for each site using the residual mean squares. In this case, horizontal variability measured variation within patches plus the variation between patches within sites. Six data sets were available for this analysis (3 shores ×1 site selected per shore ×2 modified sets of quadrats within each site), yielding 6 estimates of horizontal variation at the site scale.

To measure horizontal variation at the shore scale (variation among quadrats 100s to 1000s of m apart), we used the abundance data from the remaining site within each shore. First, we pooled the quadrats from each elevation zone across the 3 shores. Then, we randomly divided the resulting 18 quadrats into 6 groups of 3 quadrats each. We then randomly generated 6 data sets, each one including abundance data from 3 quadrats from each elevation zone. After applying one-way PERMANOVAs separately to those data sets, we obtained 6 measures of horizontal variation using the residual mean squares. In this case, horizontal variability measured variation within patches plus variation between patches within sites plus variation among shores.

We also calculated univariate variance components based on the abundance of the taxa contributing most to the spatial variation in community structure, for which we used the quadrat grouping procedure described above for the multivariate approach. For the univariate analyses, we used one-way ANOVAs to obtain estimators of variance. When negative variance components (either multivariate or univariate) were obtained for any scale, we set them to zero under the assumption that they were sample underestimates of small or zero variances [28,43]. We calculated the contribution of each taxon to the spatial variation in community structure with Similarity Percentage (SIMPER) routines. In this procedure, we calculated Bray-Curtis dissimilarity values among replicate plots, between groups, and within groups in the entire dataset of species percent covers. The average between-group dissimilarities were then broken down into separate contributions from each taxon [44]. Species contributions were calculated separately for the factors elevation, patch, site, and shore.

We tested the hypothesis that vertical variation in community structure is larger than fine-scale horizontal variation but smaller that broad-scale horizontal variation by applying a one-way ANOVA to the multivariate pseudo-variance component data obtained as explained above for each scale. In a similar way, we tested the same hypothesis for the variance component data of the taxa with highest contributions to the spatial variation in community structure. We considered "type of variation" as a fixed factor with 4 levels: vertical variation and patch, site, and shore-scale horizontal variation. We confirmed the homoscedasticity assumption by running Levene's tests after graphical exploration of residuals vs. fitted values and $\log_{10}(y + 1)$ or square-root transformation of the data. After significant differences among treatments were detected by the ANOVA, we ran Dunnett's tests to evaluate the significance of pairwise differences. We considered Dunnett's test appropriate for our goals because this test compares each group against a "control" group, which in this case corresponded to the vertical variation. SIMPER routines were conducted with PRIMER v6. We did all other statistical analyses and ordinations with *R* environment version 2.12.2 [45]. We used the vegan package to compute the CAP ordinations and PERMANOVAs, the stats package to compute the one-way ANOVAs, and the multcomp package to compute Dunnett's tests.

Results

We identified a total of 62 taxa (29 seaweeds and 33 invertebrates) among the 3 studied shores. Four sessile taxa were present in at least 30% of the quadrats, exhibiting a mean percent cover higher than 20%: the turf-forming red alga *Mastocarpus stellatus*, canopy-forming brown algae of the genus *Fucus* (*F. spiralis*, *F. vesiculosus*, and *F. serratus*), crustose algae (*Hildenbrandia rubra*, *Ralfsia* sp., and *Phymatolithon* spp.), and the acorn barnacle *Semibalanus balanoides*. The most common grazers and carnivorous invertebrates were periwinkles of the genus *Littorina* (*L. obtusata*, *L. saxatilis*, and *L. littorea*) and the green crab *Carcinus maenas*, respectively. The complete list of species appears in a related publication that utilised the abundance data to investigate vertical trends in species diversity as predicted by an environmental stress model [46]. SIMPER routines identified 16 taxa that explained up to 90% of average Bray-Curtis between-group dissimilarities within each type of variability (Table 1). However, *F. spiralis*, *F. serratus* (both species hereafter analysed together in the univariate context as *Fucus* spp.), and *M. stellatus* explained over 50% of dissimilarities across all comparisons (Table 1).

The CAP ordinations revealed evident differences in community structure among the three intertidal elevations on the studied shores (Fig. 2). Fits for each shore were statistically significant ($P<0.005$). Differences among sites within shores were less marked, although they were indeed clear in some cases. For example, at the high intertidal zone of Bunker, quadrats from sites 1 and 3 were well separated in the multivariate space (Fig. 2). At Nord-Ost-Hafen, sites 1 and 2 showed separation at the high intertidal zone and, to a lesser degree, at the middle intertidal zone (Fig. 2).

Multivariate pseudo-variance components significantly differed among the studied types of spatial variation (Fig. 3; ANOVA on $\log_{10}(y+1)$-transformed data: $F_{3,\ 20} = 39.53$, $P<0.001$). Vertical variation in community structure was significantly higher than patch- and site-scale horizontal variation and significantly lower than shore-scale horizontal variation (Fig. 3; Dunnett's tests, $P<0.05$).

The type of spatial variation also had significant effects on variance components for *Fucus* spp. and *M. stellatus* (Fig. 4; ANOVA: $F_{3,\ 20} = 14.23$, $P<0.001$ for *Fucus* spp. and $F_{3,\ 20} = 6.94$, $P = 0.002$ for *M. stellatus*). For *Fucus* spp., the vertical variation in abundance was significantly higher than patch- and site-scale horizontal variation (Dunnett's tests, $P<0.05$) and statistically similar to shore-scale horizontal variation (Fig. 4; Dunnett's tests, $P>0.05$). For *M. stellatus*, the vertical variation in abundance was

Table 1. Breakdown of average Bray-Curtis dissimilarities (untransformed data) indicating the contributions of different taxa to 90% of the average between-group dissimilarity for each studied type of variation.

	Vertical			Patch (horizontal)			Site (horizontal)			Shore (horizontal)		
Species	$\bar{\delta}$	$\bar{\delta}\%$	$\Sigma\bar{\delta}\%$	$\bar{\delta}$	$\bar{\delta}\%$	$\Sigma\bar{\delta}\%$	$\bar{\delta}$	$\bar{\delta}\%$	$\Sigma\bar{\delta}\%$	$\bar{\delta}$	$\bar{\delta}\%$	$\Sigma\bar{\delta}\%$
Mastocarpus stellatus	16.7	23.0	23.0	45.4	21.9	21.9	45.4	21.8	21.8	15.3	21.3	21.3
Fucus spiralis	14.1	19.1	42.2	19.1	18.6	40.4	18.6	18.7	40.5	14.7	19.8	41.0
Fucus serratus	10.2	14.1	56.2	26.1	14.7	55.1	26.5	14.7	55.2	10.3	14.4	55.4
Blidingia spp.	4.4	5.8	62.0	6.2	5.9	61.0	6.3	5.9	61.1	4.1	5.6	61.0
Fucus vesiculosus	3.5	4.8	66.8	5.1	5.1	66.1	5.0	5.1	66.1	3.5	5.0	66.0
Ulva sp. 2	2.3	3.3	70.0	5.6	3.3	69.4	5.6	3.3	69.4	2.4	3.2	69.2
Semibalanus balanoides	2.2	3.0	73.0	4.8	3.1	72.5	4.8	3.1	72.5	2.3	3.1	72.4
Chondrus crispus	2.0	2.7	75.7	2.5	2.6	75.0	2.5	2.6	75.0	1.8	2.5	74.8
Phymatolithon spp.	1.9	2.6	78.3	1.9	2.8	77.8	1.8	2.8	77.8	2.1	3.0	77.8
Ulva sp. 1	1.8	2.5	80.8	4.4	2.5	80.3	4.4	2.5	80.3	1.8	2.5	80.3
Flustrellidra hispida	1.6	2.3	83.1	5.1	2.3	82.6	5.2	2.3	82.7	1.8	2.6	82.8
Cladophora rupestris	1.5	2.1	85.2	3.9	2.2	84.8	4.0	2.2	84.8	1.5	2.1	85.0
Littorina littorea	1.4	1.9	87.2	1.3	2.0	86.7	1.3	1.9	86.7	1.4	1.9	86.9
Ralfsia sp.	1.4	1.8	89.0	2.5	2.0	88.7	2.5	2.0	88.7	1.4	2.0	88.8
Verruca stroemia	0.7	1.0	90.0	2.1	1.1	89.7	2.2	1.1	89.8	0.8	1.1	89.9
Hildenbrandia rubra	0.7	0.9	91.0	0.5	1.0	90.8	0.5	1.0	90.8	0.7	1.0	90.9

$\bar{\delta}$= average contribution of taxon *i* to between-group dissimilarities; $\bar{\delta}\%$ = average percentage contribution of taxon *i*; $\Sigma\bar{\delta}\%$ = cumulative average percentage contribution at the level of taxon *i*. The contribution of each taxon was averaged across all between-group pairwise comparisons.

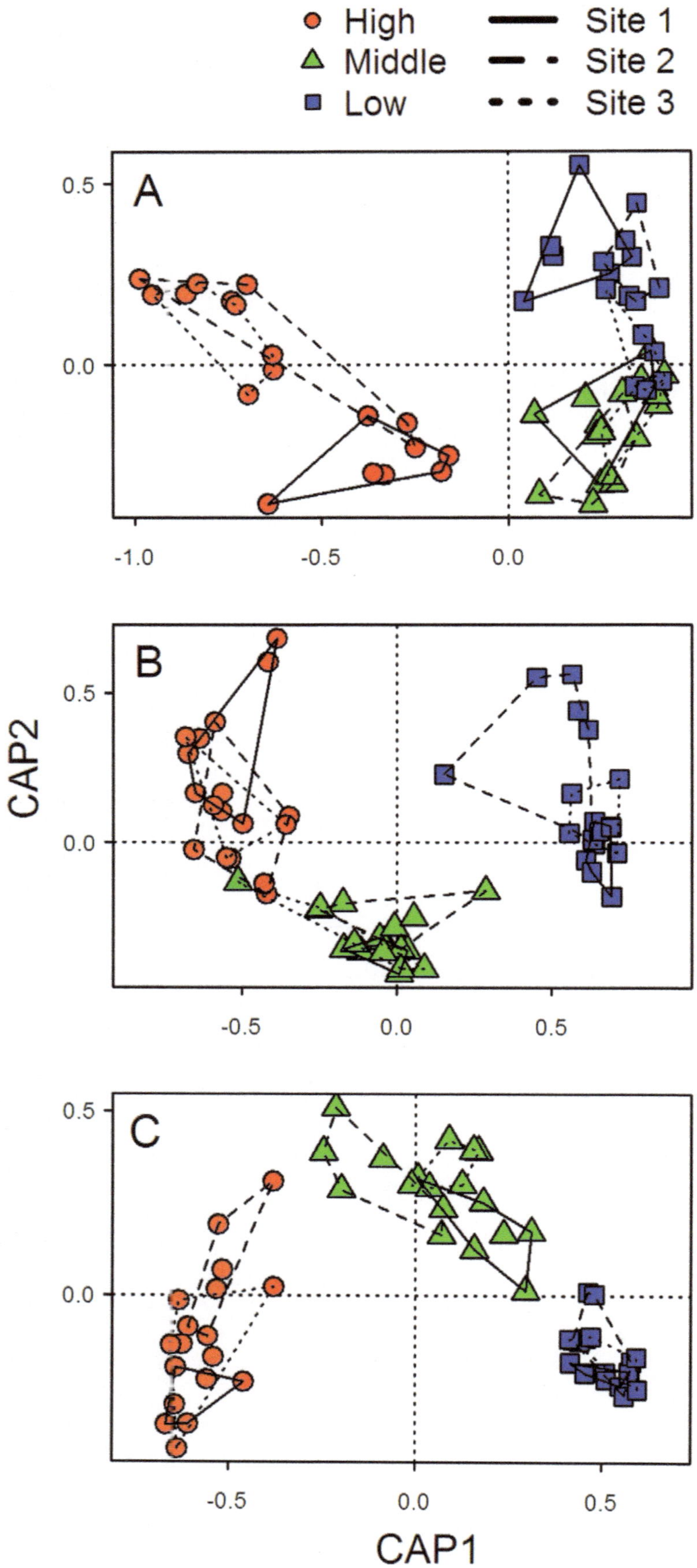
High
Middle
Low
Site 1
Site 2
Site 3
A
B
C
CAP2
CAP1
0.5
0.0
-1.0
-0.5
0.0
0.5
0.0
-0.5
0.0
0.5
0.5
0.0
-0.5
0.0
0.5

Figure 2. Canonical analysis of principal coordinates (CAP) ordinations showing the axes that best discriminate species assemblages at the 3 intertidal elevations surveyed on the 3 study shores: (A) Bunker, (B) Kringel, and (C) Nord-Ost-Hafen. The first and second canonical axes explained 63% and 14% of the total variation, respectively, in Bunker, 56% and 18% respectively in Kringel, and 62% and 16% respectively in Nord-Ost-Hafen.

significantly higher than patch- and site-scale horizontal variation and significantly lower than shore-scale horizontal variation (Fig. 4; Dunnett's tests, $P<0.05$).

Discussion

The pronounced vertical variation in community structure found on Helgoland shores corroborates previous findings for other intertidal systems [8,47]. The observed vertical differences are likely primarily a result of the marked variation in environmental stress known to occur between low and high elevations [7]. Now, fine-scale horizontal variation in species assemblages is a common attribute of marine benthic habitats. The evidence comes from studies conducted in subtidal areas [48,49], soft bottoms [50], coral reefs [51], and seagrass beds [52]. Yet, our results showed that vertical variation in community structure across elevations was higher than horizontal variation measured at fine spatial scales in terms of sampling interval (patches and sites). Conversely, vertical variation was lower than variation measured at our largest horizontal scale (shores). These findings suggest that broad-scale structuring processes along the coastline may be particularly relevant for intertidal community organization in Helgoland. In addition, broad-scale processes arise as being as important (or more) in determining community structure as the abiotic stresses that act so pervasively on marine rocky shores across the vertical elevation gradient [7].

The vertical gradient of abiotic stress that occurs in rocky intertidal habitats determines changes in the intensity of interspecific interactions that also influence vertical patterns in community structure [11,20,21]. Previous experimental work conducted in Helgoland suggested that herbivory by *Littorina littorea* and competition for space between mussels and *Fucus* spp. structured communities at middle elevations, while predation by *Carcinus maenas* shaped communities at low elevations [53]. More recently, however, the introduction to Helgoland shores of potentially dominant species, such as *Mastocarpus stellatus,* has been followed by a reduction in the abundance of mussels and an associated change in community structure [54]. Although new experimental work is therefore needed to understand what interspecific interactions are currently structuring intertidal communities in Helgoland, it is apparent that biological relationships still play a major role in determining species composition together with abiotic stresses across elevations.

It is now relevant to ask what factors might have determined the larger horizontal variation in community structure measured at the shore scale compared with vertical variation across elevations. On Helgoland Island, propagules are transported away from coastal waters [55], suggesting that propagule retention on local shores is probably low. Thus, limited connectivity among shores is a potential mechanism that may have led to the observed high horizontal variability across shores. This notion is in line with work

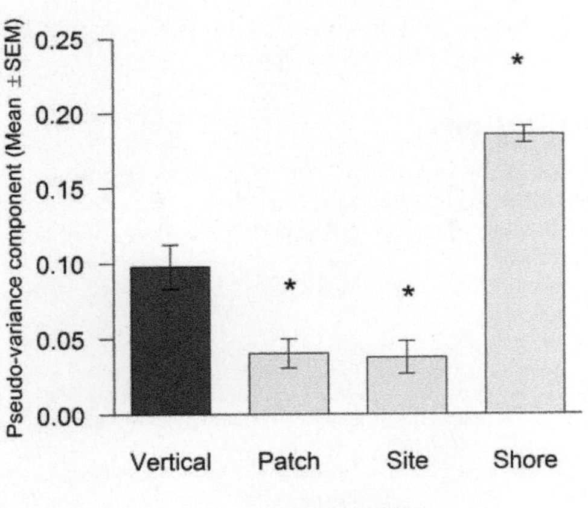

Figure 3. Vertical variation in community structure and horizontal variation in community structure for the 3 spatial scales measured in terms of sampling interval (mean ± SEM, n = 6). The asterisks over the bars for the horizontal scales indicate significant differences relative to vertical variation (Dunnett's tests, $P<0.05$).

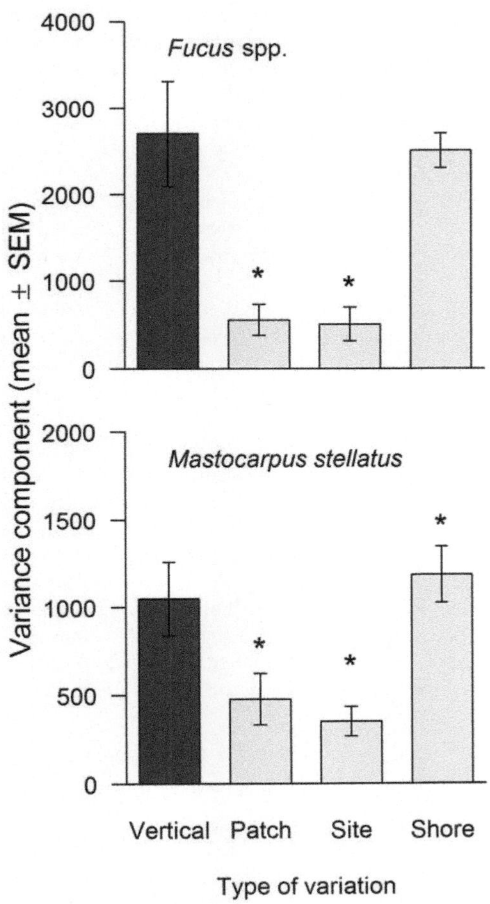

Figure 4. Vertical variation in the abundance of dominant taxa (*Fucus* spp. and *Mastocarpus stellatus*) and horizontal variation in such abundances for the 3 spatial scales measured in terms of sampling interval (mean ± SEM, n = 6). An asterisk over a bar for a horizontal scale indicates a significant difference relative to vertical variation (Dunnett's tests, $P<0.05$).

conducted in intertidal systems from other oceanic islands (e.g. the Azores) [5] and in island coral reefs [56], where dispersal limitation leads to broad-scale variation in macrobenthic assemblages and fish populations, respectively. Scale-dependent effects derived from recruitment variability have been described for rocky shores [3,57], indicating that studies on the spatial scales of propagule availability might help to explain the variation noted across shores in Helgoland.

Foundation species, such as *Fucus* spp. and *M. stellatus,* are known to have strong effects on the assembly, maintenance, and resilience of community structure [14,58,59]. Intertidal algal canopies can have positive effects on understory species by alleviating thermal and desiccation stress during low tides [60,61], but also negative effects through pre-emption of space or the action of sweeping fronds [19,62]. Moore et al. [63] suggested that broad-scale alterations in the distribution of fucoid seaweeds can have profound effects on community-level properties due to variation in shelter availability for benthic grazers. In addition, Bertocci et al. [59] showed that the consequences of removal of fucoid canopies for community structure depend on the presence or absence of additional destructive events. On the other hand, increased cover of *M. stellatus* negatively influences post-settlement mortality of *F. serratus* at Bunker [64], in a similar way as *Chondrus crispus*, morphologically similar to *M. stellatus*, affects recruitment of *F. evanescens* in eastern Canada [65]. On our study shores, both *M. stellatus* and *Fucus* spp. are often the most abundant taxa [48]. Our analyses indicated that *M. stellatus* showed the same spatial patterns of variability in abundance than the patterns of variation found for community structure, hinting for potential canopy effects of this bioengineer on many other species. *Fucus* spp. also showed a higher degree of variation in abundance at the shore scale than at the two finer horizontal scales, again suggesting that the dominant canopy-forming seaweeds may have strongly contributed to making a large broad-scale horizontal variation in community structure.

Physical attributes of marine shores may also affect broad-scale patterns in community variability. For example, differences in substrate topography and roughness may influence species composition through indirect effects on herbivore activity [66,67], which can vary across spatial scales [21]. Although the 3 studied shores in Helgoland are rocky, granite substrate is the norm at Nord-Ost-Hafen, concrete substrate predominates with sand suspended in seawater due to a nearby beach at Kringel, and concrete substrate predominates over a matrix of sandstone at Bunker. Thus, experimentally evaluating the possible effects of substrate type on community structure could be a useful next step to understanding shore-scale effects in Helgoland. Physical stresses such as wave action are also known to influence community structure at the shore scale [47]. In Helgoland, *in situ* measurements using dynamometers have revealed that wave exposure is higher at Bunker than at Nord-Ost-Hafen and Kringel, being similar between these last two shores [46]. Thus, wave exposure might explain the horizontal variation at the shore scale to a certain degree. Experiments will also be needed to elucidate the role of wave action in generating shore-scale variation in intertidal species composition in Helgoland.

Our study conducted on cold-temperate NE Atlantic shores supports the previous findings of a study done on warm-temperate NW Mediterranean shores [28]. The fact that similar results were obtained from systems largely different in terms of physical (e.g. tidal range, sea surface temperature, productivity) and biotic (e.g. species composition) conditions could be interpreted as evidence that either (1) processes producing broad-scale variability operate similarly across different environmental conditions and species pools, or that (2) different processes may lead to similar patterns in spatial variability. The identity and functional characteristics of species strongly influence ecosystem functioning and seem to determine structural processes in natural communities [68]. Therefore, it is likely that different processes may have led to strikingly similar patterns, as demonstrated in comparative studies conducted across northern Atlantic coasts [69].

In summary, ours is the second study comparing ecological variation across a well known vertical gradient of environmental stress (the intertidal elevation gradient) with variation occurring at a variety of spatial scales (in terms of sampling interval) along the coastline. Our multivariate analyses have indicated that vertical variation in community structure was larger than horizontal variation at fine scales but smaller than that at broad scales. In turn, the univariate analyses on dominant taxa suggest that variation in the abundance of canopy-forming species may have contributed to generating the large degree of variation in community structure observed at the shore scale. In other words, broad-scale processes may add an important amount of variability to the pervasive small-scale patchiness observed along marine shores. Future studies on species distribution and community structure should benefit from applying a multi-scale approach in the face of increasing frequency and intensity of climate-change related impacts on natural communities and resources.

Acknowledgments

We thank Elise A. Keppel and two anonymous reviewers for their comments on earlier versions of the manuscript.

Author Contributions

Conceived and designed the experiments: NV RAS MM. Performed the experiments: RAS MM ASK. Analyzed the data: NV. Contributed reagents/materials/analysis tools: NV RAS MM ASK. Wrote the paper: NV RAS MM.

References

1. Sarr DA, Hibbs DE, Huston MA (2005) A hierarchical perspective of plant diversity. Q Rev Biol 80: 187–212.
2. Lawton JH (1999) Are there general laws in ecology? Oikos 84: 177–192.
3. Underwood AJ, Chapman MG (1996) Scales of spatial patterns of distribution of intertidal invertebrates. Oecologia 107: 212–224.
4. Fraschetti S, Terlizzi A, Benedetti-Cecchi L (2005) Patterns of distribution of marine assemblages from rocky shores: evidence of relevant scales of variation. Mar Ecol Prog Ser 296: 13–29.
5. Martins GM, Thompson RC, Hawkins SJ, Neto AI, Jenkins SR (2008) Rocky intertidal community structure in oceanic islands: scales of spatial variability. Mar Ecol Prog Ser 356: 15–24.
6. Stephenson TA, Stephenson A (1949) The universal features of zonation between tide-marks on rocky coasts. J Ecol 37: 289–305.
7. Menge BA, Branch GM (2001) Rocky intertidal communities. In: Bertness MD, Gaines SD, Hay ME, eds. Marine community ecology: Sinauer Associates. pp 221–251.
8. Benedetti-Cecchi L, Bertocci I, Vaselli S, Maggi E (2006) Temporal variance reverses the impact of high mean intensity of stress in climate change experiments. Ecology 87: 2489–2499.
9. Connell JH (1961) The influence of interspecific competition and other factors on the distribution of the barnacle *Chathamalus stellatus*. Ecology 42: 710–723.
10. Coleman RA, Underwood AJ, Benedetti-Cecchi L, Aberg P, Arenas F, et al. (2006) A continental scale evaluation of the role of limpet grazing on rocky shores. Oecologia 147: 556–564.
11. Paine RT (1974) Intertidal community structure. Oecologia 15: 93–120.
12. Bertness MD, Callaway R (1994) Positive interactions in communities. Trends Ecol Evol 9: 191–193.
13. Burnaford JL (2004) Habitat modification and refuge from sublethal stress drive a marine plant-herbivore association. Ecology 85: 2837–2849.
14. Bruno JF, Stachowicz JJ, Bertness D (2003) Inclusion of facilitation into ecological theory. Trends Ecol Evol 18: 119–125.

15. Menge BA, Sutherland JP (1987) Community regulation: variation in disturbance, competition, and predation in relation to environmental stress and recruitment. Am Nat 130: 730–757.
16. Scrosati RA, van Genne B, Heaven CS, Watt CA (2011) Species richness and diversity in different functional groups across environmental stress gradients: a model for marine rocky shores. Ecography 34: 151–161.
17. Noda T (2009) Metacommunity-level coexistence mechanisms in rocky intertidal sessile assemblages based on a new empirical synthesis. Popul Ecol 51: 41–55.
18. Sousa WP (1979) Disturbance in marine intertidal boulder fields: the nonequilibrium maintenance of species diversity. Ecology 60: 1225–1239.
19. Kiirikki M (1996) Experimental evidence that *Fucus vesiculosus* (Phaeophyta) controls filamentous algae by means of the whiplash effect. Eur J Phycol 31: 61–66.
20. Benedetti-Cecchi L, Menconi M, Cinelli F (1999) Pre-emption of the substratum and the maintenance of spatial pattern on a rocky shore in the northwest Mediterranean. Mar Ecol Prog Ser 181: 13–23.
21. Díaz ER, McQuaid CD (2011) A spatially explicit approach to trophic interactions and landscape formation: patchiness in small-scale variability of grazing effects along an intertidal stress gradient. J Ecol 99: 416–430.
22. Scrosati R, Heaven C (2007) Spatial trends in community richness, diversity, and evenness across rocky intertidal environmental stress gradients in eastern Canada. Mar Ecol Prog Ser 342: 1–14.
23. Benedetti-Cecchi L, Bulleri F, Cinelli F (2000) The interplay of physical and biological factors in maintaining mid-shore and low-shore assemblages on rocky coasts in the north-west Mediterranean. Oecologia 123: 406–417.
24. Broitman BR, Kinlan BP (2006) Spatial scales of benthic and pelagic producer biomass in a coastal upwelling ecosystem. Mar Ecol Prog Ser 327: 15–25.
25. Blanchette CA, Wieters EA, Broitman BR, Kinlan BP, Schiel DR (2009) Trophic structure and diversity in rocky intertidal upwelling ecosystems: A comparison of community patterns across California, Chile, South Africa and New Zealand. Prog Oceanogr 83: 107–116.
26. Doherty P, Fowler T (1994) An empirical test of recruitment limitation in a coral reef fish. Science 263: 935–939.
27. Navarrete SA, Wieters EA, Broitman BR, Castilla JC (2005) Scales of benthic-pelagic coupling and the intensity of species interactions: From recruitment limitation to top-down control. P Natl Acad Sci USA 102: 18046–18051.
28. Benedetti-Cecchi L (2001) Variability in abundance of algae and invertebrates at different spatial scales on rocky sea shores. Mar Ecol Prog Ser 215: 79–92.
29. Kornmann P, Sahling P-H (1977) Meeresalgen von Helgoland: Benthische Grün-, Braun- und Rotalgen. Helg Mar Res 29: 1–289.
30. Kornmann P, Sahling P-H (1994) Meeresalgen von Helgoland: Zweite Ergänzung. Helg Mar Res 48: 365–406.
31. Luther G (1987) Seepocken der deutschen Küstengewässer. Helg Mar Res 41: 1–43.
32. Hayward PJ, Ryland JS (1995) Handbook of the marine fauna of north-west Europe. Oxford: Oxford University Press.
33. Maggs CA, Hommersand MH (1993) Seaweeds of the British Isles: Rhodophyta, part 3 A Ceramiales. London: The Natural History Museum.
34. Thompson RM, Hemberg M, Starzomski BM, Shurin JB (2007) Trophic levels and trophic tangles: The prevalence of omnivory in real food webs. Ecology 88: 612–617.
35. Konar B, Iken K, Edwards M (2009) Depth-stratified community zonation patterns on Gulf of Alaska rocky shores. Mar Ecol 30: 63–73.
36. Scrosati R (2005) Review of studies on biomass-density relationships (including self-thinning lines) in seaweeds: Main contributions and persisting misconceptions. Phycol Res 53: 224–233.
37. Dethier MN, Graham ES, Cohen S, Tear LM (1993) Visual versus random-point percent cover estimations: 'objective' is not always better. Mar Ecol Prog Ser 96: 93–100.
38. Anderson MJ, Willis TJ (2003) Canonical analysis of principal coordinates: a useful method of constrained ordination for ecology. Ecology 84: 511–525.
39. Underwood AJ, Chapman MG (1998) A method for analysing spatial scales of variation in composition of assemblages. Oecologia 117: 570–578.
40. Anderson MJ (2001) A new method for non-parametric multivariate analysis of variance. Austral Ecol 26: 32–46.
41. McArdle BH, Anderson MJ (2001) Fitting multivariate models to community data: a comment on distance-based redundancy analysis. Ecology 82: 290–297.
42. Quinn GP, Keough MJ (2002) Experimental design and data analysis for biologists. Cambridge: Cambridge University Press.
43. Fletcher DJ, Underwood AJ (2002) How to cope with negative estimates of components of variance in ecological field studies. J Exp Mar Biol Ecol 273: 89–95.
44. Clarke KR, Warwick RM (2001) Change in marine communities: an approach to statistical analysis and interpretation Plymouth: PRIMER-E Ltd .
45. R Development Core Team (2011) R: A language and environment for statistical computing. 2.12.1 ed Vienna: R Foundation for Statistical Computing.
46. Scrosati R, Knox A, Valdivia N, Molis M (2011) Species richness and diversity across rocky intertidal elevation gradients in Helgoland: testing predictions from an environmental stress model. Helg Mar Res 65: 91–102.
47. Heaven CS, Scrosati RA (2008) Benthic community composition across gradients of intertidal elevation, wave exposure, and ice scour in Atlantic Canada. Mar Ecol Prog Ser 369: 13–23.
48. Smale DA (2010) Monitoring marine macroalgae: the influence of spatial scale on the usefulness of biodiversity surrogates. Divers Distrib 16: 985–995.
49. Terlizzi A, Anderson MJ, Fraschetti S, Benedetti-Cecchi L (2007) Scales of spatial variation in Mediterranean subtidal sessile assemblages at different depths. Mar Ecol Prog Ser 332: 25–39.
50. Ysebaert T, Herman PMJ (2002) Spatial and temporal variation in benthic macrofauna and relationships with environmental variables in an estuarine, intertidal soft-sediment environment. Mar Ecol Prog Ser 244: 105–124.
51. Hughes TP, Baird AH, Dinsdale EA, Moltschaniwskyj NA, Pratchett MS, et al. (1999) Patterns of recruitment and abundance of corals along the Great Barrier Reef. Nature 397: 59–63.
52. Piazzi L, Balata D, Cinelli F, Benedetti-Cecchi L (2004) Patterns of spatial variability in epiphytes of *Posidonia oceanica* - Differences between a disturbed and two reference locations. Aquat Bot 79: 345–356.
53. Janke K (1990) Biological interactions and their role in community structure in the rocky intertidal of Helgoland (German Bight, North Sea). Helg Mar Res 44: 219–263.
54. Bartsch I, Tittley I (2004) The rocky intertidal biotopes of Helgoland: present and past. Helg Mar Res 58: 289–302.
55. Giménez L, Dick S (2007) Settlement of shore crab *Carcinus maenas* on a mesotidal open habitat as a function of transport mechanisms. Mar Ecol Prog Ser 338: 159–168.
56. Swearer SE, Caselle JE, Lea DW, Warner RR (1999) Larval retention and recruitment in an island population of a coral-reef fish. Nature 402: 799–802.
57. Methratta ET, Petraitis PS (2008) Propagation of scale-dependent effects from recruits to adults in barnacles and seaweeds. Ecology 89: 3128–3137.
58. Arenas F, Sanchez I, Hawkins SJ, Jenkins SR (2006) The invasibility of marine algal assemblages: role of functional diversity and identity. Ecology 87: 2851–2861.
59. Bertocci I, Arenas F, Matias MG, Vaselli S, Araújo R, et al. (2010) Canopy-forming species mediate the effects of disturbance on macroalgal assemblages on Portuguese rocky shores. Mar Ecol Prog Ser 414: 107–116.
60. Bertness D, Leonard G, Levine J, Schmidt P, Ingraham A (1999) Testing the relative contribution of positive and negative interactions in rocky intertidal communities. Ecology 80: 2711–2726.
61. Dayton PK (1971) Competition, disturbance, and community organization: the provision and subsequent utilization of space in a rocky intertidal community. Ecol Monogr 41: 351–389.
62. Jenkins SR, Norton TA, Hawkins SJ (2004) Long term effects of *Ascophyllum nodosum* canopy removal on mid shore community structure. J Mar Biol Assoc UK 84: 327–329.
63. Moore P, Hawkins SJ, Thompson RC (2007) Role of biological habitat amelioration in altering the relative responses of congeneric species to climate change. Mar Ecol Prog Ser 334: 11–19.
64. Hahn A (2007) Effekte einer invasiven Spezies und natürlicher Weidegänger auf die Besiedlung und das Überleben von habitat-bildenden Makroalgen an der Helgoländer Küste. Leipzig: University Leipzig. 91 p.
65. Worm B, Chapman ARO (1998) Relative effects of elevated grazing pressure and competition from a red algal turf on two post-settlement stages of *Fucus evanescens* C. Ag. J Exp Mar Biol Ecol 220: 247–268.
66. Underwood AJ (2004) Landing on one's foot: small-scale topographic features of habitat and the dispersion of juvenile intertidal gastropods. Mar Ecol Prog Ser 268: 173–182.
67. Johnson MP, Hanley ME, Frost NJ, Mosley MWJ, Hawkins SJ (2008) The persistent spatial patchiness of limpet grazing. J Exp Mar Biol Ecol 365: 136–141.
68. Hooper DU, Chapin FS, Ewel JJ, Hector A, Inchausti P, et al. (2005) Effects of biodiversity on ecosystem functioning: A consensus of current knowledge. Ecol Monogr 75: 3–35.
69. Jenkins SR, Moore P, Burrows MT, Garbary DJ, Hawkins SJ, et al. (2008) Comparative ecology of north Atlantic shores: Do differences in players matter for process? Ecology 89: S3–S23.

A Snapshot of the Population Structure of *Branchiostoma lanceolatum* in the Racou Beach, France, during Its Spawning Season

Yves Desdevises, **Vincent Maillet**, **Michael Fuentes, Hector Escriva***

CNRS UMR 7232, UPMC Univ Paris 06, Observatoire Océanologique, Banyuls-sur-Mer, France

Abstract

A methodology for inducing spawning in captivity of the lancelet *Branchiostoma lanceolatum* has been developed recently with animals collected at the Racou beach, in the southern coast of France. An increasing amount of laboratories around the world are now working on the evolution of developmental mechanisms (Evo-Devo) using amphioxus collected in this site. Thus, today, the development of new aquaculture techniques for keeping amphioxus in captivity is needed and the study of the natural conditions at which amphioxus is exposed in the Racou beach during their spawning season becomes necessary. We have investigated the amphioxus distribution, size frequency, and population structure in the Racou beach during its natural spawning season using multivariate methods (redundancy analysis and multiple regression). We found a clear preference of amphioxus for sandy sites, something that seems to be a general behaviour of different amphioxus species around the world. We have also estimated the amphioxus growth rate and we show how the animals are preferentially localized in shallow waters during April and June.

Editor: Johannes Jaeger, Centre for Genomic Regulation (CRG), Universitat Pompeu Fabra, Spain

Funding: This study was supported by L'Agence Nationale de la Recherche (ANR) BLAN 171601 and BLAN 123402 and by CNRS PEPS (Projet Exploratoire Premier Soutien) SDV-D-2009-1926 (http://www.cnrs.fr/insb/carrieresemplois/peps-2009.htm). The funders had no role in study design, data collection and analysis, decision to publish, or preparation of the manuscript.

Competing Interests: The authors have declared that no competing interests exist.

* E-mail: hector.escriva@obs-banyuls.fr

These authors contributed equally to this work.

Introduction

Amphioxus (or lancelets) from the subphylum Cephalochordata have long been considered interesting organisms due to their phylogenetic position close to the vertebrates [1,2]. Recent studies have shown that amphioxus are in fact basal chordates, urochordates being the closest extant relatives to vertebrates [3]. However, in spite of its basal position within the chordate lineage, at the anatomical, genetic, and genomic levels, amphioxus is vertebrate-like but simpler. Both amphioxus and vertebrate embryos have a dorsal hollow nerve cord, segmentally repeated trunk muscles, a notochord, and a pharynx perforated with gill slits. Moreover, the amphioxus genome includes representatives of nearly all the vertebrate gene families, but has fewer gene duplications because it diverged from the vertebrate lineage before the two rounds of complete genome duplication occurred [4,5]. These anatomical and genomic characteristics have rendered amphioxus an invaluable animal model for studies on the evolution of developmental mechanisms during the invertebrate chordate-to-vertebrate transition [1,2].

Amphioxus is widespread in tropical and temperate seas and the adult lives burrowed in the sand, gravel, or shell deposits. Traditionally, myomere counts, counts of fin chambers, the position of the atriopore, the position of the anus, and other qualitative differences in notochord and caudal fin shape, have been used to distinguish up to 23 species within the genus *Branchiostoma* [6]. However, to date, developmental studies have been carried out only on three amphioxus species (the European *Branchiostoma lanceolatum*, the East Asian *B. belcheri*, and the Floridian-Caribbean *B. floridae*). To obtain amphioxus embryos, adult animals are collected from the field during their ripe season and allowed (or stimulated) to spawn in the laboratory. In any given year, dates of laboratory spawning have been limited by two factors. First, natural populations of these three most studied species of amphioxus are ripe, at most, for only two to four months each year and, second, even when apparently ripe, animals spawn only at unpredictable intervals of several days. Induction of spawning is not possible for the Asian species (*B. blecheri*) and is possible in the American species (*B. floridae*) only on the natural spawning days (once every 7–10 days) by an electric shock [7]. However, a few years ago, a method for inducing spawning on a daily basis (during the natural spawning season) was developed for the European species *B. lanceolatum* [8,9]. Since then, several research teams have started to work on the European species and particularly on the amphioxus collected at the Racou beach in southern France.

Even if some studies on the amphioxus ecology have been performed, particularly for *B. nigeriense* on the west coast of Africa [10,11], and for *B. floridae* from Tampa Bay, Florida [12,13], very few data exist on the distribution, habitat preference and ecology of *B. lanceolatum*, particularly at the Racou beach location [8,9,14,15]. Reliable ecological information, particularly during the natural spawning season, may be instrumental for the development of adapted culture techniques in the laboratory. This work describes some aspects of the population structure, the spatial distribution, and the relationships between biotic and non-biotic characteristics for *B. lanceolatum* in the Racou beach during this important period of the amphioxus life cycle. We show a clear preference of the animals for sandy sites and we describe how the amphioxus population is structured. We also estimate the lifespan of the amphioxus as well as their growth rate both in captivity and in the wild.

Materials and Methods

Study area and amphioxus collection

The study was conducted in the Racou beach close to Argelès-sur-Mer, France, since previous studies showed the presence of an important amphioxus population at this location [9,15]. Collection was performed with a research boat from the "Laboratoire Arago, Observatoire Océanologique, Université Pierre et Marie Curie/C.N.R.S," Banyuls-sur-Mer, France. Bottom sediment grab samples containing the amphioxus were collected using a Van Venn-type dredge [9]. The maximum surface and volume covered by the Van Venn dredge was 0.1024 m^2 and 30 l. Adult amphioxus (>30 mm) were collected from the sand by sieving and small juveniles (<30 mm) by treating the sand in a known volume of sea water with a 10^{-4} dilution of clove oil (85–95% eugenol), an anaesthetic from which the animals recover when returned to sea water [16]. This treatment anesthetizes the animals allowing easy collection from the supernatant water with a 0.2 mm sieve. Once animals (both adults and juveniles) were collected from the sand they were placed in classical culture conditions until further manipulation [8,9,17]. Then, amphioxus were counted and total length was measured. The volume of sand was also measured and density of the population was calculated both as individuals/litre of sediment and as individuals/m^2.

Sampling design

The diversity and distribution of amphioxus was assessed by sampling multiple sites (a total of 50) every 100 m along transects separated by 100 m starting from the site R5.1 (42°32′54″N and 03°03′688″E) in a total surface of 360000 m^2 (see Fig. 1). Transects were oriented north-east in order to be perpendicular to the sea shore, in an angle of 62° with the azimuth. The number of transects was determined from prior knowledge of the studied area. The southern limit of the studied area was established by the presence of an area of rocks and *Posidonia* sp., and north-east limit was established by the presence of an area of mud (15 extra points were sampled in this area and confirmed the absence of amphioxus in mud, data not shown). Transects have been placed on a map and marked in the field using a global positioning system (GPS). Amphioxus sampling was conducted twice, a first sampling at the end of April 2008 (24th April) and a second sampling at the end of June 2008 (19th June) for radials R3 to R5, but only once for radial R1 and R2 (on 24th April 2008). Since data were collected twice only in radial 3, 4 and 5, we decided to perform our analyses only on these three radials (Fig. 1 shows radials where samples were obtained twice).

Amphioxus culture in captivity

Amphioxus were induced to spawn in captivity in June 2007 by the temperature shock method [9]. Amphioxus embryos were cultured in Petri dishes until they were 1 month old (8–10 gill slits) at a constant temperature of 20°C and daily feeding with 30,000–80,000 cells/ml/day of mixed algae (1/3 of *Dunaliella tertiolecta*, 1/3 of *Isochrysis galbana*, and 1/3 of *Tetraselmis suecica*). One-month old larvae were then changed to a 50 l aquarium with sand (temperature ranged between 10°C in winter to 21°C in summer) and with the same feeding conditions. 1/3 of the water volume was changed twice a week and larvae present in the changed water were counted, measured and reintroduced into the aquarium. Following metamorphosis (6–7 weeks post-fertilization), the larvae migrated into the sand. Once per month, the larvae were collected

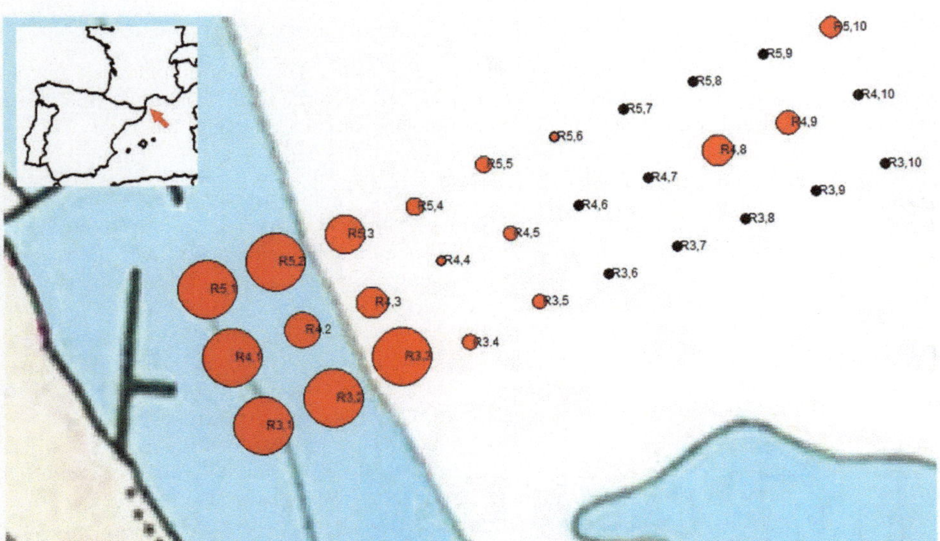

Figure 1. Study area showing the position of the sampling stations. The inset map shows the location of the sampling stations in the Racou beach, France. The size of each red circle is proportional to the density of amphioxus individuals (in ind./l). Stations predominantly composed by silt with a total absence of amphioxus individuals are shown as black dots.

from the sand (the small volume of sand in the tank allowed collection of juveniles without the need of an anaesthetic), counted, measured and reintroduced into the aquarium.

Non-biotic variables

Temperature measurements were performed with a thermo data logger (HOBO Pendant Temp/Light Intensity Data logger, ONSET Computer Corporation, Bourne, MA). This battery-powered thermometer can report 6,500 consecutive data records. The initial settings, including measurement intervals and data acquisition, were set with a computer using the software platform HOBOwares (V. 2.1.1_18), ONSET Computer Corporation. Salinity measurements were determined using a salinity refractometer (American Optical Corp., Buffalo, N.Y.). Particle size analysis was carried out with two different methods: (i) by using a Mastersizer 2000 (Malvern Co. Ltd, UK) whose analysis range is between 0.2 and 2000 µm, and (ii) by sieving dried samples (72 hrs at 65°C) when the sample size was higher than 1600 µm. Depth measurements were performed with a calibrated echosounder (Furuno FCV582L).

Data analysis

Calculation of grain size parameters was performed as proposed by Folk And Ward [18]. Four parameters were calculated: mean ($M = (\Phi_{16}+\Phi_{50}+\Phi_{84})/3$) and median value ($\Phi_{50}$ or d(0.5)), standard deviation ($\sigma_1 = (\Phi_{84}-\Phi_{16})/4+(\Phi_{95}-\Phi_5)/6.6$), skewness ($Sk_1 = (\Phi_{16}+\Phi_{84}-2\Phi_{50})/2(\Phi_{84}-\Phi_{16})+(\Phi_5+\Phi_{95}-2\Phi_{50})/2(\Phi_{95}-\Phi_5)$) and kurtosis ($Kg = (\Phi_{95}-\Phi_5)/2.44(\Phi_{75}-\Phi_{25})$). These calculations allowed us to classify sediments into three major classes, as proposed by Cailleux [19]: gravel (>1250 µm), sand (200–1250 µm) and silt (0–200 µm). Temperature and salinity were not used in our analyses due to their constant values throughout the 50 studied sites (i.e. temperature 15±0.3°C in depth; salinity 36–37 psu). Counts of individual amphioxus at sampling sites were used to build the amphioxus abundance matrix.

To identify environmental variables linked to the abundance and size of amphioxus in the study area, biotic variables (size, density, and number of individuals for juveniles and adults) were related to non-biotic variables (particle size -sand, silt, gravel- and depth) using redundancy analysis (RDA). Redundancy analysis is a direct gradient method (constrained ordination) that can be described as a series of multiple linear regressions of each response variable on explanatory variables, followed by a principal component analysis on fitted values [20]. The significance of the RDA (to assess if environmental variables account for a significant portion of the variation of the biotic variables) was carried out *via* a permutation procedure. The same procedure was used to test the significance of the canonical axes. RDA and the tests of significance were computed using the functions 'rda' and 'anova.cca' from the 'vegan' library [21] of the R statistical language [22]. Amphioxus size (on individuals) and density (on stations) were also separately studied using multiple regressions against environmental variables (depth and particle size (silt, sand, gravel)) followed by a forward selection of significant explanatory variables (see [20]). Due to non normality of variables, permutational testing were performed using the 'multRegress' and 'forward.sel' functions in R, written respectively by P. Legendre and S. Dray (available at http://www.bio.umontreal.ca/legendre/indexEn.html). Age and growth in length of cohorts of amphioxus was calculated using the successive maximum method [23]. All tests were performed using permutational procedures with 999 permutations.

Results and Discussion

Environmental characteristics

The slope of the bottom is the same for the five radials in the study area. The specific depth of each station is shown in Table S1. Tides are not significant in the Racou beach and depths are constant at each point during the day. Thus the closest points to the seashore (R1.1 to R5.1) (see Fig. 1 for R3.1 to R5.1) were at 5 m deep and the five most distant points to the seashore (R3.10 to R5.10) at 18 m deep. However the slope is not constant through the radials. Thus slope is 4% from 150 to 400 m away from the seashore and then decrease to 0.2% from 400 to 600 m away from the seashore.

A zone predominantly composed by silt ($\Phi_{50} = 100.8\pm19.8$ µm) is present from 15 m to 18 m deep (sites R1.2–1.7; R1.9–1.10; R2.5–2.10; R3.6–3.10; R4.6–4.7; R4.10; R5.7–R5.9) (see black dots in Fig. 1). Since no amphioxus were found in these silty sites, they were excluded from our analyses (see below). Sandy sediments ($\Phi_{50} = 834.1\pm65.8$ µm) are present in two zones, from the seashore (4 m deep) to 15 m deep and a second zone in the middle of the silt zone at 18 m deep (sites R1.8; R2.1–2.4; R3.1–3.5; R4.1–4.5; R4.8–4.9; R5.1–R5.6) (red dots in Fig. 1). Finally, a small gravel zone ($\Phi_{50} = 1440.6$ µm) is present in only one of the 50 studied points (R5.10) at 18 m deep. These sites have been called sandy and gravel sites because they are predominantly composed by sand and gravel respectively. However, low amounts of silt are always present at each site. The granulometry changed only in a single point between April and June, R3.5, which presented a sand profile in April ($\Phi_{50} = 891.2$ µm) and a silt profile in June ($\Phi_{50} = 138.4$ µm). This change may be explained by sediment dynamics due to short-term processes, such as floods and storms, which are quite usual through the SW Gulf of Lions, including the Racou beach [24,25].

Small variations of temperature were detected with maximum differences of 1.2°C, ranging from 14.9°C in depth to 16°C at the surface, but no variations were detected in depth between the different sites. Salinity was quite constant throughout the 50 studied points with small changes ranging from 36 to 37 psu.

Distribution and size frequency of amphioxus

Since aquaculture techniques try to reproduce the conditions present in nature close to the spawning season (which expands from mid-May to mid-July [8,9]), two quantitative samplings were performed at the Racou beach, with a total sample size of 1106 amphioxus. The first one in April 2008, just before the natural spawning season starts (626 collected animals) and the second in June 2008, within the natural spawning season (480 collected animals), (see Table S1). The geographic and quantitative distribution of the amphioxus within the studied zone is shown on Fig. 1 (for transects R3–5) with a total surface colonized by amphioxus in the Racou studied zone of 20000 m^2. In April, the highest density of animals was found at station R3.2 with 11.4 ind/l or 869 ind/m^2, and the lowest was found at R5.10 with 0.7 ind/l (48 ind/m^2). In June the population structure was similar, with the highest density of animals found at stations R5.1, R4.1 and R3.2 (15.1 ind/l or 731.5/ind/m^2), and the lowest at R5.10 with 0.85 ind/l (53.3 ind/m^2). The average density of animals in the studied zone in April and June was 5.4 ind/l or 234.5 ind/m^2 and 6.5 ind/l or 255.1 ind/m^2 respectively (density at each station is indicated in Table S1). The amphioxus density in the Racou beach has already been reported twice, by Monniot [15] and by Fuentes et al [9], with 20 ind/l and 1–5 ind/l respectively. These differences may be explained by sampling differences (both in terms of sampling techniques or collection sites), or by differences in the population between different years. Other studies on sites colonized by *B. lanceolatum* have shown important density differences, from 400 animals/m^2 in Helgoland [26] to 20 animals/m^2 in the coastal waters of Netherlands [27]. Density of other amphioxus species in different locations is also highly variable, with a highest recorded density of 9000 animals/m^2 in the west African coast for *B. senegalense* [28]. One interesting point is that our study and the study by Fuentes et al. [9], were performed at the same period of the year, between April and July, which corresponds to the natural spawning season [8,9]. In contrast, Monniot [15] did not specify the period of the year at which her study was performed. Density changes at the same location depending on the period of the year have been described for *B. caribaeum* [29] and *B. lanceolatum* at Helgoland [26]. Webb suggested that the reason for these changes was probably related to the permeability of the substrate, with amphioxus moving away if the deposit becomes unsatisfactory [26]. Since mobility and other biological activities (i.e. growth, respiration, feeding) of amphioxus change with temperature, Webb suggested that in winter, low temperature inducing low metabolic rate allows amphioxus to live in finer sands, while in summer, higher temperatures increase the amphioxus metabolic rate and induce their migration to coarse or medium grade deposits [26], but this suggestion cannot be applied to *B. floridae*, since it lives in fine sand even in summer with temperatures around 30°C. Thus, an annual study of the *B. lanceolatum* distribution in the Racou beach in order to find possible changes in their distribution would be highly informative.

Body length was measured for all the collected amphioxus. The bimodal frequency distribution (Fig. 2) suggests that the whole population can be divided into two major groups. The first group represents 39.1% and 58.8% of the whole population in April and June respectively and contains juveniles of less than 3 cm long, (our unpublished observations show that ripe animals are always longer than 3 cm [8,9]). The second group contains adults (longer than 3 cm, Fig. 2) (the number of adults/juveniles at each station is shown in Table S1). Several studies have estimated the lifespan of different amphioxus species. Wells [30], Nelson [31], and Futch & Dwinell [32] suggested a 2 to 3 yrs lifespan for *B. floridae* based on the apparent number of generations present in the sampled size-frequency histograms. Chin [33] and Chen [34] suggested a maximum age of 2 to 3 yrs for *B. belcheri*. Gosselck & Spittler [35] estimated 4–5 years life-span for *B. senegalense* off North-West Africa and a considerably longer lifespan of 8 yrs was suggested by Courtney [36] for *B. lanceolatum* living in the relatively cold waters off Helgoland. For *B. lanceolatum* in the Racou beach, we have constructed length-frequency histograms showing the presence of 5 modal lengths of 12, 36, 41, 48 and 50 mm in April (Fig. 2 blue bars). A similar distribution into five modal lengths was also found in June (Fig. 2 red bars). Since the spawning season spans from May to July and considering that it has been proposed that amphioxus grow continuously during their entire life [13], these sizes may represent the average size of animals of 1 to 5 years old. Thus, *B. lanceolatum* lifespan at the Racou can be estimated to 5 years.

An interesting point is that the juvenile population (i.e. less than 3 cm long) increased their average size between April and June from 11.9±1.1 mm to 18.4±1.9 mm (compare the first blue and red peaks in Fig. 2), indicating that the growth rate of juveniles during April/June 2008 in the wild was 3–4 mm/month. The amphioxus growth rate in captivity (Fig. 3) also shows a growth rate of 3–4 mm/month for amphioxus larvae with a similar size. Previous studies on *B. floridae* showed that growth rate for juveniles was 6 mm in 12 days from the 18th to the 30th of July [13], indicating a clear difference between these two amphioxus species. *In vitro* comparisons of embryonic development at different temperatures also showed a higher growth rate for *B. floridae* than for *B. lanceolatum* [8]. Moreover, in a study of the age structure of *B. lanceolatum* in the North Sea off Helgoland (0 to 17°C yearly temperature range) and in the warmer water at the Racou (10 to 24°C yearly temperature range), Courtney reported initial growth rates of 10 and 20 mm/yr, respectively, with faster growth during the warmest months of the year [36]. This would mean that the first cohort in Fig. 2 corresponds, as we have assumed, to the 1 year old juveniles (less than 20 mm long). The lack of growth

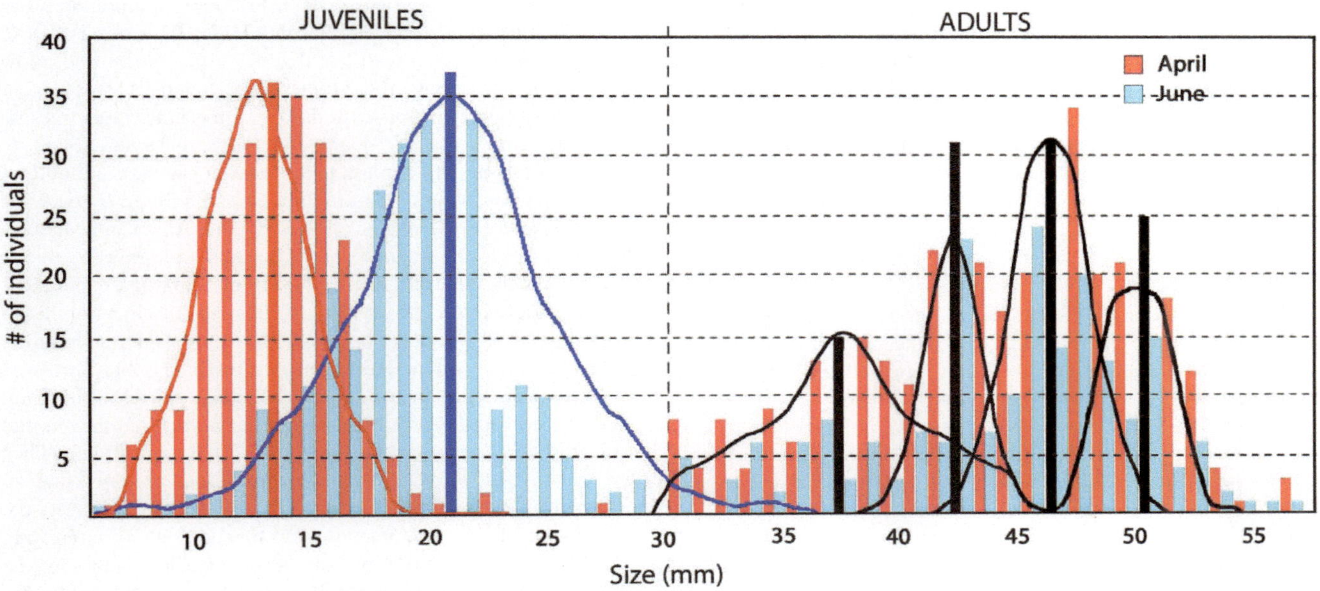

Figure 2. Size frequency distribution computed from data sampled in April (in pink) and June (in light blue). Solid bars indicate the mode of each peak, in red the first peak in April, in blue the first peak in June, in black 2nd–5th peaks in both April and June. Frequency distribution of adult animals (>3 cm long) was similar in April and June.

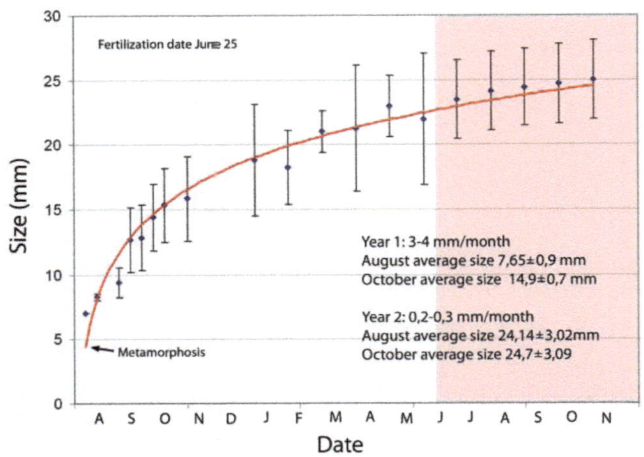

Figure 3. Growth curve of amphioxus larvae in captivity at room temperature (from 10°C in winter to 21°C in summer). The white part of the graph corresponds to the first year (animals were fertilized in late June 2007 and the graph starts when these embryos become post-metamorphic larvae in August 2007) and the pink side to the second year (2008). A minimum of 10 larvae were measured for each point (error bars = standard deviation from the mean).

between April and June in the second group of amphioxus composed of adults (longer than 3 cm, data not shown) may reflect a logarithmic decrease of growth with age, something that has been suggested for *B. lanceolatum* and other amphioxus species [35,36]. In this sense, amphioxus grown in captivity support this assumption since they show a clear decrease in the growth rate with time (from 3–4 mm/month in postmetamorphic animals to 0.2–0.3 mm/month in one year old juveniles) (see Fig. 3). Other possibilities may also explain differences in growth rate. Thus, Webb observed differences in the growth rate between nearby populations of *Branchiostoma nigeriense* [37] in Lagos, Nigeria. A population living within the lagoon grew 30 mm in three months (January–March) while two other populations closely located, one within the connexion between the lagoon and the open sea, and the second in the open sea, grew only 7 and 5 mm respectively in the same period. Webb interpreted these differences by the different food availability in these three different locations. Future laboratory experiments testing amphioxus growth at different temperatures and with different food regimes should shed light on the specific requirements for different growth rates of the same amphioxus population.

Habitat preference

Redundancy analysis (RDA) assesses how biotic factors of the amphioxus population (mean size, density, and abundances of juveniles (<30 mm), adults (>30 mm), and all individuals) are explained by environmental variables (depth and sediment granulometry). Even if temperature as well as salinity changes have been shown to be very important in the life cycle of other amphioxus species [11,38,39], we have not included these two environmental variables in the analysis because both were constant during this study. In fact, this study represents a snapshot of a precise moment in the year and it is not a dynamic view of the amphioxus population during its life cycle. Results are shown using a triplot representation (Fig. 4), displaying stations (Fig. 4, in black), response biological variables (Fig. 4, in red), and explanatory environmental variables (Fig. 4, in blue). We only used in these analyses stations were amphioxus were detected. Stations predominantly composed by silt (sites R1.2–1.7; R1.9–1.10; R2.5–2.10; R3.6–3.10; R4.6–4.7; R4.10; R5.7–R5.9), with a total absence of amphioxus, were excluded (see above). Temperature and salinity were excluded from these analyses due to their constant values at all sites. The gravel vector was also excluded from the analysis because it was almost perfectly negatively collinear to the sand vector. The RDA was globally significant (constrained (e.g. canonical) variation is 63.5%, $P = 0.001$) as well as the first canonical axis (accounting for 99.99% of the canonical variation, $P = 0.001$).

The triplot representation shows that the number of adults is strongly negatively correlated to depth (which is the main variable structuring the dataset, close to canonical axis 1), as well as the number of juveniles but in a weaker manner. The number of adults and total number of individuals are positively correlated to sand. However, while the number of adults is weakly negatively correlated to silt, juvenile abundance and density are positively correlated to silt (i.e. here it should be noted that the silt variable indicates the presence of low amounts of silt in a sample composed predominantly by sand). This is coherent with the positive link ($P = 0.01$) of size with sand (adult animals are longer and prefer sand) while size is negatively correlated to silt and density (juvenile animals are smaller and in higher density). The variable selection procedure used in multiple regressions always selected depth among the significant explanatory variables accounting for variations in body size and density, with which depth was always negatively correlated (Table 1). Silt was also selected as positively linked to size for juveniles ($P = 0.01$) and all individuals ($P = 0.02$), as well as density ($P = 0.03$). These results suggest a clear preference of amphioxus, particularly adult animals, for sandy sites in the Racou beach. Taking into account the total absence of amphioxus (both adults and juveniles) in silt, the positive link between both the number of juvenile animals and density with silt does not necessarily indicate a juvenile preference for silty sites (i.e. silt is always present in low amounts). However, the differences observed between adults and juveniles may indicate a higher tolerance for the presence of silt by juveniles than by adults. The importance of sediment type in the distribution of amphioxus has been observed repeatedly. Webb and Hill [10] showed that *B. nigeriense* preferred sand with less than 25% fine grains and a low silt content, Boschung and Gunter [40] found *B. caribaeum* in Mississippi Sound on coarse sand 90% of the time, rarely on fine sand, and never on clay, Cory and Pierce [41] also found *B. caribaeum* from South Carolina to Georgia in coarser sand and no specimens were found in silt, and Gosselck [35] found *B. senegalense* in the off-shore shelf region off North West Africa in sediments similar to gravel and mixtures of coarse sand and gravel but never in mud. Webb and Hill [10] conducted an interesting experiment by placing amphioxus (*B. nigeriense*) of different sizes in a number of graded sands (from 2 to 0.1 mm). They showed that animals placed in coarse and medium-size sands survived either burrowed in the sand or lying on the surface, but those placed in fine sand and silt died with an accumulation of sand particles in the gut and with the oral aperture blocked by a mass of grains. They concluded that amphioxus avoid fine sand bottoms due to occlusion of the oral aperture and the atriopore. Moreover, the permeability of the bottom type and the presence of circulating water between the sand grains are also controlling factors for the preference of amphioxus for sand and coarse sand. Our results also show a clear preference of *B. lanceolatum* in the Racou beach for sandy sites, suggesting that most if not all the different amphioxus species in the world need such a permeable bottom and avoid fine grains that may block their oral aperture. *Branchiostoma floridae*

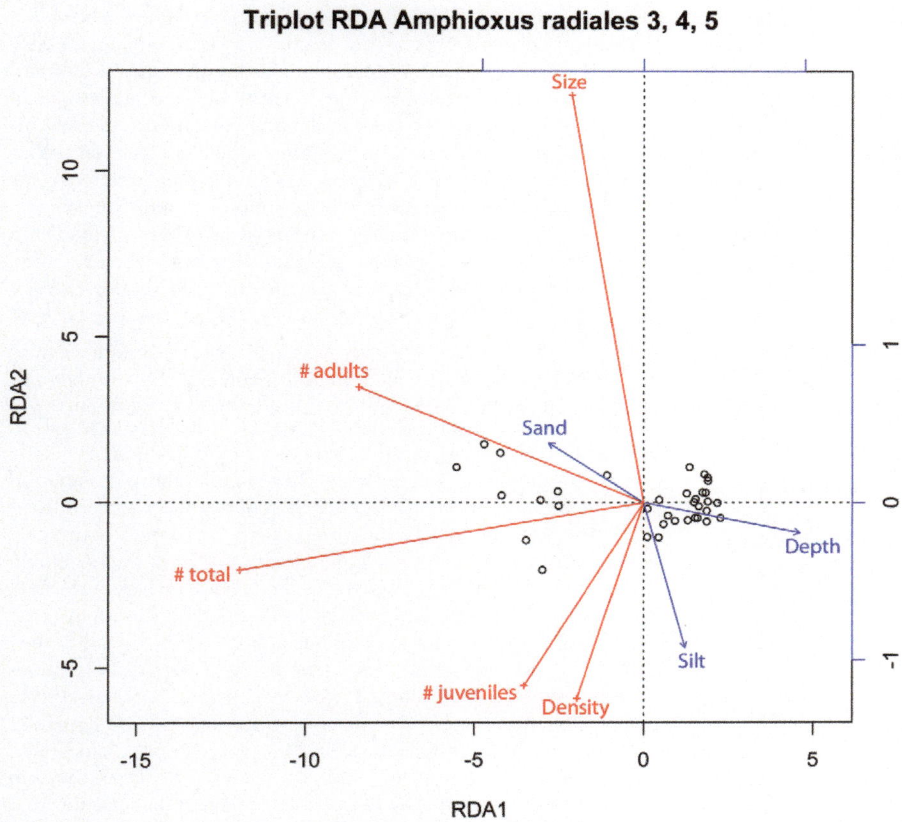

Figure 4. Triplot showing the result of the redundancy analysis of amphioxus size, number (of adults, juveniles, and all individuals) and density (in ind./l) constrained by environmental variables (percentage of sand and silt, and depth). Canonical axes 1 and 2 are represented (RDA1 and RDA2), accounting in total for 63.5% of the variation of the original dataset.

Table 1. Result of multiple regression of a response variable (body size or density) against explanatory environmental variables (depth, silt, sand, and gravel).

Response variable	Age	Sampling period	Significant explanatory variables (partial regression coefficient and p-value)	Slope of relationship with depth
		All	Depth (−1.77, 0.001), Silt (9.74, 0.02)	−1.60
	All	April	Depth (−1.90, 0.001), Silt (26.10, 0.002), Sand (36.41, 0.001)	−2.47
		June	Depth (−1.44, 0.001), Sand (−16.65, 0.008)	−0.88
		All	Depth (−0.46, 0.001), Silt (4.61, 0.01)	−0.35
Body size	Juveniles	April	Sand (3.99, 0.03)	0
		June	Depth (−0.22, 0.001)	−0.22
		All	Depth (−0.58, 0.001)	−0.58
	Adults	April	Depth (−0.61, 0.001)	−0.61
		June	Depth (−0.59, 0.001)	−0.59
		All	Depth (−0.93, 0.001), Silt (9.75, 0.03)	−0.76
Density	All	April	Depth (−0.55, 0.001)	−0.55
		June	Depth (−0.96, 0.003)	−0.96

Results were computed for the complete dataset, and for different sub-datasets, considering different combinations of age (adult and juvenile) and sampling period (April and June). Significant explanatory variables were selected using a forward selection procedure and tested by permutations.

seems to be an exception to this rule since they live in fine sand bottoms that evidently do not block their oral aperture.

As previously reported by Webb at the Racou [14], our RDA analysis shows a strong negative correlation between the presence of adults and depth but only a weak negative correlation between the presence of juveniles and depth (Fig. 4). However, bivariate linear relationships between amphioxus individual size and depth show no differences (compare green and blue lines in Fig. 5A and Table 1). This difference may be explained by the fact that bivariate linear analyses do not consider the multiplicity of interactions between all variables like RDA do. In any case, these results suggest that amphioxus are always concentrated in shallow waters close to the seashore. This is very clear on Fig. 5B showing the linear relationship between density and depth. Reasons for this distribution are not clear, thus Webb interpreted this fact by the need for a copious stream of interstitial water in adults (they pass a greater volume of water through the pharynx) but not in juveniles. Webb suggested that as depth increases, the deposit becomes finer and less permeable hampering the presence of adults [14]. Gosselck also observed higher densities of amphioxus (*B. senegalense*) in shallow waters of North West Africa [28] (never deeper than 40 m), and he attributed this distribution to the presence of detritus in deep waters that may block the water stream required by the filter feeding animals. But this distribution could also be related to the spawning season since seasonal changes in the amphioxus distribution in a given site have been described [26,29] and only sampling throughout the year will help to answer this question.

Conclusions

The present study provides some new knowledge on amphioxus distribution during their spawning season within the Racou beach in southern France, a place where amphioxus are usually collected for Evo-Devo studies by different laboratories around the world. Amphioxus is now an animal model for many researchers and the publication of recent studies concerning its life cycle, reproduction in captivity, and aquaculture, shows the growing interest of the scientific community for the biology of this animal [7,8,9,17,34,42,43,44,45,46]. In the present preliminary work, we have established that it exists a clear preference of amphioxus for sandy sites and their complete absence in silt. This result is in accordance with many other studies on other amphioxus species [10,21,28,37], suggesting that preference for coarse sediments is a general behaviour of amphioxus species, even if some exceptions exist since *B. floridae* lives in fine sand habitats. Several studies have also shown a negative correlation between amphioxus presence and the organic matter content [11,47]. In this study we did not controlled the presence of organic matter in the sand, and future studies will be necessary to confirm this correlation as a general fact for amphioxus species. We also describe a clear structure of the adult population at the Racou beach. Indeed, amphioxus are preferentially localized in shallow waters close to the seashore. This preliminary study represents an ecological snapshot of the amphioxus population in the Racou beach because the collection of animals was performed uniquely during April and June 2008, a period extremely important in the amphioxus life cycle since it lies at the beginning of the natural spawning season [9]. Since observations performed by others [26,29] showing population fluctuations during the year suggest that the population structure observed in this study may change depending on the moment of the year, further studies throughout the whole annual cycle over several successive years would therefore be highly informative to establish the dynamics of the population structure during the year. Finally, our study, even if concentrated on two months, represents a first approach towards a comprehension of possible ecological factors affecting the reproductive behaviour of amphioxus, since adults close to the spawning season (possessing ripe gonads) prefer lower depths and specific sand granulometry. Further studies on the physical and biological components of the sites where adult amphioxus preferentially live may help to further develop aquaculture facilities allowing development and maintenance of whole life cycle of *B. lanceolatum* in captivity.

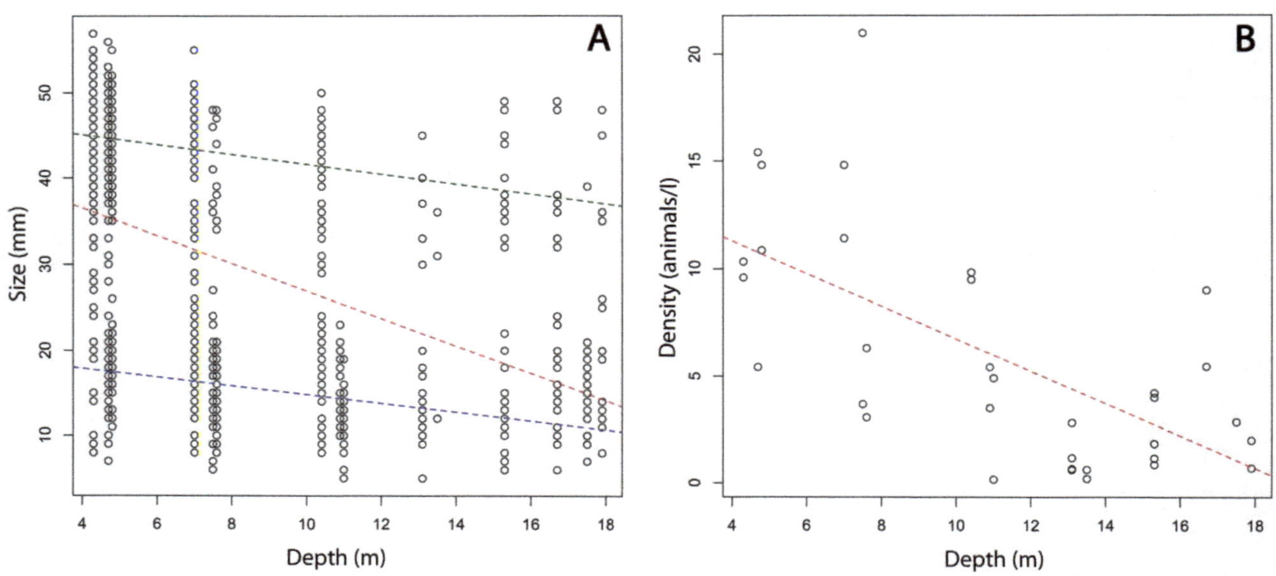

Figure 5. Bivariate linear relationships between amphioxus and depth. A: Linear relationships between amphioxus individual size and depth for juveniles (green line), adults (blue line) and all individuals (red line). B: Linear relationships between individual density and depth for all stations.

Supporting Information

Table S1 Data obtained from the different sites where amphioxus have been found. Sites containing only silt, where amphioxus were absent, have not been included. The total number of animals per site, both juveniles and adults, is indicated as well as their density (both as number of animals/m2 and number of animals/l) and average size. Numbers of adults (>3 cm long) and juveniles (<3 cm long) are also included. The percentage of silt, sand and gravel in each site is indicated as well as the depth and the collection date. ND: not determined.

Acknowledgments

We thank Stephanie Bertrand, Silvan Oulion and Yann Le Petillon for critical reading. We also thank the "Anfiox catala" association for their continuous support.

Author Contributions

Conceived and designed the experiments: HE YD. Performed the experiments: YD VM MF. Analyzed the data: YD HE. Wrote the paper: YD HE.

References

1. Bertrand PS, Camasses A, Escriva H (2007) Amphioxus: how to become a vertebrate. J Soc Biol 201: 51–57.
2. Schubert M, Escriva H, Xavier-Neto J, Laudet V (2006) Amphioxus and tunicates as evolutionary model systems. Trends Ecol Evol 21: 269–277.
3. Delsuc F, Brinkmann H, Chourrout D, Philippe H (2006) Tunicates and not cephalochordates are the closest living relatives of vertebrates. Nature 439: 965–968.
4. Dehal P, Boore JL (2005) Two rounds of whole genome duplication in the ancestral vertebrate. PLoS Biol 3: e314.
5. Putnam NH, Butts T, Ferrier DE, Furlong RF, Hellsten U, et al. (2008) The amphioxus genome and the evolution of the chordate karyotype. Nature 453: 1064–1071.
6. Poss SG, Boschung HT (1996) Lancelets (Cephalochordata: Branchiostomatidae): How many species are valid? Israel J Zool 42 suppl: 13–66.
7. Holland LZ, Yu JK (2004) Cephalochordate (amphioxus) embryos: procurement, culture, and basic methods. Methods Cell Biol 74: 195–215.
8. Fuentes M, Benito E, Bertrand S, Paris M, Mignardot A, et al. (2007) Insights into spawning behavior and development of the european amphioxus (Branchiostoma lanceolatum). J Exp Zoolog B Mol Dev Evol.
9. Fuentes M, Schubert M, Dalfo D, Candiani S, Benito E, et al. (2004) Preliminary observations on the spawning conditions of the European amphioxus (Branchiostoma lanceolatum) in captivity. J Exp Zoolog Part B Mol Dev Evol 302: 384–391.
10. Webb JE, Hill MB (1958) The ecology of Lagos Lagoon. V. Some physical properties of lagoon deposits. Phil Trans Roy Soc London, Ser B 241: 393–419.
11. Webb JE, Hill MB (1958) The Ecology of Lagos Lagoon. IV. On the Reactions of Branchiostoma nigeriense Webb to Its Environment. Philosophical Transactions of the Royal Society of London Series B, Biological Sciences 241: 355–391.
12. Stokes MD, Holland ND (1996) Life-history characteristics of the Florida lancelet, Branchiostoma floridae: some factors affecting population dynamics in Tampa bay. Israel Journal of Zoology 42: 67–86.
13. Stokes MD (1996) Larval settlement, post-settlement growth and secondary production of the Florida lancelet (= amphioxus) Branchiostoma floridae. Mar Ecol Prog Ser 130: 71–84.
14. Webb JE (1975) The distribution of amphioxus. Symp zool Soc Lond 36: 179–212.
15. Monniot F (1962) Recherches sur les graviers a amphioxus de la region de Banyuls sur Mer. Vie et Milieu XIII: 232–322.
16. Soto CG, Burhanuddin (1995) Clove oil as a fish anaesthetic for measuring length and weight of rabbitfish (Siganus lineatus). Aquaculture 136: 149–152.
17. Somorjai IML, Camasses A, Rivière B, Escrivà H (2008) Development of a semi-closed aquaculture system for monitoring of individual amphioxus (Branchiostoma lanceolatum), with high survivorship. Aquaculture 281: 145–150.
18. Folk RL, Ward WC (1957) Brazos River bar: a study in the significance of grain size parameters. Journal of Sedimentary Petrology 27: 3–26.
19. Cailleux A (1969) Quaternary Periglacial Wind-Worn Sand-Grains in URSS. Periglacial Environment Past and Present. Montréal: McGill Univ. Press. pp 285–301.
20. Legendre P, Legendre L (1998) Numerical ecology. Amsterdam: Elsevier.
21. Oksanen J, Kindt R, Legendre P, O'Hara RB (2007) vegan: Community Ecology Package. 1.8–5 ed.
22. Team RDC (2008) R: a language and environment for statistical computing. In: Computing RFfS, ed. Viena, Austria, .
23. Gueno Y, Le Guen JC (1968) Determination de l'age et croissance de Sardinellu eba (Val.) dans la region de Pointe-Noire. Cahiers ORSTOM, serie Oceanographique 6: 69–82.
24. Guillén J, Bourrin F, Palanques A, Durrieu de Madron X, Puig P, et al. (2006) Sediment dynamics during wet and dry storm events on the Têt inner shelf (SW Gulf of Lions). Marine Geology 234: 129–142.
25. Ferre B, Guizien K, Durrieu de Madron X, Palanques A, Guillen J, et al. (2005) Fine-grained sediment dynamics during a strong storm event in the inner-shelf of the Gulf of Lion (NW Mediterranean). Continental Shelf Research 25: 2410–2427.
26. Webb JE (1971) Seasonal changes in the distribution of Branchiostoma lanceolatum (Pallas) at Helgoland. Vie et Milieu 22: 827–839.
27. Eisma D, Wolff WJ (1970) The distribution of Branchiostoma lanceolatum (Cephalochordata) in the coastal waters of the Netherlands and the Southern North Sea. Netherlands Journal of Sea Research 5: 101–107.
28. Gosselck F (1975) The Distribution of Branchiostoma senegalense (Acrania, Branchiostomidae) in the Off-shore Shelf Region off North West Africa. Int Revue ges Hydrobiol 60: 199–207.
29. Franckenberg D (1968) Seasonal aggregation in amphioxus. BioScience 18: 877–878.
30. Wells MM (1926) Collecting amphioxus. Science 64: 187–188.
31. Nelson G (1969) Amphioxus in old Tampa Bay, Florida. Q J Fl Acad Sci 31: 93–100.
32. Futch CR, Dwinell SE (1977) Nearshore marine ecology at Hutchinson Island, Florida: 1971–1974: IV. lancelets and fishes. Fla mer res Publ 24: 1–23.
33. Chin TG (1941) Studies on the biology of the Amoy amphioxus Branchiostoma belcheri Gray. Philipp J Sci 75: 369–421.
34. Chen Y, Shin PKS, Cheung SG (2008) Growth, secondary production and gonad development of two co-existing amphioxus species (Branchiostoma belcheri and B. malayanum) in subtropical Hong Kong. Journal of Experimental Marine Biology and Ecology 357: 64–74.
35. Gosselck F, Spittler P (1979) Age structure, growth and weight of Branchiostoma senegalense (Acrania, Branchiostomidae) off North-West Africa. int Revue ges Hydrobiol 64: 541–550.
36. Courtney WAM (1975) The temperature relationships and age-structure of North sea and Mediterranean populations of Branchiostoma lanceolatum. Symp zool Soc Lond 36: 213–233.
37. Webb JE (1958) The Ecology of Lagos Lagoon III. The Life History Of Branchiostoma Nigeriense Webb. Philosophical Transactions of the Royal Society of London Series B, Biological Sciences 241: 335–353.
38. Webb JE (1956) A note on the lancelets of Singapore, with a description of a new species of Branchiostoma. Proceedings of the Zoological Society of London 127: 119–123.
39. Webb JE (1956) On the populations of Branchiostoma lanceolatum and their relations with the West African lancelets. Proc zool Soc Lond 127: 125–140.
40. Boschung HT, Gunter G (1962) Distribution and variation of Branchiostoma caribaeum in Missisipi Sound. Tulane Studies Zool 9: 245–257.
41. Cory RL, Pierce EL (1967) Distribution and ecology of lancelets (order Amphioxi) over the continental shelf of the southeastern United States. Limnol Oceanogr 12: 650–656.
42. Kubokawa K, Mizuta T, Morisawa M, Azuma N (2003) Gonadal State of Wild Amphioxus Populations and Spawning Success in Captive Conditions during the Breeding Period in Japan. Zoolog Sci 20: 889–895.
43. Wu XH, Zhang BL, Guo ZY, Qu YM (2000) Artificial culture of amphioxus (Branchiostoma belcheri tsingtauense). Chinese Journal of Oceanology and Limnology 18: 334–337.
44. Yasui K, Urata M, Yamaguchi N, Ueda H, Henmi Y (2007) Laboratory culture of the Oriental lancelet Branchiostoma belcheri. Zoolog Sci 24: 514–520.
45. Zhou RJ, Fang Q (2007) Artificial reproduction and batch production for Amphioxus (Branchiostoma belcheri) fry. Journal of Oceanography in Taiwan Strait 26: 121–128.
46. Theodosiou M, Colin A, Schulz J, Laudet V, Peyrieras N, et al. (2011) Amphioxus spawning behavior in an artificial seawater facility. J Exp Zool B Mol Dev Evol. In press.
47. da Silva LFB, Tavares M, Soares-Gomes A (2008) Population structure of the lancelet Branchiostoma caribaeum (Cephalochordata: Branchiostomidae) in the Baía de Guanabara, Rio de Janeiro, southeastern Brazil. Rev Bras Zool 25: 617–623.

Temporal-Spatial Variation and Controls of Soil Respiration in Different Primary Succession Stages on Glacier Forehead in Gongga Mountain, China

Ji Luo[1], Youchao Chen[1,2]*, Yanhong Wu[1], Peili Shi[3], Jia She[1,2], Peng Zhou[1,2]

1 Institute of Mountain Hazards and Environment, Chinese Academy of Sciences, Chengdu, China, **2** Graduate University of the Chinese Academy of Sciences, Chinese Academy of Sciences, Beijing, China, **3** Institute of Geographic Sciences and Natural Resources Research, Chinese Academy of Sciences, Beijing, China

Abstract

Soil respiration (SR) is an important process in the global carbon cycle. It is difficult to estimate SR emission accurately because of its temporal and spatial variability. Primary forest succession on Glacier forehead provides the ideal environment for examining the temporal-spatial variation and controlling factors of SR. However, relevant studies on SR are relatively scarce, and variations, as well as controlling factors, remain uncertain in this kind of region. In this study, we used a static chamber system to measure SR in six sites which represent different stages of forest succession on forehead of a temperate glacier in Gongga Mountain, China. Our results showed that there was substantial temporal (coefficient of variation (CV) ranged from 39.3% to 73.9%) and spatial (CV ranged from 12.3% to 88.6%) variation in SR. Soil temperature (ST) at 5 cm depth was the major controlling factor of temporal variation in all six sites. Spatial variation in SR was mainly caused by differences in plant biomass and Total N among the six sites. Moreover, soil moisture (SM), microbial biomass carbon (MBC), soil organic carbon (SOC), pH and bulk density could influence SR by directly or indirectly affecting plant biomass and Total N. Q_{10} values (ranged from 2.1 to 4.7) increased along the forest succession, and the mean value (3.3) was larger than that of temperate ecosystems, which indicated a general tendency towards higher-Q_{10} in colder ecosystems than in warmer ecosystems. Our findings provided valuable information for understanding temporal-spatial variation and controlling factors of SR.

Editor: Jack Anthony Gilbert, Argonne National Laboratory, United States of America

Funding: This study was supported by the Knowledge Innovation Project of the Chinese Academy of Science (Grants KZCX2-YW-BR-21 and KZCX2-EW-309) and the National Natural Science Foundation of China (Grant 40871042). The funders had no role in study design, data collection and analysis, decision to publish, or preparation of the manuscript.

Competing Interests: The authors have declared that no competing interests exist.

* E-mail: whchenyouchao@yahoo.com.cn

Introduction

As the implications for the climate systems, carbon (C) cycle has received growing attention in recent year [1,2]. Soils constitute the major C reserve in terrestrial ecosystems [3], and soil respiration (SR), which follows gross primary production (GPP) as the second largest C flux in the global terrestrial C cycle [4], is a critical ecosystem process that regulates C cycle in the earth system. Because of the large flux, even a small change in SR rate would have profound impact on the atmospheric CO_2 concentration [5]. Therefore, accurately quantifying SR emission is crucial for the climate change projects. However, an obstacle to estimating SR emission accurately is its inherent temporal and spatial variability within and across ecosystems [6,7,8].

Soil temperature (ST) and soil moisture (SM) are reported as two main controlling factors on temporal variation of SR [9,10]. Some biotic variables, such as microbial, root and litter biomass, may also strongly influence the seasonal variability of SR [11]. The spatial variation of SR has often been related to either the abiotic factors, such as ST, SM, soil organic carbon (SOC), soil total nitrogen (Total N), soil bulk density, soil porosity, soil pH, or biotic factors, such as plant root density, microbial biomass, litter amount [12,13]. However, the accountability of these abiotic and biotic variables on temporal and spatial variation of SR will be dependent on the type of ecosystems being studied [14].

Forest succession is a fundamental ecological process which can modify biogeochemical cycles [15], ameliorate stand conditions and alter microclimate factors [16]. As forests develop from young to mature developmental stages, the differences in soil carbon pools, root biomass, soil microbial biomass, soil chemical and physical properties, etc, may cause SR varied within and across sites [17,18,19]. Many of the controlling factors compounded in SR change during the course of forest succession, thus more knowledge of SR changes along with a forest successional gradient is required. Primary forest succession can take place on glacier forehead, with receding ice cover presenting a chronosequence of development from the bare substrate to complex plant communities [20,21]. As the forehead of a glacier is sensitive to climate change [22], forest succession on glacier forehead is the ideal environment for examining the SR and carbon cycle. However, few reports have been found on temporal-spatial variation and controlling factors of SR along forest succession on glacier forehead [23], especially for temperate glaciers.

According to the Chinese Glacier Inventory, there are 8607 monsoonal temperate glaciers in China. Hailuogou Glacier, located on the east slope of Gongga Mountain in China, is one of temperate glaciers. Our study was conducted in stands representing different stages of primary succession on the forehead of Hailuogou Glacier. The objectives of this study were to (i) identify the temporal and spatial variation of SR in different stages of primary forest succession on the forehead of a temperate glacier; (ii) distinguish the controlling factors of temporal and spatial variation of SR within and across the succession stages.

Materials and Methods

Ethics Statement

All necessary permits were obtained for the described field studies. We carried out the study based on Alpine Ecosystem Observation and Experiment Station of Gongga Mountain, which belongs to the Institute of Mountain Hazards and Environment, Chinese Academy of Sciences & Ministry of Water Conservancy. We obtained the permissions from the station and institute to use the sample plots, our study had no harm to the environment and the field study did not involve endangered or protected species.

Sites Description

Gongga Mountain (29°20′–30°20′N, 101°30′–102°15′E, 7556 m), located in the south-eastern fringe of Tibetan Plateau, is the highest peak in the eastern part of the Tibetan Plateau, and is one of the easternmost glacier areas in China (Fig. 1A). There are up to 74 glaciers distributed around Gongga Mountain, and Hailuogou Glacier, where our study was carried out, is the biggest one. Hailuogou Glacier is one of China's monsoonal temperate glaciers, located on the east slope of Gongga Mountain, with the area of 25 km^2. The regional climate is dominated by South Asian summer monsoon [24]. The mean annual air temperature is 3.8°C, the minimum mean air temperature is −4.3°C in January, and the maximum mean air temperature is 11.9°C in July. The total annual precipitation is about 2000 mm, most of which occurs from June to September.

The observed recession of the Hailuogou Glacier began in about AD 1823 [25], and the recession has accelerated markedly since the early 20th century (Fig. 1B). Within about 2 km-long forehead of Hailuogou Glacier, a series of sites, representing different stages of vegetation succession, can be easily recognized. This study was conducted in six sites of this long-term primary succession (Fig. 1C). The six sites represent a sequence of successional stages from pioneer to climax vegetation communities [26]. The newly deglaciated moraine were well-drained, and gradually colonized by pioneer species including *Astragalus adsurgens, Salix* spp., *Hippophae rhamnoides,* and *Populus purdomii* (Site 1); *P. purdomii* grew quickly thus leading to species competition, and consequently a large number of *Salix* spp. and *H. rhamnoides* disappeared from the community (Site 2); *Betula utilis, Abies fabri* and *Picea brachytula* entered into the community (Site 3); The

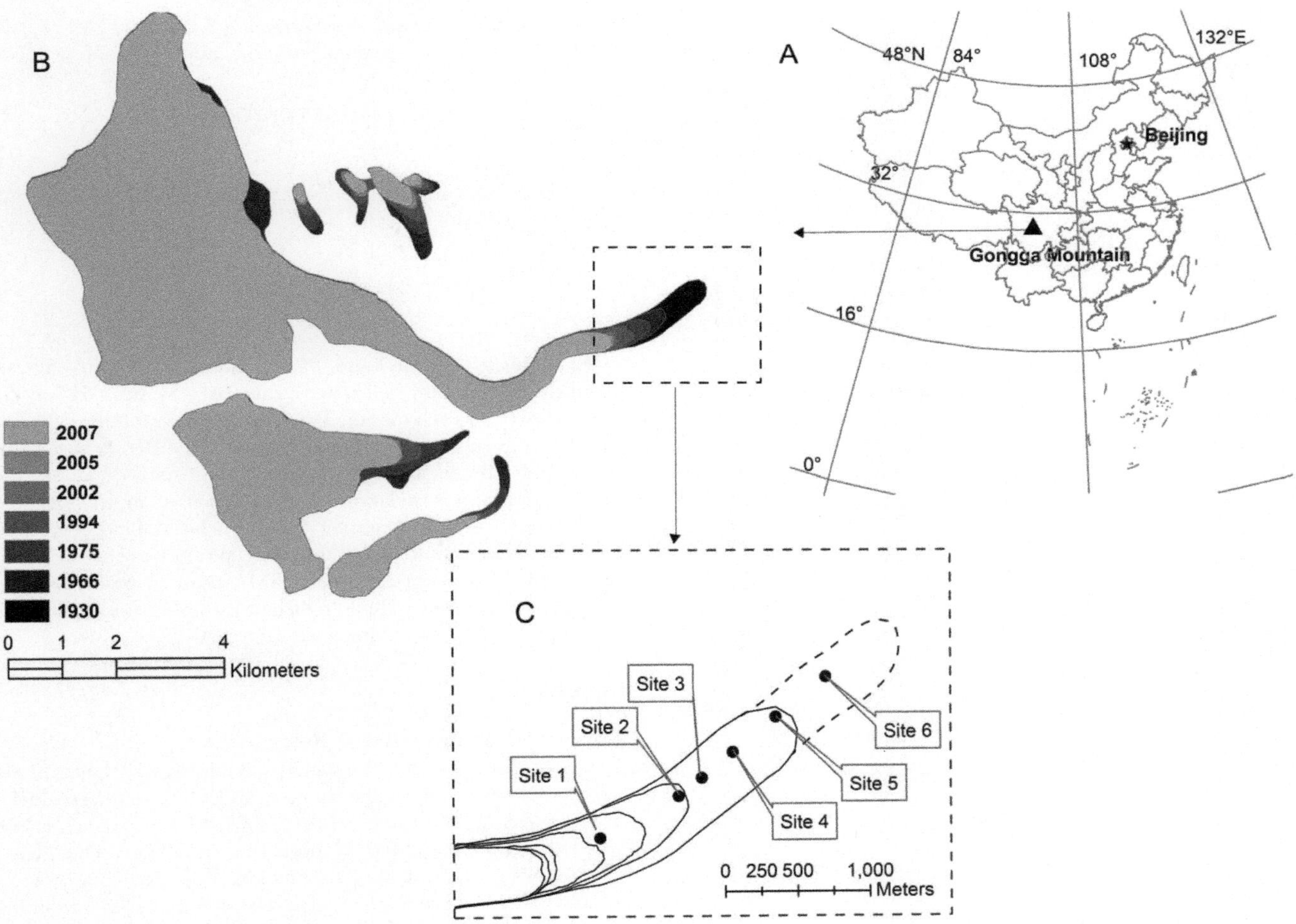

Figure 1. Location of study area. Location of Gongga Mountain (A), map of Hailougou Glacier (B) and approximate location of study sites (C). The dotted curve in panel C was adapted from Li et al. [24].

subdominant conifers increased in volume and eventually entered the canopy (Site 4); *P. purdomii* have been replaced by *A. fabri* and *P. brachytula* (Site 5); The forest has developed into a climax community dominated by *A. fabri* and *P. brachytula* (Site 6). Some characteristics of the stands are listed in Table 1.

SR, ST, SM and Litterfall Measurements

SR. Three 10 m×10 m plots were randomly located within each site. SR rates were measured about every 5 days in the year 2009, using a soil chamber (LI-6400–09, Li-Cor, Inc., Lincoln, NE) connected to a portable infrared gas analyzer (IRGA, Li-Cor, Inc., Lincoln, NE). To minimize soil surface disturbances, the chamber was mounted on polyvinyl chloride (PVC) soil collars inserted into the soil 2 or 3 cm. Three PVC collars were randomly placed in each plot one day before the measurements. A measurement consisted of placing the chamber on soil collar, scrubbing the CO_2 to sub-ambient levels, and determining soil CO_2 efflux over several 5-s periods. Data were recorded at 5-s intervals by the datalogger in the LI-Cor 6400 console. Each measurement usually took 1–3 min after placing the chamber on the collar. Concurrent measurements were made on the same day on each occasion at all six sites.

ST and SM. During the experimental period, soil temperature was monitored concurrently with SR using a copper/constantan thermocouple penetration probe (LI-6400–09 TC, LI-Cor) inserted into the soil to a depth of 5 cm in the vicinity of the SR chamber collars. The soil moisture content in the topsoil layer (0–20 cm) was measured gravimetrically and simultaneously with SR [27].

Litterfall. Five 1 m×1 m litter traps were placed randomly in each plot and we collected litter from the traps monthly in 2009 for this study. All litter samples were cleaned, oven dried at 60°C for 48 h and recorded the final dry weight.

Soil Sampling and Analysis

Three soil cores for chemical analyses, 20 cm in depth, were collected randomly from each plot in May and September 2009. The soil samples were divided into two parts: one was sieved to remove all visible plant material and air-dried for measurements of soil C and N concentration and pH value, the other was sieved and stored at 4°C until analysis for the estimation of microbial biomass carbon (MBC). SOC was determined using wet combustion as described by *Nelson and Sommers* [28], soil Total N was measured by semimicro-Kjedahl method [29], soil pH was measured with a potentiometer pH meter, MBC was determined by fumigation-extraction method [30]. Mean values were used for the later investigation.

Table 1. Some characteristics of the study sites in year 2009.

Site	1	2	3	4	5	6
Stand Age (year)	19	46	53	71	82	121
Plot size (m^2)	100	100	100	100	100	100
Mean ST (°C)	6.2	6.1	5.8	5.4	5.1	5.4
Mean SM (%)	28.4	35	31.3	35.4	36.7	39.8
Biomass ($kg \cdot m^{-2}$)	5.93	22.25	26.05	34.69	36.21	39.25
Total N (g/kg)	1.39	1.38	1.56	1.4	2.58	2.69
SOC (g/kg)	23.93	28.67	28.31	29.81	59.63	60.56
MBC (mg/kg)	43.95	118.75	131.54	244.94	281.33	191.95
pH	6.4	5.4	5.6	5.2	4.8	4.3
Bulk Density ($g \cdot cm^{-3}$)	1.41	1.44	1.36	0.92	1.1	0.94
Litterfall (g C m^{-2} $year^{-1}$)	44.14	75.26	53.34	57.12	56.28	63.91

ST means soil temperature at 5 cm depth; SM means soil moisture in the topsoil (0–20 cm); Biomass means plant biomass of stand which was extracted from unpublished data; Total N and SOC represent total nitrogen and soil organic carbon in the top 20 cm depth respectively; MBC means microbial biomass nitrogen in the top 20 cm depth soil.

Data Analysis

An exponential function was employed to describe the relationship between SR and ST [31]:

$$SR = \alpha \exp(\beta \cdot ST) \quad (1)$$

where α is the SR rate at 0°C (or a basal rate) and âis a parameter describing the temperature sensitivity (related to Q_{10} where $Q_{10} = \exp(10\beta)$).

Polynomial model was used to examine the relationship between SR and SM:

$$SR = \sum_{i=0}^{n} a_i SM^i \quad (2)$$

where a_i are constants, SM is soil moisture.

We developed an equation combining ST and SM to explain the SR. The best fits were obtained using an equation of the form:

$$SR = a \exp(b \cdot ST) \times (c \cdot SM^d) \quad (3)$$

Coefficient of variation (CV) was used to quantify the temporal and spatial variability in SR:

$$CV = \text{Standard deviation/mean} \times 100\% \quad (4)$$

We applied monthly and yearly values of SR to further investigation. Monthly values were calculated as mean SR on the observation days each month. Then we used the monthly values to calculate the sum for the year. One-way ANOVA was used to compare differences in mean values of SR, ST and SM among the six sites. Pearson correlation analysis was used to examine the relationship between SR and ST/SM in each site. Liner regression model was applied to describe the relationship between mean annual SR and mean ST, mean SM, mean litterfall, biomass, SOC, Total N, MBC, pH and bulk density among the six sites. All statistical analyses were conducted using SAS 9.0.

Results

Temporal Variation in Soil Respiration

As is shown in Fig. 2A, the patterns of temporal variation of SR were largely the same for the six sites. SR showed a similar single-peak pattern in all six sites in year 2009. SR was higher in the growing season than that in non-growing season. The lowest monthly SR rate was 0.32±0.05 µmol m^{-2} s^{-1} occurring at site 1 in January. By contrast, SR rate peaked in June or July in all the sites, with the maximum rate of 6.25±0.93 µmol m^{-2} s^{-1} in site 6.

The temporal coefficient of variation (CVs) of SR ranged from a minimum value of 39.3% for site 2 to a maximum value of 73.9%

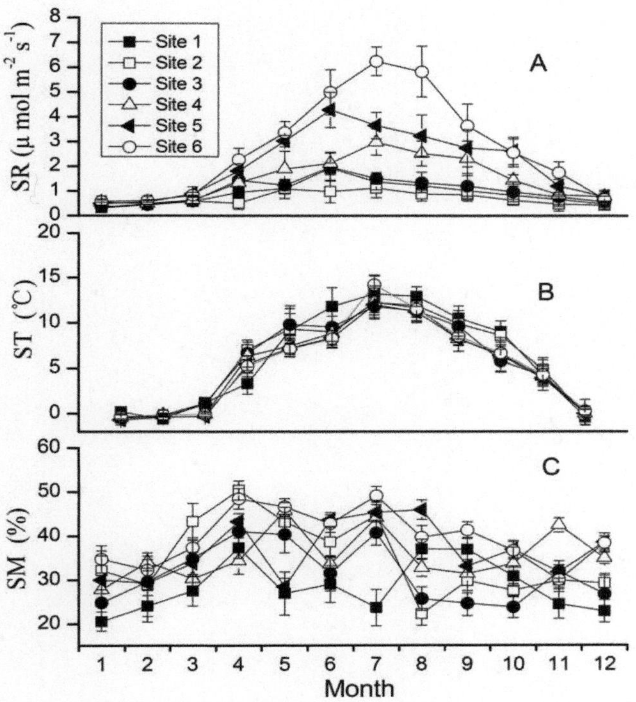

Figure 2. Temporal dynamics of SR (A), ST (B) and SM (C) for six sites. SR, ST and SM means soil respiration, soil temperature and soil moisture, respectively. The six sites represent different stages of forest succession on forehead of Hailougou Glacier in Gongga Mountain, China.

for site 6. The temporal CVs of SR appeared to be greater in the later successional stage than in the earlier stage (Fig. 3A).

Factors Controlling on Temporal Variation

SR experienced a distinct seasonal variation that paralleled the seasonality observed in ST at 5-cm depth (Fig. 2). SR increased exponentially with ST for all the sites over the whole year (Fig. 4). The ST-based regression models explained from 58.5% (site 1) to 90.3% (site 6) of the temporal variation in SR. The values of regression coefficient for all six sites were listed in Table 2.

Significantly positive correlation between SR and SM could be found in all the six sites ($p<0.001$). The polynomial functions were found to provide the best fit for the relationship between SR and SM in this study (Fig. 5). The SM-based models could explain the temporal variation in SR from 12.3% for site 4 to 32.4% for site 2. Values of R^2 here are smaller than the corresponding R^2 value using ST-based exponential model (equation (1)) for corresponding plots. When both SM and ST are used as independent variables and fitted to equation (3), we did not find a significant increase in R^2 as compared with those fitted using equation (1) for each site (Table 2).

Spatial Variation in Soil Respiration with Vegetation Succession

The mean SR rate tended to increase with the progress of succession stages for every month (Fig. 2A). The annual SR also tended to increase with the progress of primary succession: the later stages were significantly higher than the earlier stages (Fig. 6). The lowest annual SR rate was 260.1±90.2 g C m^{-2} year^{-1} in site 2, and the highest SR rate was 1030.8±184.6 g C m^{-2} year^{-1} in site 6.

The spatial CVs of SR among the six sites was higher in growing season than those in non-growing season (Fig. 3B). The

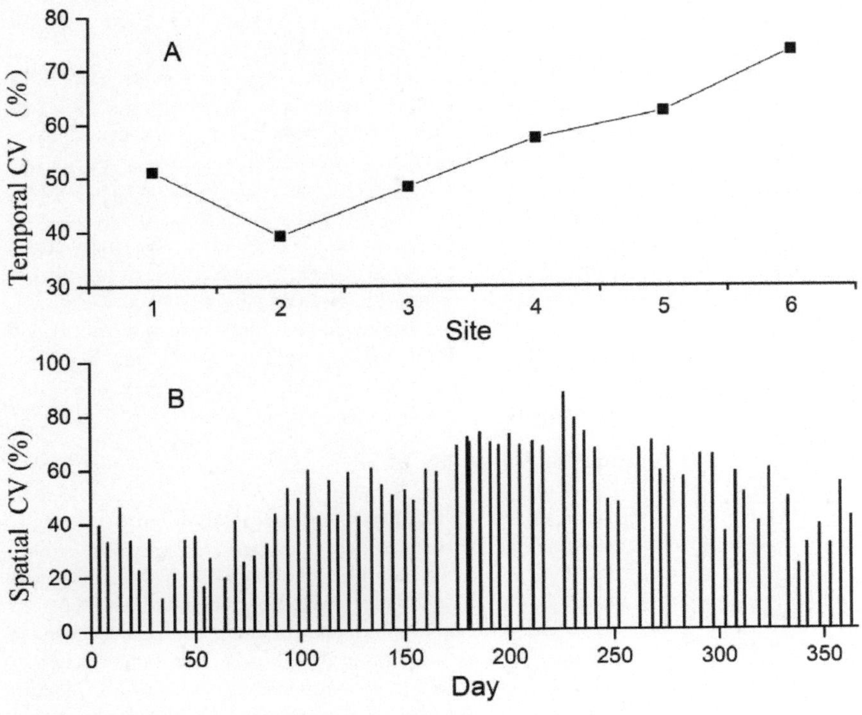

Figure 3. Temporal and spatial coefficient of variation (CV) of soil respiration. (A) Temporal CV of soil respiration for all six sites; (B) spatial CV of soil respiration during the whole year.

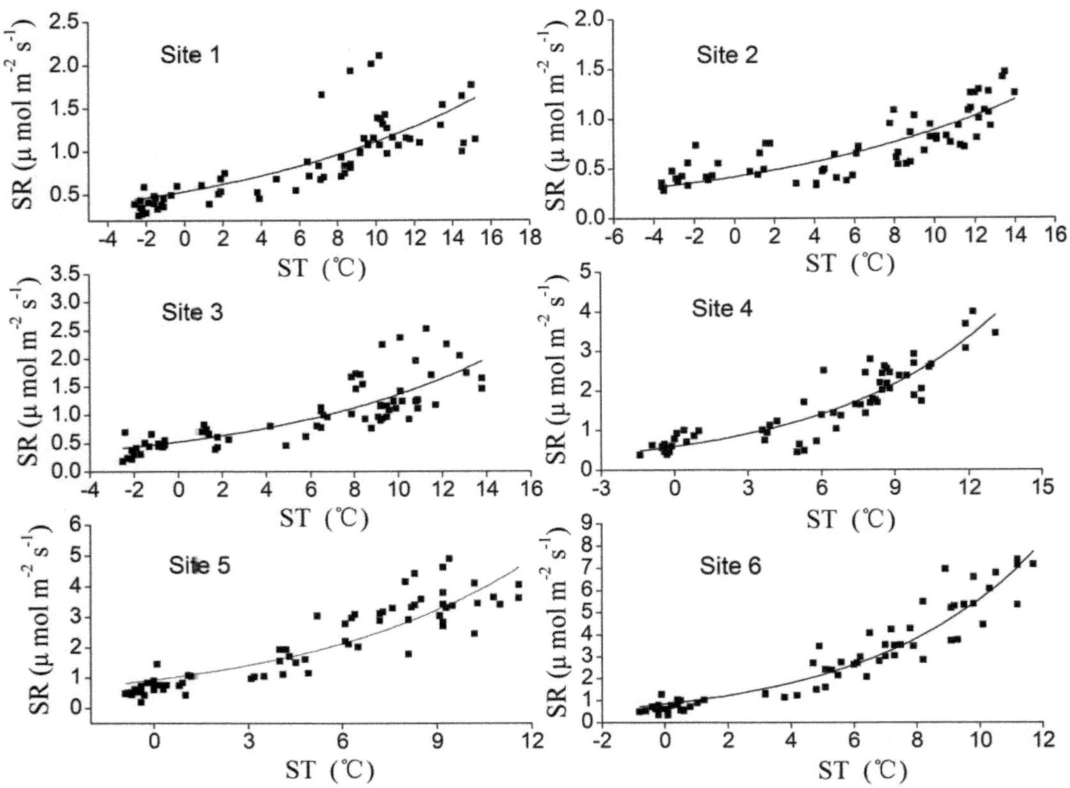

Figure 4. Relationship between SR and ST at 5 cm depth for six sites. SR and ST means soil respiration and soil temperature, respectively. The curves were fitted using equation $SR = \alpha \exp(\beta \cdot ST)$.

highest mean spatial CV among the sites occurred at August, with a maximum value of 88.6%. While the lowest SR mean spatial CV was 12.3% occurring at February.

Factors Controlling on Spatial Variation

Among the six sites, no significant correlation could be detected between mean annual SR rate and mean ST ($R^2 = 0.49$, $p = 0.09$) (Fig. 7A), which could explain much of the temporal variation in SR for each site. However, SM, which was not a controlling factor of temporal variation in SR, significantly correlated with SR among the six sites ($R^2 = 0.72$, $p = 0.02$) (Fig. 7C).

Simple regression analyses were carried out to investigate the relationship between other possible driving variables (soil properties, plant biomass, litterfall and MBC) and spatial variability of SR. The results showed that, among the six sites, mean SR rate positively correlated with plant biomass ($R^2 = 0.81$, $p = 0.008$), Total N ($R^2 = 0.80$, $p = 0.01$), SOC ($R^2 = 0.81$, $p = 0.008$) and MBC ($R^2 = 0.68$, $p = 0.02$) (Fig. 7D, 7E, 7F, 7G). In contrast, negative correlation can be detected between SR rate and pH ($R^2 = 0.66$, $p = 0.03$), bulk density ($R^2 = 0.55$, $p = 0.05$) (Fig. 7H, 7I). No significant correlation can be found between mean SR and mean yearly litterfall input ($R^2 = 0.0052$, $p = 0.89$) (Fig. 7C).

Table 2. Values of regression coefficient (α and β) and R^2 in $SR = \alpha \exp(\beta \cdot ST)$, $SR = \sum_{i=0}^{n} a_i SM^i$ and $SR = a \exp(b \cdot ST) \times (c \cdot SM^d)$.

	ST				SM	ST-SM
Sites	**á**	**â**	**R^2**	**Q_{10}**	**R^2**	**R^2**
S1	0.539	0.083	0.585***	2.3	0.303***	0.595***
S2	0.420	0.075	0.713***	2.1	0.324***	0.732***
S3	0.525	0.095	0.641***	2.6	0.132***	0.672***
S4	0.593	0.144	0.845***	4.2	0.123***	0.847***
S5	0.930	0.138	0.798***	4.0	0.269***	0.804***
S6	0.845	0.156	0.903***	4.7	0.291***	0.910***

ST is soil temperature at 5 cm depth (°C), SM means soil moisture in the topsoil (0–20 cm), SR is monthly mean soil respiration (μ mol $m^{-2} s^{-1}$), Q_{10} is calculated from equation $Q_{10} = \exp(10\beta)$, three asterisk means $p < 0.001$.

Discussion

Temporal Variability of SR

In all the six sites, a main factor controlling on temporal variation of SR is ST. The exponential relationship between SR and ST (Fig. 4) was widely reported in previous studies [4,32,33,34]. The ST-based model (equation (1)) can explain temporal variation on SR very well for each site (from 58.5% to 90.3%). The differences in ST-control along the forest succession stages may be caused by plant growth or soil microbial activities which may also contribute to temporal variations in SR [35,36]. Q_{10} was used to describe ST dependence of SR. The Q_{10} values (Table 2), with a mean of 3.3, were higher than the median value of 2.4 in temperate ecosystems [4]. There seems to be a general tendency towards higher-Q_{10} in colder ecosystems than in warmer

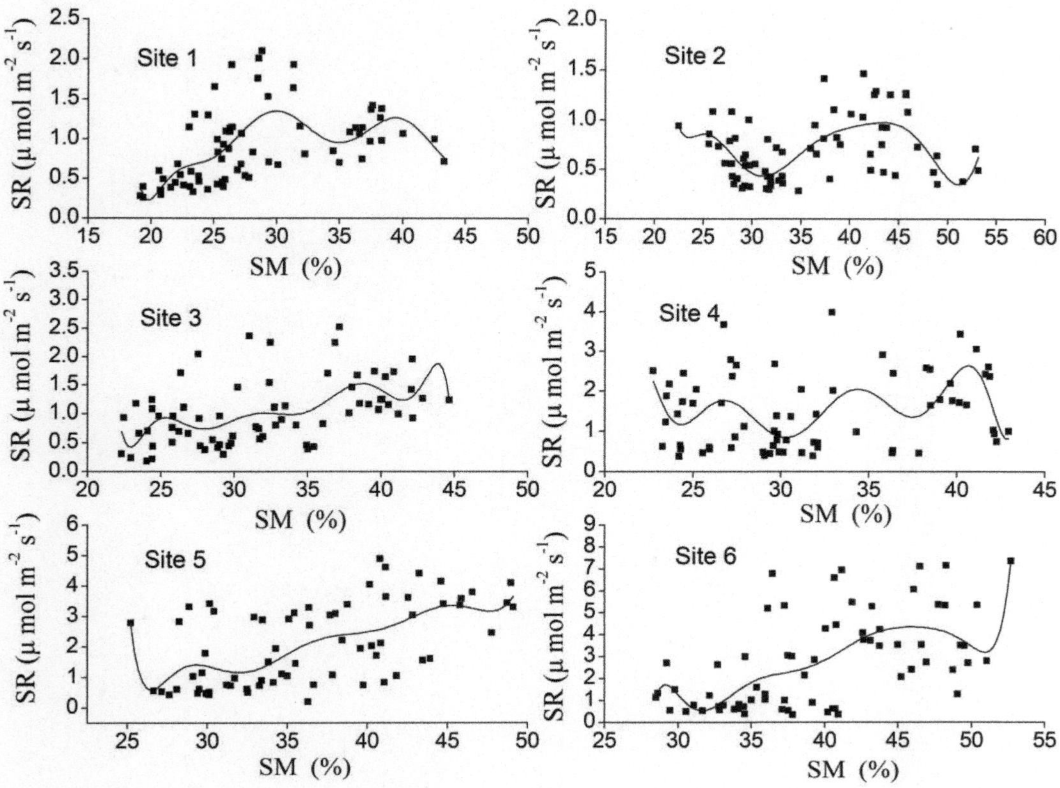

Figure 5. Relationship between SR and SM for six sites. SR and SM means soil respiration and soil moisture, respectively. The curves were fitted using equation $SR=\sum_{i=0}^{n} a_i SM^i$.

ecosystems, although the conclusion is a little up in the air for the exceptions in this study (Q_{10} in site 1 and site 2).

Apart from ST, SM is also recognized as the main factor controlling the temporal variability of SR [9,14]. In this study, positive correlation between SR and SM can be detected in all the six sites. The relationship between SR and SM is complex, and both linear and nonlinear relationship have been reported [35,37,38]. In this study, polynomial equations best fitted the SR with SM. However, the resultant fits were low (Table 2), and the combined use of ST and SM functions did not improve our capability to better explain SR compared with the model based on ST only, thus indicated that ST exerts a stronger control than SM on SR in forests on forehead of Hailuogou Glacier. The effect of SM on temporal variability of SR was negligible in this region. The reason may be that, in the forehead of Hailuogou Glacier, the atmospheric humidity and SM content were high because of the frequent precipitation, and the temporal variation in SM for each site was small (CV from 14.9% in site 6 to 23.8% in site 2). SM availability is not the major controlling factor of temporal variation of SR in forests on forehead of Hailuogou Glacier.

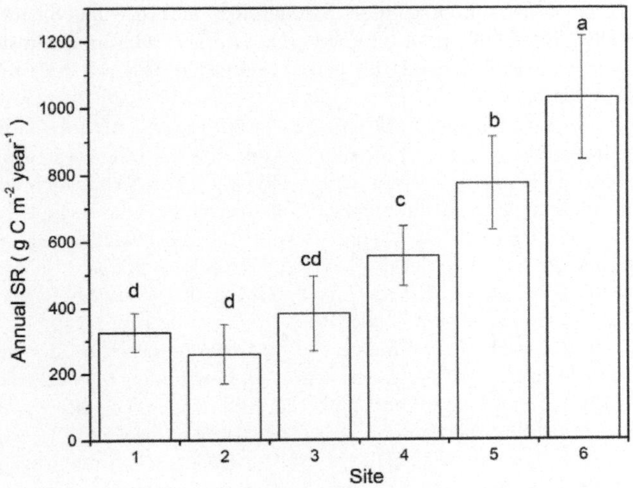

Figure 6. Annual SR for the six sites. SR means soil respiration. Different letters denote significant differences among means (á = 0.05) as determined by Turkey's HSD test.

Spatial Variability of SR

Comparing to the factors controlling on temporal variation of SR, we found SM (Fig. 7C), rather than ST (Fig. 7A), was a major factor determining spatial variability of SR in different forest succession stages on forehead of Hailuogou Glacier. In the progress of forest succession, SM content tended to increase with the succession stages (Fig. 2C), for forest succession can affect the water-holding capacity of soil by increasing the SOC, improving soil construct, receding soil bulk density and enhancing soil porosity [39]. SM can affect SR through affecting water limitation of soil microbe and site plant productivity [40]. While the differences in ST among the six sites was not significant in this

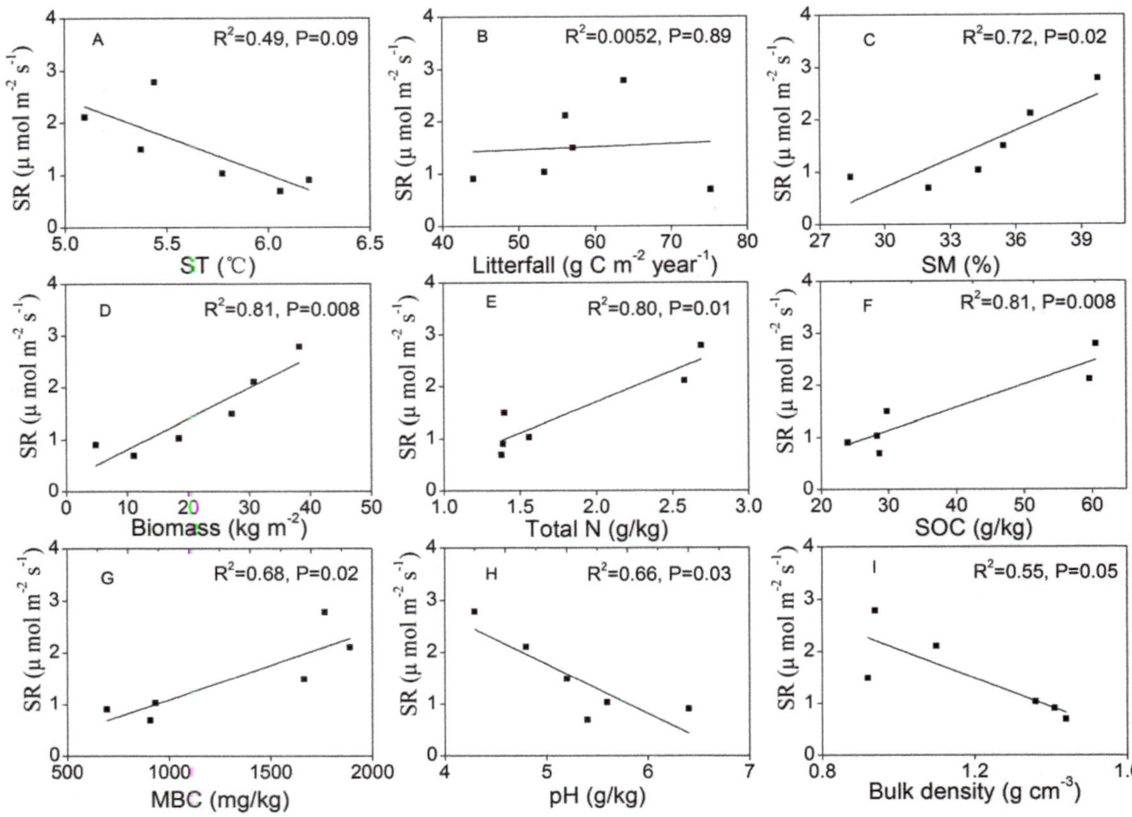

Figure 7. Relationships between mean SR and possible driving factors among sites. SR represented soil respiration; The driving factors were soil temperature (ST) at 5 cm depth (A), litterfall input (B), soil moisture (SM, C), plant biomass (D), total nitrogen (Total N, E), soil organic carbon (SOC, F), microbial biomass carbon (MBC, G), pH (H) and bulk density (I).

region ($p>0.05$), which made ST as the unimportant factor controlling on the spatial variation of SR.

We also found that plant biomass, Total N, SOC, MBC, pH and bulk density could contribute well to the spatial variation of SR among sites. Some study showed that SR rate negatively correlated with SOC and positively correlated with pH among the different stages of forest succession [41]. While in this study, we found that the annual mean SR was positively correlated with SOC (Fig. 7F), plant biomass (Fig. 7D), total N (Fig. 7E), and MBC (Fig. 7G), and negatively correlated with pH (Fig. 7H) and bulk density (Fig. 7I) among the six sites. As SR involves the process of converting organic into inorganic C, SR rate is untimely determined by the supply of C substrate [42]. The spatial dependence of SR on biomass has been reported in previous studies [4,35]. As forest developed from pioneer to climax vegetation stand, plant biomass increased with this process. Plant biomass can affect SR by affecting soil C pool size derived from litterfall, root biomass and SOC [36,40]. The correlation between SR and Total N content maybe explained by the dependence of plant growth and root activities on soil N availability. Soil pH can affect the SR through affecting activities of soil microbes. Microorganisms are generally considered as the driving force or catalyst behind the decomposition process [43], thus the magnitude of MBC may indicate potential rate of C flux [44]. The positive correlation between SR and MBC can also be found in previous studies [45,46]. Bulk density tends to decrease with forest succession (Table 1). Lower soil bulk density contributes to higher soil porosity, thus resulting in higher soil oxygen availability which facilitates microbial activities [35], leading to the increase in SR along the succession stages. SR partly results from decomposition of ground litter [47], however, in this study, we did not find a significant correlation between annual SR and yearly litterfall input among sites (Fig. 7C). In early stages, the main components of litterfall was branches and leaves of deciduous tree, and in later stages, the main component was needles of spruce and fir, the difference in litterfall amount among sites was insignificant ($p>0.05$), thus caused the poor relationship between SR and litterfall.

As showed in Fig. 7, the single factor best explaining the spatial variation of SR among the different successional stages was plant biomass or soil properties such as Total N and SOC. Stepwise multiple regression analysis with all measured variables showed that plant biomass and Total N were the only variables in the model for SR (data not shown). The exclusion of the other factors may be accounted for the fact that most of site variables were highly autocorrelated (Table 3).

In this study, our estimates of Q_{10} had a tendency to increase with the process of succession (Table 2). Our finding is in contrast to that of Yi et al. [48] and Yan et al. [49], while is consistent with the research of Wang et al. [41]. The spatial variation of Q_{10} was found to correlate with ST, SM, SOC, MBC, and ecosystem types [50,51]. In our study, SOC and MBC varied among successional stages, so the lower Q_{10} values in early stages might be ascribed to the lower SOC and MBC contents.

In summary, we investigated the temporal-spatial variation and the controlling factors of SR in different stages of forest succession

Table 3. Correlation coefficient of measured variables among the six sites.

	ST	Litterfall	SM	Biomass	Total N	SOC	MBC	pH
ST	1							
Litterfall	−0.053^	1						
SM	−0.833*	0.336^	1					
Biomass	−0.894*	0.215^	0.980**	1				
Total N	−0.700^	0.112^	0.786^	0.806*	1			
SOC	−0.755^	0.203^	0.817*	0.840*	0.988***	1		
MBC	−0.958**	0.172^	0.867*	0.935**	0.755*	0.821*	1	
pH	0.783^	−0.527^	−0.959**	−0.934**	−0.803^	−0.857*	−0.865*	1
Bulk density	0.829*	−0.050^	−0.807*	−0.890**	−0.566^	−0.621^	−0.908*	0.751^

ST, soil temperature; SM, soil moisture; Biomass, plant biomass; Total N, total nitrogen; SOC, soil organic carbon; MBC, microbial biomass nitrogen.
Caret, $p > 0.05$;
one asterisk, $p < 0.05$;
two asterisk, $p < 0.01$;
three asterisk, $p < 0.001$.

on Hailuogou Glacier forehead in Gongga Mountain. Our results indicated that SR was high in growing season and low in non-growing season for all the six sites, and the total annual SR tended to increase with the progress of succession. ST was the main controlling factor of temporal variation of SR. In contrast, spatial variation in SR could be mainly attributed to plant biomass and Total N, while the other variables (SM, MBC SOC, pH and bulk density) could affect SR by directly or indirectly influencing plant biomass and Total N. Q_{10} values increased along the forest succession, and the mean value was larger than that of temperate ecosystems, which indicated a general tendency towards higher-Q_{10} in colder ecosystems than in warmer ecosystems. Our findings show that successional stage affects microclimate, plant and microbial biomass and soil properties, thus directly and indirectly influencing SR.

Acknowledgments

We thank Qiao Liu for providing the map data of Hailuogou Glacier and Yang Cao for providing some foundation data of Alpine Ecosystem Observation and Experiment Station of Gongga Mountain.

Author Contributions

Conceived and designed the experiments: JL. Performed the experiments: JL YC. Analyzed the data: YC. Contributed reagents/materials/analysis tools: YW PS. Wrote the paper: YC. Helped in field experiments and data analysis: JS PZ.

References

1. Falkowski P, Scholes RJ, Boyle E, Canadell J, Canfield D, et al. (2000) The global carbon cycle: A test of our knowledge of earth as a system. Science 290: 291–296.
2. Piao S, Ciais P, Friedlingstein P, de Noblet-Ducoudre N, Cadule P, et al. (2009) Spatiotemporal patterns of terrestrial carbon cycle during the 20th century. Global Biogeochemical Cycles 23.
3. Dixon RK, Brown S, Houghton RA, Solomon AM, Trexler MC, et al. (1994) Carbon pools and flux of global forest ecosystems. Science 263: 185–190.
4. Raich JW, Schlesinger WH (1992) The global carbon dioxide flux in soil respiration and its relationship to vegetation and climate. Tellus Series B-Chemical and Physical Meteorology 44: 81–99.
5. Rustad L, Huntington T, Boone R (2000) Controls on soil respiration: implications for climate change. Biogeochemistry 48: 1–6.
6. Raich JW, Potter CS (1995) Global patterns of carbon-dioxide emission from soils. Global Biogeochemical Cycles 9: 23–36.
7. Soe ARB, Buchmann N (2005) Spatial and temporal variations in soil respiration in relation to stand structure and soil parameters in an unmanaged beech forest. Tree physiology 25: 1427–1436.
8. Franzluebbers K, Franzluebbers AJ, Jawson MD (2002) Environmental controls on soil and whole-ecosystem respiration from a tallgrass prairie. Soil Science Society of America Journal 66: 254–262.
9. Davidson EA, Belk E, Boone RD (1998) Soil water content and temperature as independent or confounded factors controlling soil respiration in a temperate mixed hardwood forest. Global Change Biology 4: 217–227.
10. Curiel Yuste J, Janssens IA, Carrara A, Meiresonne L, Ceulemans R (2003) Interactive effects of temperature and precipitation on soil respiration in a temperate maritime pine forest. Tree physiology 23: 1263–1270.
11. Longdoz B, Yernaux M, Aubinet M (2000) Soil CO_2 efflux measurements in a mixed forest: impact of chamber disturbances, spatial variability and seasonal evolution. Global Change Biology 6: 907–917.
12. Hanson PJ, Wullschleger SD, Bohlman SA, Todd DE (1993) Seasonal and topographic patterns of forest floor CO_2 efflux from an upland oak forest Tree physiology 13: 1–15.
13. Epron D, Nouvellon Y, Roupsard O, Mouvondy W, Mabiala A, et al. (2004) Spatial and temporal variations of soil respiration in a Eucalyptus plantation in Congo. Forest Ecology and Management 202: 149–160.
14. Saiz G, Green C, Butterbach-Bahl K, Kiese R, Avitabile V, et al. (2006) Seasonal and spatial variability of soil respiration in four Sitka spruce stands. Plant and Soil 287: 161–176.
15. De Kovel CGF, Van Mierlo A, Wilms YJO, Berendse F (2000) Carbon and nitrogen in soil and vegetation at sites differing in successional age. Plant Ecology 149: 43–50.
16. Yan JH, Zhou GY, Zhang DQ, Chu GW (2007) Changes of soil water, organic matter, and exchangeable cations along a forest successional gradient in southern China. Pedosphere 17: 397–405.
17. Cote L, Brown S, Pare D, Fyles J, Bauhus J (2000) Dynamics of carbon acid nitrogen mineralization in relation to stand type, stand age and soil texture in the boreal mixedwood. Soil Biology & Biochemistry 32: 1079–1090.
18. Jia GM, Cao J, Wang CY, Wang G (2005) Microbial biomass and nutrients in soil at the different stages of secondary forest succession in Ziwulin, northwest China. Forest Ecology and Management 217: 117–125.
19. Thuille A, Buchmann N, Schulze ED (2000) Carbon stocks and soil respiration rates during deforestation, grassland use and subsequent Norway spruce afforestation in the Southern Alps, Italy. Tree physiology 20: 849–857.
20. Nicol GW, Tscherko D, Embley TM, Prosser JI (2005) Primary succession of soil Crenarchaeota across a receding glacier foreland. Environmental Microbiology 7: 337–347.
21. Robbins JA, Matthews JA (2010) Regional Variation in Successional Trajectories and Rates of Vegetation Change on Glacier Forelands in South-Central Norway. Arctic Antarctic and Alpine Research 42: 351–361.
22. Su Z, Shi Y (2002) Response of monsoonal temperate glaciers to global warming since the Little Ice Age. Quaternary International 97–98: 123–131.
23. Yukiko SB, Atsushi K, Takehiro M, Hiroshi K, Takayuki N, et al. (2004) Soil respiration in a high arctic glacier foreland in Ny-A Òlesund, Svalbard. Polar Biosci 17: 36–46.

24. Li Z, He Y, Yang X, Theakstone WH, Jia W, et al. (2010) Changes of the Hailuogou glacier, Mt. Gongga, China, against the background of climate change during the Holocene. Quaternary International 218: 166–175.
25. Li X, Xiong SF (1995) Primary vegetation succession on deglaciated terrain at Hailuogou Glacier, Gongga Mountain (in chinese). Mountain Research 13: 109–115.
26. Cheng GW, Luo J (2002) Successional features and dynamic simulation of sub-alpine forest in the Gongga Mountain, China. Acta Ecologica Sinica 22: 1049–1056.
27. Jackson RB, Anderson LJ, T PW (2000) Measuring water avail ability and uptake in ecosystem studies. In: Sala OE, al e, editors. Methods in Ecosystem Science. New York: Springer. pp. 199–214.
28. Nelson DW, Sommers LE (1982) Total carbon, organic carbon and organic matter. In: Page ALM, R.H and Keeney, D.R., editor. Methods of Soil Analysis. Madison, Wis: American Society of Agronomy. pp. 539–579.
29. Bremner DJ (1996) Nitrogen: Total In: Sparks DL, al. e, editors. Method of soil analysis Madison, Wis.: American Society of Agronomy. pp. 539–579.
30. Brookes PC, Landman A, Pruden G, Jenkinson DS (1985) Chloroform fumigation and the release of soil N: a rapid direct extraction method to measure microbial biomass N in soil. Soil Biology & Biochemistry 17: 837–842.
31. Raich JW, Potter CS (1995) Global patterns of carbon-dioxide emissions from soils. Global Biogeochemical Cycles 9: 23–36.
32. Rustad LE, Huntington TG, Boone RD (2000) Controls on soil respiration: Implications for climate change. Biogeochemistry 48: 1–6.
33. Liu Y, Han S, Hu Y, Dai G (2005) Effects of soil temperature and humidity on soil respiration rate under Pinus sylvestriformis forest. The Journal of Applied Ecology 16: 1581–1585.
34. Yu X, Zha T, Pang Z, Wu B, Wang X, et al. (2011) Response of Soil Respiration to Soil Temperature and Moisture in a 50-Year-Old Oriental Arborvitae Plantation in China. Plos One 6.
35. Chen Q, Wang Q, Han X, Wan S, Li L (2010) Temporal and spatial variability and controls of soil respiration in a temperate steppe in northern China. Global Biogeochemical Cycles 24.
36. Raich JW, Tufekcioglu A (2000) Vegetation and soil respiration: Correlations and controls. Biogeochemistry 48: 71–90.
37. Yan JH, Wang YP, Zhou GY, Zhang DQ (2006) Estimates of soil respiration and net primary production of three forests at different succession stages in South China. Global Change Biology 12: 810–821.
38. Saiz G, Byrne KA, Butterbach-Bahl K, Kiese R, Blujdeas V, et al. (2006) Stand age-related effects on soil respiration in a first rotation Sitka spruce chronosequence in central Ireland. Global Change Biology 12: 1007–1020.
39. Chen ZJ, He JF, R TD, al e (2002) Nutrients in forest soils of eucommia ulmoides oliv and its responses to fertilization in south of Shaanxi province (in Chinese with English abstract). Journal of Soil and Water Conservation 16: 119–123.
40. Reichstein M, Rey A, Freibauer A, Tenhunen J, Valentini R, et al. (2003) Modeling temporal and large-scale spatial variability of soil respiration from soil water availability, temperature and vegetation productivity indices. Global Biogeochemical Cycles 17.
41. Wang X, Jiang Y, Jia B, Wang F, Zhou G (2010) Comparison of soil respiration among three temperate forests in Changbai Mountains, China. Canadian Journal of Forest Research-Revue Canadienne De Recherche Forestiere 40: 788–795.
42. Wan SQ, Luo YQ (2003) Substrate regulation of soil respiration in a tallgrass prairie: Results of a clipping and shading experiment. Global Biogeochemical Cycles 17.
43. Smith JL, McNeal BL, Cheng HH (1985) Estimation of soil microbial biomass: an analysis of the respiratory response of soils. Soil Biology & Biochemistry 17: 11–16.
44. Wang WJ, Dalal RC, Moody PW, Smith CJ (2003) Relationships of soil respiration to microbial biomass, substrate availability and clay content. Soil Biology & Biochemistry 35: 273–284.
45. Jenkinson DS, Powlson DS, Wedderburn RWM (1976) The effects of biocidal treatments on metabolism in soil–III. The relationship between soil biovolume, measured by optical microscopy, and the flush of decomposition caused by fumigation. Soil Biology & Biochemistry 8: 189–202.
46. Shen RF, Brookes PC, Powlson DS (1997) Effect of Long-Term Straw Incorporation on Soil Microbial Biomass and C and N Dynamics. Pedosphere 7: 297–302.
47. Bowden RD, Nadelhoffer KJ, Boone RD, Melillo JM, Garrison JB (1993) Contributions of aboveground litter, belowground litter, and root respiration to total soil respiration in a temperature mixed hardwood forest. Canadian Journal of Forest Research-Revue Canadienne De Recherche Forestiere 23: 1402–1407.
48. Yi ZG, Yi WM, Zhou GY, Zhou L, Zhang DQ, et al. (2003) Soil carbon effluxes of three major vegetation types in Dinghushan Biosphere Reserve. Acta Ecologica Sinica 23: 1673–1678.
49. Yan JH, Zhang DQ, Zhou GY, Liu JX (2009) Soil respiration associated with forest succession in subtropical forests in Dinghushan Biosphere Reserve. Soil Biology & Biochemistry 41: 991–999.
50. Zheng ZM, Yu GR, Fu YL, Wang YS, Sun XM, et al. (2009) Temperature sensitivity of soil respiration is affected by prevailing climatic conditions and soil organic carbon content: A trans-China based case study. Soil Biology & Biochemistry 41: 1531–1540.
51. Zhang JB, Song CC, Yang WY (2005) Temperature sensitivity of soil respiration and its effecting factors in the different land use. Acta Scientiae Circumstantiae 25: 1537–1542.

13

Perturbation Dynamics of the Rumen Microbiota in Response to Exogenous Butyrate

Robert W. Li[1]*, Sitao Wu[2], Ransom L. Baldwin VI[1], Weizhong Li[2], Congjun Li[1]

1 United States Department of Agriculture- Agricultural Research Service, Bovine Functional Genomics Laboratory, Beltsville, Maryland, United States of America, **2** Center for Research in Biological Systems, University of California, San Diego, California, United States of America

Abstract

The capacity of the rumen microbiota to produce volatile fatty acids (VFAs) has important implications in animal well-being and production. We investigated temporal changes of the rumen microbiota in response to butyrate infusion using pyrosequencing of the 16S rRNA gene. Twenty one phyla were identified in the rumen microbiota of dairy cows. The rumen microbiota harbored 54.5±6.1 genera (mean ± SD) and 127.3±4.4 operational taxonomic units (OTUs), respectively. However, the core microbiome comprised of 26 genera and 82 OTUs. Butyrate infusion altered molar percentages of 3 major VFAs. Butyrate perturbation had a profound impact on the rumen microbial composition. A 72 h-infusion led to a significant change in the numbers of sequence reads derived from 4 phyla, including 2 most abundant phyla, Bacteroidetes and Firmicutes. As many as 19 genera and 43 OTUs were significantly impacted by butyrate infusion. Elevated butyrate levels in the rumen seemingly had a stimulating effect on butyrate-producing bacteria populations. The resilience of the rumen microbial ecosystem was evident as the abundance of the microorganisms returned to their pre-disturbed status after infusion withdrawal. Our findings provide insight into perturbation dynamics of the rumen microbial ecosystem and should guide efforts in formulating optimal uses of probiotic bacteria treating human diseases.

Editor: David M. Ojcius, University of California Merced, United States of America

Funding: SW and WL are supported by Award Number R01RR025030 from the National Center for Research Resources (NCRR). The content is solely the responsibility of the authors and does not necessarily represent the official views of the NCRR or the National Institutes of Health. SW and WL are supported by Award Number R01RR025030 from the National Center for Research Resources (NCRR). The content is solely the responsibility of the authors and does not necessarily represent the official views of the NCRR or the National Institutes of Health. The funders had no role in study design, data collection and analysis, decision to publish, or preparation of the manuscript.

Competing Interests: The authors have declared that no competing interests exist.

* E-mail: robert.li@ars.usda.gov

Introduction

Volatile fatty acids (VFAs or short-chain fatty acids), such as acetate, butyrate and propionate, are major fermentation products of microorganisms in the rumen and hindgut. VFAs contribute up to 70% of the total metabolizable energy supply in ruminants [1]. In addition to their energetic or nutritional roles, VFAs are able to regulate animal physiology, including cholesterol synthesis and insulin and glucagon secretion. For example, butyrate is a preferred energy source for ruminal epithelial cells [2]. Most importantly, it has a multitude of cellular regulatory effects, such as modulating cell differentiation and motility, inducing apoptosis, and inhibiting cell proliferation [3,4]. Intraruminal infusion of VFAs, including butyrate, has been used to study the effect of nutrient supply on milk secretion [5] and nutrient partition, liver physiology, lipid metabolism, and rumen wall development, as well as ruminal pH maintenance [6–8]. A decreased ratio of glucogenic (propionate)/lipogenic (acetate and butyrate) in ruminal VFAs could decrease hepatic gluconeogenesis [9], leading to a reduction in milk fat secretion [10]. Steady-state ruminal butyrate concentration reflects a delicate balance between butyrate production by rumen microbes and its clearance, mainly via epithelial absorption. Many attempts have been made to identify butyrate-producing bacteria in the human gut and the rumen [11–13]. However, little is known about the ecological and physiological role of predominant butyrate-producing bacteria in the rumen microbial ecosystem. The community level response of rumen microorganisms to exogenous VFAs, such as butyrate, has yet to be understood. While biochemical processes leading to butyrate biosynthesis is well described [14], the rumen microorganisms involved in this process have yet to be fully identified. In this study, we attempted to understand the effect of butyrate perturbation on the rumen microbial community composition and dynamics.

Results

Ruminal VFA concentrations

Ruminal sodium butyrate infusion rapidly affected rumen VFA concentration as well as VFA molar percentages. In the normal rumen of dairy cows (prior to the infusion), concentrations of three major VFAs, acetate, butyrate and propionate were 377.1, 19.5 and 23.4 mM, respectively. While the total VFA concentration, which is strongly diet- and feeding time-dependent, seemed high, their molar ratio (40:2:2) was within the normal range observed in the rumen [1]. 24 h after butyrate infusion ruminal butyrate concentration increased from 19.5 mM at the baseline to 23.7 mM (Table 1). This increase was associated with simultaneous reduction in the concentration of both acetate and propionate. After the initial decrease, the concentration of acetate in the rumen remained at the same repressed level as the infusion continued. However, butyrate concentration continued to increase

Table 1. The ruminal pH and volatile fatty acid concentrations (mM) in response to butyrate infusion.

	0 h	24 h	72 h	168 h	post24 h	post168 h
Acetate	377.1±56.3	247.0±13.9*	232.1±35.1*	239.6±35.9*	348.5±103.8	286.9±21.5*
Butyrate	19.5±5.4	23.7±7.1	32.5±4.0*	38.5±18.8*	11.1±3.8	14.5±4.0
Propionate	23.4±3.2	20.7±6.2	13.7±6.1*	12.5±6.4	20.5±3.8	21.1±4.5
pH	6.70±0.39	6.95±0.43	7.36±0.44	7.11±0.71	7.03±0.29	6.60±0.48

The numbers denote mean ± SD ($N=4$).
*$P<0.05$ based on paired *t*-test. The pairs are pre- and post-infusion values of the same cow.

with infusion and peaked at 168 h infusion, while propionate concentrations continued to decrease and reached its lowest level at 168 h. Within 24 h following withdrawal of butyrate infusion, ruminal butyrate concentration experienced a drastic reduction from the baseline level (43.1% reduction down to 11.1 mM). Concomitantly, the concentration of both acetate and propionate increased from their respective repressed levels. By 168 h post infusion (post168 h), the ruminal concentration of all 3 VFAs returned to the baseline. The ruminal molar proportion (percentage) of the 3 VFAs displayed a similar trend. At the baseline, acetate accounted for 89.5% of all ruminal VFAs while butyrate and propionate accounted for 4.8% and 5.7%, respectively. A 24 h infusion resulted in a significant increase in the ruminal butyrate molar percentage from 4.8% to 8.0% ($P=0.012$; Figure 1). As the infusion progressed, the butyrate concentration continued to increase and reached a peak around 168 h post infusion (12.7%; $P=0.031$). By 24 h after infusion withdrawal, a 76% decrease in the butyrate molar percentage was observed. This sudden reduction was temporary, and butyrate molar percentage also returned to its pre-infusion baseline at 168 h post infusion. Of note, temporal changes in the butyrate concentration as well as butyrate molar percentage composition in response to exogenous butyrate among individual dairy cows were not homogeneous. The ruminal butyrate percentage as well as the concentration reached its peak value at 72 h infusion in 2 of the 4 dairy cows studied. However, the both values did not reach their peak until 168 h infusion in the remaining 2 cows. The ruminal butyrate infusion using a buffered solution (pH 7.0) led to a slight increase in the rumen pH, however, the difference was insignificant (Table 1).

The rumen microbiota composition

A total of 21 prokaryotic phyla were identified from the 16S rRNA gene sequences by RDP Classifier at a 95% confidence threshold. In the rumen microbial community of dairy cows at their mid-lactation, Bacteroidetes was predominant, represented by 70.87% of the 16S sequences in the undisrupted rumen microbial community. Firmicutes was the second most abundant phylum (22.20%), followed by Proteobacteria (3.25%), Fibrobacteres (1.52%), and Spirochaetes (1.11%). These 5 most abundant phyla accounted for ~99% of the 16S sequences. A total of 10 phyla were identified in the rumen of all cows and in all time points tested. These phyla included Actinobacteria, SR1, Synergistetes, Tenericutes, and Verrucomicrobia, in addition to the 5 most abundant phyla. In addition, the phyla such as Cyanobacteria, Euryarchaeota, Planctomycetes, and TM7 were fairly common in the rumen communities. Additionally, archaea (Euryarchaeota) were widespread in the microbial communities of dairy cows even though our primers were not specifically designed to efficiently amplify the 16S rRNA gene sequences of archaeal origin.

The mean number of genera present in the rumen community of individual animals was 54.5 (ranging from 45 to 66). 26 genera were common to all 4 cows at all time points tested, probably representing a core microbiome of the bovine rumen community fed a silage-based diet. Collectively, 454 pyrosequencing allowed us to identify 137 prokaryotic genera in the rumen microbiota of dairy cows. *Prevotella* was the most abundant genus in the rumen community, represented by 64.82% of the 16S rRNA gene sequences in the rumen prior to the perturbation. Other most abundant genera included *Succiniclasticum* (12.09%), *Fibrobacter* (4.56%), *Ruminococcus*

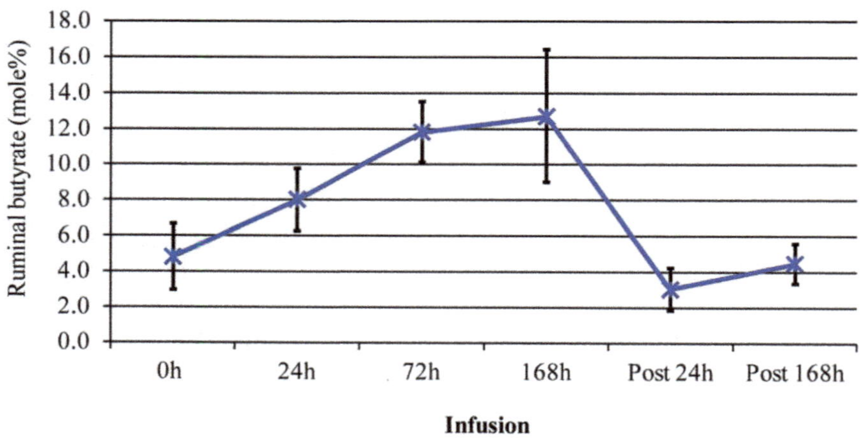

Figure 1. Butyrate molar percentages in the rumen of dairy cows in response to ruminal butyrate infusion. Error bars = SD. Post24 h and Post168 h = 24 h and 168 h after infusion withdrawal, respectively.

(2.90%), and *Treponema* (2.19%). There were a total of 9 genera with relative abundance greater than 1% in the community, such as *Butyrivibrio*, *Selenomonas*, *Moryella*, and *Succinivibrio*, in addition to the 5 most abundant genera described above. The rumen microbial community was dominated by the 10 most abundant genera, which accounted for 92.52% of the 16S rRNA gene sequences.

The microbial diversity and OTU-level composition were analyzed using CD-HIT-OTU, a novel clustering algorithm for rapid and accurate identification of microbial composition and diversity (http://weizhong-lab.ucsd.edu/cd-hit-otu/). On average, each rumen microbial community consisted of 127.25±4.42 OTUs (Operational Taxonomic Unit), ranging from 118 to 135. Collectively, 142 OTUs were identified in this study. Eighty two OTUs were present in the rumen microbial communities of all 4 cows in all time points tested and represented the core microbiome of the rumen microbial community. The lineage of the 4 most abundant OTUs, consisting of 57.07% of all 16S rRNA gene sequences in the rumen community prior to butyrate perturbation, can be traced back to the genus *Prevotella*. The 5th most abundant OTU (OTU#5) was related to the genus *Moryella* (4.33%). The rumen microbial community was dominated by 20 OTUs; and these OTUs accounted to 89.39% of the 16S rRNA gene sequences.

Dynamics of the rumen microbiota in response to butyrate infusion

Ruminal butyrate perturbation resulted in a profound change in rumen microbial composition at phylum-, genus-, and OTU-levels. Butyrate infusion led to a significant reduction in the most abundant phylum, Bacteroidetes. A 168 h ruminal butyrate infusion resulted in a reduction in its percentage composition from 70.87% prior to the infusion to 60.42%. The effect appeared instantaneous and long-lasting. The reduction was observed as early as 24 h after the perturbation was applied (reduction from 70.87% to 68.69%, $P<0.05$). However, a further reduction to 52.99% continued until 24 h post infusion. On the other hand, exogenous butyrate resulted in a significant increase in the relative abundance of Firmicutes, the second most abundant phylum. Ruminal butyrate infusion led to a significant increase in its relative abundance from 22.20% at 0 h to 29.61% ($P<0.05$) at 168 h. The increase continued and reached a plateau (35.86%; $P<0.05$) at 24 h post infusion. The relative abundance of as many as 6 phyla, including Bacteroidetes, Firmicutes, Fibrobacteres, Synergistetes, Planctomycetes, and Verrucomicrobia, was significantly impacted by exogenous butyrate at one or more time points.

A total of 19 genera were significantly impacted by ruminal butyrate infusion (Table 2) at one or more time points. *Succinivibrio* was the only genus significantly affected as early as 24 h infusion. A 72 h-infusion led to a significant change in the relative abundance of 9 genera, displaying a clear separation between the microbiota of the two groups (0 h and 72 h-fusion) (Figure 2). The maximal impact of exogenous butyrate on the rumen microbial composition was readily observable at 168 h infusion and as many as 10 genera (18% of all genera in a given rumen microbial community) were affected.

Exogenous butyrate resulted in a drastic reduction in the most abundant genus, *Prevotella* (Figure 3A). *Prevotella* consists of a group of bacteria with predominant roles in protein metabolism. Its relative abundance in the control rumen microbiota (0 h) was at 64.82%. A 72-h butyrate infusion reduced its abundance to 48.45%. However, the relative abundance of the second most abundant genus, *Succiniclasticum*, was little changed (Figure 3A). The differences in microbiota composition induced by the infusion

Table 2. Ruminal butyrate infusion resulted in a profound change in genus-level microbial composition (mean percentage ± SD; $N=4$).

Genus	0 h	24 h	72 h	168 h	post 24 h	post 168 h
Acetobacter	0.29±0.12	0.19±0.17	0.07±0.04	0.17±0.17	0.18±0.12	0.03±0.01*
Acidaminococcus	0.18±0.09	0.18±0.12	0.07±0.04	0.01±0.02*	0.02±0.01*	0.30±0.23
Anaerovorax	0.04±0.01	0.04±0.03	0.08±0.03*	0.12±0.07	0.11±0.07	0.05±0.03
Asteroleplasma	0.81±0.25	0.96±0.51	1.08±0.71	0.87±0.93	0.20±0.10*	1.37±0.66
Blautia	0.04±0.01	0.08±0.04	0.18±0.09	0.14±0.09	0.12±0.02*	0.10±0.09
Coprococcus	0.23±0.06	0.21±0.14	0.18±0.05	0.15±0.05*	0.24±0.21	0.20±0.16
Desulfobulbus	0.09±0.05	0.10±0.04	0.15±0.07	0.24±0.08*	0.18±0.11	0.13±0.06
Fusobacterium	0.00±0.01	0.02±0.01	0.08±0.05*	0.32±0.29	0.27±0.26	0.01±0.01
Holdemania	0.00±0.00	0.00±0.00	0.01±0.01	0.02±0.00*	0.02±0.02	0.00±0.01
Methylobacterium	0.20±0.03	0.13±0.07	0.12±0.07	0.12±0.03*	0.15±0.06	0.21±0.05
Paraprevotella	0.26±0.17	0.24±0.14	0.35±0.18*	0.26±0.21	0.25±0.17	0.29±0.32
Prevotella	64.82±2.61	61.52±4.54	48.45±4.77**	51.56±12.19	43.28±7.90*	59.69±8.61
Ruminobacter	0.43±0.47	1.49±1.41	8.25±3.05*	7.30±3.29*	9.76±3.41*	1.70±2.61
Ruminococcus	2.90±0.73	3.56±1.00	2.33±0.90	1.70±0.41*	2.13±0.50	3.14±0.77
Selenomonas	1.39±0.20	1.30±0.73	0.85±0.34*	0.63±0.11*	0.59±0.23*	1.29±0.40
Streptophyta	0.03±0.01	0.05±0.03	0.04±0.04	0.05±0.05	0.08±0.07	0.07±0.03*
Succinimonas	0.03±0.03	0.08±0.12	0.16±0.21	0.66±0.30*	1.40±1.51	0.07±0.10
Succinivibrio	1.11±0.73	1.95±1.10*	2.37±1.52	1.77±0.87	2.27±2.48	3.24±2.76
Treponema	2.19±0.29	3.61±1.38	5.91±2.05*	5.97±1.95*	4.04±1.73	2.57±0.90

*$P<0.05$;
**$P<0.001$.

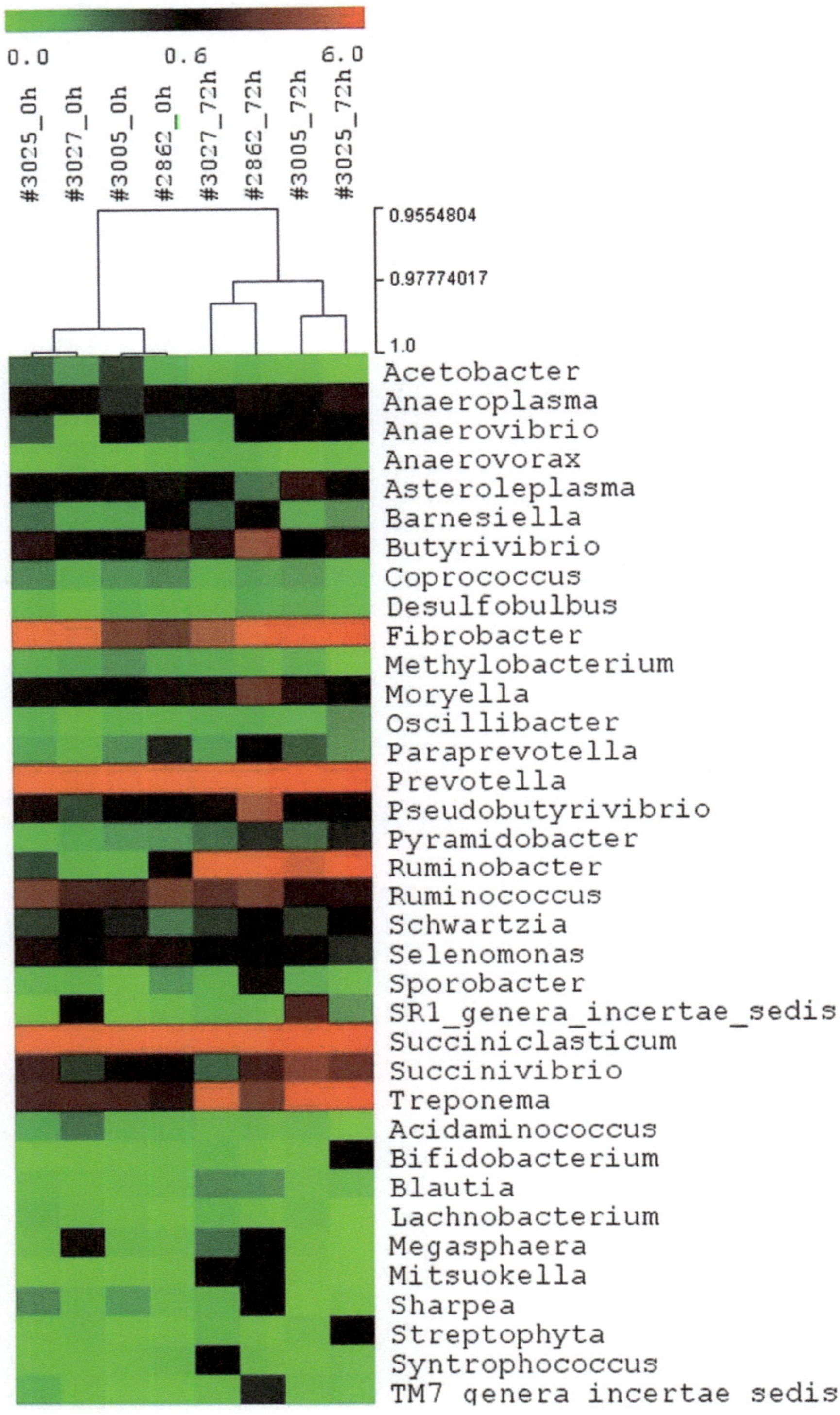

Figure 2. A heat-map of the rumen microbiota composition at a genus level. The 26 genera that shared by all samples tested (core microbiome) were displayed ($N = 4$). The scale was the percentage composition (log 2) based on the 16S sequences analyzed using RDP Classifier. A 72 h butyrate infusion induced a profound change in the rumen microbial composition, which allowed all 4 samples from this time point to be clustered using a Hierarchical Competitive Learning algorithm (HCL).

appeared to be genus-specific and did not seem to be synchronized. As Figure 3B shows, the relative abundance of *Treponema*, the 5th most abundant genus, reached its plateau between 72 h and 168 h. On the other hand, the relative abundance of *Ruminobacter* was significantly increased starting at 72 h (Figure 3B).

43 OTUs were significantly impacted by infusion at a stringent cutoff FDR<5%, representing ~52% of the core rumen

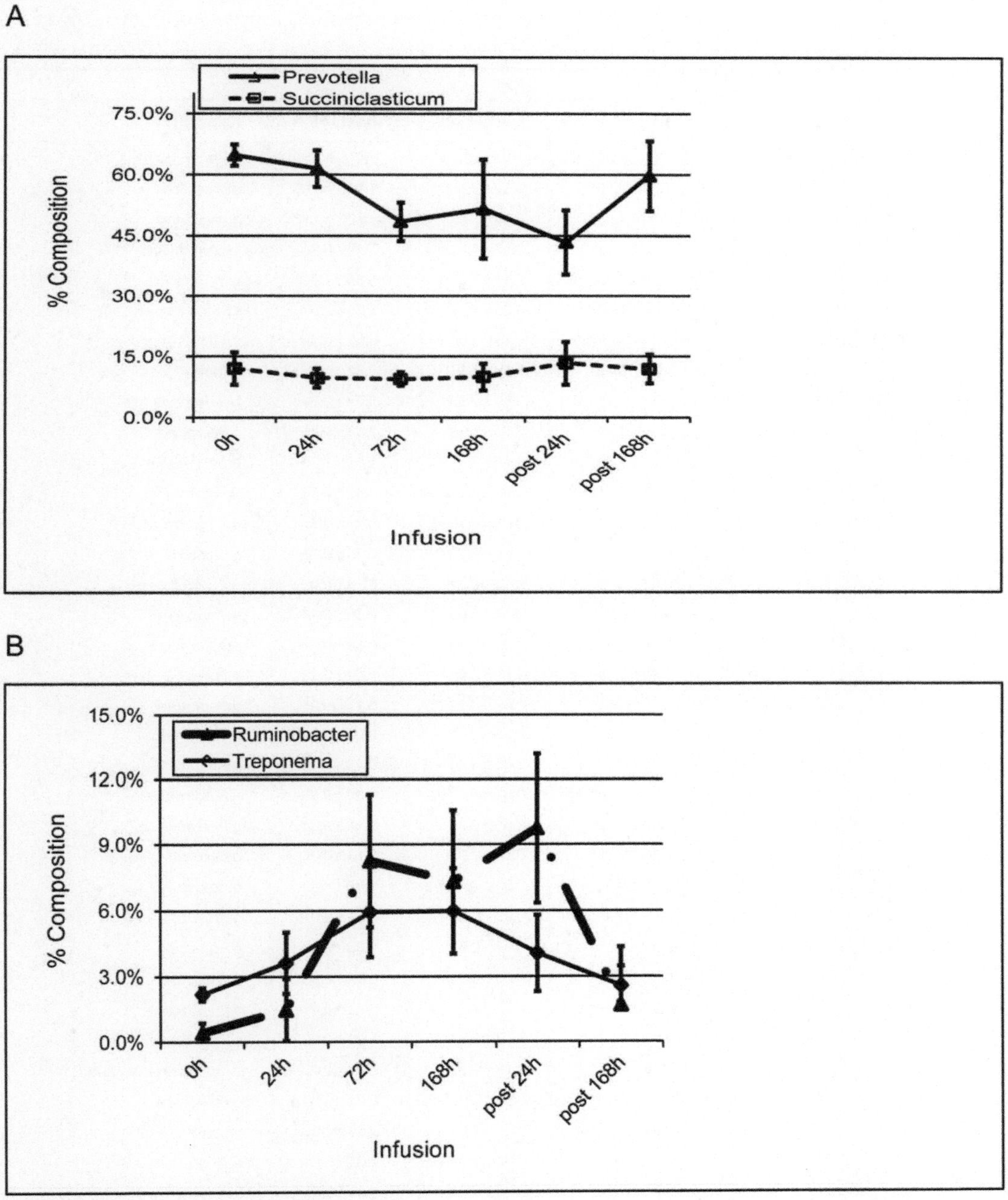

Figure 3. Temporal changes in the relative abundance (% reads) of A) 2 most dominated genera, *Prevotella* (solid line) and *Succininclasticum* (dashed line), and B) *Ruminobacter* and *Treponema*, in the rumen microbial community of dairy cows in response to butyrate infusion. Error bars represent SD of 4 samples. Normalized percentages represent the number of 16S rRNA gene sequences assigned to a given genus.

microbiome (Table 3). A Self Organizing Map (SOM) analysis, using an unsupervised learning technique to generate a type of artificial neural networks, allowed us to cluster all 82 OTUs that formed the rumen core microbiome into 4 distinct clusters. Clusters1 and 4 were the largest and contained 35 and 30 OTUs, respectively (Figure 4). These 2 clusters displayed an opposite trend in their temporal patterns in response to butyrate infusion. The relative abundance of OTUs in Cluster 1 tended to increase as infusion progressed and did not reach their peak until the 24 h post infusion (post24 h). On the other hand, the relative abundance of OTUs in Cluster 4 started to a descending slope as the infusion began; and the lowest level of the abundance was observed only at post24 h (Figure 4). OTUs in Cluster 1 were phylogenetically diverse, involving in at least 6 phyla/genera, ranging from *Moryella*, *Ruminococcus*, *Ruminobacter*, *Treponema*, to *Clostridium* and *Selenomonas*. However, the majority of OTUs belonging to *Prevotella* were concentrated in Cluster 4. OTUs in Cluster 2 increased in their relative abundance and reached their plateau at 72 h infusion and then started to decrease (Figure 5). OTU#65, annotated to *Selenomonas ruminantium*, was a typical representative of this cluster. Cluster 3, on the other hand, started its descending slope in its relative abundance and reached its

Table 3. Ruminal butyrate infusion resulted in a profound change in the species-level rumen microbial composition (mean percentage ± SD; $N=4$).

Name	0 h	24 h	72 h	168 h	post24 h	post168 h	Annotation
OTU1	20.31±1.00	18.49±2.01	13.67±1.53**	13.81±4.69	13.56±3.81*	17.19±2.50	g:Prevotella
OTU2	15.37±1.74	15.27±2.43	13.81±4.63	10.45±2.55*	9.32±3.10*	14.98±5.03	g:Prevotella
OTU3	13.20±0.76	10.96±1.00	6.96±0.98**	8.24±2.83*	8.19±3.02*	10.23±2.96	g:Prevotella
OTU6	8.19±1.39	7.39±0.95	5.41±1.19	5.19±1.53*	4.32±1.63*	7.42±0.96	s:P.ruminicola
OTU5	4.33±0.80	4.50±0.89	5.57±0.80	6.22±0.95*	6.52±1.09*	5.86±1.30	g:Moryella
OTU11	3.04±0.86	3.55±0.96	4.14±2.01	5.01±1.09	5.83±1.77*	3.25±2.47	s:R. flavefaciens
OTU7	2.69±0.61	3.51±0.92	5.38±0.66**	5.00±1.42*	4.85±0.93*	3.51±2.39	s:R.flavefaciens
OTU10	2.62±0.60	2.98±0.61	3.77±1.58	4.09±0.72*	5.58±1.51*	2.82±1.95	f:Ruminococcaceae
OTU16	2.47±0.45	2.35±0.64	2.50±0.86	3.27±0.65	3.49±0.38*	3.07±0.80	f:Lachnospiraceae
OTU12	1.50±0.39	2.13±0.58	3.10±0.39*	3.09±1.30	2.53±0.78	2.03±0.70	o:Bacteroidales
OTU9	0.54±0.41	0.94±0.60	2.43±0.48**	3.07±1.43*	4.84±3.26*	1.32±1.70	o:Aeromonadales
OTU52	0.40±0.04	0.63±0.19	0.86±0.34	0.86±0.15**	0.63±0.22	0.47±0.11	g:Treponema
OTU77	0.37±0.08	0.51±0.10	0.95±0.33	0.77±0.15*	0.48±0.12	0.46±0.17	g:Treponema
OTU71	0.36±0.15	0.30±0.06	0.63±0.12	0.76±0.18*	0.65±0.32	0.24±0.15	g:Clostridium
OTU74	0.34±0.09	0.28±0.05	0.42±0.09	0.62±0.16*	0.64±0.27	0.22±0.08	g:Selenomonas
OTU14	0.30±0.10	0.30±0.10	0.15±0.03	0.10±0.05*	0.09±0.03*	0.30±0.14	g:Prevotella
OTU56	0.21±0.08	0.26±0.13	0.26±0.18	0.24±0.32	0.05±0.03*	0.33±0.17	g:Asteroleplasma
OTU53	0.20±0.13	0.50±0.32	1.45±0.43*	1.35±0.68*	1.81±0.96*	0.49±0.50	g:Ruminobacter
OTU20	0.20±0.02	0.16±0.05	0.11±0.04	0.09±0.04*	0.08±0.02**	0.17±0.05	g:Prevotella
OTU108	0.18±0.04	0.27±0.12	0.32±0.14	0.37±0.07*	0.28±0.09	0.20±0.05	g:Treponema
OTU133	0.15±0.03	0.21±0.10	0.16±0.07	0.19±0.13	0.05±0.03*	0.15±0.06	g:Anaeroplasma
OTU35	0.14±0.10	0.26±0.20	0.55±0.08**	0.68±0.30*	0.83±0.49*	0.30±0.39	g:Ruminobacter
OTU19	0.11±0.03	0.11±0.02	0.14±0.08	0.08±0.05	0.05±0.03*	0.11±0.04	g:Prevotella
OTU76	0.09±0.01	0.10±0.04	0.09±0.04	0.26±0.09*	0.31±0.34	0.13±0.06	o:Bacteroidales
OTU126	0.09±0.02	0.05±0.02	0.06±0.02	0.05±0.01*	0.05±0.01*	0.08±0.02	g:Prevotella
OTU26	0.08±0.02	0.07±0.01	0.05±0.01	0.08±0.03	0.05±0.02*	0.05±0.01	g:Prevotella
OTU34	0.08±0.02	0.07±0.02	0.04±0.02	0.04±0.01*	0.04±0.02*	0.08±0.03	g:Prevotella
OTU50	0.06±0.03	0.04±0.02	0.01±0.01	0.00±0.00*	0.00±0.00*	0.02±0.03	g:Prevotella
OTU49	0.05±0.02	0.04±0.01	0.01±0.01	0.00±0.01*	0.00±0.00*	0.03±0.01	g:Prevotella
OTU22	0.04±0.01	0.06±0.02	0.09±0.02	0.07±0.04	0.12±0.07*	0.06±0.05	o:Clostridiales
OTU128	0.03±0.05	0.04±0.04	0.24±0.15	0.35±0.21*	0.19±0.29	0.01±0.01	c:Alphaproteobacteria
OTU67	0.03±0.01	0.02±0.02	0.01±0.01	0.01±0.00*	0.00±0.00**	0.02±0.02	g:Prevotella
OTU141	0.02±0.01	0.02±0.01	0.04±0.02	0.04±0.01*	0.01±0.00	0.02±0.01	g:RFN20
OTU139	0.02±0.02	0.02±0.01	0.07±0.01	0.07±0.02*	0.06±0.03	0.04±0.04	o:Clostridiales
OTU79	0.02±0.01	0.02±0.00	0.01±0.00	0.01±0.01	0.00±0.00*	0.02±0.02	g:Prevotella
OTU58	0.02±0.00	0.02±0.01	0.03±0.01	0.02±0.01	0.01±0.00*	0.02±0.01	g:Paludibacter
OTU91	0.01±0.01	0.01±0.01	0.00±0.00	0.01±0.00	0.00±0.00*	0.01±0.00	g:Prevotella
OTU59	0.01±0.01	0.01±0.01	0.01±0.01	0.01±0.00	0.05±0.03*	0.01±0.01	g:Ruminococcus
OTU92	0.01±0.01	0.01±0.01	0.01±0.01	0.01±0.00	0.00±0.00*	0.01±0.01	g:Prevotella
OTU110	0.01±0.01	0.02±0.01	0.02±0.01	0.04±0.01*	0.03±0.01*	0.01±0.01	g:Treponema
OTU107	0.01±0.00	0.01±0.00	0.01±0.00	0.01±0.01	0.02±0.01*	0.01±0.00	o:Clostridiales
OTU113	0.01±0.01	0.01±0.01	0.02±0.02	0.04±0.03	0.04±0.02*	0.02±0.02	g:Coprococcus
OTU131	0.00±0.01	0.01±0.01	0.01±0.01	0.02±0.02	0.02±0.01*	0.00±0.00	o:Aeromonadales

**FDR<5% and $P<0.001$.
*FDR<5% and $P<0.05$.
The consensus sequence of each OTU was annotated to the closest lineage using FR-HIT against the GreenGene database.
s: = species; g: = genus; f: = family; o: = order; and c: = class.
A stringent cutoff at a global false discovery rate or FDR<5% was used. OTUs were sorted based on their relative abundance in the rumen microbial community prior to butyrate infusion (0 h) in a descending order.

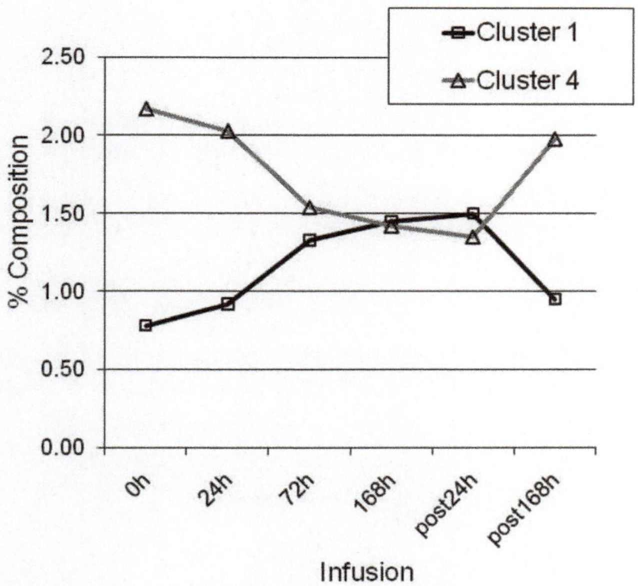

Figure 4. Temporal profiles of the relative abundance (% reads) of Operational Taxonomical Units (OTUs) of Clusters #1 and 4. Clusters were generated based on Self Organizing Maps (SOM). Percentage composition represents mean normalized percentages of OTUs of 4 samples (*N* = 4).

lowest point at 168 h infusion (Figure 5), which was different from Cluster #4, where the lowest abundance level was not reached until 24 h post infusion (Figure 4). OTUs annotated to the genus *Prevotella* were distributed in 3 clusters, Clusters#2, #3, and #4, suggesting its functional divergence. 72 h butyrate infusion induced a sufficient change in the ruminal microbial composition at both genus and OTU levels, which allowed 4 samples at this time point to be clustered using HCL (Figure 2) and PCoA analysis by Fast UniFrac (Figure 6). Infusion did not seem to affect biodiversity indices, such as ACE and Chao1, in the rumen. Of interest, infusion may have created a favorable rumen environment that allowed more microbial genera to be established as the number of genera was higher at 168 h infusion (60.3±4.9) than at the baseline (52.8±3.3), coinciding with the maximal butyrate molar percentage in the rumen (Figure 1). However, it seemed paradoxical that the number of OTUs in this time point was lower than the baseline (124.5±3.5 vs 130.5±2.4). Nevertheless, understanding temporal changes in the rumen microbiota composition in response to butyrate infusion will undoubtedly facilitate rapid identification and isolation of novel butyrate-producing bacteria and provide insight into their contributions in the rumen microbial ecosystem output.

Discussion

The rumen microbiota plays an essential role in nutrient production and utilization in ruminants [15,16]. Efficient microbial transformation of plant fibers results in production of VFAs, which are subsequently used to produce meat and milk for human consumption. Rumen fermentation is poorly understood process controlled by the interacting rumen microbiota constituents. Understanding of microbial interactions and dynamics in the rumen microbial ecosystem should provide a scientific basis for successful manipulation of ruminal fermentation for optimal outcomes. In this study, we characterized temporal changes of the rumen microbiota of dairy cows in their mid-lactation in response to an exogenous butyrate disturbance. We identified 21 prokaryotic phyla in the rumen microbial community, which were dominated by 2 phyla, Bacteroidetes and Firmicutes. Together, these 2 phyla accounted for up to 93% of all 16S rRNA gene sequences. Compared to the rumen of 12-month-old steers fed a similar hay-based diet, where 16S rRNA gene sequences can be assigned to 15 bacterial phyla [16], the rumen microbiota of mid-

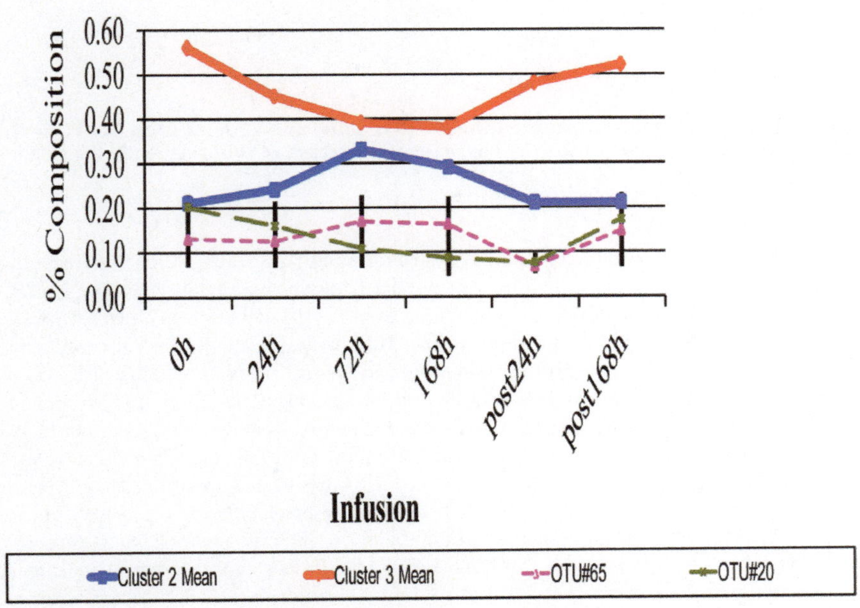

Figure 5. Temporal profiles of the relative abundance (% reads) of Operational Taxonomical Units (OTUs) of Clusters #2 and 3. Clusters were generated based on Self Organizing Maps (SOM). Group Mean represents mean normalized percentages of OTUs of 4 samples in this cluster (*N* = 4). A temporal profile of representative OTUs (OTU#65 and OTU#20) that belonged to clusters#2 and 3, respectively was shown as a reference. Error bars represent SD of 4 samples.

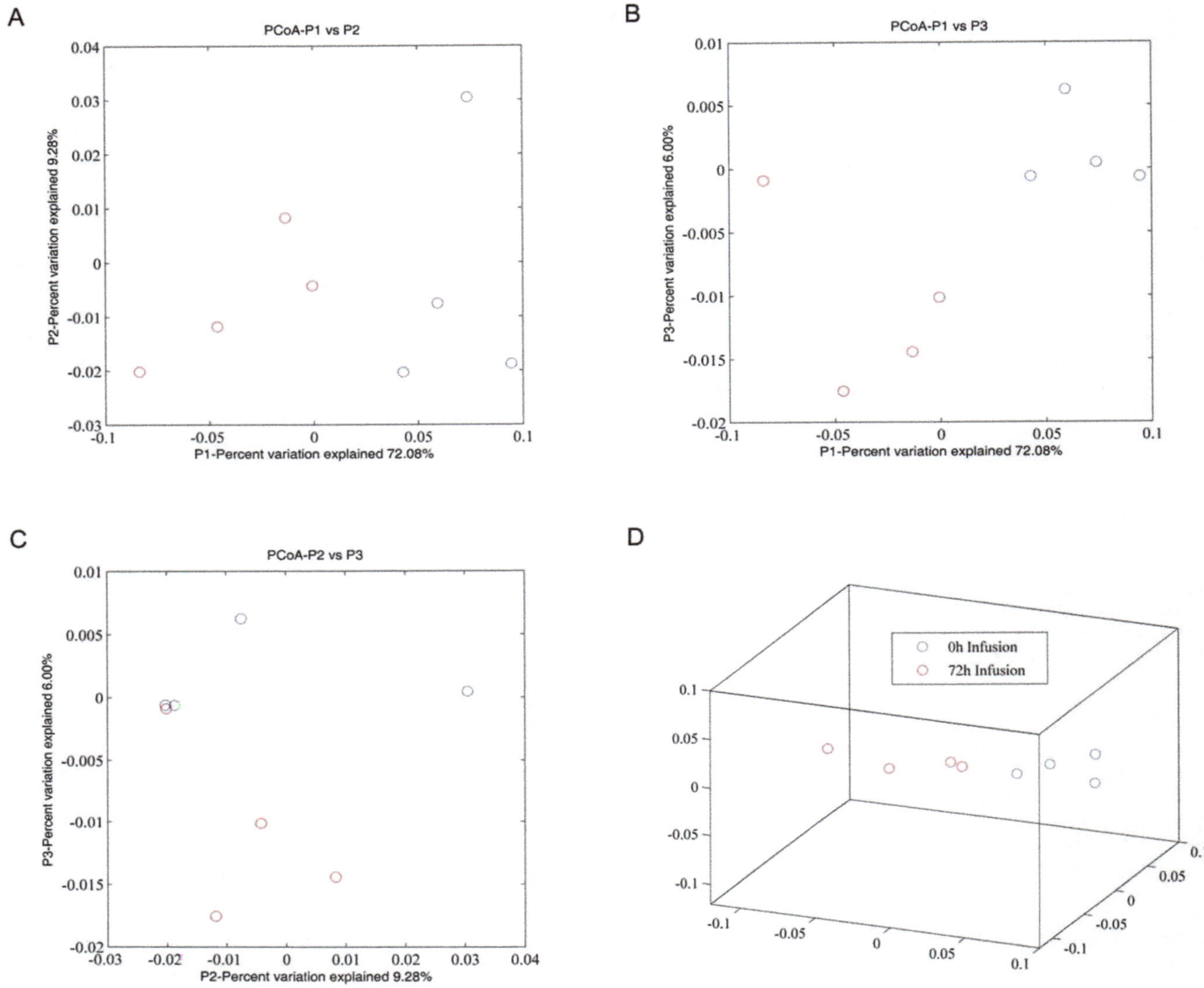

Figure 6. Principal Coordinates Analysis of weighted UniFrac values between pre-infusion (0 h; blue) and 72 h infusion (red) (*N*= 4). (A) Visualization of the first two dimensions (un-scaled); (B) Visualization of the first and third dimensions (un-scaled); (C) Visualization of the second and third dimensions (un-scaled); (D) in 3-D.

lactating dairy cows displayed some unique features. One of the salient features was prominent establishment of bacteria from a candidate phylum SR1 in the community, which was absent from the rumen of 12-month-old steers. The role of these bacteria in the rumen microbial ecosystem remains unknown.

Butyrate is a major microbial fermentation product in the forestomach (rumen) of ruminants and the lower digestive tract of humans and all animal species and contributes to approximately 70% of the daily metabolizable energy requirement of ruminants and ~10% for humans [1]. In addition to serve as a preferred energy source for the rumen epithelia and human colonocytes, butyrate is an important regulator of host physiology and acts as a signaling molecule in epithelial cells [3,17]. Butyrate has important implications in human health as well, especially in gastrointestinal disorders (colon cancer and enterocolitis) and cardiovascular diseases, including its role in development of normal colon mucosa and the prevention of colon cancer [18]. Additionally, butyrate irrigation has been used to treat colitis. Butyrate production in the gastrointestinal (GI) tract, which has recently been shown to play a special role in modulating bacterial energy metabolism in the gut ecosystem [19], depends on diets, microbial species and their relative abundance present in the gut ecosystem, and gut transit time. Over the years, extensive searches have been conducted to identify butyrate-producing bacteria in the rumen of both domestic and wild ruminants and the hindgut of humans [11,20]. In the human hindgut, butyrate-producing bacteria are mainly from the phylum Firmicutes (the clostridial clusters IV and XIVa; [21]) and 80% of butyrate-producing isolates belong to the XIVa cluster of Gram^{+} bacteria, such as species from *Butyrivibrio*, *Eubacterium* and *Roseburia* [11]. In addition, cluster IV includes several species that are butyrate-producing, such as *Faecalibacterium prausnitzii*, *Subdoligranulum variabile*, and *Anaerotruncus colihominis* [20]. In the rumen, butyrate-producing bacteria as a functional group likely had a distinct phylogenetic profile. *Butyrivibrio fibrisolvensis* is a major butyrate producing bacterium and responsible for fiber digestion and utilization in the rumen [22]. Our results show that

Butyrivibrio was among the 5 most abundant genera and accounted for 1.46% of all 16S sequences in the undisturbed rumen ecosystem. The major butyrate-producing bacteria in the colon of monogastric species including humans, such as *Eubacterium*, *Anaerostipes*, *Roseburia*, and *Faecalibacterium*, were very rare in the rumen. *Butyrivibrio* (1.46%) and *Pseudobutyrivibrio* (0.82%) were probably predominant butyrate producers in the rumen microbial ecosystem. Interestingly, exogenous butyrate seemingly had a stimulating effect on the native butyrate-producing bacterial population. A 168-h infusion resulted in an increase in the relative abundance of *Butyrivibrio* and *Pseudobutyrivibrio* by 65% and 141%, respectively, albeit not statistically significant, compared to their pre-infusion levels. Other butyrate-producing bacteria, such as *Roseburia*, also responded favorably to exogenous butyrate. If these patterns were confirmed, it is likely to isolate many previously unrecognized butyrate-producing bacteria from the rumen, especially from those belonging to Clusters#1 and 2 (Figures 4 and 5). The expansion in the relative abundance of these butyrate-producing bacteria may at least partially account for the observed expansion of the phylum Firmicutes in response to butyrate infusion.

Ruminal butyrate infusion resulted in a doubling in butyrate concentration from 19.5±5.4 mM at the basal level to 38.5±18.8 mM at 168 h infusion at the expense of both ruminal acetate and propionate concentrations (Table 1). As a result, butyrate molar percentage increased significantly from 4.8% in the undisturbed rumen (the baseline) to 12.7% at 168 h infusion ($P<0.05$). These drastic changes had a profound impact on ruminal microbial composition, community structures, and metabolic potentials because the substrate-product equilibrium in the rumen microbial ecosystem was altered. At 168 h infusion, the relative abundance of 10 genera was significantly changed ($P<0.05$), coinciding with the maximal change in both ruminal butyrate concentration and molar percentage ratios of 3 major VFAs. The perturbation also led to changes in relative abundance of as many as 26 OTUs representing 32% of OTUs in the core microbiome. A sudden withdrawal of exogenous butyrate had a repressive effect on ruminal butyrate biosynthesis and led a 43% reduction in ruminal butyrate concentration from the basal level (19.5 mM) to 11.1 mM at 24 h post infusion. Concurrent with this reduction was the changes in relative abundance of the maximal number of OTUs observed (33) at FDR<5% at 24 h post infusion, suggesting adaptation of the rumen ecosystem to substrate availability may be time-dependent. However, both ruminal butyrate concentration and butyrate molar percentage returned to the basal level at 168 h post infusion. By 168 h post infusion, the genus-level rumen microbial composition showed a high-degree of similarity (>81%) to their normal (undisturbed) status based on Bray-Curtis similarity coefficients (Table 4) while the normal (0 h) and 168 h infusion shared the least similarity at the genus level. Cluster analysis indicates that the genus-level composition of the rumen microbial communities was indistinguishable between 0 h (normal/undisturbed) and post168 h infection (data not shown). These data suggest that the rumen microbial ecosystem displayed substantial resilience to the short-term change in butyrate concentration.

Butyrate plays an important role in host physiology and gut health. A persistent elevated level of butyrate in the GI tract could have positive impacts on nutrient utilization efficiency in ruminants and the prevention and treatment of colon diseases in nonruminants. It has been known that the metabolic outputs of the gut microbial ecosystem depend on both diets and gut environmental factors, such as pH [12,21]. Factors affecting gut production of butyrate have been investigated. Higher-fiber intake in cattle tends to increase the population of major butyrate-producing bacteria, *Butyrivibrio* [22], resulting in an increase of ruminal butyrate concentration, whereas high-energy feeds lead to the suppression of these bacteria. Resistant starch is known to be butyrogenic in humans [23] and modify the composition of the gut microbiota and elevate butyrate production in human flora-associated rats [24]. In healthy obese human subjects, lower carbohydrate intakes lead to a significant reduction in fecal butyrate with a proportional decrease in the abundance of the *Eubacterium rectale/Roseburia* subgroup of the clostridial cluster XIVa [25]. Formulating a diet aiming at a selective stimulation of butyrate-producing bacteria as well as non-butyrate producing bacteria, which may indirectly affect butyrate production via metabolic cross-feeding [26], can have a desired effect on butyrate biosynthesis. The key to a successful dietary regime is to use metagenomic tools to holistically monitor dietary impacts on the microbial composition and prevent overexpansion of lactic acid producing bacteria, which may contribute to rumen acidosis in cattle.

Table 4. Bray-Curtis similarity matrix.

	0 h_#1	0 h_#2	0 h_#3	0 h_#4	168 h_#1	168 h_#2	168 h_#3	168 h_#4	p168 h_#1	p168 h_#2	p168 h_#3	p168 h_#4
0 h_#1												
0 h_#2	86.79											
0 h_#3	83.90	84.23										
0 h_#4	80.31	82.07	81.78									
168 h_#1	74.04	69.04	69.75	64.04								
168 h_#2	76.65	73.41	72.66	68.29	81.19							
168 h_#3	78.29	80.34	78.28	74.07	69.36	74.19						
168 h_#4	75.23	73.64	73.01	69.85	80.70	83.82	72.61					
p168 h_#1	86.21	81.48	79.71	75.84	77.22	73.46	72.04	75.79				
p168 h_#2	82.27	85.58	86.20	82.79	69.39	72.31	80.17	74.97	79.73			
p168 h_#3	83.25	80.68	82.77	75.18	77.25	79.18	78.47	76.89	79.73	80.30		
p168 h_#4	77.12	82.34	79.71	83.59	63.91	66.89	77.95	69.03	74.02	85.71	77.10	

The relative abundance was first square root transformed.
Bray-Curtis similarity coefficients were calculated using PRIMER v6. p168 h = post168 h.

Introducing butyrate-producing bacteria into the gut ecosystem has been suggested as a means to treat and prevent colon cancer and enterocolitis, including inflammatory bowel diseases. Indeed, an oral administration of *Butyrivibrio fibrisolvens*, a species isolated from the bovine rumen with a high capacity to produce butyrate, in mice results in a significant increase in the rate of fecal butyrate production and decreases the formation of experimentally induced aberrant crypt foci [27] and alleviated the symptoms of experimentally induced enterocolitis [28]. An ideal probiotic should possess properties able not only to colonize in the desired gut location for a prolonged period of time but also to maintain its high production potentials. Previous successes in establishing introduced bacteria into the rumen have involved species with unique metabolic capabilities that can fill empty habitat niches [29]. Butyrate producing bacteria are phylogenetically diverse [11,12,20] and host species may display their unique microbial profiles of butyrate biosynthesis. In addition, the tendency to resist colonization of foreign probiotic bacteria by the indigenous microflora and apparent host specificity of the gut bacterial community [29] require us to have a holistic understanding of dynamics of the gut microbial ecosystem in response to perturbation before a successful probiotic strategy can be optimized.

Materials and Methods

Cows and diet

Four ruminally-cannulated Holstein cows in mid-lactation were used in this study. These multiparous cows fed *ad libitum* standard lactation rations as a Total Mixed Ration (TMR; 50% corn silage and 50% concentrate at a dry matter basis). Diets were formulated to provide or exceed NRC dietary recommendations for lactation. Complete dietary chemical composition was determined on daily samples of TMR composited weekly. All animal procedures were conducted under the approval of the Beltsville Area Institutional Animal Care and Use Committee and the University of Maryland Animal Care and Use Committee. Cows were moved to a tie stall barn for adaptation and acclimation at least five days prior to the infusion experiment. Rumen contents (solids and rumen juice) of the normal dairy cows (just prior to infusion or 0 h) were collected from different locations inside the rumen. Infusion of butyrate was initiated immediately following 0 h sampling and thereafter continued for 168 h (7 days) at a rate of 5.0 L/d of a 2.5 M solution such that 12.5 moles were infused daily in a buffered saliva solution (pH 7.0; 3.8% $KHCO_3$, 7.3%$NaHCO_3$) as a continuous infusion. This amount of butyrate was selected to represent greater than 105% of daily anticipated metabolizable energy (ME) intake and ranged from 15.5 to 25.7. After 168 h infusion, cows were returned to a standard ration without ruminal infusion for an additional 168 h. Rumen contents were serially collected at 0, 24, 72, and 168 h of infusion, and 24 and 168 h post infusion through rumen fistulas. The rumen liquor pH was measured using a standard pH meter. The liquid fraction of rumen contents passed through a 300-μM metal sieve was collected and centrifuged at 16,000× *g* for 3 min. The supernatant was decanted and the remaining solid materials were snap frozen in liquid nitrogen and stored in −80C until DNA extraction.

Metagenomic DNA extraction, amplicon preparation and pyrosequencing

Metagenomic DNA extraction and sequencing were essentially same as described [16]. Briefly, a QIAamp DNA stool kit (Qiagen, Valenica, CA) with modifications was used for DNA extraction. A six-minute incubation at 95°C was used to replace the 70°C lysis recommended in the standard protocol. A 570-bp region of the 16S rRNA gene (*E. coli* position 357 to 926) containing hypervariable regions V3- V5, selected because of their high variability [30], of the 16S rRNA gene, was amplified from 40 ng of metagenomic DNA with 8-bp sample-specific barcoded primers using 2.5 units of AccuPrime Taq DNA Polymerase High Fidelity (Invitrogen, Carlsbad, CA) in a 50-μl reaction buffer containing 200 nM primers, 200 nM dNTP, 60 mM Tris-SO_4, 18 mM $(NH4)_2SO_4$, 2.0 mM MgSO4, 1% glycerol, and 100 ng/ul bovine serum albumin (New England Biolabs, Ipswich, MA). PCR was performed using the following cycling profile: Initial denaturing at 95°C for 2 minutes followed by 25 cycles of 95°C 30 s, 50°C 30 s, and 72°C 120 s. The amplicons were generated from each metagenomic DNA sample separately, purified using a Agencourt AMPure XP kit (Beckman Coulter Genomics, Danvers, MA), and quantified using a QuantiFluor fluorometer (Promega, Madison, WI). The amplicons from individual samples were pooled in equal mass (molar) ratios. The amplicon pool at the desired size (~672 bp including primers and adaptors) was excised from 1.0% agarose gel and purified using a QIAquick Gel Extraction Kit (Qiagen). The purified amplicon library was further verified and quantified using a BioAnalyzer 2000 (Agilent) and subject to 454/ Roche pyrosequencing.

Unidirectional pyrosequencing of amplicon libraries was performed according to the manufacturer's instructions with a modification (App No 001-2009, Roche Applied Science, Indianapolis, IN). This modification, using a specific fusion primer design, accommodates amplification using the GS FLX Titanium emPCR Kits (Lib-L). Pyrosequencing was conducted using a GS FLX Titanium System (Roche) following the manufacturer's protocol. A total of 24 samples (4 cows at 6 time points, 0, 24, 72, and 168 h of infusion, and 24 and 168 h post infusion) were sequenced using 454 FLX Titanium pyrosequencing. The number of raw sequences reads for each sample was 30,606.58±5,261.7 (mean ± SD) at ~450 bp/read (NCBI SRA accession# SRA043755.1).

Sequence analysis

The raw reads were first decoded based on 8-bp bar codes; their quality was checked and artifacts were removed as previously described [16]. Sequences were filtered to remove low-quality reads and initially analyzed using BLAST against the GreenGene database. Sequence reads shorter than 200 bp were excluded and resultant quality sequence reads were then annotated using RDP Classifier [31] from the Ribosomal Database Project (release 10) at a 95% confidence threshold for their taxonomic classification and phylogenetic inference. Raw read counts were normalized. A square root transformation was applied to the relative abundance data. Bray-Curtis similarity matrix was then calculated using PRIMER v6 software.

The 16S rRNA gene sequences were also analyzed using CD-HIT-OTU [32] for phylogenetic inference at the OTU level. This algorithm uses a greedy incremental clustering process to identify OTUs from 16S rRNA gene tags, which involves 3 major steps: raw read filtering and trimming, selection of error-free reads, and clustering selected representative reads into individual OTUs at a user-specific cutoff (97% identity). The program avoids over estimation of OTUs, a common problem for many existing programs, and results in a rapid and more accurate estimation of microbial diversity in complex microbial ecosystems. OTUs identified using CD-HIT-OTU were then annotated using FR-HIT [33] against the GreenGene database. Statistical analysis was carried out according to MetaStats [34]. Self-organizing maps (SOMs) were generated according to unsupervised learning.

The 16S sequences were further analyzed using Fast UniFrac [35]. Briefly, the core set of the 16S GreenGene database was first downloaded, and the input 16S sequences were analyzed using MegaBLAST. The resultant hit table was input into the Fast UniFrac server for Principal Coordinates Analysis (PCoA).

Acknowledgments

Mention of trade names or commercial products in this publication is solely for the purpose of providing specific information and does not imply recommendation or endorsement by the U. S. Department of Agriculture. The USDA is an equal opportunity provider and employer.

Author Contributions

Conceived and designed the experiments: RWL. Performed the experiments: RWL RB CL. Analyzed the data: SW WL.

References

1. Bergman EN (1990) Energy contributions of volatile fatty acids from the gastrointestinal tract in various species. Physiological reviews 70: 567–590.
2. Bugaut M (1987) Occurrence, absorption and metabolism of short chain fatty acids in the digestive tract of mammals. Comparative biochemistry and physiology B, Comparative biochemistry 86: 439–472.
3. Li RW, Li C (2006) Butyrate induces profound changes in gene expression related to multiple signal pathways in bovine kidney epithelial cells. BMC genomics 7: 234.
4. Li CJ, Li RW, Wang YH, Elsasser TH (2007) Pathway analysis identifies perturbation of genetic networks induced by butyrate in a bovine kidney epithelial cell line. Functional & integrative genomics 7: 193–205.
5. Miettinen H, Huhtanen P (1996) Effects of the ratio of ruminal propionate to butyrate on milk yield and blood metabolites in dairy cows. Journal of dairy science 79: 851–861.
6. Pitt RE, Van Kessel JS, Fox DG, Pell AN, Barry MC, et al. (1996) Prediction of ruminal volatile fatty acids and pH within the net carbohydrate and protein system. Journal of animal science 74: 226–244.
7. Lopez S, Hovell FD, Dijkstra J, France J (2003) Effects of volatile fatty acid supply on their absorption and on water kinetics in the rumen of sheep sustained by intragastric infusions. Journal of animal science 81: 2609–2616.
8. Kristensen NB, Harmon DL (2004) Splanchnic metabolism of volatile fatty acids absorbed from the washed reticulorumen of steers. Journal of animal science 82: 2033–2042.
9. Huhtanen P, Miettinen H, Ylinen M (1993) Effect of increasing ruminal butyrate on milk yield and blood constituents in dairy cows fed a grass silage-based diet. Journal of dairy science 76: 1114–1124.
10. Storry JE, Rook JA (1966) The relationship in the cow between milk-fat secretion and ruminal volatile fatty acids. The British journal of nutrition 20: 217–228.
11. Barcenilla A, Pryde SE, Martin JC, Duncan SH, Stewart CS, et al. (2000) Phylogenetic relationships of butyrate-producing bacteria from the human gut. Applied and Environmental Microbiology 66: 1654–1661.
12. Pryde SE, Duncan SH, Hold GL, Stewart CS, Flint HJ (2002) The microbiology of butyrate formation in the human colon. FEMS microbiology letters 217: 133–139.
13. Sato T, Matsumoto K, Okumura T, Yokoi W, Naito E, et al. (2008) Isolation of lactate-utilizing butyrate-producing bacteria from human feces and in vivo administration of Anaerostipes caccae strain L2 and galacto-oligosaccharides in a rat model. FEMS microbiology ecology 66: 528–536.
14. Russell JB, Rychlik JL (2001) Factors that alter rumen microbial ecology. Science 292: 1119–1122.
15. Hess M, Sczyrba A, Egan R, Kim TW, Chokhawala H, et al. (2011) Metagenomic discovery of biomass-degrading genes and genomes from cow rumen. Science 331: 463–467.
16. Li RW, Connor EE, Li C, Baldwin Vi RL, Sparks ME (2011) Characterization of the rumen microbiota of pre-ruminant calves using metagenomic tools. Environmental microbiology 2011 Sep 12: doi: 10.1111/j.1462-2920.2011.02543.x.
17. Li CJ, Li RW (2008) Butyrate induced cell cycle arrest in bovine cells through targeting gene expression relevant to DNA replication apparatus. Gene regulation and systems biology 2: 113–123.
18. Wong JM, de Souza R, Kendall CW, Emam A, Jenkins DJ (2006) Colonic health: fermentation and short chain fatty acids. Journal of clinical gastroenterology 40: 235–243.
19. Li F, Hinderberger J, Seedorf H, Zhang J, Buckel W, et al. (2008) Coupled ferredoxin and crotonyl coenzyme A (CoA) reduction with NADH catalyzed by the butyryl-CoA dehydrogenase/Etf complex from Clostridium kluyveri. Journal of bacteriology 190: 843–850.
20. Louis P, Flint HJ (2009) Diversity, metabolism and microbial ecology of butyrate-producing bacteria from the human large intestine. FEMS microbiology letters 294: 1–8.
21. Louis P, Scott KP, Duncan SH, Flint HJ (2007) Understanding the effects of diet on bacterial metabolism in the large intestine. Journal of applied microbiology 102: 1197–1208.
22. Mrazek J, Tepsic K, Avgustin G, Kopecny J (2006) Diet-dependent shifts in ruminal butyrate-producing bacteria. Folia microbiologica 51: 294–298.
23. Topping DL, Clifton PM (2001) Short-chain fatty acids and human colonic function: roles of resistant starch and nonstarch polysaccharides. Physiological reviews 81: 1031–1064.
24. Silvi S, Rumney CJ, Cresci A, Rowland IR (1999) Resistant starch modifies gut microflora and microbial metabolism in human flora-associated rats inoculated with faeces from Italian and UK donors. Journal of applied microbiology 86: 521–530.
25. Duncan SH, Belenguer A, Holtrop G, Johnstone AM, Flint HJ, et al. (2007) Reduced dietary intake of carbohydrates by obese subjects results in decreased concentrations of butyrate and butyrate-producing bacteria in feces. Applied and environmental microbiology 73: 1073–1078.
26. Belenguer A, Duncan SH, Calder AG, Holtrop G, Louis P, et al. (2006) Two routes of metabolic cross-feeding between Bifidobacterium adolescentis and butyrate-producing anaerobes from the human gut. Applied and environmental microbiology 72: 3593–3599.
27. Ohkawara S, Furuya H, Nagashima K, Asanuma N, Hino T (2005) Oral administration of butyrivibrio fibrisolvens, a butyrate-producing bacterium, decreases the formation of aberrant crypt foci in the colon and rectum of mice. The Journal of nutrition 135: 2878–2883.
28. Ohkawara S, Furuya H, Nagashima K, Asanuma N, Hino T (2006) Effect of oral administration of Butyrivibrio fibrisolvens MDT-1 on experimental enterocolitis in mice. Clinical and vaccine immunology : CVI 13: 1231–1236.
29. Weimer PJ, Stevenson DM, Mantovani HC, Man SL (2010) Host specificity of the ruminal bacterial community in the dairy cow following near-total exchange of ruminal contents. Journal of dairy science 93: 5902–5912.
30. Andersson AF, Lindberg M, Jakobsson H, Backhed F, Nyren P, et al. (2008) Comparative analysis of human gut microbiota by barcoded pyrosequencing. PloS one 3: e2836.
31. Wang Q, Garrity GM, Tiedje JM, Cole JR (2007) Naive Bayesian classifier for rapid assignment of rRNA sequences into the new bacterial taxonomy. Applied and environmental microbiology 73: 5261–5267.
32. Li W, Godzik A (2006) Cd-hit: a fast program for clustering and comparing large sets of protein or nucleotide sequences. Bioinformatics 22: 1658–1659.
33. Niu B, Zhu Z, Fu L, Wu S, Li W (2011) FR-HIT, a very fast program to recruit metagenomic reads to homologous reference genomes. Bioinformatics 27: 1704–1705.
34. White JR, Nagarajan N, Pop M (2009) Statistical methods for detecting differentially abundant features in clinical metagenomic samples. PLoS computational biology 5: e1000352.
35. Hamady M, Lozupone C, Knight R (2010) Fast UniFrac: facilitating high-throughput phylogenetic analyses of microbial communities including analysis of pyrosequencing and PhyloChip data. The ISME journal 4: 17–27.

14

Oyster Reefs as Natural Breakwaters Mitigate Shoreline Loss and Facilitate Fisheries

Steven B. Scyphers[1,2]*, Sean P. Powers[1,2], Kenneth L. Heck Jr.[1,2], Dorothy Byron[2]

1 Department of Marine Sciences, University of South Alabama, Mobile, Alabama, United States of America, **2** Dauphin Island Sea Lab, Dauphin Island, Alabama, United States of America

Abstract

Shorelines at the interface of marine, estuarine and terrestrial biomes are among the most degraded and threatened habitats in the coastal zone because of their sensitivity to sea level rise, storms and increased human utilization. Previous efforts to protect shorelines have largely involved constructing bulkheads and seawalls which can detrimentally affect nearshore habitats. Recently, efforts have shifted towards "living shoreline" approaches that include biogenic breakwater reefs. Our study experimentally tested the efficacy of breakwater reefs constructed of oyster shell for protecting eroding coastal shorelines and their effect on nearshore fish and shellfish communities. Along two different stretches of eroding shoreline, we created replicated pairs of subtidal breakwater reefs and established unaltered reference areas as controls. At both sites we measured shoreline and bathymetric change and quantified oyster recruitment, fish and mobile macro-invertebrate abundances. Breakwater reef treatments mitigated shoreline retreat by more than 40% at one site, but overall vegetation retreat and erosion rates were high across all treatments and at both sites. Oyster settlement and subsequent survival were observed at both sites, with mean adult densities reaching more than eighty oysters m^{-2} at one site. We found the corridor between intertidal marsh and oyster reef breakwaters supported higher abundances and different communities of fishes than control plots without oyster reef habitat. Among the fishes and mobile invertebrates that appeared to be strongly enhanced were several economically-important species. Blue crabs (*Callinectes sapidus*) were the most clearly enhanced (+297%) by the presence of breakwater reefs, while red drum (*Sciaenops ocellatus*) (+108%), spotted seatrout (*Cynoscion nebulosus*) (+88%) and flounder (*Paralichthys* sp.) (+79%) also benefited. Although the vertical relief of the breakwater reefs was reduced over the course of our study and this compromised the shoreline protection capacity, the observed habitat value demonstrates ecological justification for future, more robust shoreline protection projects.

Editor: Howard Browman, Institute of Marine Research, Norway

Funding: Funding for this project was provided by the National Oceanic and Atmospheric Administration Fisheries Office of Habitat Restoration through the University of South Alabama's Oyster Restoration Program and the Northern Gulf Institute. The funders had no role in study design, data collection and analysis, decision to publish, or preparation of the manuscript.

Competing Interests: The authors have declared that no competing interests exist.

* E-mail: sscyphers@disl.org

Introduction

Nearshore, biogenic habitats of estuaries support a broad spectrum of marine life and serve as nursery grounds for economically-important fishes and shellfish [1–4]. Estuarine and vegetated nearshore habitats comprise only 0.7% of global biomes, yet the value of their ecosystem services has been estimated at $7.9 trillion dollars annually, or 23.7% of total global ecosystem services [5]. Nearshore ecosystem services include disturbance resistance, nutrient cycling, habitat, food production, and recreation. Unfortunately, coastal and estuarine shorelines are among the most degraded and threatened habitats in the world because of their sensitivity to sea level rise, storms and increased utilization by man [6,7]. Many previous efforts to protect shorelines have involved the introduction of hardened structures, such as seawalls, rocks or bulkheads to dampen or reflect wave energy [8–10]. Although such structures may adequately mitigate shoreline retreat, the ecological damages that result from their presence can be great [8,10,11]. The cumulative effects of habitat alteration and losses in the nearshore have had substantial economic and ecological consequences [12,13] and threaten the sustainability of many ecosystem services. Efforts to combat degradation and loss of nearshore, biogenic habitats have increased over the last decade [7,14,15]. Unfortunately, many shoreline protection approaches still value engineering over ecology in determining mitigation and restoration efficacy.

The "engineering first" approaches, including vertical bulkheads, concrete and granite rip-rap revetments and seawalls, are often used by coastal engineers because they are viewed as permanent and non-retreating structures. Unfortunately, insufficient concern may have been given to the ecological, aesthetic or socioeconomic impacts of these hardened structures. A major concern in implementing bulkheads and seawalls for coastal property protection is that erosive wave energies are reflected back into the water body, instead of being absorbed or dampened [10]. This subjects adjacent shorelines to even greater wave energy and can cause vertical erosion down the barrier with subsequent loss of intertidal habitats [10,16].

The benthic setting adjacent to many armored shores is generally absent of complex, structured habitats [16]. Most structurally complex, natural habitats are thought to function as nurseries for many finfish and shellfish species because of their elevated faunal densities, enhanced growth or survival rates, or higher contribution of individuals that emigrate offshore to adult habitats [1,4]. Biogenic, three-dimensional structure can reduce water velocities, increase sedimentation rates and enhance propagule settlement and retention, indirectly creating a more suitable environment for many species [17–20]. Despite the known lack of ecological benefits, shoreline hardening has continued to increase for decades primarily due to a lack of practical and ecologically valuable alternatives. However, a growing initiative for sustainable shoreline protection has focused on balancing effective protection and habitat creation by a variety of new methodologies collectively termed "living shorelines" [8].

Living shoreline projects often involve the planting or restoration of naturally-occurring biogenic habitats that have numerous ecological benefits, in addition to providing a buffer for wave action. In their natural setting, oyster reefs are often found seaward of salt marshes and can attenuate erosive wave energies, stabilize sediments and reduce marsh retreat, thereby making them an attractive living shoreline approach [19,21,22]. Beyond the targeted shoreline protection, living oyster reefs may provide many ecosystem services including seston filtration, benthic-pelagic coupling, refuge from predation and abundant prey resources [2,18]. Given adequate recruitment and survival, oyster reefs could be self-sustaining elements of coastal protection [21,22] that enhance other habitats of the natural landscape, although few studies have examined the premise of restoration through facilitation [17,23].

Located on the northern Gulf of Mexico, Mobile Bay is one of the best examples of a classic estuary [3] and, like many other coastal areas, is highly developed with a large and increasing proportion of its shorelines armored by bulkheads and seawalls [10] (Figure 1). At last analysis in 1997, Douglass and Pickel estimated that over 30% of the bay's available coastline was armored with over 10–20 acres of intertidal habitat lost, a high percentage in this microtidal bay (<0.5 m tidal amplitude). The historical armoring and marsh-edge losses have already had negative fisheries consequences, with projections of further reductions of blue crab harvest if armoring continues [24].

In this study, we experimentally examined the ecological effects of constructing subtidal breakwater oyster reefs for coastal and estuarine shoreline protection. In addition to documenting changes in the physical setting near breakwaters and unaltered control treatments, we quantified the habitat value for oysters, fishes and mobile invertebrates. We focus particular attention on the potential impacts on economically-important species, as this provides insight into the economic implications of different shoreline protection alternatives. We hypothesized that the addition of breakwater reefs of oyster shell would: 1) mitigate shoreline retreat, (2) provide substrate for recruitment and survival of oysters, (3) support higher densities of small fishes, mobile macro-invertebrates and larger and transient fishes and (4) promote higher species richness and a different community structure than unaltered control areas.

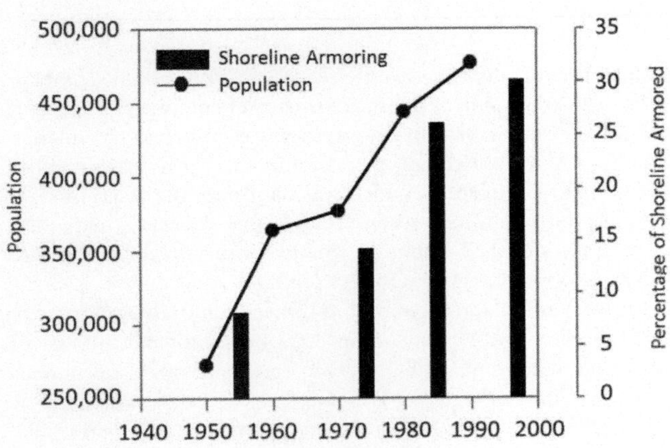

Figure 1. Population Growth and Shoreline Armoring in Mobile Bay, Alabama. Adapted with permission from Douglass and Pickel 1999, this figure depicts the rate and extent of shoreline armoring in Mobile Bay. The vertical bars in the main graph show the proportion of armoring while the line depicts the increasing population levels for Mobile and Baldwin Counties.

Methods

Ethics Statement

This study was conducted in accordance with the laws of the State of Alabama and under IACUC protocols (Permit # 05047-FSH) approved by the University of South Alabama.

Study Setting and Site Selection

To determine the ecological and physical effects of created breakwater oyster reefs, we conducted a manipulative field experiment at two sites in coastal Alabama that contained stretches of rapidly eroding coastlines. Study sites were selected within regions known to have adequate larval supply of oysters [25] and moderate wave climates [26]. At each site, we constructed two breakwater reefs of loose oyster shell and designated non-restored plots as controls in a randomized, paired design (Figure 2). The first site, known locally as Point aux Pins, received breakwater reefs in May 2007. The treatments at Point aux Pins (site center point: 30.370098, −88.308578) were located along the southern extent of a peninsula of eroding salt marsh habitat, largely comprised of fringing cordgrass (*Spartina alterniflora*) and black needlerush (*Juncus roemerianus*). Remnants of oysters (*Crassostrea virginica*) are found throughout the marsh and buried in the subtidal sediments. The second site, Alabama Port (site center point: 30.347917, −88.121338), is located along the southwestern shore of Mobile Bay, just north of the Dauphin Island bridge. The treatments at Alabama Port were located along a two kilometer stretch of eroding shoreline that has been encroached by armoring at its northern and southern extents. Small patches of *Spartina alterniflora* can be found at Alabama Port, but the most abundant vegetation is *Phragmites* sp., which is largely present in the upper intertidal zone. Both sites were selected within regions of high oyster spat settlement (40–180 spat m^{-2} day^{-1}) [25].

Breakwater Reef Dimensions

The experimental oyster reefs were designed as subtidal wave-attenuating breakwaters, a common coastal engineering approach [8]. Each reef complex was comprised of three 5 m×25 m rectangular-trapezoid sections (Figure 3B). Each section consisted of loose oyster shell, purchased from a local seafood processing plant, placed on a geo-textile fabric to prevent subsidence and secured by a plastic mesh covering (with 1 cm^2 openings) that was anchored by rebar. The purpose of the mesh covering was to help maintain the vertical relief of breakwaters until adequate recruitment of oysters cemented the loose shell in place. The initial height of each reef was slightly above MLLW (~1 m), under the assumption that the loose oyster shell would settle below that level and eventually become subtidal. The subtidal design of the reefs allowed for maximum exposure for oyster settlement and increased available substrate for foraging by transient and larger

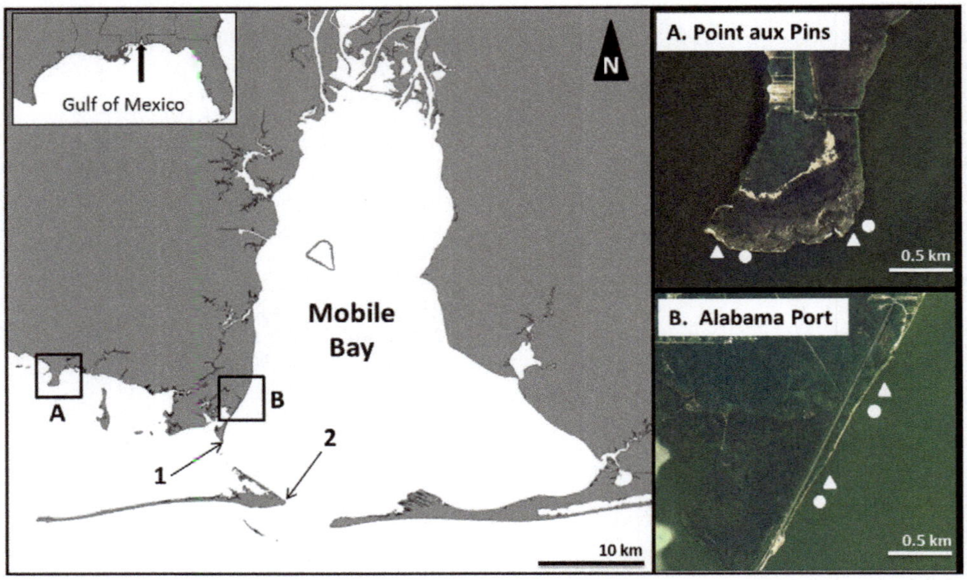

Figure 2. Map of Study Sites in Mobile Bay and Mississippi Sound, Alabama. White triangles represent breakwater reef complexes and white circles represent control treatments at the two restoration sites of (A) Point aux Pins and (B) Alabama Port. The locations of the (1) Cedar Point and (2) Dauphin Island hydrographic monitoring stations are denoted by the numbered arrows.

resident fishes, while maximizing potential capacity for wave attenuation.

Hydrographic Environment

Mean surface water temperatures, recorded by electronic thermometer, and salinity, measured by a refractometer, were recorded during each sampling event. To observe longer term patterns in salinity, we utilized publicly available data recorded by hydrographic monitoring stations located at Cedar Point and Dauphin Island, AL. The Cedar Point station is approximately 17.5 km from Point aux Pins and 4.0 km from Alabama Port site center points. The Dauphin Island station is approximately 25.5 km from Point aux Pins and 11.0 km from Alabama Port site center points. The Cedar Point station has been active since 2008 and the Dauphin Island station since 2003. To consider the effects of wave climate and dominant wind direction and magnitude on our study setting, we reviewed historical and recently published coastal engineering studies [26,27].

Shoreline and Bathymetry Change

Vegetation retreat and changes in nearshore depth profiles were monitored to evaluate the effect of the breakwaters on the nearshore setting. Bathymetry surveys were conducted at both sites during preliminary site selection and yearly following construction at Point aux Pins. Bathymetric data was collected using a Ceeducer Pro DGPS system with an integrated depth sounder mounted to a 1 m×2 m platform on pontoons. We surveyed each site manually by walking the pontoon through multiple parallel transects of the reef and control treatments. At each reef treatment, the breakwater reef footprint was delineated using the Ceeducer DGPS to measure reef spreading and consequential reduction in reef height. The width of each reef section was also measured by transect tape at reef construction and the end of the study to measure changes in reef footprint. The data collected by the Ceeducer unit was imported in ESRI's ArcView, corrected for tidal amplitude, and maps depicting depth at mean low water (MLW) were created. To measure the shoreward retreat of emergent vegetation, permanent rebar stakes were installed at 25 m intervals along the 100 m stretch of shoreline at each replicate treatment. Each 6 m rebar stake was driven into the marsh edge so that 1 m remained visible. These shoreline stakes were installed shortly after breakwater construction at both Point aux Pins and Alabama Port and were monitored periodically thereafter. During each survey, marsh retreat was measured as the distance from the rebar stake to the living vegetation line. Mean differences between vegetation retreat rates adjacent to breakwaters and controls were analyzed by repeated-measures analysis of variance (ANOVA). Because of differences in reef creation dates and sampling period, Point aux Pins and Alabama Port were analyzed separately.

Oyster Recruitment

To assess the value of the breakwater reef complexes for oysters and other sessile invertebrates, we periodically collected quadrat samples. Oyster settlement, growth and survival were quantified using a 0.25 m^2 quadrat, which was haphazardly placed at three locations on each reef section ($n=9$ per replicate reef). The exposed layer of shell within the quadrat was collected and placed in a large container. Juvenile ($\leq$3 cm) and adult oysters (>3 cm) were enumerated and measured in the field, and then returned to the reef. Mortality was quantified by enumerating dead oysters, which had both valves still articulated and were absent of fouling organisms inside the shell. We sampled the breakwater reefs at Point aux Pins in July, August and November 2007, May and October 2008 and June 2009. We sampled the reefs at Alabama Port in March, June and October 2008 and June 2009. For the final sampling period of June 2009, six 0.25 m^2 quadrats were sampled from each section ($n = 18$ per replicate reef) to account for the reef spreading and to assure a similar proportion of reef surface area was sampled.

We used univariate one-way ANOVA to test for differences in densities of live juveniles, live adults and dead oysters among

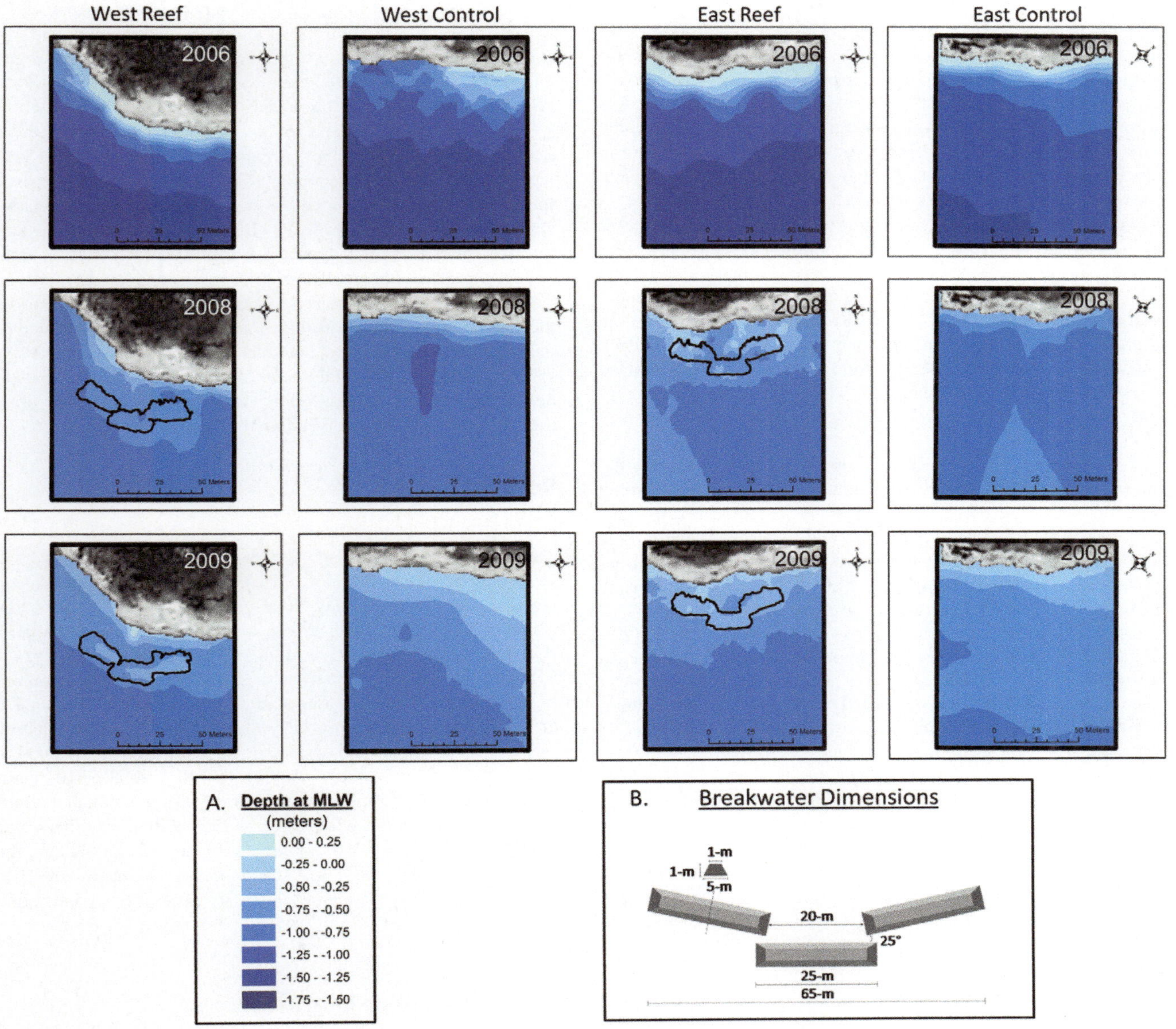

Figure 3. Bathymetry Plots from the Western Experimental Breakwater Reef and Control Treatments at Point aux Pins. The top row of 2006 plots was approximately one year prior to construction. The 2008 and 2009 plots are from one and two years post construction. Depth gradients are shown in inset (A). A schematic of the initial reef shape is depicted in (B). The crest width of each reef was approximately 1-m at MLLW.

sampling events. Point aux Pins and Alabama Port recruitment data were analyzed separately because the independent variable of sampling date was different at each site. Density estimates determined by individual quadrat samples (n = 3 all, except n = 6 for June 2009) for each reef section were averaged. The pooled values from each of the three reef sections for each of the two replicated treatments were used as replicates in a one-factor ANOVA to test the effect of sampling date. These data were tested for normality using the Kolmogorov-Smirnov test and homogeneity of variances using Bartlett's test. To meet the assumptions of ANOVA, all values were log transformed and retested. After transformation, minor violations of normality and equal variances were still present for live adults and dead oysters at both sites. Because the violations from quadrat sampling are generally minor and ANOVA is considered robust to such violation [28], we proceeded with parametric ANOVA. When ANOVA results showed significant differences, we used Tukey's HSD post-hoc test for multiple comparisons.

Fishes and Mobile Invertebrates

The response of fishes and mobile invertebrates was measured using a combination of gear types to target small and large individuals. Experimental gillnets (2 m×30 m) were used to capture larger species and individuals of coastal finfish species. Sampling occurred twice each month for one year following construction and monthly thereafter through all seasons, but was reduced to every other month during winter. Gillnets were deployed on adjacent sides of each reef or control treatment and perpendicular to shore. Each net was comprised of two 15 m panels (5 cm and 10 cm maximum opening) to broaden the size range and body shape of animals captured. Gillnets were fished for two hours starting one hour prior to sunrise. During winter months, low tides prevented crepuscular sampling so nets were fished for two hours starting one hour prior to sunset. Gillnets were retrieved in the same order they were deployed, and soak time was recorded as the time from when the net was first deployed until the time retrieval began. All specimens captured were placed in labeled bags and returned to the lab where they were identified, measured and their biomass recorded.

To quantify smaller fishes and invertebrates, we seined adjacent to each breakwater reef and control monthly, except every other month during winter. At each treatment, a 6 m wide bag seine with 6.25 mm mesh was towed three times between the treatment and shore. All seine distances were 15 m and terminated into the shore at Point aux Pins or a 4 m wide block net at Alabama Port. All captured mobile invertebrates and fishes were placed in labeled bags and returned to the laboratory where they were identified to the lowest taxonomic level possible, measured and biomass recorded.

To determine the effects of site and treatment on the communities of fishes and invertebrates, we used multivariate and univariate analyses. Differences in community structure between reef and control treatments and between Alabama Port and Point aux Pins sites were tested for each gear type using permutational analysis of variance (PERMANOVA). Multivariate PERMANOVA used Bray-Curtis similarity matrices of log (x+1) transformed abundance data with 4,999 permutations [29]. Logarithmic transformations were applied to reduce the influence of overwhelmingly abundant species. For univariate analyses on gillnet data, PERMANOVA was used to test for site and treatment effects on the total abundance, species richness and abundance of demersal fishes in an approach similar to parametric ANOVA. Univariate PERMANOVA tests were run on Euclidean distances matrices with 4,999 permutations [30]. PERMANOVA was chosen for univariate analyses because it allows for two-factor designs, considers an interaction term and does not assume a normal distribution of errors. The environmental classifications of demersal, pelagic (including benthopelagic, pelagic, and pelagic-neritic) and reef-associated fishes were acquired from FISHBASE [31]. Seine data were analyzed identically to gillnet data analyses as previously stated with the addition of a response variable containing only decapod crustaceans. All multivariate tests and univariate PERMANVOA were run in the software package PRIMER-E v6 [32] with the PERMANOVA extension.

To determine the effects of breakwater reefs on the most common demersal fishes and decapods, we analyzed these taxa separately as they include many economically-important coastal species. We used Wilcoxon signed-rank tests to compare relative abundances of each species (≥1%) between the paired breakwater reef and mudflat control treatments. This approach allowed us to test for overall treatment effects, while controlling for date and site variability through the paired experimental design but ignored their interactive effects. Certain species that were closely related or difficult to distinguish were analyzed as grouped taxa (e.g. *Menidia* sp., *Paralichthys* sp.). For all tests, we considered results of $p \leq 0.05$ to be significant. The ANOVA and Wilcoxon tests were run using the R Statistical Platform Version 10.1.1 [33].

Results

Hydrographic Environment

At Point aux Pins, mean surface water temperature over all sampling events was 21.4°C (±10.1 SD) measured by digital thermometer, and salinity averaged 23.1 PSU (±8.7 SD) measured by refractometer. Mean water temperature at Alabama Port was 21.8°C (±7.8 SD) and salinity averaged 16.1 PSU (±7.4 SD). Salinity data, shown as box and whisker plots, was acquired from hydrographic monitoring stations at Cedar Point (Figure 4A) and Dauphin Island (Figure 4B) to further investigate the salinity regime over a longer time period. Cedar Point data shows 2008 to have the highest salinity regime of the 2008–2010 years (Figure 4A). The Dauphin Island station shows a similar pattern with 2007 and 2008 having higher salinities than all other years between 2003 and 2010. In addition to higher average salinity, the outliers representing the lowest salinity measurements in 2007 and 2008 are substantially higher the other recent years indicating fewer freshets.

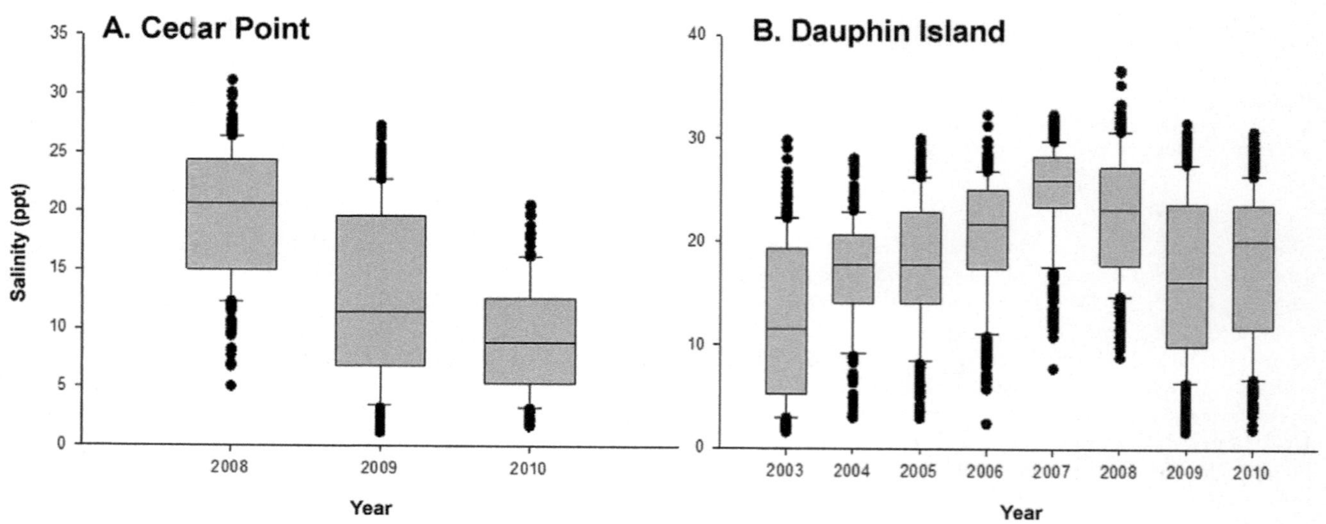

Figure 4. Salinity Ranges Recorded by Hydrographic Monitoring Stations in Coastal Alabama. Box and whisker plots of salinity data recorded by the hydrographic monitoring stations at (A) Cedar Point and (B) Dauphin Island. The Cedar Point Station has been active since 2008 and the Dauphin Island Station since 2003.

Most wind-driven wave energies along coastal Alabama shorelines are generated by dominant south to southeasterly winds from spring through early fall and north-oriented winds from late fall throughout most of winter [26,27]. Fetch at Point aux Pins averages approximately 15 km with a longest fetch of 32 km. The erosion rate at this site has potentially increased in recent years after Hurricane Katrina opened a one mile gap termed "Katrina Cut" in Dauphin Island, a protective barrier island located due south of Point aux Pins. The wave climate near Alabama Port is strongly affected by prevalent southeast winds, as well as the wakes of ships utilizing the Mobile shipping channel less than ten kilometers to the East. At Alabama Port, average fetch is approximately 21 km with a longest fetch of 34 km. For more detailed discussion of wind and wave climates, erosion, sediment sizes, Keddy exposure values and Knutson et al.'s vegetation success scores, refer to Roland and Douglass (2005) [26].

Shoreline and Bathymetry Changes

Changes in the nearshore and shoreline environments of reef and control sites were observed from measuring vegetation retreat and bathymetric surveys. Bathymetric surveys at Point aux Pins found that, in addition to a general trend of decreasing depth, areas inshore of breakwater reefs appeared to gain more sediments than areas inshore of control plots (Figure 3). The footprint of East and West breakwaters expanded approximately 300% over the course of the study, and reef crest height was reduced from approximately 1 m to 0.3 m. The living vegetation line at Point aux Pins retreated nearly 6 m on average in slightly over two years (Figure 5A). Repeated measures ANOVA found no differences in the vegetation retreat rates between treatments, a strong effect of time and no interaction between the two factors (Table S1). At Alabama Port, breakwater reefs mitigated vegetation retreat by more than 40% over two years (Figure 5B). Repeated measures ANOVA found a marginally-significant treatment effect ($p = 0.089$) and a strong effect of time with no interaction (Table S1).

Oyster Recruitment

Point aux Pins reefs were constructed in May 2007 and first sampled for oyster recruitment the following July. Densities of juvenile oysters continually increased until peaking at greater than 700 oysters m^{-2} in November 2007, but were much lower the following year with ranges between 50 and 150 m^{-2} ($F_{5,30} = 28.15$, $p \leq 0.001$, Figure 6A). Adult oysters were found in highest densities during November 2007 and May 2008 sampling with approximately 35 oysters m^{-2} ($F_{5,30} = 38.29$, $p \leq 0.001$, Figure 6B). The highest mortality was observed during the October 2008 sampling event ($F_{5,30} = 22.492$, $p \leq 0.001$, Figure 6C) and 88% of measured dead oysters were juveniles (≤3 cm).

Alabama Port reefs were constructed in October 2007 and were first sampled in March 2008. Live juveniles densities at Alabama Port were between 70 and 140 m^{-2} in the last three sampling events and higher than the first sampling event in March 2008 ($F_{3,20} = 47.40$, $p \leq 0.001$, Figure 6D). Adult oysters were observed first and at a maximum in October 2008 (~75 oysters m^{-2}) and found in lower densities in June 2009 (~20 oysters m^{-2}) ($F_{3,20} = 18.82$, $p \leq 0.001$, Figure 6E), although October and June were not significantly different. The first and highest mortality (~70 oysters m^{-2}) was recorded in October 2008 ($F_{3,20} = 114.29$, $p \leq 0.001$, Figure 6F), and juvenile oysters accounted for 80% of the total dead.

Fishes and Mobile Invertebrates

Gillnet and seine sampling near breakwater reefs and controls captured a diverse assemblage of fishes and mobile macro-invertebrates. From the use of multiple gears, over 100 species of fish and invertebrates were collected during the 30 month sampling period. Gillnet sampling collected nearly 8,000 individuals of 45 different species in 5 cm mesh panels while larger 10 cm panels captured over 1,500 individuals of 44 different species. Seines captured 71,640 individuals that represented 88 species or grouped taxa. Demersal fishes appeared to be the most broadly enhanced by the oyster reef structure when the overall percent difference in CPUE between oyster reefs and mudflat controls was calculated across both sites and all sampling events (Table S2). The dominant pelagic and reef-associated species did not appear strongly affected by oyster reef presence. Of the twelve species that comprised at least 1% of the 5 cm gillnet catch, six were categorized as demersal species. Four of these six demersal taxa were more abundant on breakwater reefs than controls. Spotted seatrout were 38% more abundant near breakwater reefs, and displayed the strongest trend of enhancement among 5 cm captured fishes. Twenty species comprised at least one percent of the 10 cm gillnet catch, and eleven of these were demersal fishes. Fourteen of the twenty species were captured more often

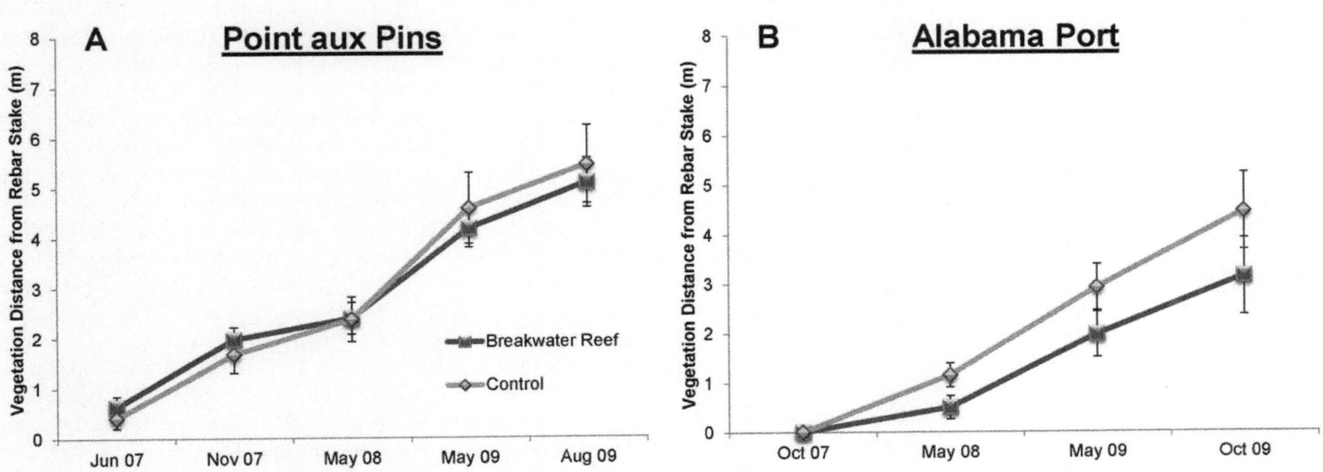

Figure 5. Shoreline Vegetation Retreat. Mean retreat (± SE) of living vegetation shoreward of each treatment at (A) Point aux Pins and (B) Alabama Port.

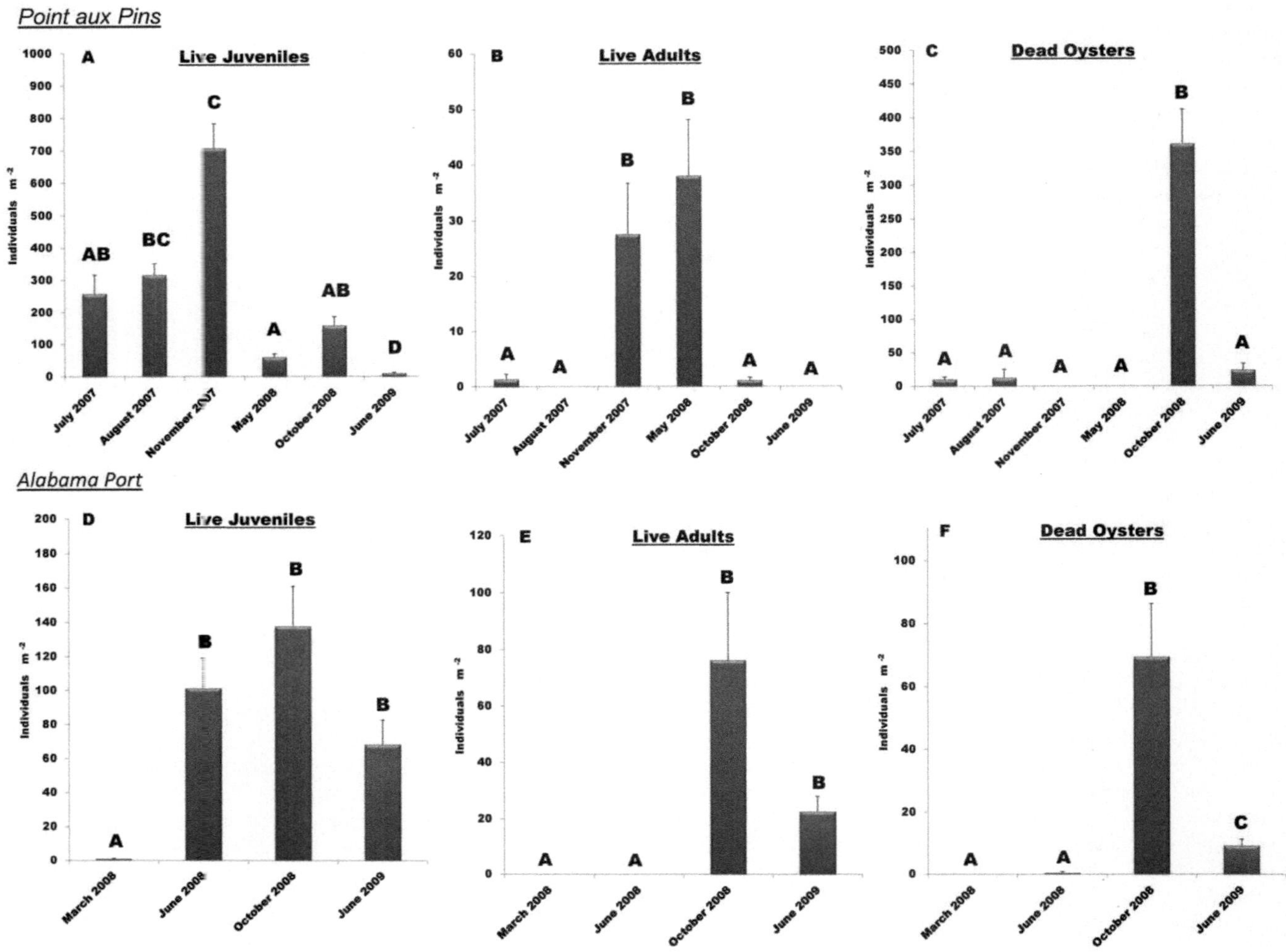

Figure 6. Oyster Recruitment and Survival. Mean oyster densities (+SE) of live juvenile, live adult and dead oysters at Point aux Pins (A–C) and Alabama Port (D–F). Different letters indicate statistical differences ($p<0.05$) from Tukey's HSD post-hoc tests.

near breakwater reefs than controls. Nine species or grouped taxa comprised at least one percent of seine catches, seven of which were more frequently captured near breakwater reefs. Included in these seven were three demersal fishes and three decapod crustaceans.

We used multivariate PERMANOVA to test for differences in the community structure between breakwater reef and control treatments. PERMANOVA tests on 5 cm gillnet catches found that site and the site-treatment interaction were both significant factors (Table S3). There were no community-level differences between our breakwater and control treatments with 5 cm captured fishes. The communities of larger fishes captured by 10 cm gillnets differed significantly by site, treatment and the interaction of the two factors (Table S3). The community structure of smaller and juvenile fishes and mobile invertebrates captured by seines were different between sites and treatments, with no interaction between the two factors (Table S3).

We used univariate PERMANOVA tests on total abundance, species richness and demersal and decapod abundances to detect differences between breakwater reef and control treatments and between Alabama Port and Point aux Pins. For 5 cm total abundance, a significant interaction between site and treatment was observed (Table S4) because total abundance was higher near breakwaters at Point aux Pins but higher near controls at Alabama Port (Figure 7A). For 10 cm gillnet catch, total abundance was higher adjacent to oyster reefs than controls (Table S4). For both 10 cm and seine data, abundances were significantly higher at Point aux Pins than Alabama Port, and no interaction was observed between site and treatment (Table S4). The PERMANOVA tests on species richness found significant differences between sites across all gear types, between treatments only for 10 cm catches and no significant interactions (Table S4). For 10 cm catches, species richness was significantly higher near reefs than controls (Table S4) and higher at Point aux Pins than Alabama Port. Demersal fishes showed no differences between reef and mudflat treatments for 5-cm catches (Table S3), but again there was a significant interaction between site and treatment(Figure 7B). For 10 cm, demersal fishes were more abundant near breakwater oyster reefs (Figure 8A) and higher at Point aux Pins. From seine catches, demersal fish abundance showed no differences, but decapod crustacean abundance was higher near reefs than mudflat controls (Table S4 and Figure 8B).

The relative abundance of each demersal fish and decapod species ($\geq 1\%$) between breakwater and control treatments was tested using paired Wilcoxon signed-rank tests. For 5 cm gillnet samples, six demersal species contributed $\geq 1\%$ of the total catch

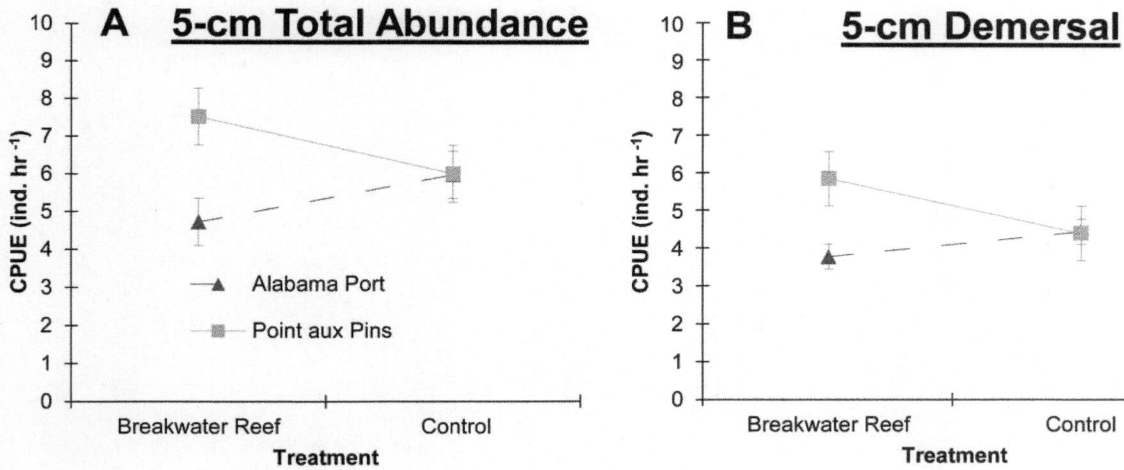

Figure 7. Relative Demersal Fish and Decapod Crustacean Abundance. Mean ±1 SE CPUE of (A) demersal fishes separated by collection method and (B) decapod crustaceans collected by seines near breakwater reefs and controls. Significant differences at P≤0.05 from univariate PERMANOA tests are indicated by asterisks.

(Table S5). Of those, only sand seatrout abundance was significantly enhanced by breakwater reefs (Figure 9A). Silver perch, spotted seatrout and southern kingfish showed positive trends of enhancement, but not statistically significant. Eleven demersal fishes were analyzed from the 10 cm catches, seven of which were significantly enhanced by reefs including sand seatrout, spotted seatrout, red drum and black drum (Table S5 and Figure 9B). Only finetooth shark abundance in 10 cm gillnets was significantly greater on controls than breakwater reef treatments. Seine samples had nine species or taxa that comprised ≥1% of the total catch, including three demersal fish species and three decapods. Of the demersal fishes, which were silver perch, Atlantic croaker and juvenile sciaenids, only silver perch showed a significant difference and were more common near breakwater reefs (Figure 9C). All three decapods, caridean shrimp, penaeid shrimp and blue crabs were present in significantly higher densities near breakwater reefs.

Discussion

Our study found that breakwater reefs constructed of loose oyster shell provided substrate for oyster recruitment and harbored a more diverse community of fishes and mobile invertebrates than control areas without reefs. This habitat enhancement is uncommon among shoreline protection schemes and could be a vast improvement over traditional armoring techniques, many of which have detrimental impacts on nearshore species [16]. While our experimental breakwaters were an "ecology-first" approach

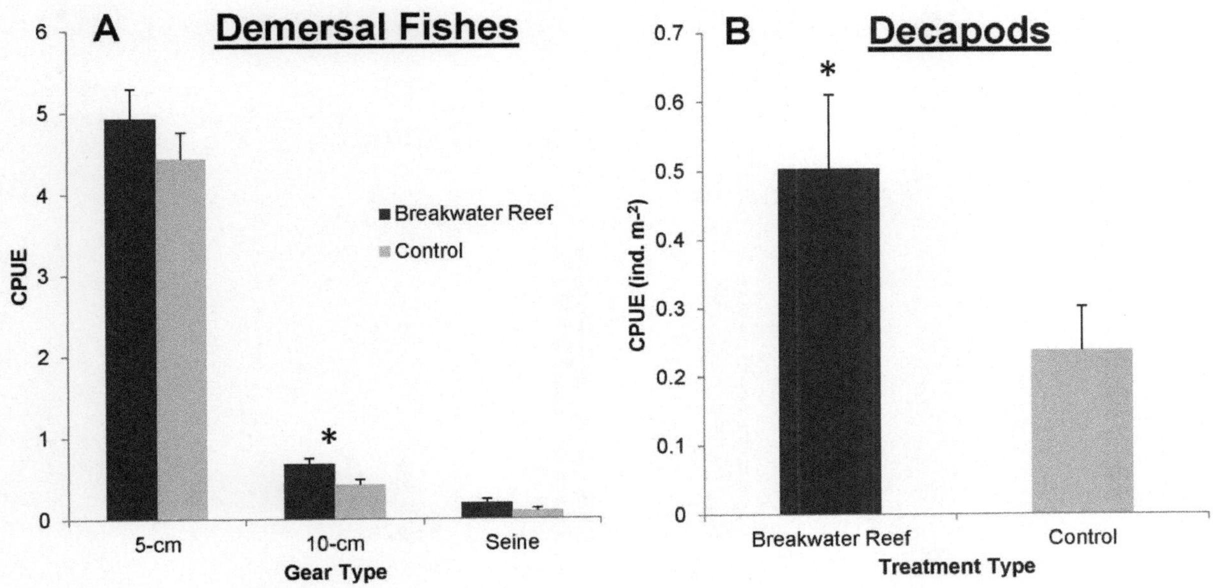

Figure 8. Total Abundance and Demersal Fish Abundance Separated by Site. Mean+1 SE catch per unit effort of (A) total fish and invertebrate abundance and (B) demersal fish abundance collected by 5 cm gillnets. CPUE is presented as the total individuals captured for each hour of soak time.

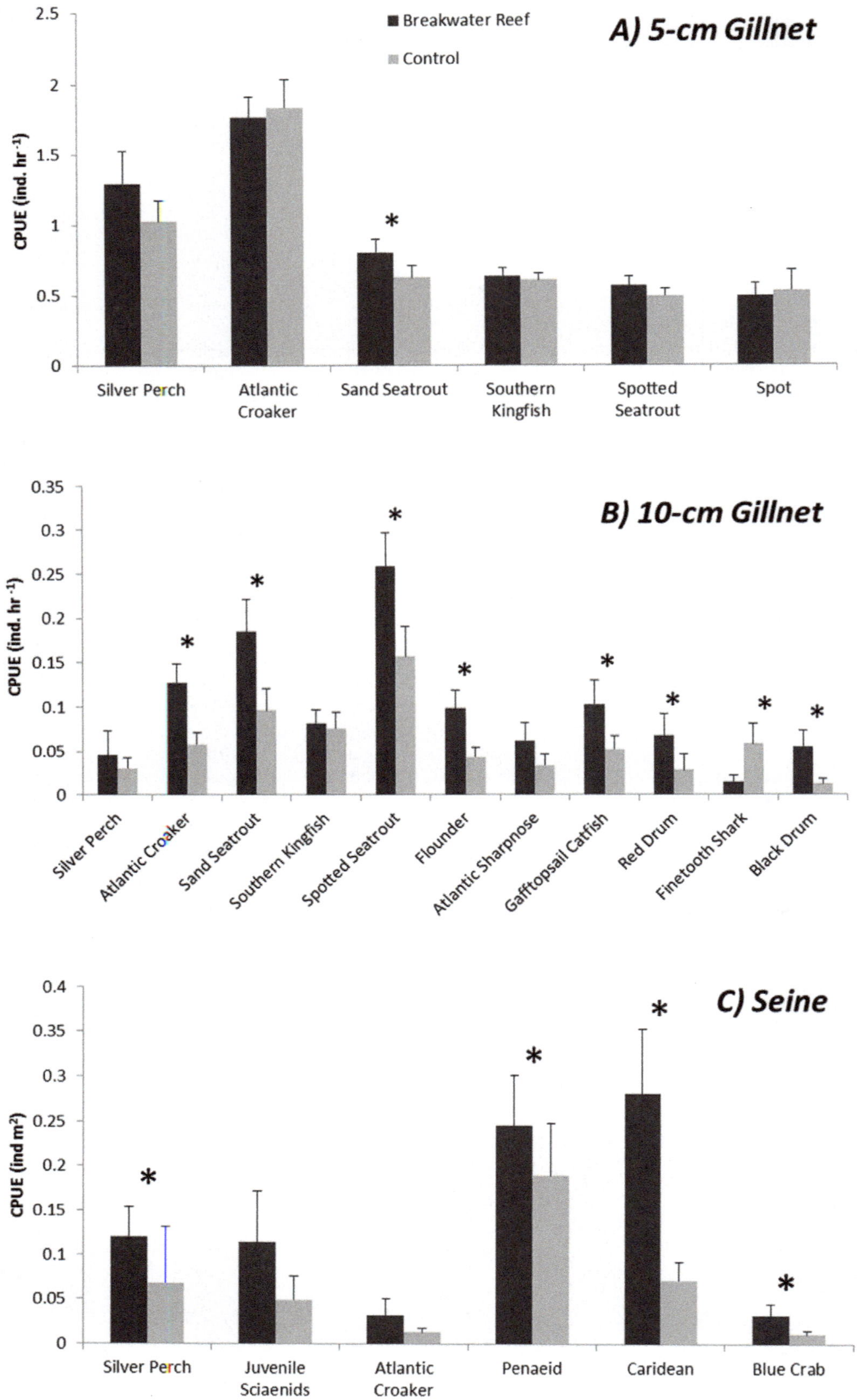

Figure 9. Relative Abundance of Dominant Demersal Fish and Decapod Taxa. Mean+1 SE CPUE of dominant demersal and decapod species or grouped taxa between treatments. Significant differences at $P \leq 0.05$ from Wilcoxon signed rank tests comparing paired breakwater reef and control treatments.

and were successful in creating valuable habitat, they did not provide the amount of protection that could be offered by well-engineered methodologies. This shortcoming highlights the need for coastal protection philosophies that balance ecology and engineering. However, an approach similar to ours could serve as an immediate solution to the habitat losses experienced along many sheltered coasts. In these settings, breakwater oyster reefs that were installed seaward of already armored shorelines could mitigate losses of fish and shellfish habitat.

Roland and Douglass (2005) found that many stretches of Alabama's shoreline are faced with wave energies well above critical limits where vegetation can naturally persist and proposed breakwaters as a potential mechanism to reduce wave energies [26]. The wave-attenuating capacity of the breakwaters in our study was compromised because the loose shell reefs expanded and flattened prior to the cementing together that could result from oyster settlement and survival. The mesh covering used in our study to maintain the breakwater reefs' integrity was not rigid enough to withstand the wave energy of our sites, but an improvement in this aspect of the breakwater design could allow for better shoreline protection and less disturbance of the reef. To mitigate reef spreading and flattening, we suggest the introduction of a more rigid structure as a temporary backbone which would deteriorate or could be removed after reef cementing occurred.

At both Alabama Port and Point aux Pins, we documented oyster recruitment and survival to reproductive size, but substantial mortality limited reef cementing and success. The high mortality recorded at both sites during October 2008 sampling appeared to be caused by predation or physical disturbance, such as wave energy. During this sampling period, very few exposed oysters were observed to be alive. In contrast, nearly all live oysters observed were found sheltered inside of dead, but still hinged oyster shells. This suggests that it is unlikely disease was the cause of mortality, since structurally-protected oysters would have no reprieve. Another factor that frequently affects oyster survival is reef height as tall reefs escape the poor water quality sometimes found near the sediment [34]. As previously discussed, the vertical relief of our reefs did decline over time, but again it is unlikely that sheltered oysters would survive if water quality caused the observed mortality. Physical disturbance could have caused many of the oyster shells that were on the surface and available for settlement to be buried under other shells, also explaining the lowered densities of live oysters. Predation is likely the most plausible explanation for the differential mortality between sheltered and exposed oysters. We frequently observed black drum, southern oyster drills (*Stramonita haemastoma*) and several species of crabs near the reefs. Stomach content analysis of the black drum collected in gillnets usually found oyster shell remains and dead oysters often showed signs of predation (S Scyphers, Pers. Observ.). A recent mark and recapture study of subtidal oyster reefs in coastal Alabama waters also documented drills as the most prevalent cause of mortality due to visible scarring on dead spat shells [35]. The high salinities and absence of freshets observed during the drought conditions 2007 and 2008 were likely beneficial for the oyster drill predators which thrive in higher salinity conditions [36,37].

The communities of fishes and mobile invertebrates that benefit from oyster reefs have been well-described, but very few studies have examined the enhancement from oyster reefs designed for protecting shorelines. The elevated species richness and densities that we observed during our study concur with most literature describing oyster reef habitats [2]. From our seines, we found blue crabs, penaeid and caridean shrimp, and juvenile silver perch were more abundant near oyster reefs than mudflat controls. Higher blue crab densities near reefs were likely due to the refuge value, as their recruitment and survival is largely augmented by structured habitats [38]. Blue crabs support an important commercial fishery throughout Gulf and Atlantic estuaries and, along with caridean and penaeid shrimp, are commonly found in the diets of several of the larger fishes. From our 10 cm gillnet sampling, we found that spotted seatrout, drum and flounder were substantially enhanced by oyster reefs. The paradigm of abundance, biomass and species richness being higher in structured areas and further increasing with habitat complexity is a pattern observed in nearly all nearshore ecosystems [20,39–41], but the relative importance of food versus refuge within structured habitats remains unresolved [42,43].

Landscape attributes, such as adjacent habitats or bathymetric features, commonly influence community composition [44–46] and are probably responsible for the interaction between site and treatment for the total abundance and demersal abundance of 5 cm gillnet catches. The interaction was driven by demersal fishes (Figure 7) and these catches were dominated by Atlantic croaker and silver perch, both which are recognized to predominately feed in non-structured habitats [47]. Geraldi et al. (2009) found very little evidence of enhancement by oyster reefs restored in marsh tidal creeks and concluded that the area was not limited by complex structure and therefore the addition of oyster shell was functionally redundant. Grabowski et al. 2005 concluded that small or few reefs may not measurably enhance transient predators. Interestingly, the broad enhancement we observed occurred in a similar setting with each reef located near structurally-complex saltmarsh habitat and of moderate size ($\sim$225 m^2).

It has proven quite challenging to predict the ecosystem services to be expected from restoring reefs at different scales or in different settings [34,41,42]. Ecosystem services provided by shallow marine habitats have received considerable attention from natural and social scientists seeking to quantify and predict potential benefits from protection or restoration [5,48,49]. Historically, most of these studies have focused on wetlands, seagrass meadows, coral reefs and mangroves [5,50,51], all habitats that receive considerable protection because of their productivity. Oyster reefs also provide important ecosystem services [18], but are more challenging to protect and manage because they are an exploited fishery [2]. A long history of excessive and destructive harvesting coupled with natural stressors like disease and storms have left shellfish populations in global demise [52–54]. Most large or landscape scale oyster reef restoration efforts have primarily targeted the re-establishment of harvestable oysters, many of which failed to achieve previous population levels. Some recent studies have detailed shortcomings of oyster restoration and cast serious doubts on the ability to achieve restoration success in subtidal and often large-scale efforts [55]. However, other recent studies have documented restored reefs that have persisted over decades [56] and on unrivaled spatial scales [57]. Attempts to quantify the economic benefits from restoring oyster reefs are very recent and forthcoming and could provide more support for protecting and restoring oyster reefs for the goods and services they provide [58,59].

Awareness of the detrimental impacts of shoreline armoring has increased in recent years, but movement towards more ecologically-responsible methods has been limited by the lack of cost-effective alternatives. "Living shoreline" approaches, including breakwater reefs, that protect coastal uplands could provide a more ecologically-responsible alternative to traditional armoring and not only mitigate coastal erosion, but also enhance certain economically-valuable fish stocks. However, as our study demonstrated, efforts to sustainably and responsibly protect coastal shoreline habitats must balance both engineering and ecology.

Supporting Information

Table S1 Results of Repeated-Measures ANOVA on Vegetation Retreat.

Table S2 Relative Abundance of the Most Abundant Fishes and Mobile Invertebrates.

Table S3 Results from Multivariate PERMANOVA Tests.

Table S4 Results from Univariate PERMANOVA Tests.

Table S5 Results from Wilcoxon Signed-Rank Tests on Single Species or Grouped Taxa.

Acknowledgments

Advice from Loren Coen, Joel Fodrie, Marti Anderson, Matt Ajemian and Michelle Brodeur and helpful comments from Howard Browman and two anonymous reviewers greatly improved our study and manuscript. We are grateful for the help provided by Carly Canion, numerous technicians, interns and graduate students that assisted in the field and lab components of this study. We thank Scott Douglass for permission to modify and include Figure 1. Caitlin Bovery and Mallory Scyphers lent their expertise on bathymetry plots and data presentation. We acknowledge the Dauphin Island Sea Lab for the hydrographic data, and we appreciate the assistance of Laure Carassou for analysis.

Author Contributions

Conceived and designed the experiments: SS SP KH DB. Performed the experiments: SS SP KH DB. Analyzed the data: SS SP KH DB. Contributed reagents/materials/analysis tools: SS SP KH DB. Wrote the paper: SS SP KH DB.

References

1. Beck MW, Heck KLJ, Able KW, Childers DL, Eggleston DB, et al. (2001) The identification, conservation, and managment of estuarine and marine nurseries for fish and invertebrates. Bioscience 51: 633–641.
2. Coen LD, Luckenbach MW, Breitburg DL (1999) The role of oyster reefs as essential fish habitat: a review of current knowledge and some new perspectives. American Fisheries Society Symposium 22: 438–454.
3. Gunter G (1967) Some Relationships of Estuaries to the Fisheries of the Gulf of Mexico. In: Lauff GH, ed. Estuaries. Washington, D.C.: American Association for the Advancement of Science. pp 621–638.
4. Heck KL, Hays G, Orth RJ (2003) Critical evaluation of the nursery role hypothesis for seagrass meadows. Marine Ecology Progress Series 253: 123–136.
5. Costanza R, d'Arge R, de Groot R, Farber S, Grasso M, et al. (1997) The value of the world's ecosystem services and natural capital. Nature 387: 253–261.
6. Syvitski JPM, Vorosmarty CJ, Kettner AJ, Green P (2005) Impact of humans on the flux of terrestrial sediment to the global coastal ocean. Science 308: 376–380.
7. Vitousek PM, Mooney HA, Lubchenco J, Melillo JM (1997) Human domination of Earth's ecosystems. Science 277.
8. NRC (2007) Mitigating shore erosion along sheltered coasts. Council NR, ed. Washington, D.C.: National Academies Press. 174 p.
9. Pilkey OH, Wright W (1988) Sea-walls vs. Beaches. Journal of Coastal Research. pp 41–64.
10. Douglass SL, Pickel BH (1999) The tide doesn't go out anymore - the effect of bulkheads on urban shorelines. Shore and Beach 67: 19–25.
11. Kennish MJ (2002) Environmental threats and environmental future of estuaries. Environmental Conservation 29: 78–107.
12. Chambers JR (1991) Coastal Degradation and Fish Population Losses. In: Stroud R, ed. Stemming the Tide of Coastal Fish Habitat Loss: National Coalition for Marine Conservation. pp 45–51.
13. Jackson JBC, Kirby MX, Berger WH, Bjorndal KA, Botsford LW, et al. (2001) Historical overfishing and the recent collapse of coastal ecosystems. Science 293: 629–638.
14. Minton MD (1999) Coastal Wetland Restoration and Its Potential Impact on Fishery Resources in the Northeastern United States. American Fisheries Society Symposium 22: 405–420.
15. Thayer GW, McTigue TA, Bellmer RJ, Burrows FM, Merkey DH, et al. (2003) Science-based restoration monitoring of coastal habitats, Volume One: A framework for monitoring plans under the Estuaries and Clean Waters Act of 2000 (Public Law 160–457). NOAA Coastal Ocean Program Decision Analysis Series No 23 : 35 plus appendices.
16. Bilkovic D, Roggero M (2008) Effects of coastal development on nearshore estuarine nekton communities. Marine Ecology Progress Series 358: 27–39.
17. Bruno JF, Kennedy CW (2000) Reduction of wave-related disturbance in cobble beach habitats by Spartina alterniflora: Importance of patch size. Oecologia 122: 98–108.
18. Coen LD, Brumbaugh RD, Bushek D, Grizzle R, Luckenbach MW, et al. (2007) AS WE SEE IT: Ecosystem Services related to oyster restoration. Marine Ecology Progress Series 341: 303–307.
19. Dame RF, Patten BC (1981) Analysis of energy flows in an intertidal oyster reef. Marine Ecology Progress Series 5: 115–124.
20. Orth RJ (1977) The importance of sediment stability in seagrass communities. In: Coull B, ed. Ecology of Marine Benthos. ColumbiaSC: University of South Carolina. pp 281–300.
21. Meyer DL, Townsend EC, Thayer GW (1997) Stabilization and Erosion Control Value of Oyster Cultch for Intertidal Marsh. Restoration Ecology 5: 93–99.
22. Piazza BP, Banks PD, La Peyre MK (2005) The potential for created oyster shell reefs as a sustainable shoreline protection strategy in Louisiana. Restoration Ecology 13: 499–506.
23. Bruno JF, Stachowicz JJ, Bertness MD (2003) Inclusion of facilitation into ecological theory. Trends in Ecology and Evolution 18: 119–125.
24. Jordan SJ, Smith LM, Nestlerode JA (2009) Cumulative effects of coastal habitat alterations on fishery resources: toward prediction at regional scales. Ecology and Society 14: 16.
25. Kim C-K, Park K, Powers SP, Graham WM, Bayha KM (2010) Oyster larval transport in coastal Alabama: Dominance of physical transport over biological behavior in a shallow estuary. Journal of Geophysical Research 115: 1–16.
26. Roland RM, Douglass SL (2005) Estimating wave tolerance of Spartina alterniflora in coastal Alabama. Journal of Coastal Research 21: 453–463.
27. Schroeder WW, Wiseman WJJ (1985) An Analysis of the Winds (1974–1984) and Sea Level Elevations (1973–1983) in Coastal Alabama. Mississippi-Alabama Sea Grant Consortium Publication MASGP-84-024. pp i-102.
28. Underwood AJ (1981) Techniques for analysis of variance in experimental marine biology and ecology. Oceanography and Marine Biology: An Annual Review 19: 513–603.
29. Anderson MJ (2001) A new method for non-parametric multivariate analysis of variance. Austral Ecology 26: 32–46.
30. Anderson MJ (2001) Permutation tests for univariate or multivariate analysis of variance and regression. Canadian Journal of Fisheries and Aquatic Sciences 58: 626–639.
31. Froese R, Pauly D (2010) Editors, Fishbase. World Wild Web Electronic Publication. www.fishbase.org, version (11/2010).
32. Clarke KR, Gorely RN (2006) Primer v6: User Manual/Tutorial. PRIMER-E, Plymouth.
33. CRAN (2010) R: A language and environment for statistical Computing. R Foundation for Statstical Computing, Vienna, Austria ISBN 3-900051-07-0, URL http://wwwR-projectorg.
34. Lenihan HS (1999) Physical-biological coupling on oyster reefs: how habitat structure influences individual performance. Ecological Monographs 69: 251–275.
35. Gregalis KC, Powers SP, Heck KLJ (2008) Restoration of oyster reefs along a bio-physical gradient in Mobile Bay, AL. Journal of Shellfish Research 27: 1163–1169.
36. Wells HW (1961) The fauna of oyster beds, with special reference to the salinity factor. Ecological Monographs 31: 239–266.
37. Garton D, Stickle WB (1980) Effects of salinity and temperature on the predation rate of Thais haemastoma on Crassostrea virginica spat. Biological Bulletin 158: 49–57.
38. Moksnes P-O, Heck KL (2006) Relative importance of habitat selection and predation for the distribution of blue crab megalopae and young juveniles. Marine Ecology Progress Series 308: 165–181.
39. Goreau T, Goreau N, Goreau T (1979) Corals and coral reefs. Scientific American 241: 124–136.
40. Rozas LP, Odum WE (1987) Fish and macrocrustacean use of submerged plant beds in tidal freshwater marsh creeks. Marine Ecology Progress Series 38: 101–108.

41. Wittman J (1985) Refuges, biological disturbance, and rocky subtidal community structure. New England Ecological Monographs 55: 421–445.
42. Bostrom C, Mattila J (1999) The relative importance of food and shelter for seagrass-associated invertebrates: a latitudinal comparison of habitat choice by isopod grazers. Oecologia 120: 162–170.
43. Grabowski JH, Powers SP (2004) Habitat complexity mitigates trophic transfer on oyster reefs. Marine Ecology Progress Series 277: 291–295.
44. Geraldi NR, Powers SP, Heck KL, Cebrian J (2009) Can habitat restoration be redundant? Response of mobile fishes and crustaceans to oyster reef restoration in marsh tidal creeks. Marine Ecology Progress Series 389: 171–180.
45. Grabowski JH, Hughes AR, Kimbro DL, Dolan MA (2005) How habitat setting influences restored oyster reef communities. Ecology 86: 1926–1935.
46. Micheli F, Peterson CH (1999) Estuarine vegetated habitats as corridors for predator movements. Conservation Biology 13: 869–881.
47. Peterson CH, Grabowski JH, Powers SP (2003) Estimated enhancement of fish production resulting from restoring oyster reef habitat: quantitative valuation. Marine Ecology Progress Series 264: 249–264.
48. Hobbs RJ, Harris LE (2001) Restoration ecology: repairing Earth's ecosystems in the new millenium. Restoration Ecology 9: 239–246.
49. Peterson CH, Lipcius RN (2003) Conceptual progress towards predicting quantitative ecosystem benefits of ecological restorations. Marine Ecology Progress Series 264: 297–307.
50. Moberg F, Folke C (1999) Ecological goods and services of coral reef ecosystems. Ecological Economics 29: 215–233.
51. Ronnback P (1999) The ecological basis for economic value of seafood production supported by mangrove ecosystems. Ecological Economics 29: 235–252.
52. Beck MW, Brumbaugh RD, Airoldi L, Carranza A, Coen LD, et al. (2011) Oyster reefs at risk and recommendations for conservation, restoration, and management. Bio Science 61: 107–116.
53. Kirby MX (2004) Fishing down the coast: historical expansion and collapse of oyster fisheries along continental margins. Proceedings of the National Academy of Sciences 101: 13096–13099.
54. Rothschild B, Ault J, Goulletquer P, Heral M (1994) Decline of Chesapeake Bay oyster population: a century of habitat destruction and overfishing. Marine Ecology Progress Series 111: 29–39.
55. Mann R, Powell EN (2007) Why oyster restoration goals in the Chesapeake Bay are not and probaly cannot be acheived. Journal of Shellfish Research 26: 905–917.
56. Powers SP, Peterson CH, Grabowski JH, Lenihan HS (2009) Success of constructed oyster reefs in no-harvest sanctuaries: implications for restoration. Marine Ecology Progress Series 389: 159–170.
57. Schulte DM, Burke RP, Lipcius RN (2009) Unprecedented restoration of a native oyster metapopulation. Science 325: 1124–1128.
58. Brumbaugh RD, Torropova C (2008) Economic valuation of ecosystem services: A new impetus for shellfish restoration? Basins and Coasts 2: 8–15.
59. Grabowski JH, Peterson CH (2007) Restoring Oyster Reefs to Recover Ecosystem Services. In: Cuddington K, Byers JE, Wilson W, Hastings A, eds. Ecosystem Engineers, 4: Academic Press.

15

Ecological Implications of Extreme Events: Footprints of the 2010 Earthquake along the Chilean Coast

Eduardo Jaramillo[1]*, Jenifer E. Dugan[2], David M. Hubbard[2], Daniel Melnick[3], Mario Manzano[1], Cristian Duarte[4], Cesar Campos[1], Roland Sanchez[1]

1 Instituto de Ciencias Ambientales y Evolutivas, Universidad Austral de Chile, Valdivia, Chile, **2** Marine Science Institute, University of California Santa Barbara, Santa Barbara, California, United States of America, **3** Institut für Geowissenschaften, Universität Potsdam, Potsdam, Germany, **4** Instituto de Ciencias Marinas y Limnológicas, Universidad Austral de Chile, Valdivia, Chile

Abstract

Deciphering ecological effects of major catastrophic events such as earthquakes, tsunamis, volcanic eruptions, storms and fires, requires rapid interdisciplinary efforts often hampered by a lack of pre-event data. Using results of intertidal surveys conducted shortly before and immediately after Chile's 2010 M_w 8.8 earthquake along the entire rupture zone (*ca.* 34–38°S), we provide the first quantification of earthquake and tsunami effects on sandy beach ecosystems. Our study incorporated anthropogenic coastal development as a key design factor. Ecological responses of beach ecosystems were strongly affected by the magnitude of land-level change. Subsidence along the northern rupture segment combined with tsunami-associated disturbance and drowned beaches. In contrast, along the co-seismically uplifted southern rupture, beaches widened and flattened increasing habitat availability. Post-event changes in abundance and distribution of mobile intertidal invertebrates were not uniform, varying with land-level change, tsunami height and coastal development. On beaches where subsidence occurred, intertidal zones and their associated species disappeared. On some beaches, uplift of rocky sub-tidal substrate eliminated low intertidal sand beach habitat for ecologically important species. On others, unexpected interactions of uplift with man-made coastal armouring included restoration of upper and mid-intertidal habitat seaward of armouring followed by rapid colonization of mobile crustaceans typical of these zones formerly excluded by constraints imposed by the armouring structures. Responses of coastal ecosystems to major earthquakes appear to vary strongly with land-level change, the mobility of the biota and shore type. Our results show that interactions of extreme events with human-altered shorelines can produce surprising ecological outcomes, and suggest these complex responses to landscape alteration can leave lasting footprints in coastal ecosystems.

Editor: Jack Anthony Gilbert, Argonne National Laboratory, United States of America

Funding: This work was supported by grants from CONICYT (National Commision for Science, Chile; www.conicyt.cl) and Dirección de Investigación y Desarrollo, Universidad Austral de Chile (http://investigacion.uach.cl/direccion/index.php) to EJ. Support was provided by Santa Barbara Coastal Long Term Ecological Research program (National Science Foundation Award # OCE-0620276) (www.nsf.gov) for JED's participation. DM was supported by grant ME 3157/2-1, German Science Foundation (www.dfg.de/en). The funders had no role in study design, data collection and analysis, decision to publish, or preparation of the manuscript.

Competing Interests: The authors have declared that no competing interests exist.

* E-mail: ejaramillo@uach.cl

Introduction

Chile overlies the convergent boundary between the Nazca plate and the South American continent (Fig. 1a), one of the most seismically active areas on Earth. Along the Chilean margin, plate convergence at ~70 mm/yr [1] results in great earthquakes (M_w~8.0–8.5) on average every ~100–150 years, with occasional giant events (M_w>9.0) every ~300 years [2], like the 1960 earthquake that reached a moment magnitude (M_w) of 9.5. Such earthquakes result in substantial land-level changes of coastal and inland regions [3,4,5,6]. The plate boundary off central Chile (*ca.* 34–38°S) ruptured last in 1835 and then on February 27, 2010. The 1835 event reached M_w~8.5 [7], and its effects were documented by Fitz Roy and Darwin, who measured coseismic uplift of 2.4–3.0 m at Isla Santa Maria, southwest of Concepcion [8]. The recent 2010 rupture, known as the Maule earthquake, reached M_w 8.8 becoming the sixth largest event recorded by modern seismology [9], and third in the era of space geodesy following Sumatra 2004 and Japan 2011 [9]. Plate-boundary slip associated with the Maule earthquake reached 20 m, localized mostly in two patches to the north and south of the epicentre [10,11,12] (Fig. 1b). The response of the coastline to co-seismic slip depends on its position with respect to these areas of high slip release; thus, subsidence of ~1 m occurred along the coastline landward of the northern portion (34–35°S), whereas the coast above the southern portion, particularly the Arauco Peninsula (37°S), was uplifted up to 2.5 m [4,5,6]. The Maule event triggered a devastating tsunami that killed nearly 500 people [13], destroyed coastal infrastructure, and further reshaped coastal landscapes [4,5,6].

Ecological footprints of large-scale extreme events, such as earthquakes, tsunamis, volcanic eruptions, fires and climatic phenomena, can persist for years, decades, centuries and even millennia [14,15,16,17,18,19,20]. For example, the gradation of muddy sedimentary and peaty layers representing coastal land types, suggest that subsidence related to large earthquakes occurred during the Holocene along the outer coast of Washington State [15]. For the same coastal area, analyses of sand

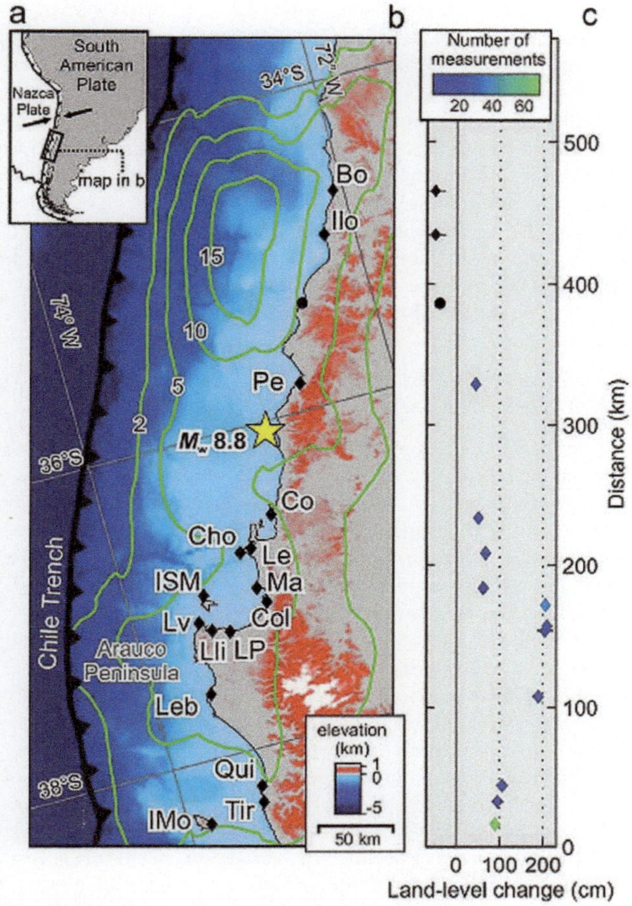

Figure 1. Index maps and land-level changes from the 2010 Maule earthquake. a, Plate-tectonic setting of Chilean margin. Arrows indicate convergences between Nazca and South American plates at *circa* 70 mm yr^{-1}. b, Maule earthquake setting. Epicentre indicated by star, green contours show plate boundary slip (m) (10). Beach sites: Boyeruca* (Bo), Iloca* (Ilo), Pelluhue* (Pe), Lenga (Le), Colcura (Col), Punta Lavapie* (Lv), Llico* (Lli), Las Peñas (LP) and Lebu* (Leb). Land-level change estimated at sites with * plus Cocholhue (Co), Chome (Cho), Maule (Ma), Isla Santa María (ISM), Quidico (Qui), Tirúa (Tir) and Isla Mocha (IMo) (Table S2). c, Geographic variability in estimated land-level changes (means ±1 standard deviation). Black dot denotes coseismic subsidence at permanent GPS site CONT [51]; black diamonds are estimates of subsidence based on drowned coastlines.

deposited over wetland peats point to a tsunami following a large earthquake which probably occurred between 500 and 1700 years ago [17]. The massive wetland created by land subsidence on the Río Cruces after the 1960 M_w 9.5 Chilean earthquake [21], is now recognised as a Ramsar wetland site [22] for its high biodiversity [23,24]. Despite the dramatic effects of these events, a lack of pre-event and comparative data from diverse ecosystems limits our understanding of their lasting impacts. Moreover, the difficulty of mounting interdisciplinary field investigations immediately following extreme events often prevents collection of ecological information that could lead to important discoveries [25,26,27,28,29,30]. For example, results of research initiated nearly 1.5 years after a large disturbance, the severe 1988 fires in Yellowstone National Park in northwest Wyoming, USA, indicating that vegetation in the park was quite resilient surprised ecologists [20,27]. Similarly, early post-disturbance research following the 1980 Mount St Helen's volcanic eruption, revealed another ecological surprise: vegetation recovery was initiated by propagules which originated within the disturbed areas, rather than outside sources [25].

We provide the first quantification of ecologically relevant effects of a large earthquake and its tsunami, on mobile intertidal invertebrates of sandy beaches by taking advantage of unique results obtained from ecological surveys conducted weeks before and after, as well as over the 10 months following, the Maule earthquake. By incorporating measurements of uplift, subsidence and estimated tsunami heights in our analyses, we provide a broader and more mechanistic perspective on the implications of extreme events for an important coastal ecosystem.

Modification of shorelines and beaches by the construction of seawalls and rock revetments, has for centuries been a common societal response to threats from shoreline erosion, extreme storms and tsunamis [31,32]. Coastal armouring covers beach habitat, reflects waves and constrains landward migration of the shoreline, leading to decreases in beach width and intertidal habitat [33]. However, despite widespread use of armouring, its ecological impacts to sandy beach ecosystems are only beginning to be considered. Field studies have supported a recent hypothesis suggesting that as beach width narrows in front of armouring structures, habitat is lost disproportionately from the upper and mid-intertidal resulting in loss of biodiversity and ecosystem function [34]. The location of seawalls or revetments on the beach profile and the degree of interaction with waves and tides, are important considerations in predicting impacts; *i.e.* the lower the structure is in the intertidal, the greater both the physical [35] and ecological effects. Importantly, the ecological implications of interactions between coastal armouring and extreme events have never been studied for sandy beaches. For this reason, anthropogenic coastal armouring was included as a key design factor in our study.

Although sandy beaches represent nearly 80% of the ice-free coasts of the world [36], ecological studies (including research on the effects of major disturbances) of this valuable coastal ecosystem have been widely neglected [37]. Thus, our study represents an outstanding opportunity to better understand responses of this economically, culturally and ecologically important coastal ecosystem to extreme events.

Methods

Ethics statement

No specific permits were required for the described intertidal field studies. The sandy beaches we studied in Chile are unrestricted to public access and use, and are not privately owned or designated as protected areas (reserves or parks). No protected or endangered species were involved in this study.

In a study designed to evaluate the responses of intertidal beach fauna to armouring we surveyed intertidal zones and animals at nine sandy beaches along the coasts of Maule and Bíobío (*ca.* 34–38°S, Fig. 1b) during late January 2010. We compared biological and physical characteristics of unarmoured sites to those located in front of adjacent seawalls and rocky revetments (Table S1). To quantitatively sample invertebrate macrofauna (defined as animals retained on a 1 mm sieve [38]) we set up four replicated shore-normal transects 5 m apart from each other, extending from the upper intertidal to the low tide level of sites located in front of seawalls or rocky revetments and in unarmoured areas at each beach (see Table S1). Along these transects we sampled each of the three intertidal faunal zones typical of sandy beaches of south central Chile, which are dominated by crustaceans [39,40]: i) the upper zone occupied primarily by talitrid amphipods (*Orchestoidea tuberculata*), usually extending from the toe of the dunes or natural cliffs to the drift line or high tide level, ii) the mid zone occupied by cirolanid isopods (*Excirolana braziliensis* and *Excirolana hirsuticauda*), extending from the drift line to the effluent line, and iii) the lower zone occupied primarily by hippid crabs (*Emerita analoga*), extending from the effluent line to the lowest tide level or bore collapse line of incoming waves. In each of the three zones present on each replicate transect, five core samples of sediments were collected with a metal cylinder (10 cm in diameter) to a depth of 30 cm at equally spaced levels across the zone, for a total sampling area of 0.04 m^2 surface area per zone on each transect. The five core samples from each zone in a transect were pooled and sieved through a 1000 μm sieve; afterwards, the collected organisms were stored in 10% formalin in sea water until laboratory sorting. Beach width was measured as the distance between the landward boundary of the beach defined by the toe of foredune or armouring structures and the low tide level during spring tides at each of the four transects at each type of site (armoured, unarmoured). The beach face slope was measured at each of the four transects of each site using the method of Emery [41]. In this study, beach face slope is expressed as 1/x, where x was the distance in meters at which a height difference of 1 m between two consecutive intertidal levels is reached. Thus, higher values of 1/x correspond to flatter beaches.

Following the Maule earthquake and tsunami in February 2010, intertidal surveys as described above were repeated at the nine study sites during March 2010 and used to evaluate responses to the earthquake and tsunami. Comparisons of the abundance of macrofaunal invertebrates between our pre- and post-earthquake and tsunami surveys were made using one way ANOVA on data that were log (n+1) transformed to meet basic assumptions. Changes in beach widths and beach face slopes of sites located in front of seawalls and revetments and in natural unarmoured areas, were calculated as the ratio between the values collected after and before the Maule earthquake (March 2010 and January 2010, respectively). The latitudinal scope of the event and its widespread effects on the coastline, made it impossible to find locations suitable for use as reference sites unaffected by the event. Three beaches from the uplifted zone (Lebu, Llico and Punta Lavapie) were selected for additional surveys during 2010 (July, September and December) based on the intensities of interaction of existing armouring with waves and tides; *i.e.* armouring structures which are located low enough on the beach profile to be reached by waves during high tides experience higher interaction intensity than those located higher on the beach profile.

We measured and compared the magnitude of land-level changes along the study coast (Fig. 1b and Table S2). The magnitude of post-earthquake uplift was determined by measuring the difference in height between the upper growth limit of living rocky shore sessile calcareous algae of the genus *Lithothamnion* (pink coloured) to that of dead organisms (white coloured) which were uplifted above mean sea level [4,5,6]. Coastal subsidence estimates were based on measurements of the depths of flooded vegetation in nearby estuaries. We also estimated tsunami height by measuring water marks left on walls and trees; those measurements were corrected for tide level at the time of tsunami arrival.

Results

Coastal uplift showed a strong geographic pattern, with highest values on the Arauco Peninsula (Punta Lavapie, Llico and Lebu) and Isla Santa María (mean ranges = 189–209 cm). Coastal uplift decreased southward (down to mean values of 96 and 90 cm in Tirúa and Isla Mocha, respectively) and northward (down to 44 cm in Pelluhue). Coastal subsidence (*ca.* −50 cm) affected the two northernmost study sites (Boyeruca and Iloca) (Fig. 1c) drowning the beaches.

The height of the tsunami ranged from ~1.5 m to >10 m across the study beaches. Post-earthquake and tsunami changes in beach width and beach slope ratios were significantly correlated with coseismic land-level change for armoured and unarmoured sites (Fig. 2 e, f). In uplifted areas, beaches became wider and flatter by March 2010 and remained so throughout 2010 (Fig. 2 b, c, d). After the earthquake and tsunami we observed significant changes in population abundances of representative taxa of mobile crustacean macrofauna in all intertidal zones, including new appearances, increases, declines and some local extinctions (Fig. 3). These ecological responses varied with land-level change, tsunami height and coastal development. On beaches that subsided and were not armoured, the abundance of all taxa declined in all intertidal zones after the earthquake (Fig. 3a). On armoured beaches with seawalls that subsided, the responses differed among taxa: an upper intertidal beach species (*Orchestoidea tuberculata*) disappeared at Iloca and a lower intertidal beach species (*Emerita analoga*) increased in abundance at the same beach while responses of mid-intertidal taxa varied after the event (Fig. 3a).

The response and direction of change in distribution and abundance of mobile crustacean macrofauna varied with intertidal zone. On uplifted beaches that were not armoured (n = 6) (Fig. 3b, c), declines observed in abundance after the event were more prevalent in the lower intertidal while increases in abundance were more prevalent in the upper intertidal; *i.e.* mean abundance of upper shore taxa increased more than two fold at 50% of sites (Pelluhue, Colcura and Lebu; $p<0.05$ for the last two sites), while that of lower shore taxa declined by half at 50% of sites (Pelluhue, Las Peñas and Llico; <0.05 for Pelluhue and Llico). In the mid-intertidal, ecological responses were split between increases and declines in abundance, but included disappearance of at least one species (*Excirolana braziliensis*) at two of the sites (Pellhue and Lenga). On uplifted beaches with armouring, increases or no changes in abundance were prevalent compared to declines for upper, mid and lower intertidal zone taxa after the event, especially for sites with revetments (Fig. 3b,c). However, uplifted sites with bedrock in the shallow subtidal, like Punta Lavapie, lost sandy intertidal zones and macroinvertebrate fauna (*Emerita analoga*) due to uplift of rocky substrate into the low intertidal. This is somewhat analogous to ecological effects of armouring on upper and mid-intertidal zones (34); *i.e.* the uplifted bedrock replaced the swash zone habitat, and the lower intertidal sand crab *Emerita analoga* disappeared. Ecological effects of the tsunami, such as those from washover and scour, were less clear; however, the lower intertidal sand crab, *Emerita analoga* declined significantly ($p<0.05$) on both unarmoured sites where tsunami heights were ≥10 m (Pelluhue and Llico; Fig. 3 b,c). In addition, the tsunami height was 10.5 m at the only site (Pelluhue) where disappearance of both mid-intertidal taxa (*Excirolana braziliensis* and *Excirolana hirsuticauda*) was observed (Fig. 3b).

Results of continuing surveys of uplifted beaches during 2010 indicated different levels of recolonization, estimated by variability in species composition and population abundance of macroinvertebrates of the intertidal zones, on sites located in front of armouring. These patterns were related to the intensity of interaction of the armouring structures with waves and tides before the earthquake. In front of a seawall that did not interact with waves and tides (Lebu), post-earthquake species composition was the same as pre-event (Fig. 4). For all except one species (*Emerita analoga*), population abundances were similar or even

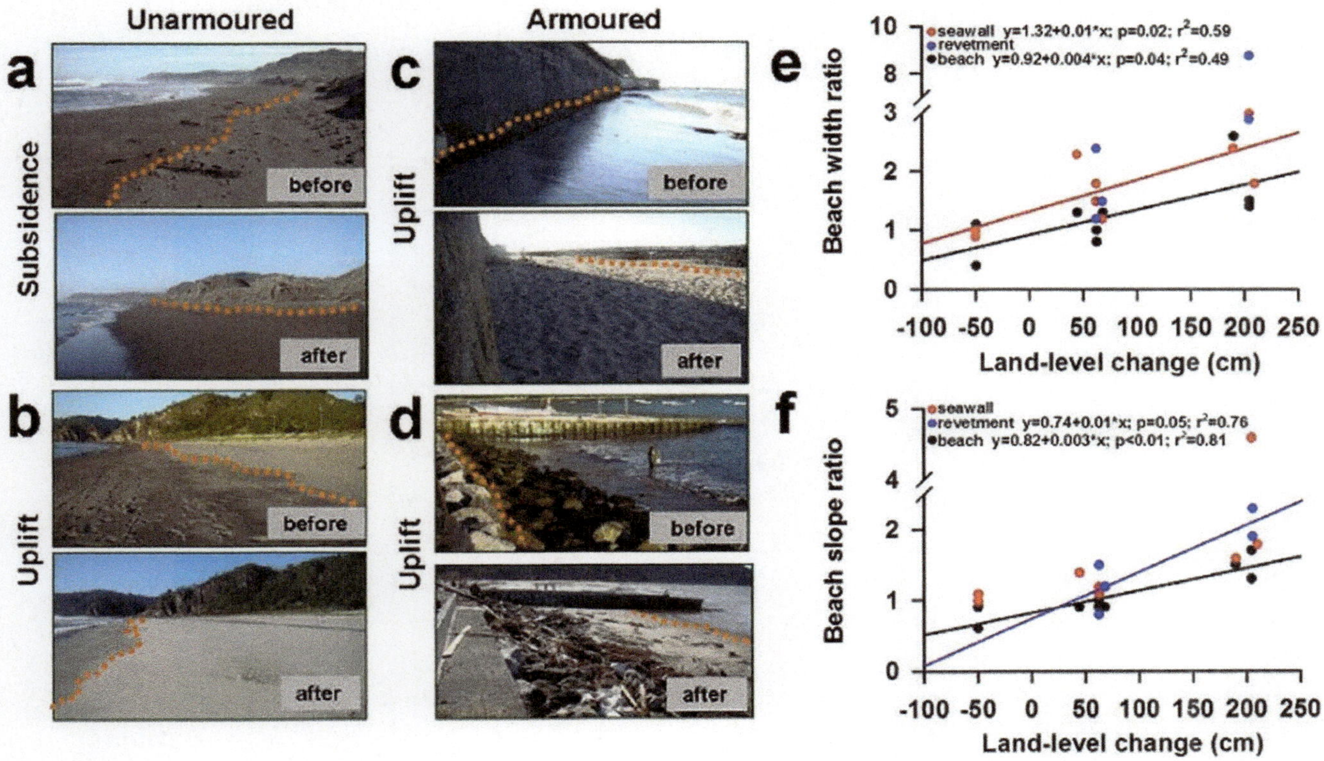

Figure 2. Beach characteristics as a function of land-level changes. a–d, Photos of study sites taken before and after the 2010 Maule earthquake. Orange dotted lines indicate 24 hour spring high tide line; a–b, unarmoured sites: a, subsidence (narrower): Boyeruca, b, coastal uplift (wider): Lebu, c–d, armoured sites: c, seawall with uplifted wider intertidal: Punta Lavapie, d, revetment with uplifted wider intertidal: Llico. Note that dry sand areas above high tide line decreased at subsided beaches (a) and increased at uplifted beaches (b, c and d), e–f, Relationships between the magnitude of land-level changes and e, beach widths and f, beach face slopes for sites with seawalls (red dots), sites with rocky revetments (blue dots) and unarmoured sites (black dots). Only statistically significant regressions are shown.

higher than before February 27th and remained similar for 10 months following the earthquake and tsunami at Lebu (Fig. 4). Although when the tsunami hit the seawall it appeared to cause high mortality of the larger size classes of the lower intertidal crab *Emerita analoga* (Fig. 5) (the most abundant species prior to the earthquake), by the time of our post-event survey in March, the overall abundance of this species was higher than before the earthquake. However, in all subsequent months the abundance of *Emerita analoga* was nearly 50% lower than before the earthquake and tsunami at that site (Fig. 4). At the unarmoured site on this beach, although post-earthquake species composition remained the same as before the earthquake for the rest of 2010, abundance varied considerably in two of the four taxa we studied (*Orchestoidea tuberculata* and *Excirolana braziliensis*; Fig. 4). In front of a formerly interacting seawall (Llico), species composition was the same before and immediately after the earthquake (Fig. 4). However, population abundance of *Excirolana braziliensis* was lower immediately following the earthquake (March), and later this species disappeared (Fig. 4). At the same site, population abundance of *Orchestoidea tuberculata* was higher in surveys five and seven months after the earthquake (July and September) but disappeared during December 2011 (Fig. 4) after the beach was cleared of tsunami debris by heavy equipment. In front of a revetment at Llico, only one species, the lower intertidal *Emerita analoga*, was present before the earthquake (Fig. 4). One month after the earthquake, the intertidal habitat restored in front of this revetment by uplift was colonized by upper and mid-intertidal species (*Orchestoidea tuberculata* and *Excirolana hirsuticauda*), and then *Excirolana braziliensis* by July (Fig. 4). Thus, five months after the earthquake this armoured site had regained the typical species composition of unarmoured sandy beaches of south central Chile (38,39). In front of the armouring with the highest interaction with waves, the seawall at Punta Lavapie, again only one species, the lower intertidal *Emerita analoga* was present before the earthquake (Fig. 4). Five to seven months after (July and September), mid-intertidal isopods (*Excirolana braziliensis* and *Excirolana hirsuticauda*) appeared for the first time on the restored intertidal habitat at this site. In contrast, formerly abundant lower intertidal sand crabs *Emerita analoga* disappeared right after 27 February (Fig. 3) coincident with the uplift of subtidal rocky substrate; very low numbers were observed during September and this species disappeared again in December 2010 (Fig. 4).

Discussion

An upward trend in the frequency of major natural disturbances across different ecosystems around the world has been suggested by a number of models and sources [42,43,44,45]. In fact, the annual number of major natural disturbances that have significantly affected society increased four fold from 1975 to 2008 [29,46]. Major earthquakes (recently in Haiti, Chile and Japan) causing tragic loss, economic devastation and destruction during the last few years, emphasize the high vulnerability of the world's heavily populated and ecologically critical coasts. Pressing needs for societal recovery can overshadow the urgency of understanding

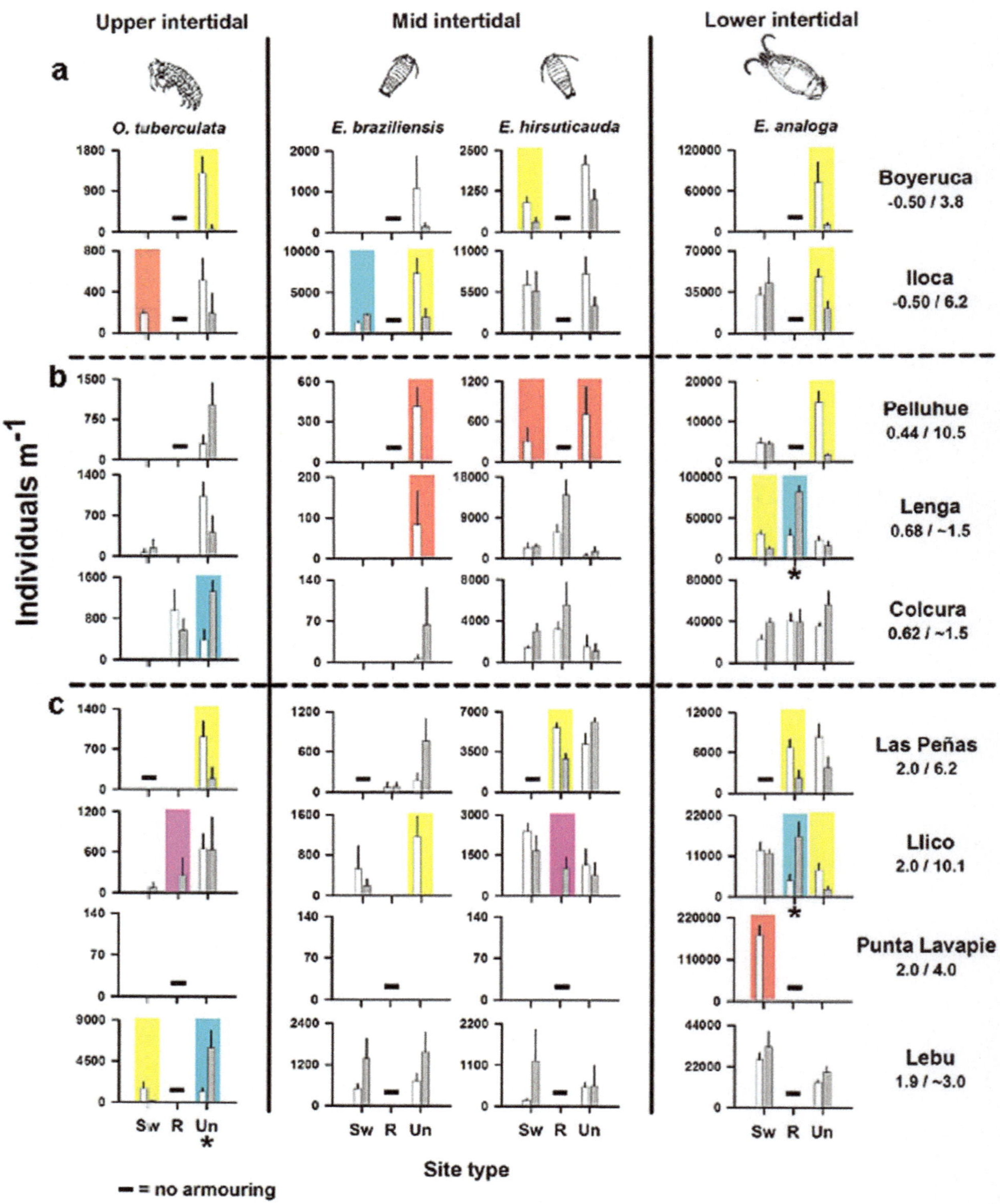

Figure 3. Responses of sandy beach macrofauna to the 2010 Maule earthquake. a, subsided beaches, b, beaches with <1.0 m uplift, c, beaches with 2 m uplift. Mean abundance (+1 standard error) of mobile crustaceans representing upper, mid and lower intertidal zones for before (white bars) and after (grey bars) the Maule earthquake at sites located seaward of seawalls (Sw) and rocky revetments (R), and at unarmoured sites (Un). Shaded rectangles over bars indicate: significant decrease (yellow), disappearance (red), significant increase (light blue) and first appearance of species at a site (purple). The three intertidal zones are represented by 1) the talitrid amphipod *Orchestoidea tuberculata* (upper), 2) the cirolanid isopods *Excirolana braziliensis and Excirolana hirsuticauda* (mid) and 3) the sand crab *Emerita analoga* (lower) as illustrated on the figure. Abundances are presented as individuals m^{-1} to accommodate beach width changes; trends for individuals m^{-2} were similar except three cases indicated by *. Note: low intertidal seaward of the seawall at Boyeruca was bedrock before and after the earthquake, excluding *Emerita analoga*, and the beach at Punta Lavapie is completely armoured with a seawall. Numbers below the names of sites on the right side of the figure indicate land-level changes and tsunami heights in meters (m/m, respectively) for each study beach.

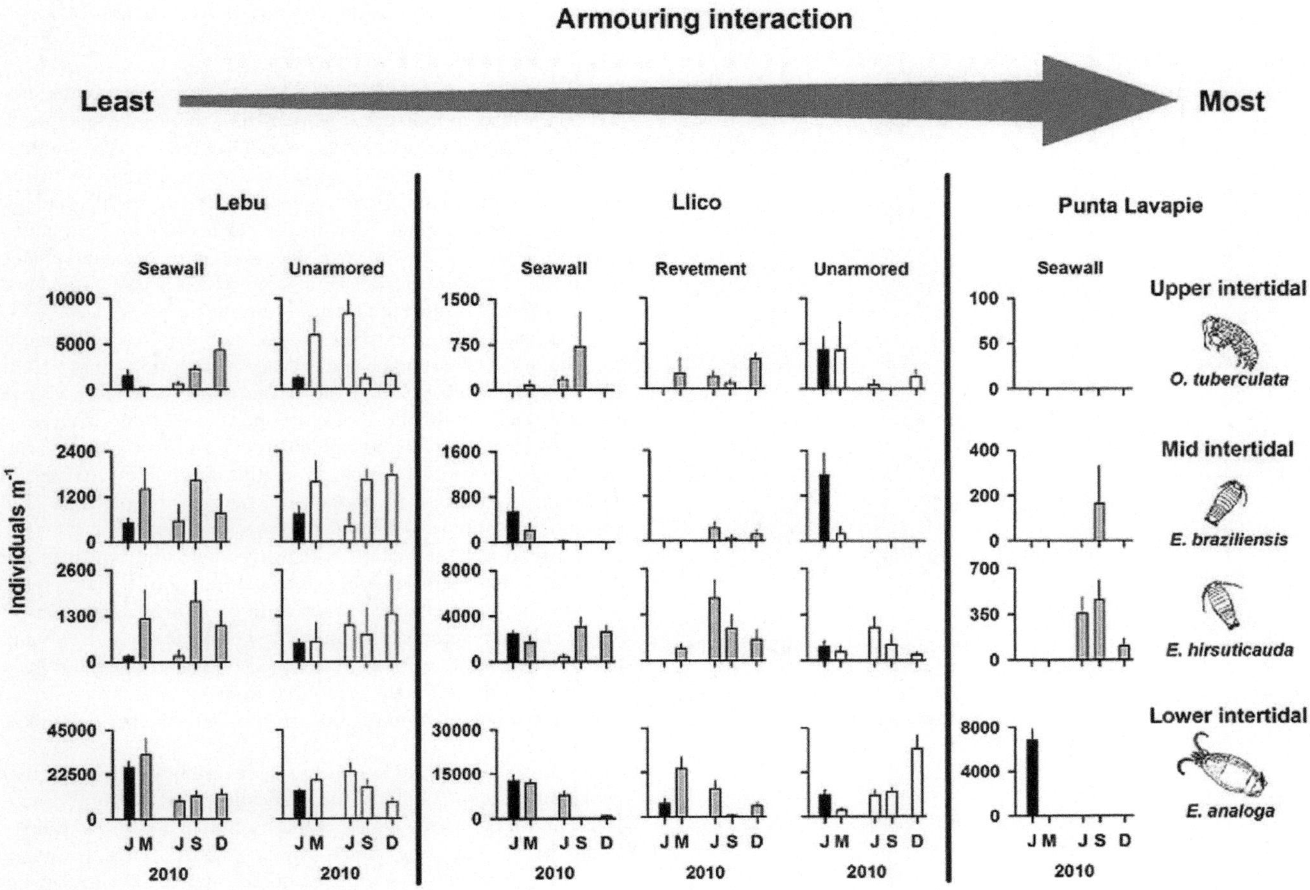

Figure 4. Responses of sandy beach macrofauna to uplift and coastal armouring over time. Temporal patterns in population abundance of crustaceans from the three intertidal zones listed in Fig. 3 for armoured sites that experienced different amounts of interaction between the structure and local waves and tides before the 2010 earthquake and tsunami, and for unarmoured sites. Mean values (+1 standard error) of abundance before the event are indicated by black bars, mean values of post-event abundance for armoured sites are grey bars and unarmoured sites are white bars. Strength of pre-event armouring and wave interaction (see Methods) increases from left to right as indicated by the grey arrow at the top of the figure. Data are presented as individuals m^{-1} to accommodate changes in beach widths; trends for individuals m^{-2} were similar.

complex effects of large-scale disasters on coastal ecosystems. In this study, we took advantage of the unique opportunity to document the complex and surprising ecological responses of an important and widespread coastal ecosystem, sandy beaches, to a major earthquake and tsunami for the first time. Our novel results indicate that the extreme earthquake and tsunami of 27th February 2010, resulted in significant and lasting changes in physical and ecological attributes of sandy beaches on the affected coast. The responses of ecologically important species that are characteristic of the upper, mid and lower intertidal zones of beaches in the region suggest the entire intertidal zone was affected on some beaches. The changes we observed in the distribution and abundance of mobile intertidal crustaceans after the Maule event (including new appearances, increases, declines and some local extinctions), were much greater than would be expected during mid-summer when sandy beaches are generally most stable and in the season of sediment accretion. Our findings also illustrate how current management of coastal urbanized zones can result in significant deterioration of sandy beach ecosystems [33,34,38,47].

The most unexpected changes observed in sandy beach invertebrates after the earthquake and tsunami, occurred in front of artificial shore armouring along the uplifted coast of the Arauco Peninsula. Where coastal armouring structures constrained landward extent of the sandy beach, increases in intertidal width created new habitat that was rapidly colonized by upper-intertidal and mid-intertidal species via migration from adjacent unarmoured beach areas. Thus, restoration of habitat in front of seawalls and rocky revetments by uplift had positive effects on mobile upper-intertidal and mid-shore macrofauna: *Orchestoidea tuberculata* and *Excirolana hirsuticauda* appeared for the first time in front of the revetment at Llico right after the Maule earthquake while *Excirolana braziliensis* also recolonized the area, but after five months (Fig. 4). Similarly, both species of *Excirolana* colonized the uplifted intertidal zone in front of the seawall at Punta Lavapie, a zone formerly occupied by just the lower intertidal crab *Emerita analoga* (Fig. 4). Nearly eight months after the Maule earthquake (October 2010), native and exotic coastal strand vegetation began to colonize the new supralittoral zones formed by uplift on sandy beaches with armouring structures (Jaramillo, unpublished). The appearance of perennial vegetation that can potentially trap wind-blown sand to form hummocks and embryonic dunes, could further transform these coseismically uplifted coastal habitats to provide an important ecotone between intertidal and terrestrial ecosystems that was eliminated by armouring prior to the earthquake. Such coastal strand zones have become increasingly

Figure 5. Dead sand crabs, *Emerita analoga*, stranded in life position at the upper-shore levels of the armoured site at Lebu, illustrate the direct mortality of a dominant lower intertidal invertebrate species observed immediately after the Maule earthquake and tsunami.

rare on developed coastlines and are threatened by sea level rise [47,48].

Although ecological effects of the tsunami on sandy beach invertebrates were challenging to isolate from the effects of uplift or subsidence, high mortality of intertidal animals were associated with large tsunami heights (>10 m), for example at Pelluhue where mid-intertidal species disappeared (*Excirolana braziliensis* and *Excirolana hirsuticauda*) or decreased significantly in abundance (*Emerita analoga*) after the Maule earthquake. Thus, our results suggest that even in the absence o coseismic uplift or subsidence, the ecological effects of tsunamis on beach ecosystems could be significant as been shown for subtidal soft bottom macrofauna [30]. How lasting those effects may be, will depend on the species assemblage affected and proximity of source populations, as well as shoreline evolution and human responses to the disaster. For example, weeks after the Indian Ocean tsunami of December 24th 2004, which removed large amounts of sediments along sandy beaches and dunes of the coast of northwestern Sumatra, a new coast begun to form with similar features to the pre-event period. However, beach recovery was hindered by coastal development, including the installation of new armouring structures in the form of rocky revetments [49,50].

As predicted [34], our pre-earthquake surveys found that the upper and mid-intertidal species (*Orchestoidea tuberculata*, *Excirolana braziliensis* and *Excirolana hirsuticauda*) of sandy beaches are more strongly affected by armouring than lower shore species (*Emerita analoga*) and that the degree of interaction with waves and tides influenced these ecological impacts on sandy intertidal animals. After the earthquake, these upper and mid-intertidal mobile species, all of which have direct development and thus low dispersal, quickly expanded distributions and increased in abundance in response to uplift that altered beach profiles and reduced interaction of the armouring structures with waves and tides. We hypothesise that this rapid response was possible due to the proximity of resident source populations living on adjacent unarmoured sections of the study beaches. For coastal areas where armouring is more extensive and for beaches that are isolated from potential source populations, responses of these intertidal species with low dispersal would be expected to be considerably slower. In contrast, the sand crab *Emerita analoga*, a species with free swimming larvae and high potential for dispersal, had not recovered from mortality due to the earthquake and tsunami at some sites after nearly a year due to extreme habitat change, specifically the uplift of rocky substrate which replaced the sandy habitat in the low intertidal zone of Punta Lavapie.

Ecological effects of extreme events, such those we observed for the Maule event, are expected to vary in duration. Shorter-term effects on beaches included direct mortality associated with the tsunami (Fig. 5) and the indirect bottom up effects of increased inputs of algal wrack from uplifted rocky shores on upper shore invertebrate consumers such as talitrid amphipods (*Orchestoidea tuberculata*). However, in areas with significant uplift (*ca.* 2 m), locations of armouring structures were shifted higher on the beach profile, reducing interaction with waves and tides and restoring intertidal zones for biota and ecological function. For this reason we expect positive changes observed in these beach ecosystems to persist, altering intertidal community composition and dynamics over the long term, even in front of existing coastal armouring. In contrast, for the subsided armoured areas, community composition and population abundances are expected to remain depressed over time.

Responses of mobile sandy beach invertebrates to the earthquake and tsunami differed dramatically from that observed on adjacent rocky shores where uplift caused massive mortality of sessile biota, including macrophytes [4,5,6]. Thus, ecological impacts of extreme events on coastlines appear to vary strongly with magnitude and direction of land-level change, across shore types and with the mobility of the biota. Our results emphasize the value of and need for baseline information on coastal ecosystems, and illustrate how interactions of extreme events with human-altered coasts can produce unexpected ecological outcomes. Understanding how these complex and habitat-specific responses to episodic extreme landscape alteration can create lasting ecological signatures, provides important new insights on processes and dynamics of coastal ecosystems.

Supporting Information

Table S1 Geographic coordinates of the sandy beaches studied indicating the types of sites sampled at each beach.

Table S2 Geographic coordinates of the rocky shore sites visited to estimate land-level changes.

Acknowledgments

We thank M. González for sorting the invertebrate macrofauna samples collected in this study. We are grateful to the anonymous reviewers for constructive suggestions that improved this manuscript.

Author Contributions

Conceived and designed the experiments: EJ CD. Performed the experiments: EJ CD MM CC RS. Analyzed the data: EJ JED DMH DM MM. Wrote the paper: EJ JED DMH DM.

References

1. Ruegg JC, Rudloff A, Vigny C, Madariaga R, de Chabalier JB, et al. (2009) Interseismic strain accumulation measured by GPS in the seismic gap between Constitucion and Concepcion in Chile. Phys Earth Planet In 175: 78–85.
2. Cisternas M, Atwater BF, Torrejon F, Sawai Y, Machuca G, et al. (2005) Predecesssors of the giant 1960 Chile earthquake. Nature 437: 404–407.
3. Plafker G, Savage JC (1970) Mechanism of the Chilean earthquake of May 21 and 22, 1960. Geol Soc Am Bull 81: 1001–1030.
4. Farías M, Vargas G, Tassara A, Carretier S, Baise S, et al. (2010) Land-level changes produced by the 2010 Mw 8.8 Chile earthquake. Science 329: 916.
5. Castilla JC, Manríquez PH, Camaño A (2010) Effects of rocky shore coseismic uplift and the 2010 Chilean mega-earthquake on intertidal biomarker species. Mar Ecol Prog Ser 418: 17–23.
6. Vargas G, Farías M, Carretier S, Tassara A, Baize S, et al. (2011) Coastal uplift and tsunami effects associated to the 2010 M_w8.8 Maule earthquake in Central Chile. Andean Geol 38: 219–238.
7. Lomnitz C (2004) Major earthquakes in Chile: a historical survey, 1535–1960. Seismol Res Lett 75: 368–378.
8. Darwin C (1851) Geological observations on coral reefs, volcanic islands and on South America. London, Smith, Elder and Co. 768 p.
9. USGS Available http://earthquake.usgs.gov/earthquakes/world/historical_mag.php Accessed 2012 March 19.
10. Moreno M, Rosenau M, Oncken O (2010) Maule earthquake slip correlates with pre-seismic locking of Andean subduction megathrust. Nature 467: 198–202.
11. Tong XP, Sandwell D, Luttrell K, Brooks B, Bevis M, et al. (2010) The 2010 Maule, Chile earthquake: downdip rupture limit revealed by space geodesy. Geophys Res Lett 37.
12. Vigny C, Socquet A, Peyrat S, Ruegg JC, Metois M, et al. (2011) The 2010 Mw 8.8 Maule megathrust earthquake of Central Chile, monitored by GPS. Science 332: 1417–1421.
13. Fritz HM, Petroff CM, Catalán PA, Cienfuegos R, Winckler P, et al. (2011) Field survey of the 27 February 2010 Chile Tsunami. Pure Appl Geophys 168: 1989–2010.
14. Foster DR, Knight DH, Franklin JF (1998) Landscape patterns and legacies resulting from large, infrequent forest disturbances. Ecosystems 1: 497–510.
15. Atwater BF (1987) Evidence for great Holocene earthquakes along the outer coast of Washington State. Science 236: 942–944.
16. Nelson AR, Manley W (1992) Holocene coseismic and aseismic uplift of Isla Mocha, south-central Chile. Quatern Int 15/16: 61–76.
17. Atwater BF, Moore AL (1992) A tsunami about 1000 years ago in Puget Sound, Washington. Science 258: 1614–1617.
18. Franklin JF, MacMahon (2000) Messages from a mountain. Science 288: 1183–1185.
19. Turner M (2005) Landscape ecology: what is the state of the science: Annu Rev Ecol Evol Syst 36: 319–344.
20. Turner M (2010) Disturbance and landscape dynamics in a changing world. Ecology 91: 2833–2849.
21. Reinhardt EG, Nairn RB, Lopez G (2010) Recovery estimates for the Rio Cruces after the May 1960 Chilean earthquake. Mar Geol 269: 18–33. (2010).
22. The Ramsar Convention on Wetlands. Carlos Anwandter Sanctuary Ramsar site no. 222 Available: http://www.ramsar.org. Accessed 2012 March 19.
23. Corti P, Schlatter R (2002) Feeding ecology of the blacknecked swan *Cygnus melancoryphus* in two wetlands of southern Chile. Stud Neotrop Fauna Environ 37: 9–14.
24. Lagos N, Paolini P, Jaramillo E, Lovengreen C, Duarte C, et al. (2008) Environmental processes, water quality degradation, and decline of waterbird populations in the Rio Cruces wetland, Chile. Wetlands 28: 938–950.
25. Franklin JF, Lindenmayer D, MacMahon JA, McKee A, Magnuson J, et al. (2000) Threads of continuity: ecosystem disturbances, biological legacies and ecosystem recovery. Conserv Biol Pract 1: 8–16.
26. Turner MG, Baker WL, Peterson CJ, Peet RK (1998) Factors influencing succession: lessons from large, infrequent natural disturbances. Ecosystems 1: 511–523.
27. Turner MG, Romme WH, Tinker DB (2003) Surprises and lessons from the 1988 Yellowstone fires. Front Ecol Environ 1: 351–358.
28. Lindenmayer DB, Likens GE, Franklin JF, Muntz R (2009) Opportunity in the wake of natural "disasters". Science 324: 463.
29. Lindenmayer DB, Likens GE, Franklin JF (2010) Rapid responses to facilitate ecological discoveries from major disturbances. Front Ecol Environ 8: 527–532.
30. Lomovasky BJ, Firstater FN, Salazar AG, Mendo J, Iribarne OO (2011) Macrobenthic community assemblage before and after the 2007 tsunami and earthquake at Paracas Bay, Peru. J Sea Res 65: 205–212.
31. Nordstrom KF (2000) Beaches and Dunes on Developed Coasts Cambridge University Press, Cambridge. 338 p.
32. Charlier RH, Chaineux MCP, Morcos S (2005) Panorama of the history of coastal protection. J Coast Res 21: 79–111.
33. Hall MJ, Pilkey OH (1991) Effects of hard stabilization on dry beach widths for New Jersey. J Coast Res 7: 771–785.
34. Dugan JE, Hubbard DM, Rodil I, Revell DL, Schroeter S (2008) Ecological effects of coastal armoring on sandy beaches. Mar Ecol 29: 160–170.
35. Weigel RL (2002) Seawalls, seacliffs, beachrock: what beach effects? Part 1. Shore & Beach 70: 17–27.
36. Brown AC (2001) Biology of sandy beaches. In: Steele JH, Thorpe SA, Turekian KK, eds. Encyclopedia of Ocean Sciences Volume 5: 2496–2504 Academic Press, London.
37. Dugan JE, Defeo O, Jaramillo E, Jones AR, Lastra M, et al. (2010) Give beach ecosystems their day in the sun. Science 329: 1146.
38. Schlacher TA, Schoeman DS, Dugan J, Lastra M, Jones A, et al. (2008) Sandy beach ecosystems: key features, sampling issues, management challenges and climate change impacts. Mar Ecol 29: 70–90.
39. Jaramillo E (2001) The sand beach ecosystem of Chile. In: Seeliger U, B. Kjerfve, eds. Coastal Marine Ecosystems of Latin America. Ecological Studies 144: 219–227 Springer Verlag, Berlin.
40. McLachlan A, Jaramillo E (1995) Zonation on sandy beaches. Oceanog Mar Biol Annu Rev 33: 305–335.
41. Emery K (1961) A simple method of measuring beach profiles. Limnol Oceanogr 6: 90.
42. Byrnes JE, Reed DC, Cardinale BJ, Cavanaugh KC, Holbrook SJ, et al. (2011) Climate-driven increases in storm frequency simplify kelp forest food webs. Glob Change Biol 17: 2513–2524. doi:10.1111/j.1365-2486.2011.02409.x.
43. Emanuel KA (1987) The dependence of hurricane intensity on climate. Nature 326: 483–485.
44. Emanuel KA (2005) Increasing destructiveness of tropical cyclones over the past 30 years. Nature 436: 686–688.
45. Westerling AL, Hidalgo HG, Cayan DR, Swetnam TW (2006) Warming and earlier spring increase western US forest wildfire activity. Science 313: 940–943.
46. International Federation of Red Cross and Red Crescent Societies (2009) World Disasters report 2009: focus on early warning and early action. Available: www.ifrc.org/Global/WDR2009-full.pdf Accessed 2011 July 20.
47. Dugan JE, Hubbard DM (2010) Loss of coastal strand habitat in southern California: the role of beach grooming. Estuaries Coasts 33: 67–77.
48. Feagin RA, Sherman DJ, Grant WE (2005) Coastal erosion, global sea-level rise, and the loss of sand dune habitats. Front Ecol Environ 3: 359–364.
49. Wong PP (2009) Impacts and recovery from a large tsunami: coast of Aceh. Polish J Environ Stud 18: 5–16.
50. Liew SC, Gupta A, Wong PP, Kwoh LK (2010) Recovery from a large tsunami mapped over time: the Aceh coast, Sumatra. Geomorphology 114: 520–529.
51. Lorito S, Romano F, Atzori S, Tong X, Avallone A, et al. (2011) Limited overlap between the seismic gap and coseismic slip of the great 2010 Chile earthquake. Nature Geosci 4: 173–177.

Anthropogenic Disturbance Can Determine the Magnitude of Opportunistic Species Responses on Marine Urban Infrastructures

Laura Airoldi[1,2]*, Fabio Bulleri[3,4]

1 Dipartimento di Biologia Evoluzionistica Sperimentale and Centro Interdipartimentale di Ricerca per le Scienze Ambientali in Ravenna, Università di Bologna, Ravenna, Italy, **2** Department of Earth and Environmental Sciences, Wesleyan University, Middletown, Connecticut, United States of America, **3** Dipartimento di Biologia, Università di Pisa, CoNISMa, Pisa, Italy, **4** Dipartimento di Science Botaniche, Ecologiche e Geologiche, Università di Sassari, Sassari, Italy

Abstract

Background: Coastal landscapes are being transformed as a consequence of the increasing demand for infrastructures to sustain residential, commercial and tourist activities. Thus, intertidal and shallow marine habitats are largely being replaced by a variety of artificial substrata (e.g. breakwaters, seawalls, jetties). Understanding the ecological functioning of these artificial habitats is key to planning their design and management, in order to minimise their impacts and to improve their potential to contribute to marine biodiversity and ecosystem functioning. Nonetheless, little effort has been made to assess the role of human disturbances in shaping the structure of assemblages on marine artificial infrastructures. We tested the hypothesis that some negative impacts associated with the expansion of opportunistic and invasive species on urban infrastructures can be related to the severe human disturbances that are typical of these environments, such as those from maintenance and renovation works.

Methodology/Principal Findings: Maintenance caused a marked decrease in the cover of dominant space occupiers, such as mussels and oysters, and a significant enhancement of opportunistic and invasive forms, such as biofilm and macroalgae. These effects were particularly pronounced on sheltered substrata compared to exposed substrata. Experimental application of the disturbance in winter reduced the magnitude of the impacts compared to application in spring or summer. We use these results to identify possible management strategies to inform the improvement of the ecological value of artificial marine infrastructures.

Conclusions/Significance: We demonstrate that some of the impacts of globally expanding marine urban infrastructures, such as those related to the spread of opportunistic, and invasive species could be mitigated through ecologically-driven planning and management of long-term maintenance of these structures. Impact mitigation is a possible outcome of policies that consider the ecological features of built infrastructures and the fundamental value of controlling biodiversity in marine urban systems.

Editor: Tamara Natasha Romanuk, Dalhousie University, Canada

Funding: Support comes from the EU projects THESEUS (EU - FP7 - ENV2009-1, grant 244104) and DELOS (EVK3-CT-2000-00041), The MIAT project MED RED and from the MIUR projects COFIN (ex 40%) and PRIN (BIORES). F.B. was further supported by Assegni di Ricerca of the University of Pisa and University of Sassari, respectively, and L.A. was supported by an Eastern College Consortium (E.C.Co) Scholarship at Wesleyan University. The funders had no role in study design, data collection and analysis, decision to publish, or preparation of the manuscript.

Competing Interests: The authors have declared that no competing interests exist.

* E-mail: laura.airoldi@unibo.it

Introduction

Marine landscapes have been altered globally by the introduction of a variety of man-made infrastructures, such as seawalls, dykes, breakwaters, groynes, jetties, pilings, bridges, artificial reefs, offshore platforms, and marine energy installations [1–4]. In Europe, 22000 km^2 of the coastal zone is covered in concrete or asphalt, and about 50% of the Mediterranean shorelines bordering Spain, France, and Italy are dominated by artificial infrastructures (more than 1500 km), most of which are developed for harbours and ports ([5] and references therein). In the USA, armouring covers more than 50% of the coastline in some estuaries and bays ([4] and references therein); overall, about 21% of the 759 km coastline of Florida and 12% of the 1763 km coastline of California have been altered by armouring, or addition of bulkheads, revetments, or other coastal infrastructures. Similarly, in the Western Pacific, 27% of the coastline in Japan [6] and more than 50% of the shores of Sydney Harbour [7] have been altered by either coastal infrastructure or armouring. It is expected that marine infrastructures will further proliferate in response to burgeoning coastal populations, expansion of coastal cities, and greater threats from climate change, storm surges and sea level rise [3,4,8].

As marine artificial substrata support many species colonising epibiota, it has been suggested that these artificial substrata may represent adequate mimics of natural hard-bottom habitats [8–9], or valuable surrogates for the habitats that they replace [10–11].

However, studies from different regions of the world and from different types of artificial infrastructures suggest that artificial surfaces do not function as natural rocky habitats [12–15], and often introduce surfaces and species that are extraneous to the natural environments [16–18]. These studies document differences in the structure of assemblages inhabiting artificial infrastructures compared to nearby natural rocky shores [15,19–22], including low species and genetic diversity [23–25], rarity of particular functional groups, such as large grazers and predators [24], different ecological processes [15,26–30], and assemblages representative of the early stages of succession, comprising opportunistic, weedy, and invasive species [15,17,31–34].

In terrestrial systems, the prevalence of opportunistic and invasive forms in urban areas has often been attributed to severe disturbances, typical of human-dominated systems [35–39], such as those linked with soil use, deforestation, fires, construction activities, gardening, and recreation, among others. Disturbances (both natural and anthropogenic) are also implicated in the prevalence of opportunistic and non-indigenous species in a variety of marine systems [e.g. 40–44]. However, the role of anthropogenic disturbances in facilitating the spread of opportunistic, and invasive species on artificial infrastructures is relatively unexplored [32]. Marine artificial infrastructures tend to be subjected to high levels of disturbances from both natural factors [e.g. storms and sediment scour, which are especially intense on infrastructures constructed for protection against erosion and flooding [21,45] and anthropogenic factors [e.g. harvesting and trampling, which are intense on many artificial infrastructures due to their accessibility from highly tourist beaches [46–47].

Another common and particularly severe form of disturbance, which is unique to artificial infrastructures, is represented by maintenance works. The failure rates of coastal infrastructures as a result of scour, undermining, outflanking, overtopping, and battering by storm waves are relatively high, and there is an ongoing need for repair and maintenance during the lifetime of the structure [48]. The ecological consequences of repair and maintenance for the biota of artificial infrastructures are currently unknown.

We analysed how the expansion of opportunistic and invasive forms on coastal defence infrastructures along the Italian side of the north Adriatic Sea (Italy) could be influenced by the continued repair and maintenance, We measured (1) the extent, frequency, and timing of occurrence of periodical maintenance, and (2) quantified the response trajectories of assemblages following maintenance. Since breakwaters introduce both sheltered (i.e. on the landward sides) and exposed (on the seaward sides) substrata that support different assemblages [17,31–32], we also tested (3) whether maintenance has different effects in these habitats. Finally, since recovery of assemblages is influenced by the mix of propagules, spores, and larvae present in the water column at the time at which free space becomes available [49–50], we experimentally tested(4) if disturbance imposed at different times of the year could influence the recovery of assemblages, and control the abundance of opportunistic and invasive forms. Results are used to identify possible management strategies in an effort to improve the ecological outcomes of artificial marine infrastructures.

Materials and Methods

Study area

The study was conducted along ~50 km of coast of the north-east Adriatic Sea (Italy), from Punta Marina (44°45′N, 12°29′E) to Cesenatico (44°20′ N, 12°40′ E). The area is characterised by flat sandy substrate, with moderate exposure to wave action and an average tidal amplitude of ~80 cm. Average surface sea temperature varies between 8°C in the winter and 24°C in the summer.

This region is severely urbanised [5,51]. Over the past 60 years, a wide variety of marine artificial infrastructures have been built along >60% of this sedimentary coastline (Fig. 1A), including >100 km of breakwaters and groynes, >60 km of seawalls and >40 km of jetties. Breakwaters (Fig. 1A), built with large blocks of quarried rock (1–3 m across), are deployed at ~100–250 m from the shore, have an average length of 100–150 m, and extend ~2–3 m above and below the Mean Low Water Level (MLWL). These breakwaters thus provide both subtidal and intertidal surfaces for colonisation by benthic organisms [17,32], along both wave-sheltered (landward) and wave-exposed (seaward) habitats.

The study focused on assemblages at low-shore levels (−10 to +20 cm relative to MLWL). At these levels on the shore, ephemeral and weedy macroalgae (*Ulva* spp.), filamentous forms, the introduced green alga *Codium fragile* ssp. *tomentosoides*), and biofilms (comprising coatings of microalgae, juvenile stages of macroalgae and silt) are the dominant floral taxa on many breakwaters. Other dominant faunal taxa include mussels (*Mytilus galloprovincialis*), oysters (*Ostrea edulis* and the introduced *Crassostrea gigas*), barnacles (*Chthamalus* spp. and *Balanus perforatus*), and the limpet *Patella caerulea*. Detailed descriptions of these assemblages can be found in [17,32,46,52–53].

Disturbance regime

Many coastal infrastructures are constructed using quarried rocks that move during sea storms [54]. Mobile rocks are subjected to stresses that can cause breakage, size reduction, or dislodgement. Maintenance, therefore, often involves the addition of new quarried rocks over large portions of the defence structures to repair damages from storms (Fig. 1B). We monitored maintenance interventions to breakwaters over 3 years, from January 2001 to December 2003, at 25 reference breakwaters. The breakwaters were selected at random along 50 km of coast. We recorded: (1) the occurrence of maintenance interventions; (2) the time of the year at which interventions were carried out, and (3) measured the extension of the damage caused by interventions, as the proportion (% of the total length) of the breakwater affected by the addition of new rocks.

Effects of maintenance

In April 2002, extensive maintenance was carried out at Cesenatico (Fig. 1A). About one third of the breakwaters at this locality were maintained through the addition of new blocks over >70% of the surface. We randomly selected 4 breakwaters among those that had undergone maintenance; 4 additional breakwaters, that were not maintained, were randomly selected as controls. Sampling started in May 2002, after a storm stabilised the new blocks and made sampling safe from risks associated with the potential overturning of blocks. Eight replicate quadrats (20 × 20 cm), randomly placed at least 1 m apart, were sampled visually at low-shore levels on both the landward and seaward sides of each breakwater, by using a frame with 25 sub-quadrats. A score from 0 (i.e. absence) to 4% (i.e. occupation of the entire surface) was given to each taxon in each sub-quadrat, and the total cover was obtained by summing over the entire set of sub-quadrats [55]. Organisms were generally identified to species level, or grouped into higher taxonomic or morphological groups when unequivocal identification in the field was not possible. Sampling was repeated in August 2002, and January and May 2003 to identify the recovery time trajectory. Although the study was planned to run until complete recovery, 2 control breakwaters were maintained in May 2003 and as a result, the experiment was terminated.

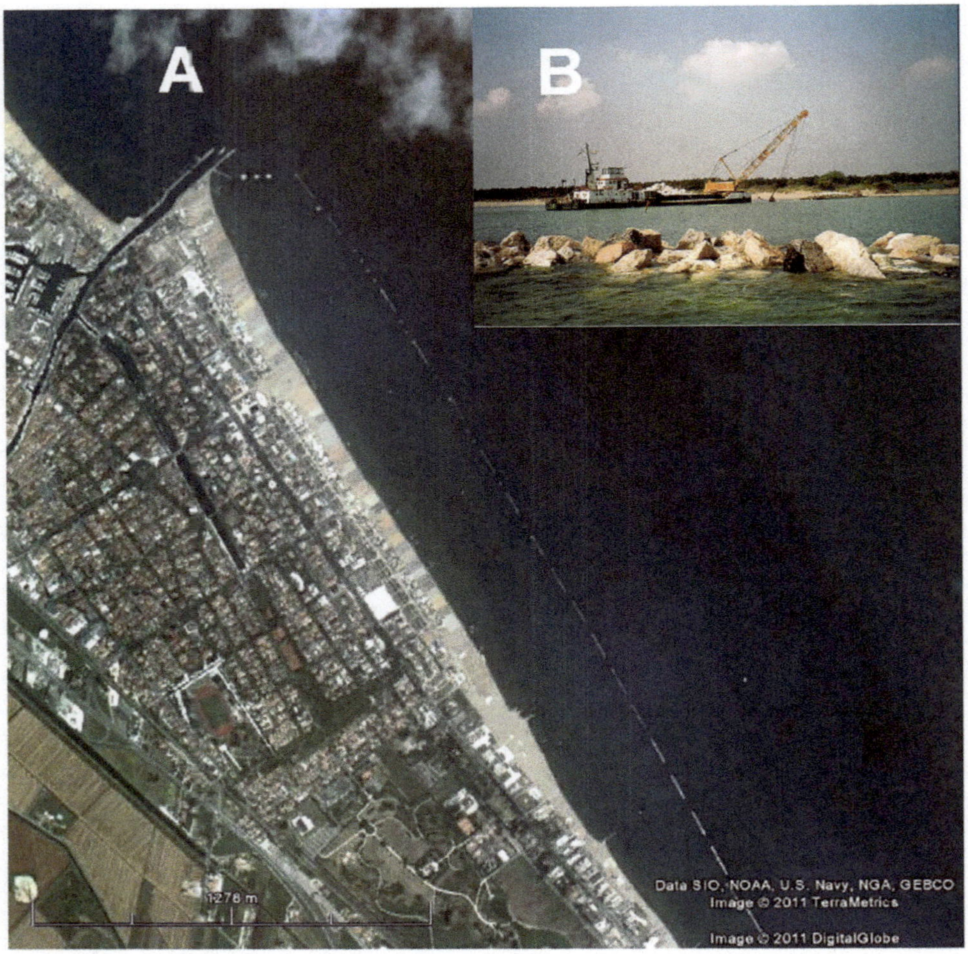

Figure 1. Human-made, coastal defence infrastructures at Cesenatico, along the Italian shores of the north Adriatic sea. A) View of the study area and breakwaters from Google Earth. B) View of a breakwater under maintenance (photo L. Airoldi).

Effects of maintenance in relation to the exposure and location of the assemblages were analysed using PERMANOVA [56] on Bray-Curtis similarity coefficients, calculated using fourth-root transformed data to preserve information on relative covers of species, while reducing differences in scales among variables. For the analysis, 9999 unrestricted random permutations of residuals were used to generate P-values. The analysis included the factors: date (random, 4 levels), treatment (fixed, 2 levels: maintained *vs* non-maintained control), exposure (fixed, 2 levels: landward *vs* seaward), and breakwater (random, 4 levels, nested within treatment). A metric multi-dimensional scaling (MDS) plot, calculated on a matrix of centroids in Bray-Curtis space (PCO; [56]) for each combination of date, treatment, breakwater, and exposure, was used to visualise patterns in multivariate data. Effects on the most abundant taxa were also analysed individually using Analysis of Variance (ANOVA), including the same factors as in the multivariate analysis. Cochran's C test was used to assess the assumption of homogeneity of variances and data were transformed when necessary. Student – Newman – Keuls (SNK) tests were used for a posteriori comparisons of means (57).

Effects of the timing of disturbance

A manipulative experiment was carried out at Cesenatico to test whether the timing at which maintenance, or other extensive disturbances are imposed affects the recovery of the assemblages and the dominance by weedy macroalgae, and whether any such effects varied between the landward and seaward sides of breakwaters and among breakwaters. In March 2003, 16 blocks >3 m apart, were randomly selected on both the landward and seaward sides of 3 randomly chosen breakwaters (100s m apart), that had not been maintained during the past 3 years. Subsets of 4 blocks were randomly assigned to each of 3 different disturbance times (April 2003, August 2003, January 2004), and controls (undisturbed blocks). Assemblages were removed from the entire surface of each block by means of paint-scrapers and brushes, in order to simulate the effects of large disturbances such as those from maintenance. Blocks assigned to different times of disturbance were marked for later relocation.

Assemblages on treatment and control blocks were sampled in May 2004, as most of the macroalgae on these breakwaters are annual species, characterised by a peak in abundance during the spring-summer. Four quadrats (20 × 20 cm) were randomly placed on each block and sampled visually, using the same technique described in the previous section. No further sampling was possible after this date, as breakwaters were subjected to unscheduled maintenance works.

The effects of disturbances applied at different times, and in relation to the exposure and location of the assemblages, were analysed using PERMANOVA. The analysis included the factors treatment (4 levels, fixed), breakwater (3 levels, random), and exposure (fixed, seaward *vs* landward), crossed to each other, and the factor block (4 levels, random), nested in the interaction treatment × exposure × breakwater. A MDS plot, calculated on a matrix of centroids in Bray-Curtis space for each combination of treatment, breakwater, and exposure, was used to visualise patterns in multivariate data. The same design was used to evaluate the response of weedy macroalgae to manipulative conditions using ANOVA.

Results

From 2001 to 2003, approximately 52% of the monitored breakwaters were maintained only once, 16% were maintained twice, and 32% were not maintained. Almost 80% of the repairs were made during spring months, from mid-March to mid-June. Maintenance opened a substantial amount of bare space, through both the addition of new surfaces and the disruption of extant assemblages by overturning blocks (Fig. 1 B). The extent of bare space just after maintenance varied from ~30–40% on the seaward sides of the breakwaters to >70% on the landward sides.

Effects of maintenance

There were significant effects of disturbance due to maintenance works that varied among dates of sampling (Table 1). Assemblages on maintained breakwaters were different from those on non-maintained breakwaters after 1 (May 2002), and 4 months (August 2002) after the time of disturbance (Fig. 2, Table 1). Despite the lack of significant effects of maintenance works in January 2003, low-shore assemblages on maintained and non-maintained breakwaters differed in May 2003, more than a year after maintenance had occurred (Fig. 2, and Table 1). Disturbance had consistently greater effects on assemblages on the landward than on the seaward sides, as shown by the significant interaction treatment × exposure (Table 1).

Table 1. Effects of maintenance in relation to the exposure and location of the assemblages on the breakwaters.

Source	df	MS	Pseudo-F	Pair-wise tests
Date = D	3	24310	13.49***	
Treatment = T	1	68385	5.52**	
Exposure = E	1	69816	6.49**	
Breakwater (T) = B (T)	6	5415	3.01***	**D x T**
D x T	3	7912	4.39***	May02: M ≠ C, Aug02: M ≠ C,
D x E	3	7223	3.02**	Jan03: M = C, May03: M ≠ C
T x E	1	21632	3.98**	**T x E**
D x B (T)[a]	16	1802	1.93***	Landward: M ≠ C, Seaward: M = C
E x B (T)	6	4408	1.84*	
D x T x E	3	1925	0.80	
D x E x B (T)[a]	16	2392	2.56**	
Residual	420	934		

The analysis is a PERMANOVA (17 variables, 4th-root transformed data) comparing assemblages between maintained (= M) and non-maintained, control (= C) breakwaters (Treatment, fixed factor), among dates of sampling (May and August 2002, January and May 2003, random factor; in May 2003, two control breakwaters were missing), between landward and seaward sides (Exposure, fixed factor) and among Breakwaters (4 levels, random factor nested in Treatment). [a] Term has one or more empty cells. * = $P<0.05$; ** = $P<0.01$, *** = $P<0.001$.

On the landward sides, maintenance caused the immediate loss of mussels and oysters and an increase in the availability of bare space, much of which was immediately occupied by a biofilm coating and, to a lesser extent, by macroalgae (Fig. 3; see Supporting Information S1 for details of the univariate analyses). In May 2002, 1 month after the maintenance, bare rock and biofilm together comprised >80% of primary substrata, compared to <30% on control breakwaters. Bare rock and biofilm cover decreased in May 2003, when macroalgae became the dominant taxa on maintained breakwaters (Fig. 3). Macroalgae, which comprised ephemerals, such as *Ulva* spp., filamentous forms, and the invasive species *Codium fragile* ssp. *tomentosoides*, were virtually absent from the landward sides of non-maintained breakwaters.

On the seaward sides, maintenance had less severe effects on mussels, initially increasing the cover of the biofilm by only 10–20% (Fig. 3). However, similar to results observed on the landward sides, macroalgae also increased significantly on the seaward sides following the maintenance (Fig. 3), reaching up to 50% cover on maintained breakwaters, compared to <20% on non-maintained breakwaters by May 2003. The increase in macroalgae on both the landward and seaward sides of breakwaters was particularly interesting because it occurred not only during the spring-summer, immediately following the repairs, but also during the following spring, one year later.

Effects of the timing of disturbance

In May 2004, the landward sides of breakwaters were extensively covered by a thin layer of sediment, possibly as a consequence of nearby beach nourishment. Increased sediment deposition probably negatively affected filter-feeding organisms (i.e. mussels and oysters), reducing their% cover compared to the beginning of the experiment. Nonetheless, it was still possible to clearly detect differences between blocks that had been disturbed at different times.

In May 2004, assemblages on blocks disturbed in April and August 2003 were still significantly different from those on control blocks (Table 2, Fig. 4). In contrast, there were no detectable differences between assemblages disturbed in January 2004 and the controls (Table 2, Fig. 4), despite the shorter period of time which had elapsed. These patterns did not vary among breakwaters or between the landward and seaward sides, despite the large natural variation at this scale, and between these habitats (Table 2).

The abundance of weedy macroalgae varied according to the time at which the disturbance was applied ($F_{3,72} = 4.58$, $P = 0.054$, see Supporting Information S2 for details of the univariate analyses). The cover of macroalgae (mostly *Ulva* spp.) on blocks disturbed in April 2003 was approximately double that on controls, on both the landward and seaward sides (Fig. 5). Blocks disturbed in August 2003 also supported a greater cover of macroalgae (mostly the invasive *Codium fragile* ssp. *tomentosoides*) than controls, but this was only significant on the landward sides (Fig. 5). Conversely, blocks disturbed in January 2004 developed macroalgal coverage comparable to undisturbed controls. The cover of other species greatly varied among replicates and blocks, and it was not possible to identify a clear response to the treatment (Supporting Information S2).

Discussion

Our results clearly indicate that intense human disturbance can be a major determinant of the spread of opportunistic species on

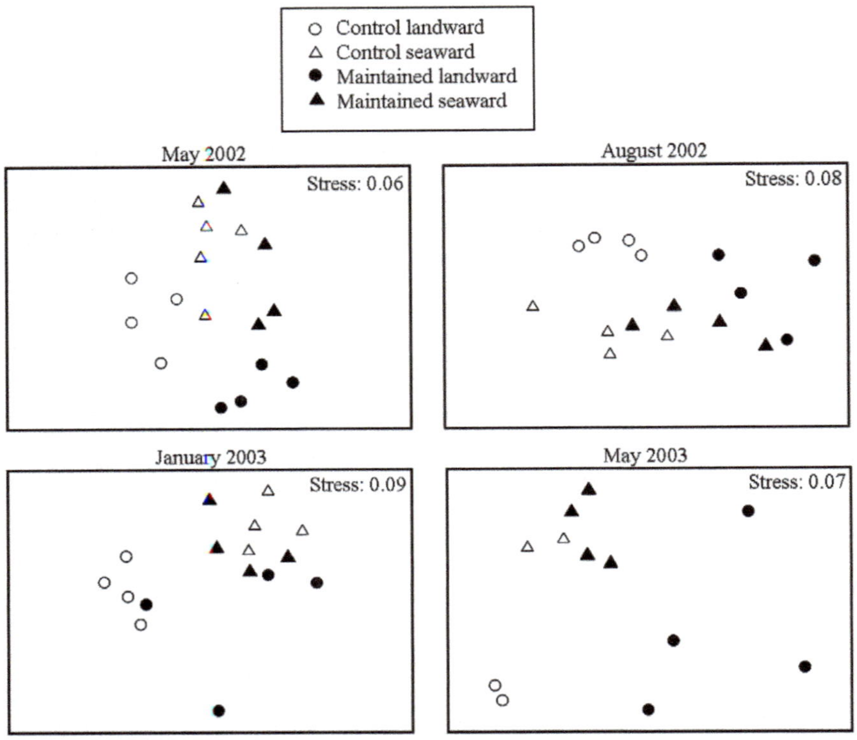

Figure 2. Two-dimensional plots of principal coordinate axes (unconstrained metric multi-dimensional scaling, MDS). The MDS shows ordination of centroids of assemblages at the landward and seaward sides of replicated control (non-maintained) and maintained (March 2002) breakwaters at Cesenatico in May 2002, August 2002, January 2003 and May 2003. There were 4 control and 4 maintained breakwaters, except for May 2003, when there were only 2 control breakwaters. Analyses were based on Bray-Curtis dissimilarities after 4th-root transformation of cover data. Stress values lower than 0.10 indicate that the ordination is good and that the interpretation of patterns in 2 dimensions is reliable.

artificial coastal infrastructures. Maintenance - which is an extreme disturbance - caused a significant decrease in the cover of dominant species such as mussels and oysters and enhanced the cover of opportunistic organisms such as biofilms, and weedy and invasive macroalgae. These effects were particularly prevalent on the landward (sheltered) sides of breakwaters, while assemblages on the seaward (exposed) sides were only moderately affected by the maintenance.

The different responses to disturbance observed between the landward and seaward sides of the breakwaters can be interpreted as a result of a combination of factors. Firstly, disturbance *per se* was more severe and persistent on the landward sides. The new blocks added to the breakwaters (e.g. Fig. 1B) were initially unstable and were overturned by wave action, rolling most frequently towards the landward sides. Secondly, the structure of dominant mussel beds also differed between the landward and seaward sides of breakwaters [32,46]: on the landward sides, mussels were generally larger in size and formed a multi-layered matrix, whilst on the seaward sides, smaller individuals formed mono-layered beds [58], which tend to be less susceptible to dislodgement by mechanical disturbances [59–60]. Finally, on the seaward sides, mussels, which were the dominant space occupiers, recovered more quickly. Efficient colonisation of space in exposed habitats may result from greater rates of recruitment of larvae or faster growth of individuals, due to greater provision of food particles in enhanced water flow [61–62].

The manipulative experiment confirmed that the high cover of weedy macroalgae, often observed on artificial infrastructures in the study region, can result from severe disturbances, such as those from maintenance. In addition, it showed that the cover of these species varied according to the timing at which the disturbance was applied. The cover of macroalgae increased by ~100% on all substrata disturbed in April, and on landward substrata disturbed in August, while no increase was observed when the disturbance was applied in January. The macroalgal assemblage was mostly composed of opportunistic forms, representing an early stage of succession, and invasive species. In the study area, *Ulva* spp. are amongst the earliest colonisers on new infrastructures, sometimes attaining up to 100% cover 2 months post-construction [17].

Although the experiment was not specifically designed to test if recovery trajectories differed depending on the timing of disturbance, our results suggest that the time at which the disturbance is applied could also influence the duration of recovery to reference conditions. Assemblages disturbed in January 2004 were similar to controls after only 4 months, while assemblages disturbed earlier (in April and August 2003) still differed from controls, despite the longer period of time which had elapsed. This suggests a non - linear relationship between recovery and time elapsed since disturbance. A possible reason why substrata disturbed in January recovered more quickly than substrata disturbed at other times could be related to the different recruitment peaks of mussels and macroalgae in this region. Mussels tend to recruit in late winter to early spring [63], while macroalgae recruit in the spring to late summer [17]. Therefore, the effects of disturbances occurring in winter, when macroalgae are less likely to monopolise space and mussel recruitment is about to start, could be short-lasting in this system. In contrast, disturbances occurring in spring-summer could result in a

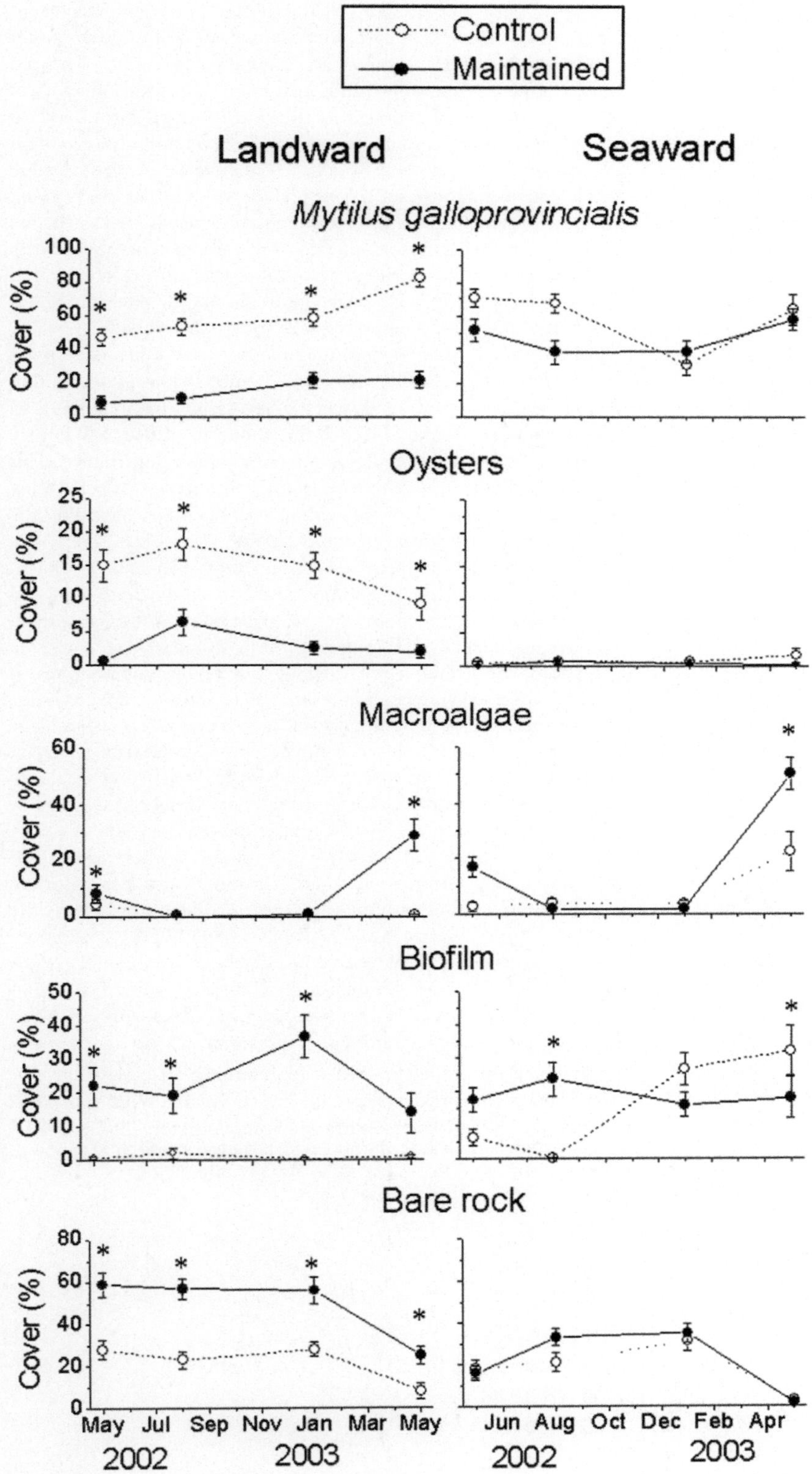

Figure 3. Abundance of most abundant taxa on the study breakwaters at Cesenatico. Data are average percent covers ±1 SE ($n=32$, 8 replicate plots for each of 4 breakwaters; in May 2003 $n=16$ for controls, 8 replicate plots for each of 2 breakwaters) of *Mytilus galloprovincialis*, oysters (a mixture of *Ostrea edulis* and *Crassostrea gigas* difficult to separate by visual sampling), macroalgae (mainly *Ulva* spp., *Codium fragile* ssp. *tomentosoides* and filamentous forms), biofilm (a coating of microalgae, juvenile stages of macroalgae and silt) and bare rock (rock non occupied by visible macroscopic forms) at the landward and seaward sides of control (non-maintained) and maintained (April 2002) breakwaters, in May 2002, August 2002, January 2003 and May 2003. Asterisks indicate significant differences between assemblages on maintained and control breakwaters as indicated by *a posteriori* SNK tests.

Table 2. Effects of disturbance applied at different times and in relation to the exposure and location of the assemblages on the breakwaters.

Source of variation	df	MS	Pseudo-F
Treatment = T	3	5404.1	9.11**
Exposure = E	1	2.95	11.14*
Breakwater = B	2	33544	16.50**
T x E	2	1745.5	0.94
T x B	6	593.1	0.29
E x B	2	26496	13.03
T x E x B	6	1844.3	0.91
Block (T x E x B)	72	2033.6	4.27**
Residual	288	475.75	
pair-wise tests: Treatment			
Comparison	*t*	*P*	
D1, D2	18.355	0.001	
D1, D3	1.6832	0.118	
D1, C	2.6654	0.032	
D2, D3	4.7585	0.005	
D2, C	2.9879	0.021	
D3, C	1.6012	0.141	

The analysis is a PERMANOVA (14 variables, 4th-root transformed data) comparing assemblages among breakwaters, between landward and seaward sides, among treatments (removal in April 2003 = D1, August 2003 = D2 and January 2004 = D3; unmanipulated plots = C) and among blocks, 10 months after the initiation of the experiment. * = $P<0.05$; ** = $P<0.01$.

longer-term persistence of macroalgae. The mortality of mussels in spring 2004 (prior to our sampling in May), probably caused by the large deposition of sediments, may have increased the similarity between assemblages on blocks disturbed shortly before (January) and those on control blocks. However, a similar response on the sheltered sides of breakwaters, little affected by the enhanced deposition of sediments, indicates that this was not the primary mechanism responsible for the patterns observed.

Our results are consistent with recent work showing how the spatio-temporal variation in the effects of disturbance is a key factor in determining the success of opportunistic and invasive species in marine systems [41,44,64]. This has important implications for the design and management of hard coastal infrastructures and, more broadly, for the conservation of coastal areas, under increasing anthropogenic pressures. Ecological considerations in the design of marine infrastructures tend to focus on construction materials, surface texture, and habitat complexity, as engineering options to enhance the ecological value of these artificial substrata [65–68]. Here we show that, in the long run, the project lifetime and required maintenance are clearly one of the most crucial factors affecting the composition, abundance, and distribution of species that colonise the infrastructures. For any new infrastructure introduced into the marine environment, it will take time for mature assemblages to develop [69–71]. Furthermore, any following maintenance, or other analogous severe disturbances, will lead to the assembly of communities that represent early successional stages. Importantly, the magnitude of this impact and the capability of the system to recover will largely depend on the scale (spatial and temporal) of the disturbance. This has some significant policy implications, in terms of incorporating aspects of the ecology of the system into decisions regarding the timing of major maintenance and repairs.

Marine urban infrastructures are often located in harsh environments and their lifetime would be significantly reduced without routine maintenance or periodic repairs. At the same time, our results suggest that approaches to maintenance could be improved, for example, by carrying out repair interventions at specific times or in a way that reduces their impacts. For example, in the study area, the impact of maintenance on assemblage dynamics would be reduced if interventions are carried out over the winter rather than spring. Higher operational costs would be largely offset by both environmental and economical benefits. Optimising maintenance would reduce the development of weedy macroalgal species. Macroalgal blooms and their associated detritus washed up on beaches are a common problem in urban coastal areas, causing nuisance to recreational and tourist activities and leading to expensive removal operations [72–74]. Additionally, sound management of maintenance activities would reduce the likelihood of the establishment and/or spread of non-indigenous marine species, with indirect benefits for fisheries and aquaculture. Marine man-made infrastructures are particularly sensitive to invasions by non-indigenous species [31-34]. Our results, in accordance with those of previous studies [32,46], suggest that the great invasibility of artificial infrastructures may be due to the severe disturbances to which they are generally exposed.

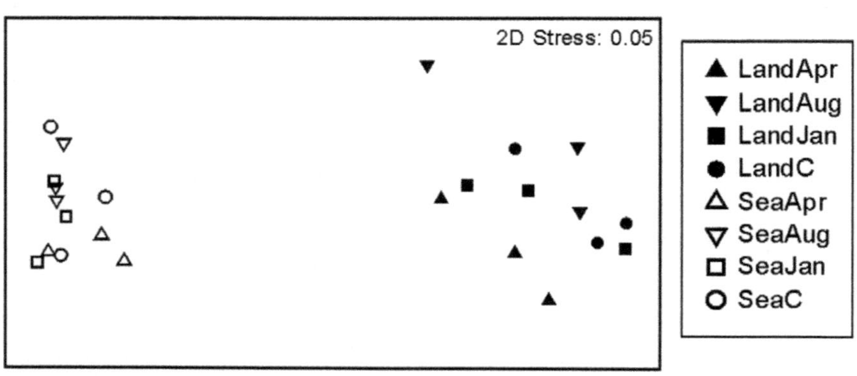

Figure 4. Two-dimensional plots of principal coordinate axes (metric multi-dimensional scaling, MDS). The MDS shows ordination of centroids of assemblages at the landward and seaward sides of each of 3 breakwaters for each time of disturbance (D1 = April 2003, D2 = August 2003, D3 = January 2004, C = unmanipulated plots). Analyses were based on Bray-Curtis dissimilarities after 4th-root transformation of data collected in May 2004. A Stress value of 0.05 indicates that the ordination is excellent and that the interpretation of patterns in 2 dimensions is highly reliable.

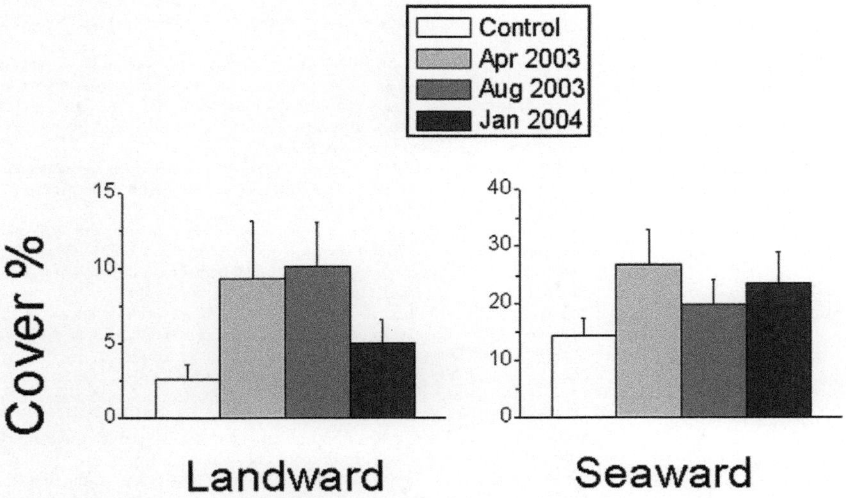

Figure 5. Abundance of macroalgae in the experimental treatments at Cesenatico. Data are average percent covers ±1 SE ($n = 36$, 4 replicate plots for each of 4 blocks for each of 3 breakwaters) of macroalgae (mainly *Ulva* spp. and *Codium fragile* ssp. *tomentosoides*) at the landward and seaward sides of breakwaters for each time of disturbance (April 2003, August 2003, January 2004, Control = unmanipulated plots).

Indeed, prolonged availability of unoccupied space, or other resources generated by human disturbances are considered one of the main factors facilitating the establishment of non-indigenous species [75].

Conclusions

Mitigation of impacts of marine infrastructures is a possible outcome of policies that explicitly consider the ecological features of artificial habitats and recognise the fundamental value of managing biodiversity in urban settings. We advocate that there is a need for ecological based planning and management of urban artificial infrastructures. Understanding the functioning of these novel habitats will be key to improving their design and management and, hence, to mitigate their negative impacts and enhance their contribution to marine biodiversity and ecosystem functioning.

Supporting Information

Supporting Information S1 ANOVAs on the effects of date, maintenance works, exposure and breakwater on the percentage cover of unoccupied space, biofilm, macroalgae and invertebrates.

Supporting Information S2 ANOVAs on the effects of Breakwater, Exposure and Treatment on the percentage cover of unoccupied space and dominant taxa.

Acknowledgments

We thank F. Bacchiocchi for her valuable contribution to various phases of the work. We also thank J.M. Anderson, N. Liverani, G.M. Branca for help with the fieldwork. Valuable comments by L. Firth and T. Romanuk improved the final version of the paper.

Author Contributions

Conceived and designed the experiments: LA FB. Performed the experiments: LA FB. Analyzed the data: LA FB. Contributed reagents/materials/analysis tools: LA. Wrote the paper: LA FB.

References

1. Glasby TM, Connell SD (1999) Urban structures as marine habitats. Ambio 28: 595–598.
2. Inger R, Attrill MJ, Bearhop S, Broderick AC, Grecian WJ, et al. (2009) Marine renewable energy: potential benefits to biodiversity? An urgent call for research. J Appl Ecol 46: 1145–1153.
3. Bulleri F, Chapman MG (2010) The introduction of coastal infrastructure as a driver of change in marine environments. J Appl Ecol 47: 26–35.
4. Dugan JE, Airoldi L, Chapman MG, Walker S, Schlacher T (2011) Estuarine and coastal structures: environmental effects. A focus on shore and nearshore structures. In: Elliott M, Dugan J, eds. Treatise on Estuarine and Coastal Science. New York: Elsevier Press. In press.
5. Airoldi L, Beck MW (2007) Loss, status and trends for coastal marine habitats of Europe. Oceanogr Mar Biol Annu Rev 45: 345–405.
6. Koike K (1996) The countermeasures against coastal hazards in Japan. GeoJournal 38: 301–312.
7. Chapman MG, Bulleri F (2003) Intertidal seawalls - new features of landscape in intertidal environments. Landsc Urban Plan 62: 159–172.
8. Thompson RC, Crowe TP, Hawkins SJ (2002) Rocky intertidal communities: past environmental changes, present status and predictions for the next 25 years. Environ Cons 29: 168–191.
9. Pister B (2009) Urban marine ecology in southern California: the ability of riprap structures to serve as rocky intertidal habitat. Mar Biol 156: 861–873.
10. Iannuzzi TJ, Weinstein MP, Sellner KG, Barrett JC (1996) Habitat disturbance and marina development: an assessment of ecological effects. I. Changes in primary production due to dredging and marina construction. Estuaries 19: 257–271.
11. Iverson ES, Bannerot SP (1984) Artificial reefs under marine docks in southern Florida. N Am J Fish Manage 4: 294–299.
12. Airoldi L, Abbiati M, Beck MW, Hawkins SJ, Jonsson PR, et al. (2005) An ecological perspective on the deployment and design of low-crested and other hard coastal defence structures. Coast Eng 52: 1073–1087.
13. Perkol-Finkel S, Shashar N, Benayahu Y (2006) Can artificial reefs mimic natural reef communities? The roles of structural features and age. Mar Environ Res 61: 121–135.
14. Burt J, Bartholomew A, Usseglio P, Bauman A, Sale PF (2009) Are artificial reefs surrogates of natural habitats for corals and fish in Dubai, United Arab Emirates? Coral Reefs 28: 663–675.
15. Miller MW, Valdivia A, Kramer KL, Mason B, Williams DE, et al. (2009) Alternate benthic assemblages on reef restoration structures and cascading effects on coral settlement. Mar Ecol Prog Ser 387: 147–156.

16. Davis JLD, Levin LA, Walther SM (2002) Artificial armored shorelines: sites for open-coast species in a southern California bay. Mar Biol 140: 1249–1262.
17. Bacchiocchi F, Airoldi L (2003) Distribution and dynamics of epibiota on hard structures for coastal protection. Est Coast Shelf Sci 56: 1157–1166.
18. Connell SD (2000) Floating pontoons create novel habitats for subtidal epibiota. J Exp Mar Biol Ecol 247: 183–194.
19. Glasby TM (1999) Differences between subtidal epibiota on pier pilings and rocky reefs at marinas in Sydney. Est Coast Shelf Sci 48: 281–290.
20. Bulleri F, Chapman MG, Underwood AJ (2005) Intertidal assemblages on seawalls and vertical rocky shores in Sydney harbour, Australia. Austral Ecol 30: 655–667.
21. Moschella PS, Abbiati M, Åberg P, Airoldi L, Anderson JM, et al. (2005) Low-crested coastal defence structures as artificial habitats for marine life: using ecological criteria in design. Coast Eng 52: 1053–1071.
22. Gacia E, Satta MP, Martin D (2007) Low crested coastal defence structures on the Catalan coast of the Mediterranean Sea: how they compare with natural rocky shores. Sci Mar 71: 259–267.
23. Johannesson K, Warmoes T (1990) Rapid colonization of Belgian breakwaters by the direct developer, *Littorina saxatilis* (Olivi) (Prosobranchia, Mollusca). Hydrobiologia 193: 99–108.
24. Chapman MG (2003) Paucity of mobile species on constructed seawalls: effects of urbanization on biodiversity. Mar Ecol Prog Ser 264: 21–29.
25. Fauvelot C, Bertozzi F, Costantini F, Airoldi L, Abbiati M (2009) Lower genetic diversity in the limpet *Patella caerulea* on urban coastal structures compared to natural rocky habitats. Mar Biol 156: 2313–2323.
26. Ivesa L, Chapman MG, Underwood AJ, Murphy RJ (2010) Differential patterns of distribution of limpets on intertidal seawalls: experimental investigation of the roles of recruitment, survival and competition. Mar Ecol Prog Ser 407: 55–69.
27. Bulleri F (2005) Role of recruitment in causing differences between intertidal assemblages on seawalls and rocky shores. Mar Ecol Prog Ser 287: 53–64.
28. Moreira J, Chapman MG, Underwood AJ (2006) Seawalls do not sustain viable populations of limpets. Mar Ecol Prog Ser 322: 179–188.
29. Perkol-Finkel S, Benayahu Y (2009) The role of differential survival patterns in shaping coral communities on neighboring artificial and natural reefs. J Exp Mar Biol Ecol 369: 1–7.
30. Martins GM, Amaral AF, Wallenstein FM, Neto AI (2009) Influence of a breakwater on nearby rocky intertidal community structure. Mar Environ Res 67: 237–245.
31. Vaselli S, Bulleri F, Benedetti-Cecchi L (2008) Hard coastal-defence structures as habitats for native and exotic rocky-bottom species. Mar Environ Res 66: 395–403.
32. Bulleri F, Airoldi L (2005) Artificial marine structures facilitate the spread of a non-indigenous green alga, *Codium fragile* ssp *tomentosoides*, in the North Adriatic Sea. J Appl Ecol 42: 1063–1072.
33. Glasby TM, Connell SD, Holloway MG, Hewitt CL (2007) Nonindigenous biota on artificial structures: could habitat creation facilitate biological invasions? Mar Biol 151: 887–895.
34. Dafforn KA, Johnston EL, Glasby TM (2009) Shallow moving structures promote marine invader dominance. Biofouling 25: 277–287.
35. Godefroid S, Koedam N (2003) Distribution pattern of the flora in a peri-urban forest: an effect of the city–forest ecotone. Landsc Urban Plan 65: 169–185.
36. Rebele F (1994) Urban ecology and special features of urban ecosystems. Global Ecol Biogeogr Lett 4: 173–187.
37. D'Antonio CM, Vitousek PM (1992) Biological invasions by exotic grasses, the grass/fire cycle, and global change. Annu Rev Ecol Syst 23: 63–87.
38. Lozon J, MacIsaac H (1997) Biological invasions: are they dependent on disturbance? Environ Rev 5: 131–144.
39. McKinney ML (2006) Urbanization as a major cause of biotic homogenization. Biol Cons 127: 247–260.
40. Airoldi L (1998) Roles of disturbance, sediment stress, and substratum retention on spatial dominance in algal turf. Ecology 79: 2759–2770.
41. Norkko A, Rosenberg R, Thrush SF, Whitlatch RB (2006) Scale- and intensity-dependent disturbance determines the magnitude of opportunistic response. J Exp Mar Biol Ecol 330: 195–207.
42. Stachowicz JJ, Byrnes JE (2006) Species diversity, invasion success, and ecosystem functioning: disentangling the influence of resource competition, facilitation, and extrinsic factors. Mar Ecol Prog Ser 311: 251–262.
43. Lohrer AM, Chiaroni LD, Hewitt JE, Thrush SF (2008) Biogenic disturbance determines invasion success in a subtidal soft-sediment system. Ecology 89: 1299–1307.
44. Clark GF, Johnston EL (2011) Temporal change in the diversity-invasibility relationship in the presence of a disturbance regime. Ecol Lett 14: 52–57.
45. Burcharth HF (1993) The design of breakwaters. In: Abbott MB, Price WA, eds. Coastal, estuarial and harbour engineer's book. London: Chapman & Hall. pp 381–424.
46. Airoldi L, Bacchiocchi F, Cagliola C, Bulleri F, Abbiati M (2005) Impact of recreational harvesting on assemblages in artificial rocky habitats. Mar Ecol Prog Ser 299: 55–66.
47. Guidetti P, Bussotti S, Boero F (2005) Evaluating the effects of protection on fish predators and sea urchins in shallow artificial rocky habitats: a case study in the northern Adriatic Sea. Mar Environ Res 59: 333–348.
48. Gregory PT, ed (2004) Port engineering: planning, construction, maintenance and security. Hoboken: John Wiley & Sons. 867 p.
49. Denley EJ, Underwood AJ (1979) Experiments on factors influencing settlement, survival and growth of two species of barnacles in New South Wales. Journal of Exp Mar Biol Ecol 36: 269–293.
50. Sousa WP (1979) Experimental investigations of disturbance and ecological succession in a rocky intertidal community. Ecol Monogr 49: 227–254.
51. Cencini C (1998) Physical processes and human activities in the evolution of the Po delta, Italy. J Coast Res 14: 774–793.
52. Bulleri F, Airoldi L, Branca GM, Abbiati M (2006) Positive effects of the introduced green alga, *Codium fragile* ssp *tomentosoides*, on recruitment and survival of mussels. Mar Biol 148: 1213–1220.
53. Bulleri F, Branca MG, Abbiati M, Airoldi L (2007) Development of reproductive structures in the introduced green alga, *Codium fragile* spp. *tomentosoides*, in the northern Adriatic Sea. Eur J Phycol 42: 137–144.
54. Van der Meer JW (1988) Rock slopes and gravel beaches under wave attack. Delft, The Netherlands: Delft Hydraulics Publication. 214 p.
55. Benedetti-Cecchi L, Airoldi L, Abbiati M, Cinelli F (1996) Estimating the abundance of benthic invertebrates: a comparison of procedures and variability between observers. Mar Ecol Prog Ser 138: 93–101.
56. Anderson MJ, Gorley RN, Clarke KR (2008) PERMANOVA+ for PRIMER: Guide to software and statistical methods. Plymouth: PRIMER-E. 214 p.
57. Underwood AJ (1997) Experiments in ecology. Their logical design and interpretation using analysis of variance. Cambridge, United Kingdom: Cambridge University Press. 504 p.
58. Bacchiocchi F (2004) Distribution, structure and variability of intertidal assemblages on human-made structures in the North Adriatic Sea. Parma: University of Parma, Phd Thesis. 198 p.
59. Harger JRE, Landenberger DE (1971) The effect of storms as a density dependent mortality factor on populations of sea mussels. Veliger 14: 195–201.
60. Paine RT, Levin SA (1981) Intertidal landscapes: disturbance and the dynamics of patterns. Ecol Monogr 51: 145–178.
61. Seed R (1976) Ecology. In: Bayne BL, ed. Marine mussels: their ecology and physiology. Cambridge: Cambridge University Press. pp 13–65.
62. Seed R, Suchanek TH (1992) Population and community ecology of *Mytilus*. In: Gosling E, ed. The mussel *Mytilus*: ecology, physiology, genetics, and culture. New York: Elsevier. pp 87–169.
63. Ceccherelli VU, Rossi R (1984) Settlement, growth and production of the mussel *Mytilus galloprovincialis*. Mar Ecol Prog Ser 16: 173–184.
64. Clark GF, Johnston EL (2009) Propagule pressure and disturbance interact to overcome biotic resistance of marine invertebrate communities. Oikos 118: 1679–1686.
65. Burt J, Bartholomew A, Bauman A, Saif A, Sale PF (2009) Coral recruitment and early benthic community development on several materials used in the construction of artificial reefs and breakwaters. J Exp Mar Biol Ecol 373: 72–78.
66. Brown CJ (2005) Epifaunal colonization of the Loch Linnhe Artificial Reef: Influence of substratum on epifaunal assemblage structure. Biofouling 21: 73–85.
67. Burcharth HF, Hawkins SJ, Zanuttigh B, Lamberti A, eds (2007) Environmental design guidelines for low-crested coastal defence structures. Amsterdam: Elsevier. 448 p.
68. Chapman MG, Blockley DJ (2009) Engineering novel habitats on urban infrastructure to increase intertidal biodiversity. Oecologia 161: 625–635.
69. Pinn EH, Mitchell K, Corkill J (2005) The assemblage of groynes in relation to substratum age, aspect and microhabitat. Est Coast Shelf Sci 62: 271–282.
70. Perkol-Finkel S, Shashar N, Barneah O, Ben-David-Zaslow R, Oren U, et al. (2005) Fouling reefal communities on artificial reefs: Does age matter? Biofouling 21: 127–140.
71. Burt J, Bartholomew A, Sale PF (2011) Benthic development on large-scale engineered reefs: a comparison of communities among breakwaters of different age and natural reefs. Ecol Eng 37: 191–198.
72. Morand P, Briand X (1996) Excessive growth of macroalgae: a symptom of environmental disturbance. Bot Mar 39: 491–516.
73. Phillips JA (2006) Drifting blooms of the endemic filamentous brown alga *Hincksia sordida* at Noosa on the subtropical East Australian coast. Mar Pollut Bull 52: 962–968.
74. Iskander MM, Frihy OE, El Ansary AE, El Mooty MMA, Nagy HM (2007) Beach impacts of shore-parallel breakwaters backing offshore submerged ridges, Western Mediterranean Coast of Egypt. J Environ Manage 85: 1109–1119.
75. Davis MA, Grime JP, Thompson K (2000) Fluctuating resources in plant communities; a general theory of invisibility. J Ecol 88: 528–534.

17

Cleft, Crevice, or the Inner Thigh: 'Another Place' for the Establishment of the Invasive Barnacle *Austrominius modestus* (Darwin, 1854)

Sally A. Bracewell*, Matthew Spencer, Rob H. Marrs, Matthew Iles, Leonie A. Robinson

School of Environmental Sciences, University of Liverpool, Liverpool, Merseyside, United Kingdom

Abstract

The proliferation of anthropogenic infrastructure in the marine environment has aided the establishment and spread of invasive species. These structures can create novel habitats in areas normally characterised as void of suitable settlement sites. The habitat requirements of the invasive acorn barnacle *Austrominius modestus* (Darwin, 1854) were assessed using a novel sampling site at Crosby Beach, Liverpool. *Austrominius modestus* has spread rapidly around the UK since its initial introduction, becoming locally dominant in many estuarine areas including the Antony Gormley art installation, 'Another Place', at Crosby Beach. The installation consists of 100 replicate solid cast-iron life-size human figures, located at a range of heights on the shore. We recorded the distribution and abundance of *A. modestus* present on all of the statues at various positions during the summer of 2006. The positions varied in location, exposure, direction, and rugosity. Although parameters such as rugosity and exposure did influence patterns of recruitment, they were less important than interactions between shore height and direction, and specific location on the beach. The addition of a suitable substrate to a sheltered and estuarine region of Liverpool Bay has facilitated the establishment of *A. modestus*. Understanding the habitat requirements of invasive species is important if we are to make predictions about their spread and the likelihood of invasion success. *Austrominius modestus* has already become locally dominant in some regions of the UK and, with projections of favourable warming conditions and the global expansion of artificial structures, the continued spread of this species can be expected. The implications of this on the balance between native and invasive species dominance should be considered.

Editor: Simon Thrush, National Institute of Water & Atmospheric Research, New Zealand

Funding: This work was partly funded by the Esmée Fairbairn Foundation (www.esmeefairbairn.org.uk/) under the URBANE project (Urban Research on Biodiversity on Artificial and Natural coastal Environments) and the British Ecological Society (www.britishecologicalsociety.org) through a Small Ecological Project Grant to LA Robinson. The funders had no role in study design, data collection and analysis, decision to publish, or preparation of the manuscript. No additional external funding was received for this study.

Competing Interests: The authors have declared that no competing interests exist.

* E-mail: sally_bracewell@hotmail.com

Introduction

Alien, non-native, introduced, or non-indigenous species (NIS) have become commonplace across many of the worlds ecosystems [1]. If a NIS is able to establish in a region distant from its native range and maintain self-sustaining populations in large numbers it is generally regarded as invasive, elevating to pest status once it has caused significant ecological or economic damage [2]. Successful invasions are contingent on numerous factors (e.g. suitability of habitat [3], interactions with native species [4]), and it has been suggested that approximately only 10% of any established introductions of NIS will become invasive [5], although this value is difficult to quantify [6]. Nevertheless, the number of biological invasions recorded continues to increase [7,8] and the implications of such increases for levels of biodiversity and ecosystem functioning are now widely acknowledged [9,10], causing concern amongst conservation biologists and resource managers.

Why some introductions of NIS result in invasions, whilst others do not, is a key question in invasion ecology, the answer to which is multi-faceted and dependent on traits related to both the species in question [11] and those related to the recipient habitat [12]. Greater dispersal abilities, faster growth rates and generation times (i.e. increased propagule pressure) [13], higher stress tolerances [14] and greater capacities for evolutionary change due to higher phenotypic plasticity [15] have emerged as traits related to invasiveness. Likewise, it has been hypothesised that certain types of habitats are more susceptible to invasions than others, with factors such as disturbance [16,17] and levels of native species diversity [12] being influential.

Recent studies indicate that artificial structures such as piers, pilings, seawalls and other sea defences are particularly vulnerable to invasion by non-native species; however, their contribution as drivers of ecological change has received limited attention [18]. These structures are often located in disturbed habitats, such as ports and estuaries, areas characterised by high shipping traffic and thus an increased abundance of NIS [19,20]. Artificial structures can create novel environmental conditions (e.g. vertical surfaces, lack of microhabitats), and do not closely resemble native habitats [18]. As such, they are often characterised by low native species diversity [21], and have been found to support assemblages of organisms that greatly contrast those of nearby natural sites [22]. In addition, if NIS are able to become naturalized on these structures, they can act as supplementary recruitment sites that may aid their spread, effectively acting as stepping-stones in areas of otherwise unsuitable habitat [23]. With the rising concern over the consequences of global climate change, the number of artificial structures is likely to increase in the hope that they will provide the necessary protection against such threats as sea level rise and increased storm activity [18]. The implications of this for the management of invasive species should be considered carefully [24,25].

Understanding the factors that contribute to, or inhibit, the establishment of invasive species is critical for developing effective management techniques and predicting future range expansions [26]. Studies monitoring abundance patterns of NIS can provide valuable information about their distribution and rate of spread [27,28]. Sessile marine invertebrates are common fouling organisms on artificial structures and their abundance and distribution on these structures could offer insights into the habitat requirements of these species, which in turn could help assess the likelihood of establishment in novel locations and on novel structures. Although patterns in abundance of marine invertebrates with a pelagic larval stage are inextricably linked to variations in larval supply, factors relating to the abiotic features of the recipient habitat or structure are also important in shaping the distribution of organisms [29].

Incorporating all potential factors that influence abundance patterns within one study is often infeasible; as such, most work has tended to focus on only one or two [30], despite evidence that it is the combination of multiple factors and the interactions between them that are often responsible for structuring resident communities of organisms [31]. Here we used variations in the distribution and abundance of an invasive intertidal acorn barnacle, *Austrominius modestus*, on a novel artificial substrate matter to develop a model that can incorporate all factors thought to be most important in determining abundance distributions of barnacle species, as well as the interactions between them. By simultaneously assessing multiple indices, using a unique field study opportunity, we aimed to gain a greater insight into the habitat requirements of *A. modestus* on a local scale and hypothesised that no one factor would be solely responsible for any patterns observed.

Methods

Study Site

In July 2005, 100 life-size cast-iron human figures (Fig.1) were distributed at various tidal heights along approximately 3 kilometres (km) of the foreshore at Crosby Beach, Liverpool (Fig.2) to form the art installation, 'Another Place', by sculptor Antony Gormley. A preliminary investigation undertaken in 2006, one year post installation, revealed the statues to be dominated by

Figure 1. Examples of the life-size cast-iron statues at Crosby Beach, Liverpool. Images show two of the 100 statues that form the art installation 'Another Place'; one at the higher end of shore height sampled (left) and one at a lower shore height (right). The statues stretch over approximately 3 km of the foreshore and are distributed at a range of tidal heights.

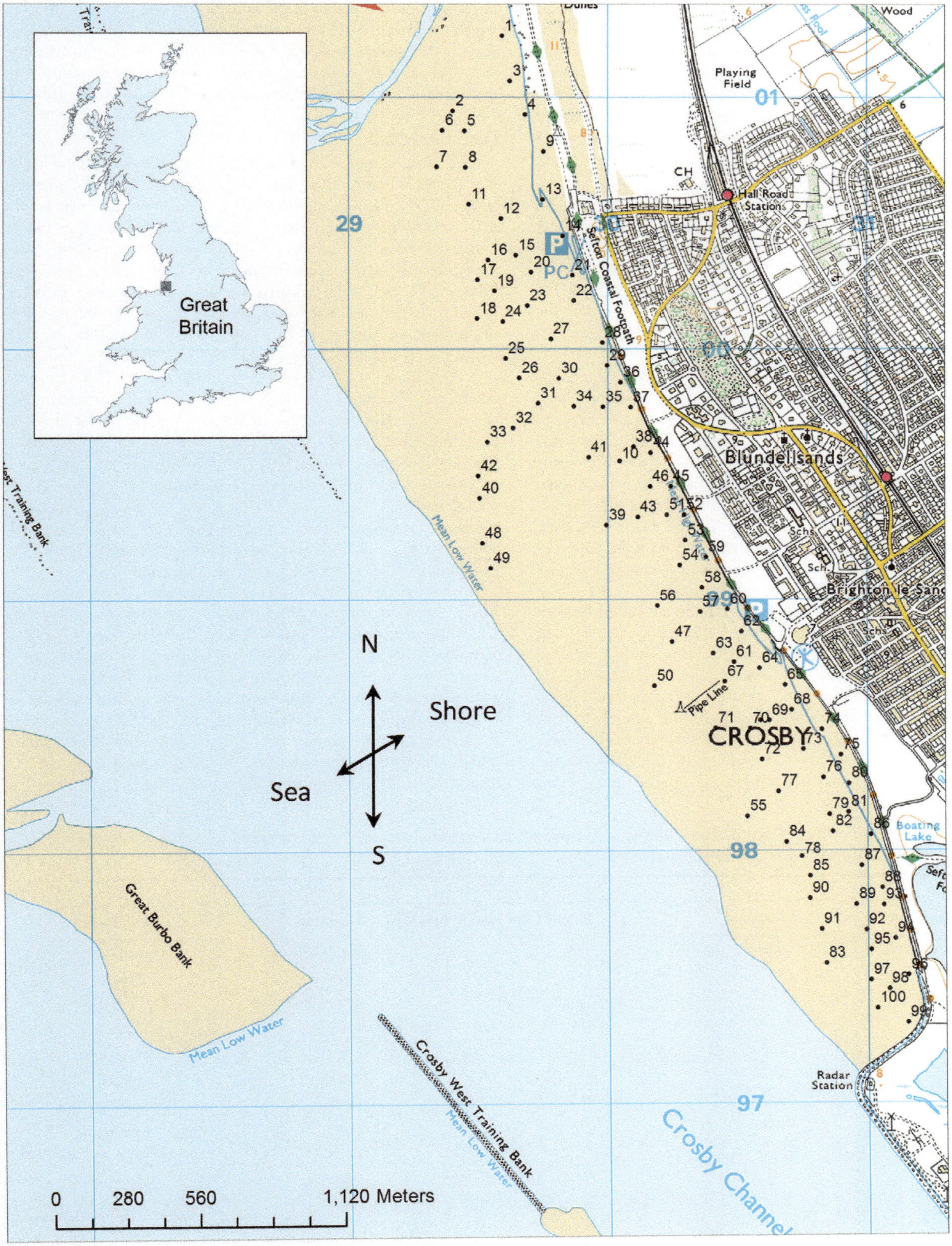

Figure 2. Location of the 100 Antony Gormley statues (numbered) at Crosby Beach, Liverpool. Locations of the statues are shown in relation to Great Britain. Also showing the direction of sampling positions used to assess barnacle abundance on the statues (north, south, sea, or shore) (Ordnance Survey © Crown copyright 2011).

the invasive barnacle, *Austrominius modestus*, to the exclusion of all other species of barnacle (L. A. Robinson, personal observation). Before installation, the Gormley statues were blasted clean; as such they provided a pristine, hard substrate onto which organisms could settle in an otherwise sandy (unsuitable) environment. The statues are placed at a range of tidal heights, and different regions of their body differ in exposure (whether or not they are sheltered by another body part – e.g. the inner thigh versus the outer thigh; we do not treat exposure as a direct measure of wave fetch), direction (north, south, sea, and shore), and rugosity (the influence of body contours, where an area such as the groin might be more complex than the torso) (Fig.1, Table 1). As a contemporary art installation, the statues were never intended as an ecological study; however they present a unique opportunity to investigate the influence of these factors, and their interactions, on the distribution and abundance of *A. modestus* in the area.

Study Species

Austrominius modestus, previously known as *Elminius modestus*, originates from Australasia, and whilst it was once considered to be a strictly southern genus [32], it has become commonplace and abundant in many European estuaries and other sheltered marine areas [33,34]. In its native habitat, *A. modestus* is a prominent fouling organism of harbours and estuaries, capable of growing 'on any sort of substratum' [35] in the upper limit of tidal ranges [32]. In the UK and Ireland, the species occupies a similar environmental niche, being predominantly found in sheltered areas of varying salinity [27,36]. However, in comparison to its native range, *A. modestus* is not limited to upper tidal limits and occupies a wide range of tidal heights, being predominant in the low-mid region in some locations [37]. *A. modestus* appears to occupy a similar niche as the native barnacles *Balanus balanoides* and *Semibalanus balanoides*, however its superior tolerance of fluctuating salinity and ability to reproduce all year round has enabled *A. modestus* to occupy a greater intertidal range, outcompeting these species in numerous sheltered estuarine habitats [33,36–38]. In the Mersey Estuary, *A. modestus* is common and abundant on other intertidal hard substrates in the estuary near the study site (most of which are artificial) [39], but both *B. balanoides* and *S. balanoides* also occur (M. Spencer, personal observation). *A. modestus* and *Balanus* spp. (including *B. crenatus* and *B. improvisus*) have been recorded on nearby subtidal artificial substrates in the Liverpool docks [40–42].

Data Collection

The abundance of *Austrominius modestus* was quantified using 10×10 cm quadrats placed at 14 positions on each of the 100 statues (Fig. 2) during August 2006, one year after their installation following permission granted by Sefton Council. Positions were chosen to represent a range of heights, rugosity, and exposure, and were placed in the same location for each statue (Table 1). Quadrats, which were constructed from garden wire to allow flexibility, were subdivided into 5×5 cm sections and the numbers of individuals present in the upper left hand corner of each quadrat were counted. Rugosity ranged from the highly contoured face and groin to the relatively uniform torso. This range in rugosity was assessed by using a piece of string to measure the distance from one side of the 10×10 cm quadrat to the other, to allow the extra distance of cracks and crevices to be incorporated. The ratio of string length to quadrat width was then used as a rugosity index, with higher values indicating a greater level of complexity and a value of 1 being a flat surface. The sampling positions faced four different directions; sea, shore, north and south, and exposure was assessed according to whether or not the sampling position was sheltered by another body part.

Each statue measured 191 cm from the base of the feet to the top of the head, and sampling positions varied from between 61 and 180 cm from the base of the feet. The shore height of each sampling position on each statue was calculated as the sum of the distance from the base of the feet and the shore height of the base of the feet (estimated from the time at which the tide reached the feet of the statues using tidal curves for Liverpool). Shore heights of the statues ranged from just over 1 m to 10 m above chart datum and the heights of the sampling positions ranged from 2 m to almost 12 m. Many of the shore heights were replicated by two or

Table 1. Description of the 14 sampling positions, and their associated variables, that were sampled for coverage of the barnacle *Austrominius modestus* on each of the 100 statues at Crosby Beach, Liverpool.

Sampling position	Rugosity	Height from base of feet (cm)	Direction	Exposure
Head front	1.62	180	Sea	Exposed
Head back	1.51	180	Shore	Exposed
Torso front	1.46	135	Sea	Exposed
Torso back	1.44	135	Shore	Exposed
Groin	1.70	97	Sea	Exposed
Buttocks	1.47	94	Shore	Exposed
Lower leg front	1.43	61	Sea	Exposed
Lower leg back	1.47	65	Shore	Exposed
Inner thigh left	1.42	71	North	Sheltered
Inner thigh right	1.42	71	South	Sheltered
Outer thigh left	1.51	71	South	Exposed
Outer thigh right	1.51	71	North	Exposed
Under arm left	1.45	134	South	Sheltered
Under arm right	1.45	134	North	Sheltered

Under the index of rugosity, higher values indicate a greater level of complexity with 1 being a flat surface; direction describes the direction in which each sampling position faced: sea, shore, north and south; and exposure was assessed according to whether or not the sampling position was sheltered by another body part.

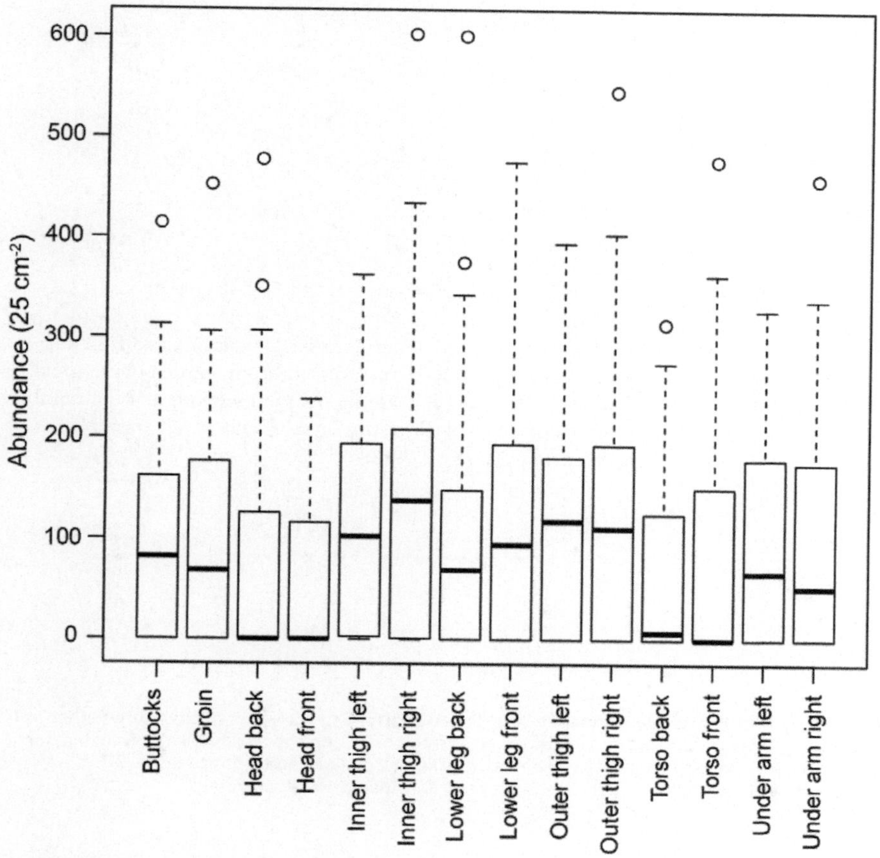

Figure 3. Mean total abundance of barnacles (25 cm^{-2}) per sampling position on each of the 100 statues. In total 14 sampling positions were chosen to represent a range of environmental conditions experienced by the statues and were located at the same points on each of the 100 statues sampled at Crosby Beach, Liverpool (for statue locations see Fig. 2). Boxplots show the medians (thicker black line) and upper and lower quartiles of abundance values at each sampling point, with whiskers extending to the extremes of data points not considered to be outliers.

more statues. The latitude and longitude of each statue was obtained using a Garmin etrex© Global Positioning System.

Statistical Analysis

Generalized linear mixed models (GLMMs) were used to investigate the effects of location, shore height, exposure, direction, and rugosity, and some of their interactions on recruitment of *Austrominius modestus* to the statues. GLMMs are useful as they can be applied to non-normal data that includes a mix of both fixed and random effects whilst also allowing for co-variation among samples [43]. We assumed that *A. modestus* counts followed a Poisson log-normal distribution, conditional on the values of the explanatory variables. The Poisson log-normal is one of several possible models for over-dispersed count data, but is particularly convenient for GLMMs [44]. We treated statue as a random effect, to account for variation among statues. We also added an additional random effect of observation, to account for unexplained variation within statues. Location (the first principal component of latitude and longitude), shore height, exposure, rugosity, and direction were treated as fixed effects. We considered models including combinations of these fixed effects and their two-way interactions (all models included both the random effects). However, due to the physical layout of sampling positions on the statues, some of these interactions (direction and rugosity, exposure and direction) were not identifiable, in the sense that models with different parameter values could fit the data equally well. We excluded any model containing such interactions. We selected the model having the lowest Akaike's Information Criterion (AIC) among this subset of possible models.

It is likely that there is spatial autocorrelation among sampling positions within a statue, because positions close together in space may experience similar environmental conditions. However, because the same set of positions is sampled on every statue, these patterns can be captured by the fixed effects in the model. It is also likely that if there are many individuals at a given position on a statue, other positions on the same statues will also have many individuals, due to similarities in environmental conditions. Such patterns will be captured by the random effect of statue. The absence of post-settlement dispersal of barnacles means that we do not expect spatial autocorrelation arising from the direct influence of numbers at one position on numbers at other positions. We checked the model assumptions by visual inspection of residuals. All statistical analysis was performed using the R software package "lme4" version 0.999375-40 [45].

Results

Substantial variations in the abundance of *Austrominius modestus* were observed over the sampling positions ($n = 1400$) analysed in the study (Fig. 3). The overall mean abundance ($\pm$ SE) was 96 ± 3 individuals 25 cm^{-2} with a maximum abundance of 603 individuals 25 cm^{-2}. Zero counts were numerous and observed

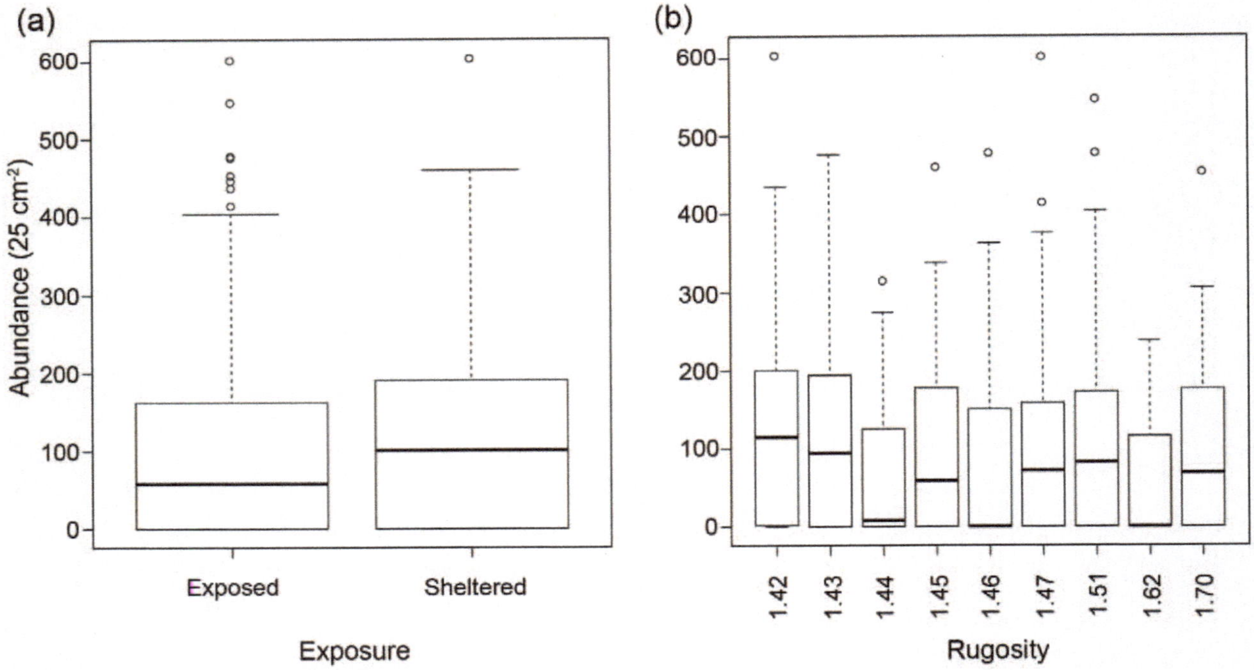

Figure 4. Total abundance of barnacles (25 cm^{-2}) in relation to (a) exposure and (b) rugosity. Exposure was defined as whether or not the sampling position was sheltered by another sampling position and a rugosity index was used to describe the complexity of each sampling position, with higher numbers indicating greater complexity. Boxplots show the medians (thicker black line) and upper and lower quartiles of abundance values for both indices, with whiskers extending to the extremes of data points not considered to be outliers.

at a range of tidal heights, although all sampling positions with a height of 9.1 m and greater had zero counts except for the inner and outer thigh of two of the statues (98 and 99). The greatest mean abundance was on the inner thigh (139±12 individuals 25 cm^{-2}) and the lowest on the front and back of the head (53±8 and 66±10, respectively). However, there was no strong trend in abundance in relation to sampling position, with no one position standing out as being most suitable for *A. modestus* recruitment. There did appear to be a difference between the numbers of barnacles in relation to exposure, with greater mean abundances observed on sampling positions described as sheltered (55±2 individuals 25 cm^{-2}) than on those described as exposed (45±1 individuals 25 cm^{-2}) (Fig.4a). There seemed to be no preference for more topographically complex sampling positions over less complex ones (Fig.4b).

The most parsimonious adequate generalized linear mixed model (GLMM) was chosen from a series of GLMMs based on its complexity and AIC value (Table 2). The final GLMM was used to understand how interactions between the various fixed effects could be affecting the distribution of barnacles on the statues. Visual inspections of the residuals indicated zero inflation, with a sharp downward trend in the smoothed residuals at low fitted values (Fig. 5 red line). Whilst the model was adequate at predicting large fitted values, smaller fitted values were more variable and there was an excess of zero counts.

Table 2. Generalized linear mixed models (GLMMs) of barnacle abundance (numbers 25 cm^{-2}) in relation to important environmental variables at Crosby Beach, Liverpool.

GLMM (Fixed effects)		d.f	*l*	AIC	Δ AIC
(a)	Location	4	−3334	6676	482
(b)	Exposure	4	−3332	6672	478
(c)	Direction	6	−3298	6608	414
(d)	Shore Height	4	−3164	6335	141
(e)	Rugosity	4	−3336	6679	485
(f)	Location+ Exposure + Direction + Shore height + Rugosity	10	−3140	6300	106
(g)	**(f) + Shore height:(Location + Direction)**	**14**	**−3083**	**6194**	**0**

Model selection was based on Akaike's Information Criterion (AIC) with additional interactions being kept if they reduced AIC by >2. The most parsimonious adequate model on this basis is shown in bold. n = 1400 quadrats; d.f., degrees of freedom; *l*, log-likelihood; Δ AIC, the difference in AIC from that of the most parsimonious adequate model.

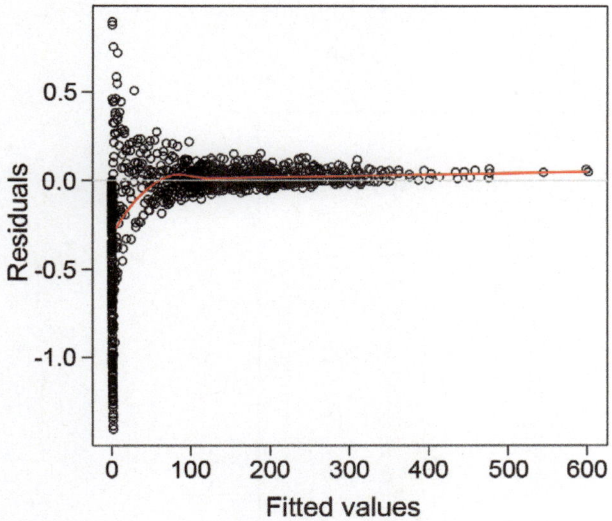

Figure 5. Goodness of final model fit used to assess barnacle abundance on the statues at Crosby. Goodness of fit of the final generalized linear mixed model is illustrated through assessment of the fitted values of the selected final model against the residuals of the model (red line indicates loess smoother).

The selected model included two different interactions between fixed effects that could help explain the distribution of barnacles observed on the statues (Table 3). The model showed an interaction between the locations of the sampling position along the beach and shore height. The value of location (the principal component of latitude and longitude) increases towards the southerly end of the beach. The interaction between these two fixed effects indicated a more negative effect of increasing shore height on the abundance of barnacles on sampling positions located towards the southerly end of the beach (Fig. 6). Clear differences in mean abundances and mean shore heights were observed when statues were compared based on their location: northern (statues 1–49) or southern (statues 50–100). Those statues located at the northern end of the beach were characterised by higher mean abundances and lower mean shore heights (2000±147 individuals 25 cm^{-2} at mean shore height of 4.7±0.2 m) when compared to those at the southern end (739±123 individuals 25 cm^{-2} at a mean shore height of 7.2±0.3 m).

The highest (mean) abundances (>200 individuals 25 cm^{-2}) were predicted at the northerly end of the beach in the mid tidal height region, and the lowest (<50 individuals 25 cm^{-2}) in most of the high tidal height region, particularly towards the southerly end (Fig. 6). The second interaction observed was between the direction of the sampling position and shore height. Increasing shore height was predicted to decrease the abundance of barnacles in all directions but was most acute on seaward and shoreward facing positions with a lesser effect observed for north and south facing positions, where abundances decreased at an almost identical rate (Fig. 7, see Fig. 2 for actual positions of statues).

Discussion

Within a year, the invasive barnacle *Austrominius modestus* was found to have thrived on the Antony Gormley statues along the sandy beach in Crosby, Liverpool (individual abundances being in their hundreds per 25 cm^{-2} in suitable positions), opportunistically colonising and dominating the man-made installation in an environment that is otherwise void of settlement sites. Using a generalized linear mixed model approach, we successfully modelled the distribution of *A. modestus* across these statues, with high abundances predicted with greater precision. We found that while individual parameters, such as rugosity, had some influence on barnacle distribution, they were less important than the interactions between shore height and direction of the sampling position, and specific location on the beach. In these conditions, *A. modestus* was only really affected by position on the statue itself (and thus 'design') at the extremes of shore height tolerated. In other words, having 'another place' offering suitable substrate within its tidal range was enough to lead to widespread colonisation and dominance.

Table 3. Parameter estimates for the final generalized linear mixed model (GLMM) assessing the influence of various factors on the abundance of barnacles.

Fixed effects	Estimate	Std error
(Intercept)	13.358	1.198
Location	4.482	0.606
Exposure(Sheltered)	0.339	0.138
Direction(Sea)	1.765	0.426
Direction(Shore)	0.979	0.420
Direction(South)	0.188	0.422
Rugosity	0.121	0.067
Shore height	−1.795	0.118
Location:Shore height	−0.592	0.071
Direction(Sea):Shore height	−0.360	0.067
Direction(Shore):Shore height	−0.192	0.065
Direction(South):Shore height	−0.017	0.067

There is increasing concern that artificial structures could aid the spread of non-native species [23]. It is clear from this study that the addition of suitable substrate to a sheltered estuarine environment, conditions in which *A. modestus* is commonly found [34,37], has facilitated the establishment of this species to the area. *A. modestus* has a longer reproductive period than any other barnacle in British waters [46] and can breed almost all year round. This particular physiological trait could have given it a competitive advantage over other barnacle species when colonising the statues initially, particularly when considering peak settlement in this species has been observed in the summer and autumn months [37,47], around the time the statues were installed. However, a more recent study has found that *A. modestus* still dominates the statues, to the exclusion of all other barnacle species, and that it rapidly re-colonises areas of the statues that become available through disturbance [48].

The influence of shore height on the distribution of *A. modestus* in this area is unsurprising given the well-described zonation patterns observed in barnacle species [49]. The negative effects of shore height on abundances were most acute at the southerly end of the beach. This was most likely a result of the comparative narrowness of the beach in this region and the greater mean shore height of the statues (see Fig. 2 for statue locations). In its native range, *A. modestus* is distributed in the upper tidal region [32,35]. In the UK, however, the species occupies a much wider range of tidal heights [38], and its distribution appears to be highly variable dependent on the relative abundances and distribution of co-occurring native species, primarily; *Cthamalus montagui*, *Cthamalus stellatus*, *Semibalanus*

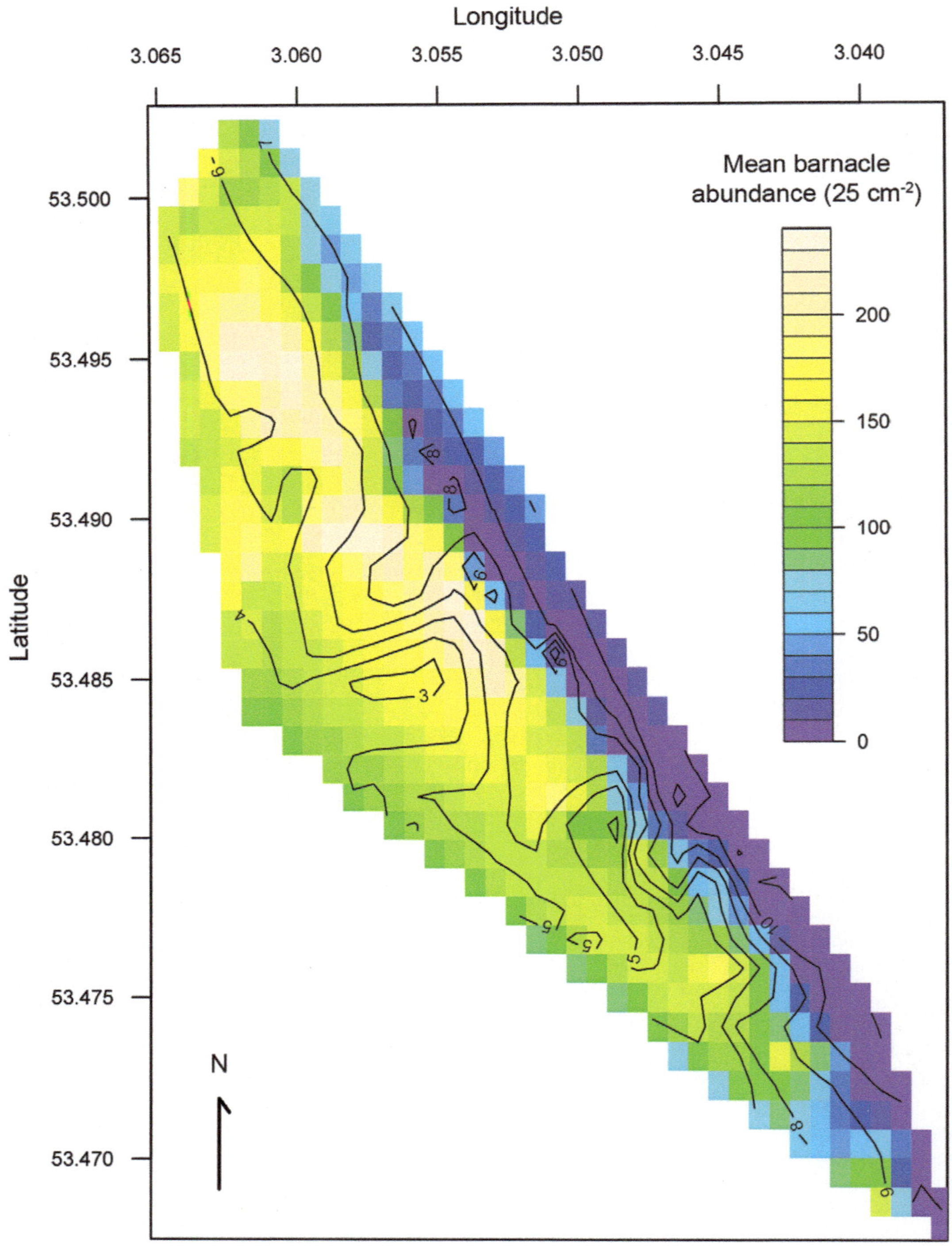

Figure 6. Predicted mean abundances of barnacles (25 cm^{-2}) per statue position in relation to their location. Predicted mean abundances are based on predicted values from a generalized linear mixed model output, with contours indicating actual shore height of the statues.

balanoides, Balanus balanoides and Balanus crenatus [37]. The species appears to occupy a similar ecological niche as both *B. balanoides* and *S. balanoides*, and its competitive superiority has allowed it to become dominant over these species in numerous areas [33,37,38]. In the absence of any competitors or predators at Crosby Beach, *A. modestus* colonised a wide range of tidal heights, only really being absent at sampling positions greater than nine metres above chart datum. Given that the spring tidal range in Liverpool Bay (in excess of 10 m), is one of the largest in the world [50], this suggests that tidal height is of little constraint for this species where other conditions are favourable.

Numbers of *A. modestus* on north and south facing positions were least affected by increasing shore height. These positions made up six of the 14 sampling positions assessed on each statue, four of

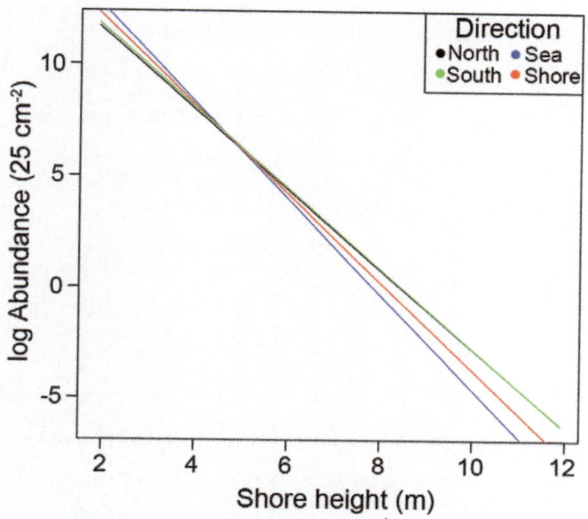

Figure 7. Predicted changes in log barnacle abundance (25 cm^{-2}) in relation to shore height and direction. Sampling positions faced four different directions: north, south, sea, shore. Values are based on those predicted from a generalized linear mixed model output using the median value of Location (first principal component of latitude and longitude), the mean level of Rugosity of the sampling positions and exposed or sheltered levels of Exposure.

which were the only positions to be described as sheltered by another body part (inner thigh and under arm, see Table 1 for full descriptions). *A. modestus* is known to recruit preferentially to sheltered shores and is often only sparsely distributed on shores that experience greater levels of exposure [27,37]. Even though Crosby Beach itself would be described as sheltered, the added protection provided by other body parts may have promoted survival of individuals as shore height increased. An additional factor that could have influenced this relationship is the direction of prevailing currents in this area, as the transport of propagules to settlement sites is greatly dependent on coastal currents and oceanographic processes on large and small scales [51,52].

The final model used in our study included rugosity as a fixed effect, indicating that rugosity did have some bearing on barnacle abundance. However, it was not as important as the interactions of other factors, and no perceivable effect was observed. Variations in substrate rugosity and its interaction with other environmental parameters influences the spatial distribution of barnacles, as the provision of refuges can promote post-settlement survival [31,53], although this effect is not always observed (see [54]). The results of this study suggested that the distribution of *A. modestus* at this study site were not greatly regulated by habitat complexity. Previous studies indicate that the influence of cracks and crevices on settlement in *A. modestus* may be greater on smaller scales of approximately 1 cm [55,56], and it is possible that the scale of rugosity used in this study was too great to generate an observable response. Alternatively, except for increasing shore height, there may be little need for refuge at this study site as wave exposure is limited and no evidence of potential predators was observed; two factors known to influence the importance of rugosity on distribution of barnacles [57,58].

The effect of individual statue location on abundance was partly explained by the interaction with shore height variation along the beach, but the locational effect may also be dependent on local current patterns within the beach and/or proximity to seeding sites (both factors we were unable to test in this study). Whilst the current patterns in Liverpool Bay have been studied extensively [50], at present no near shore oceanographic data exist on such a small scale for the area around Crosby Beach. The statues themselves could also potentially alter water flow speeds and current patterns, as has been observed for other artificial structures [59] encouraging or discouraging recruitment to particular areas. A multi-disciplinary study integrating information on very local (within beach), alongshore and regional hydrography, with information on the distribution and genetic profiles of individuals of *A. modestus* within and at sites beyond this beach, would help to further our understanding of the ecological connectivity of this species, and how artificial structures contribute to the spread of invasive species.

Identifying the variables that influence patterns of abundance in benthic invertebrates is complex, as numerous factors and interactions between them may be responsible at both the recruitment and post recruitment stage [30]. However, if we are to understand the factors responsible for the establishment and spread of invasive species, so that predictions of invasion success to new habitats can be made, a sound knowledge of the variables that determine population success is essential [60]. In this study, we have outlined the environmental parameters that influence the patterns of abundance observed in the invasive barnacle *Austrominius modestus* on an artificial substrate in an area of otherwise unsuitable habitat. This species has already displaced native barnacle species in some regions of the UK [37] and its ability to colonise and survive in large numbers in environments inhospitable to other species, combined with its rapid rate of spread, are a cause for concern [33]. With studies predicting a positive association between *A. modestus* recruitment and milder winter conditions [61], in addition to an ever-increasing presence of artificial structures in marine environments, the continued spread and dominance of this species can be expected.

Acknowledgments

We would like to thank Sefton Council for granting permission to undertake the study. We would also like to acknowledge Dr Dave Wilson whose discussions with LAR first prompted the idea to survey the statues. LAR developed the concept, MI collected the data, SAB analysed the data with assistance from MS and RHM, and wrote the paper with LAR. We also gratefully acknowledge helpful comments from Richard Osman and three anonymous reviewers on an earlier version of this manuscript.

Author Contributions

Conceived and designed the experiments: LAR. Performed the experiments: MI. Analyzed the data: SAB. Contributed reagents/materials/analysis tools: RS RHM. Wrote the paper: SAB.

References

1. Molnar JL, Gamboa RL, Revenga C, Spalding MD (2008) Assessing the global threat of invasive species to marine biodiversity. Frontiers in Ecology and the Environment 6: 485–492.
2. Richardson DM, Pyšek P, Rejmánek M, Barbour MG, Panetta FD, et al. (2000) Naturalization and invasion of alien plants: concepts and definitions. Diversity and Distributions 6: 93–107.
3. Rajakaruna H, Strasser C, Lewis M (2012) Identifying non-invasible habitats for marine copepods using temperature-dependent R-0. Biological Invasions 14: 633–647.
4. Chun YJ, van Kleunen M, Dawson W (2010) The role of enemy release, tolerance and resistance in plant invasions: linking damage to performance. Ecol Lett 13: 937–946.

5. Williamson M, Fitter A (1996) The Varying Success of Invaders. Ecology 77: 1661–1666.
6. Jeschke J, Gómez Aparicio L, Haider S, Heger T, Lortie C, et al. (2012) Support for major hypotheses in invasion biology is uneven and declining. NeoBiota 14: 1–20.
7. Hulme PE, Pyšek P, Nentwig W, Vilà M (2009) Will Threat of Biological Invasions Unite the European Union? Science 324: 40–41.
8. Mead A, Carlton JT, Griffiths CL, Rius M (2011) Revealing the scale of marine bioinvasions in developing regions: a South African re-assessment. Biological Invasions 13: 1991–2008.
9. Pysek P, Jarosik V, Hulme PE, Pergl J, Hejda M, et al. (2012) A global assessment of invasive plant impacts on resident species, communities and ecosystems: the interaction of impact measures, invading species' traits and environment. Global Change Biology 18: 1725–1737.
10. Strayer DL (2012) Eight questions about invasions and ecosystem functioning. Ecol Lett 15: 1199–1210.
11. van Kleunen M, Weber E, Fischer M (2010) A meta-analysis of trait differences between invasive and non-invasive plant species. Ecol Lett 13: 235–245.
12. Levine JM, D'Antonio CM (1999) Elton Revisited: A Review of Evidence Linking Diversity and Invasibility. Oikos 87: 15–26.
13. Angert AL, Crozier LG, Rissler LJ, Gilman SE, Tewksbury JJ, et al. (2011) Do species' traits predict recent shifts at expanding range edges? Ecol Lett 14: 677–689.
14. Allred BW, Fuhlendorf SD, Monaco TA, Will RE (2010) Morphological and physiological traits in the success of the invasive plant Lespedeza cuneata. Biological Invasions 12: 739–749.
15. Davidson AM, Jennions M, Nicotra AB (2011) Do invasive species show higher phenotypic plasticity than native species and, if so, is it adaptive? A meta-analysis. Ecol Lett 14: 419–431.
16. Chytry M, Jarosik V, Pysek P, Hajek O, Knollova I, et al. (2008) Separating habitat invasibility by alien plants from the actual level of invasion. Ecology 89: 1541–1553.
17. Clark GF, Johnston EL (2005) Manipulating larval supply in the field: a controlled study of marine invasibility. Marine Ecology Progress Series 298: 9–19.
18. Bulleri F, Chapman MG (2010) The introduction of coastal infrastructure as a driver of change in marine environments. Journal of Applied Ecology 47: 26–35.
19. Briggs JC (2012) Marine species invasions in estuaries and harbors. Marine Ecology Progress Series 449: 297–302.
20. Floerl O, Inglis GJ, Dey K, Smith A (2009) The importance of transport hubs in stepping-stone invasions. Journal of Applied Ecology 46: 37–45.
21. Glasby T, Connell S, Holloway M, Hewitt C (2007) Nonindigenous biota on artificial structures: could habitat creation facilitate biological invasions? Marine Biology 151: 887–895.
22. Edwards RA, Smith SDA (2005) Subtidal assemblages associated with a geotextile reef in south-east Queensland, Australia. Marine and Freshwater Research 56: 133–142.
23. Bulleri F, Airoldi L (2005) Artificial marine structures facilitate the spread of a non-indigenous green alga, *Codium fragile* ssp. *tomentosoides*, in the north Adriatic Sea. Journal of Applied Ecology 42: 1063–1072.
24. Airoldi L, Bulleri F (2011) Anthropogenic disturbance can determine the magnitude of opportunistic species responses on marine urban infrastructures. Plos One 6.
25. Forrest BM, Gardner JPA, Taylor MD (2009) Internal borders for managing invasive marine species. Journal of Applied Ecology 46: 46–54.
26. Bax N, Carlton JT, Mathews-Amos A, Haedrich RL, Howarth FG, et al. (2001) The control of biological invasions in the world's oceans. Conservation Biology 15: 1234–1246.
27. Allen BM, Power AM, O'Riordan RM, Myers AA, McGrath D (2006) Increases in the abundance of the invasive barnacle *Elminius modestus* Darwin in Ireland. Biology and Environment: Proceedings of the Royal Irish Academy 106B: 155–161.
28. Bulleri F, Abbiati M, Airoldi L (2006) The colonisation of human-made structures by the invasive alga *Codium fragile* ssp tomentosoides in the north Adriatic Sea (NE Mediterranean). Hydrobiologia 555: 263–269.
29. Rittschof D, Forward RB, Cannon G, Welch JM, McClary M, et al. (1998) Cues and context: larval responses to physical and chemical cues. Biofouling 12: 31–44.
30. Pineda J, Reyns N, Starczak V (2009) Complexity and simplification in understanding recruitment in benthic populations. Population Ecology 51: 17–32.
31. Munroe DM, Noda T, Ikeda T (2010) Shore level differences in barnacle (*Chthamalus dalli*) recruitment relative to rock surface topography. Journal of Experimental Marine Biology and Ecology 392: 188–192.
32. Darwin CR (1854) A monograph on the subclass Cirripedia, with figures of all the species. The Balanidae (or sessile cirripedes); the Verrucidae, etc. etc. etc. The Ray Soceity, London 2.
33. Crisp DJ (1958) The spread of *Elminius modestus* Darwin in North-West Europe. Journal of the Marine Biological Association of the United Kingdom 37: 483–520.
34. Crisp DJ, Southward AJ (1959) The further spread of *Elminius modestus* in the British Iles to 1959. Journal of the Marine Biological Association of the United Kingdom 38: 429–437.
35. Pope EC (1945) A simplified key to the sessile barnacles found on the rocks, boats, wharf piles and other installations in Port Jackson and adjacent waters. Records of the Australian Museum 21(6): 351–372.
36. Gomes-Filho JGF, Hawkins SJ, Aquino-Souza R, Thompson RC (2010) Distribution of barnacles and dominance of the introduced species Elminius modestus along two estuaries in South-West England. Marine Biodiversity Records 3: 1–11.
37. Lawson J, Davenport J, Whitaker A (2004) Barnacle distribution in Lough Hyne marine nature reserve: a new baseline and an account of invasion by the introduced Australasian species *Elminius modestus* Darwin. Estuarine, Coastal and Shelf Science 60: 729–735.
38. Witte S, Buschbaum C, van Beusekom JEE, Reise K (2010) Does climatic warming explain why an introduced barnacle finally takes over after a lag of more than 50 years? Biological Invasions 12: 3579–3589.
39. Mills DJL (1998) Liverpool to the Solway (Rhos-on-Sea to the Mull of Galloway) (MNCR Sector 11) In: Hiscock K, editor. Marine Nature Conservation Review Benthic marine ecosystems of Great Britain and the north-east Atlantic. Peterborough: Joint Nature Conservation Committee. 315–338.
40. Allen JR, Wilkinson SB, Hawkins SJ (1995) Redeveloped docks as artificial lagoons: The development of brackish-water communities and potential for conservation of lagoonal species. Aquatic Conservation-Marine and Freshwater Ecosystems 5: 299–309.
41. Russell G, Hawkins SJ, Evans LC, Jones HD, Holmes GD (1983) Restoration of a Disused Dock Basin as a Habitat for Marine Benthos and Fish. Journal of Applied Ecology 20: 43–58.
42. Wilkinson SB, Zheng W, Allen JR, Fielding NJ, Wanstall VC, et al. (1996) Water Quality Improvements in Liverpool Docks: The Role of Filter Feeders in Algal and Nutrient Dynamics. Marine Ecology 17: 197–211.
43. Bolker BM, Brooks ME, Clark CJ, Geange SW, Poulsen JR, et al. (2009) Generalized linear mixed models: a practical guide for ecology and evolution. Trends in Ecology & Evolution 24: 127–135.
44. Elston DA, Moss R, Boulinier T, Arrowsmith C, Lambin × (2001) Analysis of aggregation, a worked example: numbers of ticks on red grouse chicks. Parasitology 122: 563–569.
45. R Development Core Team (2011) R: a language and environment for statistical computing. Vienna, Austria: R Foundation for Statistical Computing.
46. Crisp DJ, Chipperfield PNJ (1948) Occurrence of *Elminius modestus* (Darwin) in British waters. Nature 161: 64–64.
47. Muxagata E, Williams JA, Sheader M (2004) Composition and temporal distribution of cirripede larvae in Southampton Water, England, with particular reference to the secondary production of Elminius modestus. ICES Journal of Marine Science 61: 585–595.
48. Bracewell SA (2011) Recruitment of the invasive barnacle *Elminius modestus*, Darwin, in Liverpool Bay [MRes Thesis]: University of Liverpool, Liverpool, UK.
49. Connell JH (1972) Community interactions on marine rocky intertidal shores. Annual Review of Ecology and Systematics 3: 169–192.
50. Polton JA, Palmer MR, Howarth MJ (2011) Physical and dynamical oceanography of Liverpool Bay. Ocean Dynamics 61: 1421–1439.
51. Gaines S, Roughgarden J (1985) Larval settlement rate: a leading determinant of structure in an ecological community of the marine intertidal zone. Proceedings of the National Academy of Sciences, USA 82: 3707–3711.
52. Pineda J (1994) Spatial and temporal patterns in barnacle settlement rate along a southern California rocky shore. Marine Ecology Progress Series 107: 125–138.
53. Raimondi PT (1990) Patterns, mechanisms, consequences of variability in settlement and recruitment of an intertidal barnacle. Ecological Monographs 60: 283–309.
54. Munroe DM, Noda T (2010) Physical and biological factors contributing to changes in the relative importance of recruitment to population dynamics in open populations. Marine Ecology Progress Series 412: 151–162.
55. Crisp DJ, Barnes H (1954) The orientation and distribution of barnacles at settlement with particular reference to surface contour. Journal of Animal Ecology 23: 142–162.
56. Wright JR, Boxshall AJ (1999) The influence of small-scale flow and chemical cues on the settlement of two congeneric barnacle species. Marine Ecology Progress Series 183: 179–187.
57. Burrows MT, Harvey R, Robb L (2008) Wave exposure indices from digital coastlines and the prediction of rocky shore community structure. Marine Ecology Progress Series 353: 1–12.
58. Connell JH (1970) A predator-prey system in the marine intertidal region. I. *Balanus glandula* and several predatory species of *Thais*. Ecological Monographs 40: 49–78.
59. Svane I, Petersen JK (2001) On the problems of epibioses, fouling and artificial reefs, a review. Marine Ecology 22: 169–188.
60. Kulhanek SA, Leung B, Ricciardi A (2011) Using ecological niche models to predict the abundance and impact of invasive species: application to the common carp. Ecological Applications 21: 203–213.
61. Broitman BR, Mieszkowska N, Helmuth B, Blanchette CA (2008) Climate and recruitment of rocky shore intertidal invertebrates in the eastern North Atlantic. Ecology 89: S81–S90.

18

Seasonal Carbonate Chemistry Covariation with Temperature, Oxygen, and Salinity in a Fjord Estuary: Implications for the Design of Ocean Acidification Experiments

Jonathan C. P. Reum[1]*, Simone R. Alin[2], Richard A. Feely[2], Jan Newton[3], Mark Warner[4], Paul McElhany[1]

1 Conservation Biology Division, Northwest Fisheries Science Center, National Marine Fisheries Service, National Oceanic and Atmospheric Administration, Seattle, Washington, United States of America, **2** Pacific Marine Environmental Laboratory, National Oceanic and Atmospheric Administration, Seattle, Washington, United States of America, **3** Applied Physics Laboratory, University of Washington, Seattle, Washington, United States of America, **4** School of Oceanography, University of Washington, Seattle, Washington, United States of America

Abstract

Carbonate chemistry variability is often poorly characterized in coastal regions and patterns of covariation with other biologically important variables such as temperature, oxygen concentration, and salinity are rarely evaluated. This absence of information hampers the design and interpretation of ocean acidification experiments that aim to characterize biological responses to future pCO_2 levels relative to contemporary conditions. Here, we analyzed a large carbonate chemistry data set from Puget Sound, a fjord estuary on the U.S. west coast, and included measurements from three seasons (winter, summer, and fall). pCO_2 exceeded the 2008–2011 mean atmospheric level (392 µatm) at all depths and seasons sampled except for the near-surface waters (< 10 m) in the summer. Further, undersaturated conditions with respect to the biogenic carbonate mineral aragonite were widespread ($\Omega_{ar}<1$). We show that pCO_2 values were relatively uniform throughout the water column and across regions in winter, enriched in subsurface waters in summer, and in the fall some values exceeded 2500 µatm in near-surface waters. Carbonate chemistry covaried to differing levels with temperature and oxygen depending primarily on season and secondarily on region. Salinity, which varied little (27 to 31), was weakly correlated with carbonate chemistry. We illustrate potential high-frequency changes in carbonate chemistry, temperature, and oxygen conditions experienced simultaneously by organisms in Puget Sound that undergo diel vertical migrations under present-day conditions. We used simple calculations to estimate future pCO_2 and Ω_{ar} values experienced by diel vertical migrators based on an increase in atmospheric CO_2. Given the potential for non-linear interactions between pCO_2 and other abiotic variables on physiological and ecological processes, our results provide a basis for identifying control conditions in ocean acidification experiments for this region, but also highlight the wide range of carbonate chemistry conditions organisms may currently experience in this and similar coastal ecosystems.

Editor: Moncho Gomez-Gesteira, University of Vigo, Spain

Funding: This research was supported by the NOAA Ocean Acidification Program, NOAA Pacific Marine Environmental Laboratory, the University of Washington including the PRISM program, and the Washington State Department of Ecology. Work in Hood Canal was partially funded through the Hood Canal Dissolved Oxygen Program via Naval Sea Systems Command Contract #N00024-02-D-6602 Task 50. J.C.P.R. was supported by a National Research Fellowship award. This work was co-sponsored by the National Oceanic and Atmospheric Administrationâ€™s Northwest Fisheries Science Center, Pacific Marine Environmental Laboratory, Ocean Acidification Program, and Integrated Ocean Observing System Program; the University of Washington including Northwest Association of Networked Ocean Observing Systems (NANOOS), Puget Sound Regional Synthesis Model (PRISM), Applied Physics Laboratory, and School of Oceanography; and the Washington State Department of Ecology. Work in Hood Canal was partially funded through the Hood Canal Dissolved Oxygen Program via Naval Sea Systems Command Contract #N00024-02-D-6602 Task 50. The funders had no role in study design, data collection and analysis, decision to publish, or preparation of the manuscript.

Competing Interests: The authors have declared that no competing interests exist.

* E-mail: Jonathan.Reum@noaa.gov

Introduction

Global surface ocean pH levels have decreased by approximately 0.1 units since the Industrial Revolution and will potentially decline an additional 0.3 to 0.4 units over the next century due to anthropogenic CO_2 emissions [1–4]. This change in ocean chemistry, termed "ocean acidification" (OA), results in a reduction in the saturation state of carbonate minerals such as calcite and aragonite, elevates pCO_2, and lowers pH [5]. A growing body of research indicates that such changes may impact a diversity of species, with some organisms benefiting from increased pCO_2 availability (e.g., photosynthesizers), while others (e.g., echinoderms, molluscs, some fishes) may be adversely impacted by enhanced shell dissolution rates, reductions in metabolic scope, or neurological impairment [2,6–8]. A major challenge to inferring the ecological implications of these studies, however, is an incomplete understanding of present-day patterns of carbonate chemistry variability, particularly in coastal habitats [9-14].

In continental shelf and estuarine systems, complex physical processes and high rates of biological production act to perturb

pCO_2 levels away from conditions under air-sea gas equilibrium [9] and may exhibit considerable variability over interannual, seasonal, and daily time scales [10,14–17]. In contrast, researchers studying organisms from these systems often use IPCC projections of global average surface ocean carbonate chemistry conditions that correspond to assumptions of air-sea gas equilibrium to set control and experimental carbonate chemistry conditions [11,18]. Recognition of the potential disparity between conditions considered "controls" in OA experiments and actual carbonate chemistry conditions experienced by organisms in the habitats they occupy has been highlighted in a suite of recent publications [9,10,18–20]. Consequently, there is increasing awareness that researchers should attempt to simulate control conditions that correspond to carbonate chemistry conditions experienced by organisms in natural habitats and interpret experimental outcomes accordingly [21,22].

However, largely missing from this discussion is the observation that carbonate chemistry commonly covaries with other biologically relevant variables including temperature and oxygen.

In some regions, the strength of these associations has permitted the development of semi-mechanistic predictive models of carbonate chemistry in coastal environments based on field measurements of temperature and oxygen [15,23]. Yet for researchers focused on laboratory OA experiments, these relationships are rarely incorporated into experimental designs. This oversight poses important potential drawbacks. Foremost, if carbonate chemistry strongly covaries with temperature and oxygen, researchers risk running experiments with control water characteristics that are atypical of the habitat that a focal organism / life stage occupies. Physiological parameters including aerobic capacity and metabolic scope (the amount of energy that can be allocated to activities beyond those for basic existence) are strongly influenced by temperature, and adversely impacted by reductions in ambient oxygen availability or increases in CO_2 concentrations [24]. Under extreme warm and cool temperatures defining the upper and lower boundaries of the thermal window an organism survives within, the negative effects of oxygen scarcity or high concentrations of CO_2 are potentially exacerbated [24]. Because of the individual and interactive effects that temperature, oxygen, and pCO_2 may have on physiology and other metrics related to fitness and survival [6,24,25], knowledge of their covariability is essential for designing OA experiments that adequately characterize biological performance under contemporary relative to future acidified conditions.

Here, we have analyzed a large carbonate chemistry data set from Puget Sound, a large fjord estuary on the U.S. west coast [26]. Puget Sound is a highly altered ecosystem that suffers from multiple stressors stemming from extensive urbanization in the surrounding watershed. Habitat loss, shoreline development, pollution, alteration of freshwater flows, and overfishing are pervasive problems [27]. Despite these impacts, the Sound remains a biologically diverse system that directly supports economically important industries including aquaculture, commercial and recreational fisheries, and tourism [27]. Located on the eastern boundary of the NE Pacific, the Sound is supplied with CO_2- and nutrient-rich waters that are upwelled onto the continental shelf [26,28]. Further CO_2 additions resulting from high rates of respiration act to make Puget Sound a highly vulnerable region for acidification [26]. Due to the reduced carbonate buffering capacity of the seawater, increases in pCO_2 and reductions in pH resulting from ocean uptake of anthropogenic CO_2 will occur more rapidly compared to open-ocean waters [26]. Similarly, the saturation state of seawater with respect to biogenic calcium carbonate minerals (calcite and aragonite) will also decline at a faster rate, suggesting that calcifying species may be adversely affected sooner in Puget Sound relative to open ocean populations [26].

We sought to evaluate seasonal and regional variability in carbonate chemistry conditions as well as patterns of covariation with temperature, oxygen, and salinity. Our primary rationale was to provide an evaluation of carbonate chemistry variability in this system as part of a larger effort to develop OA experimental designs with improved ecological relevance. Our second objective was to evaluate the potential range of carbonate chemistry conditions experienced by members of the Puget Sound pelagic food web and how that range may change with additional CO_2. In particular, we focused on potential simultaneous changes in temperature, oxygen, and carbonate chemistry conditions that are likely to be experienced by pelagic biota undergoing diel vertical migrations across regions and seasons in Puget Sound. In doing so, we aim to draw attention to the more general issue of identifying temperature, oxygen, and carbonate chemistry combinations that reflect plausible control and experimental conditions that are seasonally appropriate and correspond to the habitat occupied by the focal study organism or assemblage. We analyzed water chemistry data collected over three seasons (winter, summer, and fall), which afforded the opportunity to assemble seasonal snapshots of spatial variation in carbonate chemistry. Further, we collected data over two summers (August 2008 and September 2009) and two fall periods (October 2010 and October 2011). The inclusion of summer and fall sampling permitted an evaluation of interannual variation over a particularly dynamic season in Puget Sound. During the transition from late summer to early fall, intrusions of denser oceanic waters at depth result in the upward movement of the most oxygen-depleted waters in southern Hood Canal (a major sub-basin of Puget Sound) and southerly winds over Hood Canal result in episodic, localized upwelling events that bring oxygen-depleted, cooler waters to the surface at the southern end of Hood Canal [29]. We show that carbonate chemistry in Puget Sound covaries strongly with temperature and oxygen, but that the strength and sign of the relationship can change seasonally. Our findings offer insight into patterns of carbonate chemistry variability in this system, and we highlight their implications for the design and interpretation of OA laboratory experiments that simulate conditions representative of this and other coastal systems.

Materials and Methods

We collected water samples in Puget Sound over five cruises that spanned winter (February 2008), summer (August 2008, September 2009) and fall (October 2010 and 2011; Table 1) as part of the Puget Sound Regional Synthesis Model Program (PRISM) conducted by the University of Washington in collaboration with the Washington Department of Ecology. The summer surveys correspond to conditions prior to the intrusion of denser oceanic waters at depth while the fall surveys capture this process. The full survey includes transect lines with stations in all four major sub-basins of Puget Sound as well as the Strait of Juan de Fuca (Fig. 1). For the current study, we focused on stations south of the sill at the entrance to Hood Canal (herein referred to as 'Hood Canal') and those spanning the northern entrance of Admiralty Inlet to northern Central Puget Sound (herein 'Admiralty Inlet'). The bottom depth of the station near the sill in Hood Canal is ~100 m while the deepest stations near the center of the basin are ~170 m. Bottom depths of stations spanning Admiralty Inlet range from 70 m to 160 m. We focused our analysis on these regions because they possess contrasting

Table 1. Dates and stations sampled in Hood Canal and Admiralty Inlet for carbonate chemistry during PRISM cruises from 2008 to 2011.

Cruise	Sampling dates	Stations sampled
Feb. 2008	5 - 7 Feb.	7–22, 24, 401
Aug. 2008	12 - 13 Aug.	7–22, 24, 401
Sep. 2009	30 Sep. - 1 Oct.	7–11, 15–19, 21, 22
Oct. 2010	15 - 30 Oct.	7–22, 24, 401
Oct. 2011	12 - 13 Oct.	7–22, 24, 401

mixing patterns and were sampled consistently across cruises. All stations were located in Washington State waters and no special permission was required to collect water samples. Field activities did not involve endangered or protected species.

At all stations, full water column profiles of temperature, salinity, and oxygen were obtained using a SeaBird (SBE) 911*plus* conductivity-temperature-density (CTD) instrument equipped with an SBE 43 oxygen sensor (SeaBird Electronics Inc., Bellevue, WA). Data were binned at 0.5-m intervals. To correct oxygen sensor measurements for potential offsets, discrete water samples were analyzed for oxygen by modified Winkler titration [30]. The salinity sensor on the CTD was factory-calibrated annually; the sensor was checked against discrete values collected on the cruise at a subset of stations, analyzed with a Guildline Autosal by the University of Washington, and calibrated, if necessary. Cruise CTD data are available through the NANOOS Visualization Server (http://nvs.nanoos.org/CruisePrism).

Discrete water samples were collected and analyzed for carbonate chemistry following previously published methods [12]. Briefly, water samples were obtained with Niskin-type bottles and analyzed in the laboratory for dissolved inorganic carbon (DIC), total alkalinity (TA), oxygen, and nutrients. DIC was analyzed using coulometric titration [31–33]. TA was measured by the potentiometric titration method [33]. Certified Reference Materials (CRMs) were analyzed with both the DIC and TA samples as an independent verification of instrument calibrations [33]. The DIC and TA data have a combined precision and accuracy of $\sim$1 μmol kg^{-1} and $\sim$2 μmol kg^{-1}, respectively, based on replicate sample analysis and the use of CRMs. Carbonate chemistry measurements from all surveys are currently being prepared for submission and archival at the Carbon Dioxide Information Analysis Center (http://cdiac.ornl.gov/oceans/).

For our analysis, we focused on carbonate chemistry variation as conveyed by changes in pCO_2 and the saturation state of aragonite (Ω_{ar}). We focused on pCO_2 because this is the most commonly used treatment variable in OA experiments and the parameter directly changed by anthropogenic CO_2 emissions. We evaluated variation in Ω_{ar} because of its direct relationship to the stability of aragonite material in calcifying species. In supersaturated conditions ($\Omega_{ar} > 1$), the formation of aragonite is energetically favored, while dissolution is favored in undersaturated conditions ($\Omega_{ar} < 1$). We calculated pCO_2 and Ω_{ar} using the R library 'seacarb' [34] with dissociation constants from Lueker et al. [35]. The pressure effect on the solubility was applied as described from the equation of Mucci [36], incorporating the adjustments to the constants recommended by Millero [37]. Based on the uncertainties in the DIC and TA measurements, uncertainty in the calculated aragonite saturation state and pCO_2 is approximately 0.015–0.04 and 5–10 μatm, respectively.

We evaluated covariation of both pCO_2 and Ω_{ar} with temperature, oxygen, and salinity within each survey for Hood Canal and Admiralty Inlet stations using simple linear regression separately for each variable. We evaluated the degree of association based on the variance explained by the regression models (R^2). To comply with assumptions of normality in the response variable, both pCO_2 and Ω_{ar} were $\log_{10}$-transformed prior to fitting the regression models. The fitted regressions offer information on linear trends between variables but are not suited for the prediction of carbonate chemistry variables using temperature, oxygen concentration, or salinity measurements beyond the region and time period sampled for each cruise.

Next, we sought to visually compare seasonal and spatial patterns in the vertical distribution of pCO_2 and Ω_{ar}. To facilitate comparability, we predicted pCO_2 and Ω_{ar} across depths at each station using continuous depth profiles of temperature, oxygen, and salinity based on relationships estimated with the discrete carbonate chemistry samples. The predicted pCO_2 and Ω_{ar} values offered a higher vertical resolution of carbonate chemistry variability at each station than measurements from the discrete water samples alone. In preliminary stages of model development, we first considered multiple regression methods to predict pCO_2 and Ω_{ar}, but significant collinearity in the predictor variables was observed in some regions and time periods (absolute value of Pearson's correlation coefficient greater than 0.90), which hampered their application [38]. To overcome this issue, we instead utilized partial least squares regression (PLSR) to model pCO_2 and Ω_{ar}. PLSR is a dimension reduction technique that is well-suited to data sets that exhibit multicollinearity among predictor variables [39]. Essentially, latent factors (or component variables) are obtained from linear combinations of the predictor variables. The approach is similar to principal components regression except that in PLSR, component variables are obtained by maximizing the explained variance in the response variables rather than within the predictor variables [39]. We selected the optimal number of components by calculating the root mean square error of prediction (RMSEP) for each component using leave-one-out cross validation. We retained only those components without which the RMSEP did not show a significant decrease [40]. One PLSR model was estimated for pCO_2 and a second model was estimated for Ω_{ar}. We used CTD temperature, salinity and oxygen concentration as predictor variables in both models. However, to predict relationships for all stations across time periods and regions, we also included a categorical variable that denoted the region and cruise from which a sample was obtained along with first-order interactions between the continuous CTD variables and the categorical variable. Doing so permitted the relationships between the carbonate chemistry parameters and temperature, salinity and oxygen to vary between regions and seasons. This was deemed necessary because preliminary examination of the data sets suggested differing relationships between the explanatory and response variables across seasons and regions. As with the simple linear regression models, both pCO_2 and Ω_{ar} were $\log_{10}$-transformed to improve normality in the residual error structure. To evaluate model adequacy, we evaluated the explained variance in the response variable (R^2), RMSEP, and checked for homogeneity in the residual error structure. We used the fitted model to predict pCO_2 and Ω_{ar} values across depths using CTD and oxygen sensor measurements and visually compared patterns of carbonate chemistry variation between regions and seasons; we employed PLSR purely as a means to estimate high-resolution vertical profiles of pCO_2 and Ω_{ar}.

We note that although previous studies have developed semi-mechanistic predictive multiple regression models of carbonate

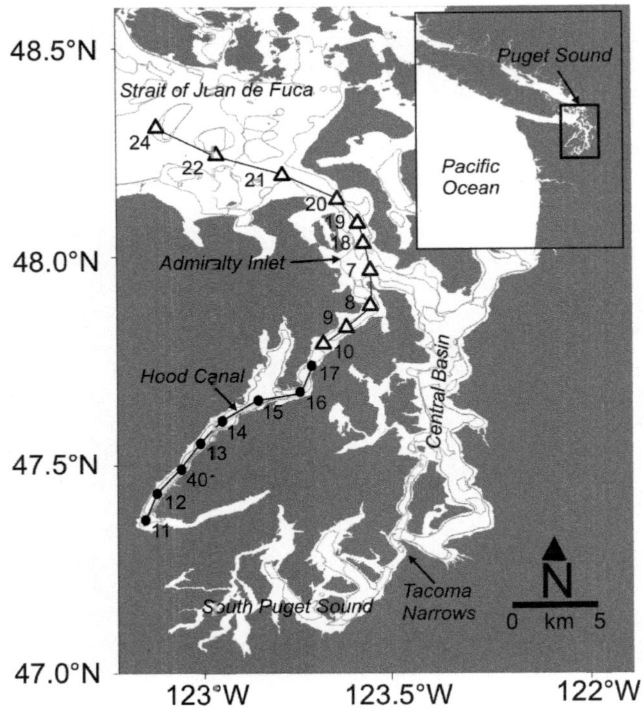

Figure 1. Map of the study area and stations sampled for carbonate chemistry in Hood Canal (denoted by solid circle symbol) and Admiralty Inlet (open triangle) from Puget Sound, Washington, USA. Sampling took place February and August 2008, September 2009, and October 2010 and 2011. Station numbering follows the convention of PRISM cruises (1998–2011, http://nvs.nanoos.org/CruisePrism).

chemistry using these variables in the California Current system [15,23], the application of similar methods to Puget Sound was hampered by several considerations beyond the multi-collinearity we observed among predictor variables. Namely, a key assumption of these models is that the stoichiometry of the CO_2:O_2 relationship in constituent water masses is primarily controlled by aerobic respiration acting over the time span since last exposure to the atmosphere. Puget Sound receives waters upwelled onto the continental shelf that are entrained in subsurface flows through the Strait of Juan de Fuca, but that mix to the surface and combine with low salinity, seaward-flowing surface waters due to the presence of shallow sills at Admiralty Inlet and the entrance to Hood Canal [41]. As a result, CO_2:O_2 relationships in the Sound reflect a combination of respiration in inflowing upwelled coastal waters, differential diffusion rates of CO_2 and O_2 between seawater and the atmosphere, and variable mixing with surface waters where CO_2:O_2 relationships are strongly influenced by biological production and freshwater inputs. We expect that semi-mechanistic models may exist with the capacity to estimate at least some carbonate chemistry variables in Puget Sound (e.g. TA), but our primary aim here was to interpolate pCO_2 and Ω_{ar} values between depths where discrete samples were collected at stations within each survey. The development of models to extrapolate carbonate chemistry parameters across time periods and locations outside our spatial and temporal sampling domain is beyond the scope of the present study but remains an area of active research.

In Puget Sound as elsewhere, many fish and zooplankton, including euphausiids, mysids, copepods, pteropods, and a large variety of fish and invertebrate larvae, undergo diel vertical migrations (DVM) and may swim vertical distances in excess of 50 m in less than an hour [42–46]. In stratified waters, diel vertical migrators may pass through water masses with drastically differing temperature, oxygen, and carbonate chemistry conditions [45]. The prevalence of DVM in Puget Sound fish and invertebrates and our need to understand the response of these groups to acidification motivated our interest in describing and comparing potential carbonate chemistry, temperature, and oxygen conditions experienced by diel vertical migrators across seasons and regions. In particular, we evaluated how these variables may change simultaneously as organisms move between near-surface and subsurface waters and compared contemporary carbonate chemistry conditions experienced by diel vertical migrators with those assuming an increase in DIC due to future ocean uptake of anthropogenic CO_2.

To estimate pCO_2 and Ω_{ar} under elevated DIC conditions, we used an equilibrium approach based on the thermodynamic equations that describe the carbonate system [5]. From the outset, we recognized the complexity of physical (vertical and horizontal mixing, air-sea CO_2 fluxes, freshwater inputs) and biological processes (photosynthesis, respiration) that potentially regulate carbonate chemistry conditions in Puget Sound and therefore did not seek to generate precise predictions of pCO_2 and Ω_{ar} corresponding to specific IPCC scenarios of CO_2 emissions. Rather, we considered a scenario in which the present-day anthropogenic DIC load in Puget Sound (~30 μmol kg^{-1}; [26]) was quadrupled to 120 μmol kg^{-1}. That is, we added 90 μmol kg^{-1} to observed in situ DIC measurements from the discrete water samples, and recalculated the carbonate system to estimate future seawater pCO_2 and Ω_{ar} (hereafter noted as pCO_2* and Ω_{ar}*). This DIC addition approximates potential anthropogenic CO_2 loads in mid-latitude surface ocean waters after equilibration with an 788 μatm pCO_2 atmosphere as predicted for the year 2100 based on the Intergovernmental Panel on Climate Change (IPCC) IS92a 'continually increasing' emissions scenario [1,5]. We

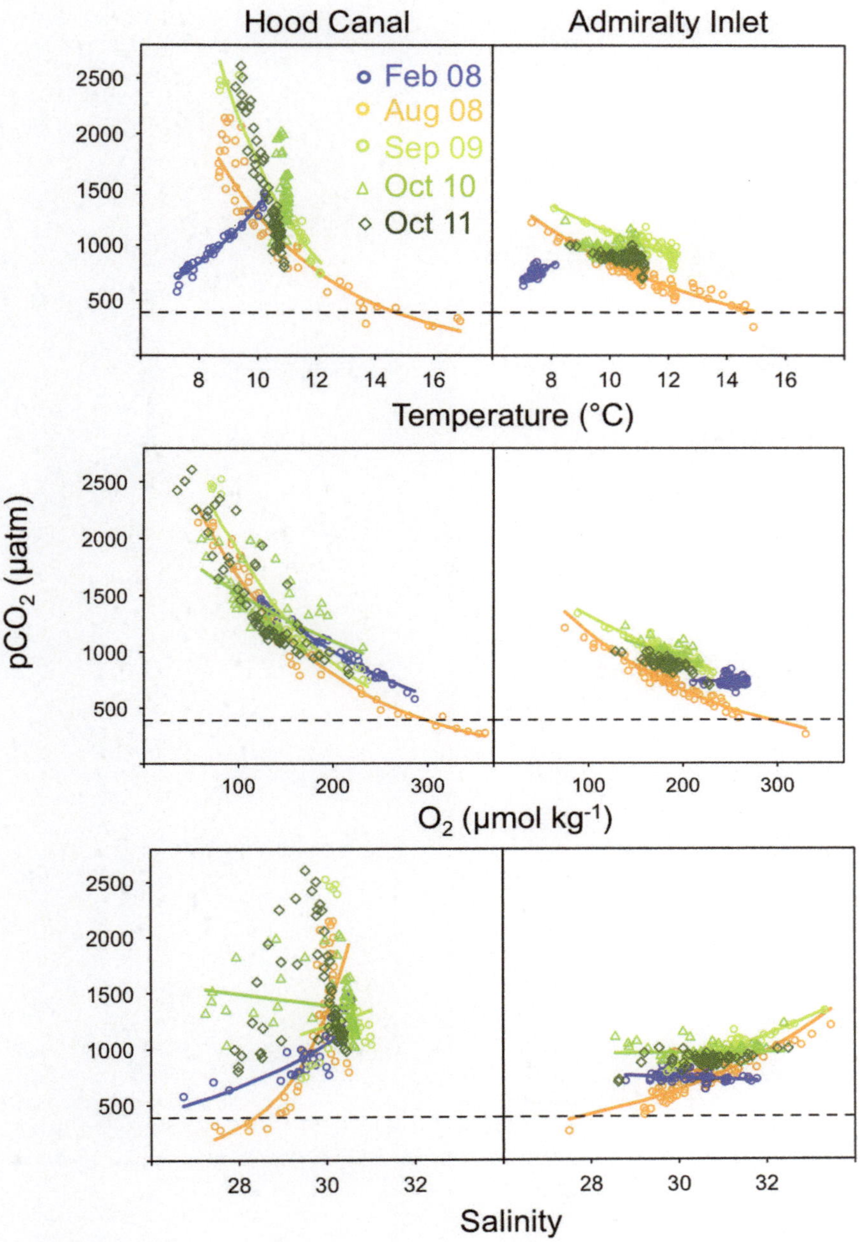

Figure 2. Patterns of covariation between pCO_2 and temperature (upper), oxygen (middle), and salinity (lower panels) in Hood Canal and Admiralty Inlet, Washington. For reference, the average atmospheric pCO_2 level (2008–2011 mean: 392 μatm) is noted by a dashed line. Trend lines correspond to relationships fitted using log_{10}-transformed pCO_2.

derived our DIC increment value from potential changes to inflowing marine waters to Puget Sound because river discharge comprises only 5% of the annual water exchange, and the median water residence time is ~80 days for Hood Canal and less than 60 days for the remaining sub-basins of Puget Sound [47]. Consequently, inflowing marine waters largely govern temperature, salinity, and nutrient loads in Puget Sound [28,41,48] and set initial conditions for carbonate chemistry in bottom waters entering Hood Canal as well [26]. The method carries the implicit assumption that alkalinity, respiration rates, and circulation patterns remain invariant. In summer, strong upwelling winds on the Washington coast supply deep, dense waters to the Sound that are thought to have last been in contact with the atmosphere approximately 50 years prior and that acquired their anthropogenic CO_2 burden at that time [12]. The anthropogenic CO_2 burden in Puget Sound can therefore lag behind conditions in open-ocean surface waters by a few decades, and future DIC loads will depend partly on interannual changes to the seasonal upwelling regime. Recognizing these complexities, our application of a uniform increase in DIC across seasons and locations is meant to simplify and facilitate general comparisons of possible future carbonate chemistry variability. As with pCO_2 and Ω_{ar}, we used CTD temperature, salinity, and oxygen values corresponding to pCO_2* and Ω_{ar}* from the discrete water samples to predict pCO_2*

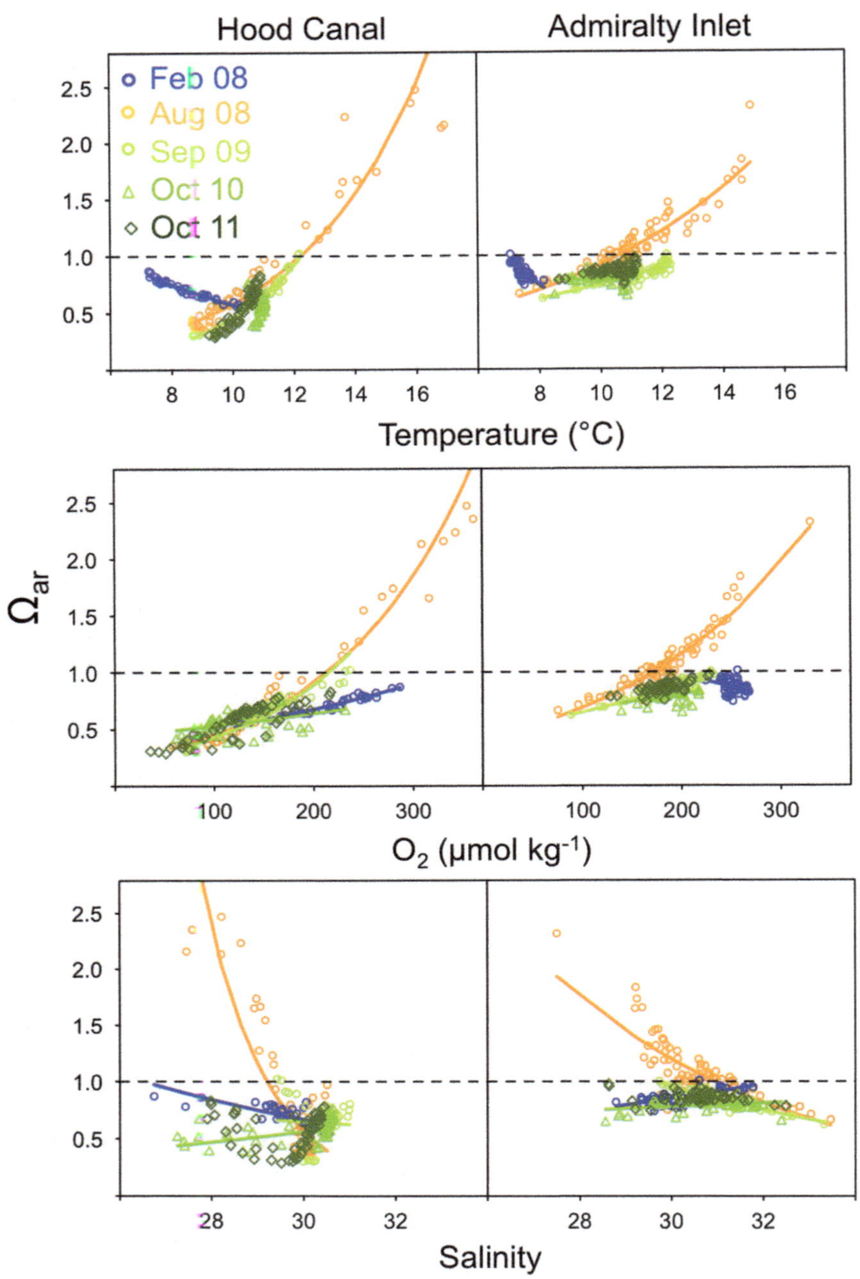

Figure 3. Patterns of covariation between aragonite saturation state (Ω_{ar}) and temperature (upper), oxygen (middle), and salinity (lower panels) in Hood Canal and Admiralty Inlet, Washington. Values below 1 (indicated by dashed line) correspond to corrosive conditions, those above are supersaturated with respect to aragonite. Trend lines correspond to relationships fitted using $\log_{10}$-transformed Ω_{ar}.

and Ω_{ar}* across all depths using PLSR; model selection proceeded in a manner identical to that used for pCO_2 and Ω_{ar}.

Based on typical DVM patterns observed in Puget Sound zooplankton and fish [42,45], we assumed that zooplankton descended from 3 to 50 m at a rate of 1 m min^{-1} starting at 0600 hr, and ascended from 50 to 3 m beginning at 1800 hr at the same rate. Given this depth pattern over a 24-hour period, we extracted the corresponding CTD data and pCO_2, pCO_2*, Ω_{ar}, and Ω_{ar}* values predicted across depths from each station in the survey to generate time traces. We compared time traces across seasons and between Admiralty Inlet and Hood Canal stations to evaluate potential high-frequency variation in carbonate chemistry resulting from diel movements, explore how carbonate chemistry changes simultaneously with other biologically important variables, and examine how carbonate chemistry may differ under CO_2-enriched conditions as experienced by diel vertical migrators. To aid comparability, we used the same DVM across seasons and locations, but acknowledge that seasonal, regional, and interspecific variations in DVM may in fact be significant.

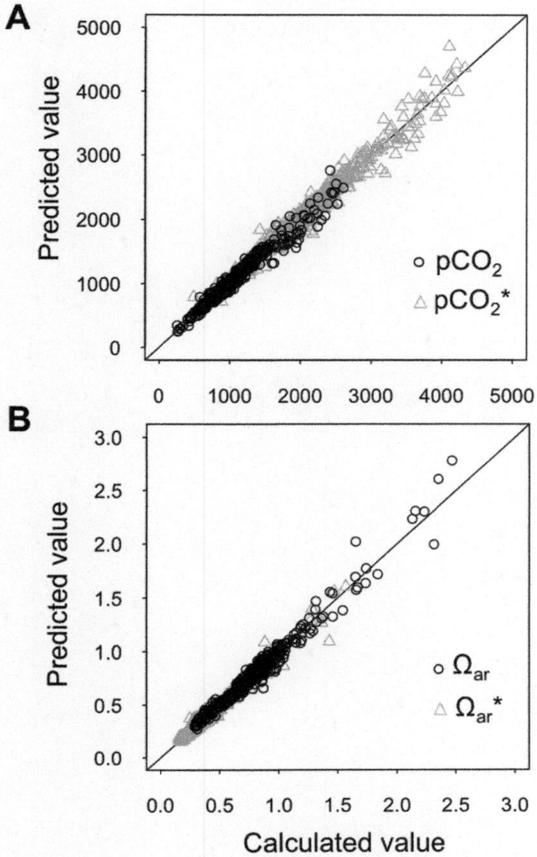

Figure 4. (a) Predicted versus calculated pCO_2 and pCO_2* and (b) Ω_{ar} and Ω_{ar}* based on partial least squares regression models for Puget Sound water samples obtained from 2008 to 2011. pCO_2* and Ω_{ar}* values correspond to estimates based on an the addition of 90 μmol kg^{-1} DIC across all samples.

Results

In Hood Canal, pCO_2 estimates obtained from discrete water samples ranged from 230 to 2610 μatm, and the majority of water samples exceeded atmospheric pCO_2 levels (mean atmospheric pCO_2 from 2008 to 2011: 389 μatm; Fig. 2). Regression intercept and slope coefficients and the explained variance (based on the $\log_{10}$-transformed response variables) associated with the trend lines depicted in Fig. 2 are presented in Table S1. The Hood Canal August relationship between pCO_2 and temperature was negative and spanned the largest range of temperatures (8.5–17°C) relative to the other months surveyed (Fig. 2). In fall, the range of temperatures observed in both years was narrower relative to summer (~9–12°C; Fig. 2). Temperatures in winter ranged from 8 to 11°C in Hood Canal but showed the opposite relationship with pCO_2 as observed in the summer and fall (Fig. 2). Similar temperature-pCO_2 relationships were evident for Admiralty Inlet, but for all cruises the range of temperatures and pCO_2 levels were narrower. In general, variation in pCO_2 was more tightly associated with temperature in Hood Canal relative to Admiralty Inlet based on the higher R^2 values observed in Hood Canal across all seasons (Table S1).

Negative relationships between pCO_2 and oxygen were evident in all seasons and years in both Hood Canal and Admiralty Inlet with the exception of Admiralty Inlet in February (Fig. 2, Table S1). As with temperature, the larger range of O_2 values were found in Hood Canal relative to Admiralty Inlet (Fig. 2). In general, the negative relationship between pCO_2 and O_2 was stronger in Hood Canal than Admiralty Inlet based on R^2 values (Table S1). Salinities, in contrast, spanned a relatively narrow range of values (~27 to 33; Fig. 2). In both regions, relationships between pCO_2 and salinity were positive in August. In the remaining seasons, pCO_2 and salinity relationships were weaker. Overall, pCO_2 was less tightly associated with salinity relative to temperature and oxygen across all seasons and both regions (Fig. 2; Table S1).

As expected, patterns of Ω_{ar} were nearly inverse to those observed for pCO_2 (Fig. 3; Table S1). With the exception of August, when *in situ* Ω_{ar} values ranged from 0.2 to 2.8, Ω_{ar} values were ~1 or lower in both Hood Canal and Admiralty Inlet in all remaining months (Fig. 3). Similar to pCO_2, the range of Ω_{ar} values was highest in Hood Canal and lowest in Admiralty Inlet (Fig. 3). Patterns of association between Ω_{ar} and temperature, oxygen, and salinity based on R^2 values were similar to those of pCO_2 between regions and across seasons (Table S1).

Partial least square regression models of pCO_2 and Ω_{ar} explained greater than 95 and 94% of variance in each survey, respectively (Fig. 4). RMSEPs (based on the original response variable scale) ranged between 4 and 50 μatm and 0.02 and 0.05, for pCO_2 and Ω_{ar}, respectively. The model fits suggested a sufficient level of predictive accuracy for estimating values across depths based on CTD depth profiles of oxygen, temperature, and salinity. We predicted pCO_2 and Ω_{ar} across depths and examined temperature and oxygen depth profiles as well, but did not evaluate salinity due to the narrow range of values observed in the survey.

Overall, seawater pCO_2 values were above mean atmospheric levels (392 μatm) at all depths and across surveys with the exception of shallow waters (< 15 m depth) at several stations sampled in August 2008 (Fig. 5). Waters were also undersaturated with respect to aragonite ($\Omega_{ar}<1$) at all depths sampled in February 2008 and October 2010 and 2011 (Fig. 5). Ω_{ar} in near-surface waters (< 15 m) at some stations in Hood Canal in August 2008 and September 2009, however, were supersaturated ($\Omega_{ar} >1$), while several stations at Admiralty Inlet sampled in August 2008 were supersaturated throughout the water column (Fig. 5).

In general, large differences in vertical gradients in carbonate chemistry, temperature, and oxygen conditions were apparent across seasons and regions. In February 2008, CTD depth profiles of temperature and oxygen and predicted values of pCO_2 and Ω_{ar} indicated waters in Admiralty Inlet were well-mixed, and showed little vertical or inter-station variability (Fig. 5). In Hood Canal, several stations were also well-mixed, though vertical gradients were evident at several stations: deep waters were warmer, oxygen depleted, rich in pCO_2, and exhibited lower Ω_{ar} values relative to surface waters (Fig. 5). In August 2008 and September 2009, the water column was strongly stratified at several stations in Hood Canal, with warm, oxygen-rich, low pCO_2, supersaturated waters occurring in near-surface waters. At depth, the opposite conditions prevailed. Variation in water conditions was comparatively lower at Admiralty Inlet both across stations and vertically (Fig. 5).

Stratification was again low at Admiralty Inlet stations in October 2010 and 2011 with near-uniform temperature, oxygen, and carbonate chemistry conditions throughout the water column and little variability between stations. However, in Hood Canal oxygen-poor, CO_2-rich, and low Ω_{ar} waters occurred in the top 50 m of the water column, while the opposite conditions were observed at depth (Fig. 5). Hypoxic conditions (oxygen concentrations less than ~60 μmol kg^{-1}) were observed in waters at depths shallower than 20 m at stations in southern Hood Canal in

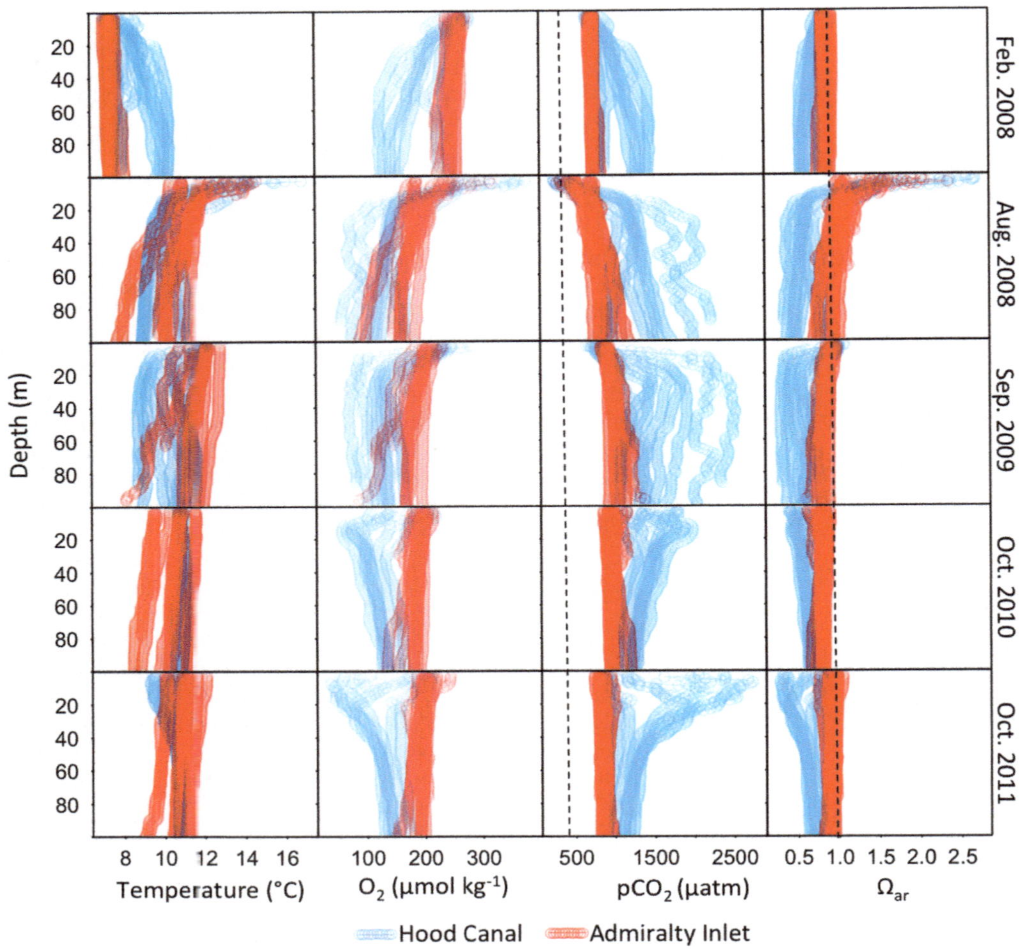

Figure 5. Temperature and oxygen CTD depth profiles at stations sampled in Hood Canal and Admiralty Inlet. pCO_2 and Ω_{ar} values are model-based estimates obtained using partial least squares regressions. To aid interpretation, vertical dashed lines are overlaid denoting the average atmospheric pCO_2 (2008 to 2011 mean: 392 μatm) and the equilibrium aragonite saturation state (Ω_{ar} = 1). Profiles are truncated at 100 m to aid comparability.

October 2011. Across all five surveys, the highest and lowest estimated pCO_2 and Ω_{ar} values, respectively, were observed at these same stations and depths: estimates of pCO_2 exceeded 2400 and Ω_{ar} values were less 0.4 (Fig. 5).

Time traces of temperature, oxygen, pCO_2, and Ω_{ar} concentrations corresponding to conditions experienced by hypothetical diel vertical migrators are depicted in Fig. 6. As suggested by the depth profiles, diel vertical migrators occupying the sampled stations likely experience simultaneous and abrupt changes in temperature, oxygen, and carbonate chemistry conditions in strongly stratified locations, and these were most prevalent in Hood Canal and Admiralty Inlet in August 2008 and September 2009 (Fig. 6). Time traces in August 2008 in Hood Canal, for instance, suggest that mean temperature, oxygen concentration, Ω_{ar}, and pCO_2 experienced by diel vertical migrators in near-surface waters (3 m) at night and deep waters (50 m) during the day differed by 5.1°C, 214 μmol kg^{-1}, 1.68, and 1100 μatm, respectively (Fig. 6). Importantly, the ranges across stations in temperature, oxygen, and carbonate chemistry conditions in both surface and deep waters were also notable during this time period in Hood Canal. The larger range in carbonate chemistry variables occurred in deep relative to near-surface waters: minimum and maximum pCO_2 values in near-surface waters were estimated to be 250 and 480 μatm and in deep waters values were 980 and 2030 μatm, respectively. Ω_{ar} ranged from 1.52 to 2.68 in near-surface waters and 0.37 to 0.77 in deep waters (Fig. 6).

In all seasons and regions, pCO_2* values were not only higher relative to pCO_2, but the difference between near-surface and deep-water values increased (Fig. 6). In Hood Canal August 2008, the mean difference between near-surface and deep-water pCO_2 widened from 1100 to 2090 μatm with the uniform addition of DIC (Fig. 6). Correspondingly, Ω_{ar} decreased across all depths as well, but the mean difference between near-surface and deep water was lower (1.68 vs. 0.98) relative to Ω_{ar}, as the decline in the carbonate ion concentration slows as it approaches zero (Fig. 6).

In contrast to August 2008 and September 2009, the mean differences in temperature, oxygen, and carbonate chemistry variables between near-surface and deep waters were lower in October 2010 and 2011 (Fig. 6). However, the upward displacement of the waters with the lowest oxygen concentrations toward the surface in Hood Canal resulted in time traces in which diel vertical migrators encountered higher pCO_2 and lower Ω_{ar} during the night rather than during the day (Fig. 6). Further, the range in pCO_2 and Ω_{ar} conditions in Hood Canal was larger among

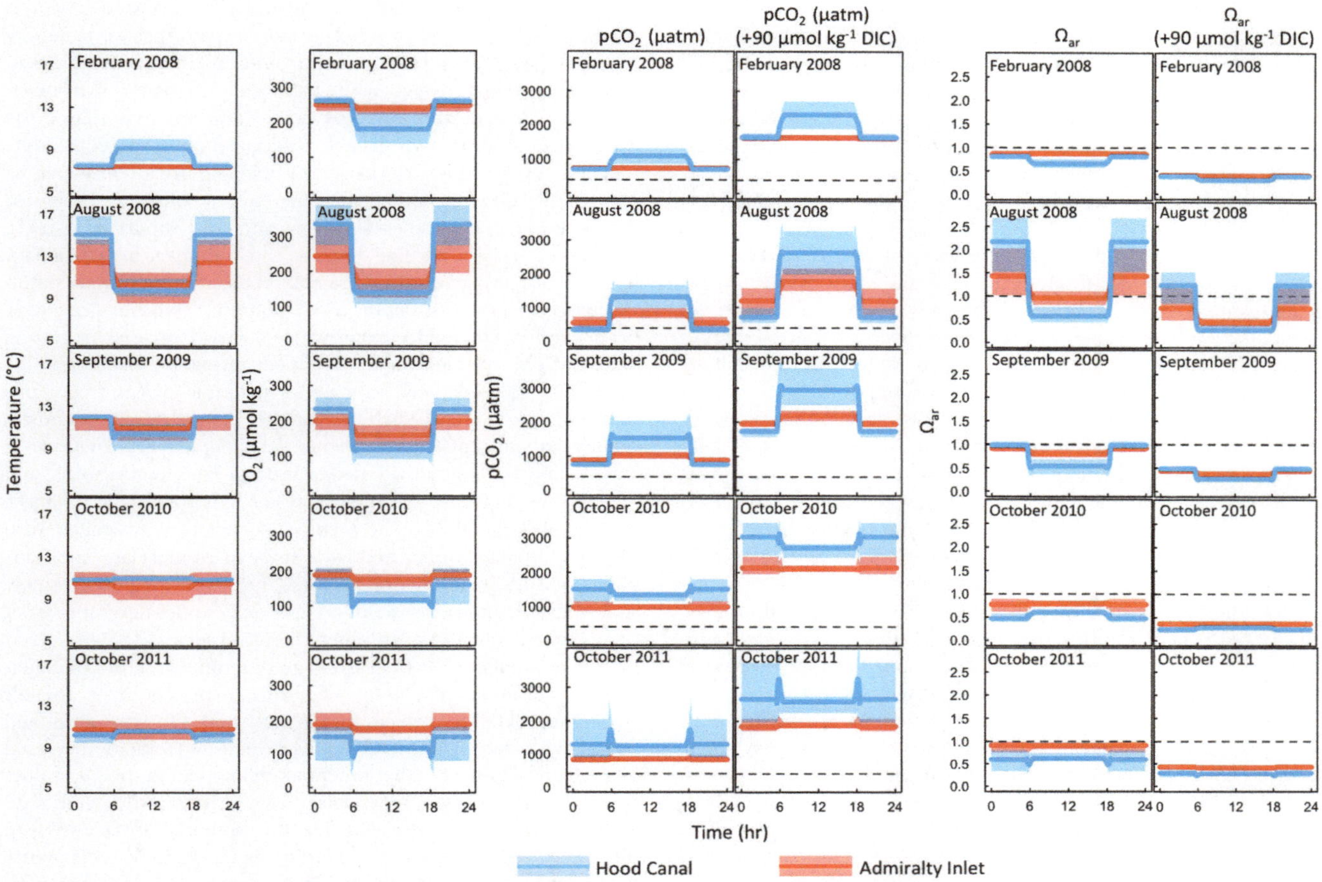

Figure 6. Time traces of temperature, oxygen, pCO_2 and Ω_{ar} conditions potentially experienced by hypothetical diel vertical migrators at stations sampled in Puget Sound, Washington, U.S.A. Organisms ascend from 50 to 3 m at a rate of 1 m min^{-1} beginning at 1800 hr and descend back to 50 m at 0600 hr. Colored bands correspond to maximum and minimum pCO_2 and Ω_{ar} values and pCO_2* and Ω_{ar}* as estimated assuming a 90 µmol kg^{-1} increase in DIC in Hood Canal and Admiralty Inlet stations. The solid lines correspond to mean values. For comparison, atmospheric pCO_2 (2008–2011 average: 392 µatm) is indicated by a dashed line. Ω_{ar} values below 1 (denoted by dashed line) indicate undersaturated conditions with respect to aragonite.

stations in near-surface relative to deep waters, and these ranges increased and decreased, respectively, with an increase in DIC (Fig. 6).

Overall, the difference in near-surface and deep temperatures, oxygen concentrations, and carbonate chemistry conditions experienced by diel vertical migrators in February 2008 were lower than in the remaining surveys (Fig. 6). Similar to August 2008 and September 2009, oxygen concentrations and Ω_{ar} were lower and pCO_2 higher for diel vertical migrators at depth during the day in Hood Canal, but unlike the remaining months surveyed the mean temperature was warmer by 1.7°C relative to near-surface waters experienced during the night. Further, the potential range of water conditions experienced by diel vertical migrators at a given depth across stations was narrower than observed in the remaining surveys (Fig. 6).

Discussion

Our study expands on previous research on Puget Sound carbonate chemistry that noted waters greatly enriched in CO_2 relative to air-sea equilibration conditions in the winter and summer [26]. Here, we show that CO_2-enriched conditions persist through the fall. We also show that carbonate chemistry conditions covary with temperature and oxygen but that the direction and strength of these correlations change with season and between nearby regions characterized by differing levels of water mass mixing. Our findings suggest that fish and invertebrates that exhibit DVM behavior in stratified locations are likely to experience large, rapid changes in ambient carbonate chemistry conditions over the course of a single day during the summer and early fall months. Further, we show through simple calculations that a uniform increase in DIC not only elevates pCO_2 and lowers Ω_{ar}, but also widens the absolute difference in these values between near-surface and subsurface waters. Together, our findings highlight the physicochemical complexity of coastal systems and the importance such data hold for the design of ecologically relevant OA experiments.

Our observations in September 2009 and October 2010 and 2011 add an additional layer of information to our general understanding of seasonal variability in carbonate chemistry in Hood Canal. Due to sluggish circulation and the availability of nutrients in inflowing marine and terrestrial waters, hypoxic conditions commonly occur in Hood Canal during the summer months [49]. As evidenced by oxygen and pCO_2 depth profiles in August 2008 and September 2009, aerobic respiration simultaneously draws down O_2 at depth and increases pCO_2.

However, we show that bottom-water intrusions into Puget Sound and Hood Canal during October 2010 and 2011 resulted in the upward movement of low-oxygen waters as observed previously [45] but that these near-surface waters were CO_2-replete and exhibited the most extreme pCO_2 and Ω_{ar} values observed across all surveys. Fall bottom-water intrusions into the Sound occur during neap tides and their onset is influenced by storm-related fluctuations in the horizontal density gradient spanning the double-silled entrance at Admiralty Inlet [50,51]. Although such intrusions play a critical role in the renewal of deep water layers in Puget Sound, they also result in the surfacing of low oxygen, high pCO_2 waters which, if upwelled suddenly to the surface due to favorable southerly winds, can result in extensive fish kills [29]. Our observations suggest that organisms that reside in shallow waters to avoid low-oxygen, CO_2-rich waters at depth in the summer may be unable to avoid exposure at some locations in Hood Canal as these waters circulate upward in the fall. Combined, our carbonate chemistry data set indicates that pCO_2 values are spatially and seasonally heterogeneous in Puget Sound as in other estuarine systems [52] and further underscores the need for researchers to consider natural variability in the design and interpretation of OA experiments.

Importantly, we show that carbonate chemistry conditions also covary with temperature and oxygen but that the strength and slope of the relationships change with season. Failure to consider covariation in carbonate chemistry with temperature or oxygen may result in experimental designs with 'control' treatments that have little correspondence to conditions that animals may have adapted to in nature and 'acidified' treatments that do not account for naturally high pCO_2 levels in many estuarine and coastal waters. For instance, strong negative relationships between pCO_2 and temperature in Hood Canal for August 2008 suggest that appropriate control pCO_2 values for experiments on a widely distributed animal from this region would depend on the temperatures considered in the experiment. Temperatures of 10 and 15°C occur within the top 50 m of the water column and have corresponding control pCO_2 values of 1270 and 360 µatm, respectively, based on the linear regression between pCO_2 and temperature. Assuming a uniform increase in 90 µmol kg^{-1} in DIC, the respective acidified treatments are 2150 and 760 µatm, again based on the regression between pCO_2* and temperature. In contrast, experimental controls based on winter Hood Canal conditions that use temperatures of 8 and 10°C would have pCO_2 values corresponding to 850 and 1360 and pCO_2* values of 1380 and 1930 µatm, respectively. Although there is increasing recognition that researchers should consider control treatments that reflect carbonate chemistry conditions experienced by the organism under study [11,18,21], little attention has been given to understanding patterns of carbonate chemistry covariability with temperature and oxygen. Given the combined effect environmental temperature, oxygen, and pCO_2 have on metabolic scope [53] and evidence indicating non-additive, interactive effects of these variables in ecological studies [6,54], we recommend care be taken in selecting appropriate values that correspond to those likely to be experienced by organisms in the field. Doing so may facilitate inferences from experimental results to population-level effects.

High-resolution carbonate chemistry time series have revealed considerable diel variability in coral reef [55,56], kelp forest [10,57], coastal upwelling [58], and shallow estuarine systems [59] that is primarily driven by oscillations in photosynthesis and respiration rates. For animals that undergo DVM in vertically heterogeneous regions, however, exposure to water masses with differing carbonate chemistry, temperature, and oxygen characteristics is also possible but has received only limited consideration in the context of OA [60]. In Puget Sound, DVM behavior is common in a diversity of fish and invertebrate species, including zooplankton occurring in stratified regions of Hood Canal [42,45]. Developing experiments with temporally dynamic carbonate chemistry, temperature, and oxygen conditions that mimic the experience of organisms in the wild may offer more accurate baselines upon which responses to acidified treatments can be compared. Our hypothetical time traces offer a basis for developing such experimental designs and offer important insights. Foremost, we show that during the summer in particular, temperature, oxygen, and carbonate chemistry conditions simultaneously covary with depth, a consideration that has previously been overlooked in OA experiments focused on diel migrating animals from stratified waters [60]. Organismal performance (e.g., growth) may potentially improve in dynamic pCO_2 treatments that emulate natural diel oscillations in carbonate chemistry relative to static pCO_2 treatments [61], but we are unaware of studies that have simultaneously varied pCO_2, temperature, and oxygen to replicate the full suite of environmental changes associated with DVM. The technical challenges associated with implementing dynamic, multi-stressor experiments are considerable but OA experimental facilities are becoming increasingly sophisticated and such systems are currently under development at several academic and government laboratories.

Our recalculation of the carbonate system to generate estimates of potential future pCO_2 and Ω_{ar} values experienced by animals undergoing DVM showed that pCO_2 and Ω_{ar} increased and decreased, respectively, at all depths as expected with an increased CO_2 load. However, the potential difference in pCO_2 values between near-surface and deep waters increased relative to contemporary observations, while the potential surface-to-deep difference for Ω_{ar} decreased. This pattern results from nonlinearity in the carbonate system at near-surface ocean pH values: as DIC is added in the form of CO_2 (either from anthropogenic CO_2 emissions or respiration), the carbonate ion concentration of seawater declines, while the CO_2 concentration and pCO_2 forms of inorganic carbon increase exponentially [62,63]. In waters already burdened with high concentrations of CO_2 due to respiration, an increase in DIC results in an increase (decrease) in pCO_2 (Ω_{ar}) that will be larger (smaller) relative to less burdened waters [62,64]. Our estimates of potential change in carbonate chemistry variability as experienced by DVM fauna are qualitatively similar to simulations in other studies that have shown an amplification of natural variability in carbonate chemistry associated with increased DIC [65]. These findings suggest more broadly that changes in the mean as well as the variance of pCO_2 will accompany increased DIC loadings. The influence of changes to pCO_2 variance on organisms is poorly understood and warrants further attention.

In addition, we show that although spatial and vertical variation in carbonate chemistry is considerable in Puget Sound, additional DIC may place pCO_2 and Ω_{ar} at most stations and depths outside the range of values observed presently across depths and stations. This possibility was most evident in winter where there was no overlap between present-day and future pCO_2 and Ω_{ar} as experienced by fauna undergoing DVM. During winter, the water column is less stratified and conditions are more uniform across stations and depths [41]. Consequently, low pCO_2, supersaturated refugia are absent and carbonate chemistry is relatively homogenous. In contrast, some near-surface waters in August remained supersaturated (Ω_{ar}) and the pCO_2 at levels low, despite the DIC addition. These same near-surface waters initially possessed low DIC concentrations due to biological drawdown through photosynthesis and thus higher buffering capacity.

The general seasonal patterns observed suggest that species with CO_2-sensitive life history stages that occur in winter are even less likely to find depths or regions with favorably low pCO_2 conditions in the future. For species with CO_2-sensitive life history stages that occur in summer, the potential for carbonate chemistry refugia persisting in the future may be better because high rates of primary production can locally drive pCO_2 levels below air-sea equilibrium in the surface layer, though the benefits of such refugia depend on the suitability of other biological variables and their persistence in time and space. Our analysis also suggests that CO_2-sensitive species may be more fully excluded from regions such as Admiralty Inlet that are well-mixed and therefore lack low pCO_2 refugia at most depths and stations relative to more heterogeneous regions like Hood Canal.

We limited our ocean acidification scenario to a uniform change in DIC. Our estimates therefore do not account for processes that may further alter carbonate chemistry and the vertical and spatial distribution of CO_2, which may include climate variables (upwelling winds, circulation patterns, river discharge), changes in watershed use, eutrophication [9], and atmospheric deposition of nitrogen and sulfur from shipping vessels [66] or urban pollution [67]. Processes that reduce salinity (and thus TA) or that increase respiration (and DIC) increase pCO_2 [64], while warming directly affects carbonate system thermodynamics, resulting in solubility-related increases in pCO_2, but more importantly influences ecosystem metabolism, which strongly affects carbonate chemistry [68]. Biogeochemical models of coastal and estuarine systems are increasingly utilized to identify the potential processes that dampen or exacerbate acidification [69,70] and may help clarify the contribution of climate variability and different human activities to future changes in Puget Sound's carbonate chemistry. We also note that our data set does not permit an evaluation of diurnal carbonate chemistry patterns, which can be pronounced in coastal surface waters, and is limited to stations in open-water habitats. We still lack information on shallower habitats such as mudflats, eelgrass meadows, pocket estuaries, and oyster beds, which are important components of the Puget Sound ecosystem. We recommend targeted sampling in these key habitats over seasonal cycles and the development of long-term monitoring programs in order to better identify baseline carbonate chemistry variability and to facilitate the attribution of the contribution of anthropogenic CO_2.

Puget Sound receives waters upwelled along the Pacific Coast that are already burdened with anthropogenic CO_2 [26], and concentrations will continue to increase over the proceeding decades regardless of possible changes to atmospheric pCO_2. This is because upwelled waters were last in contact with the atmosphere ~50 years prior to surfacing and therefore carry CO_2 burdens that reflect air-sea equilibration at that time [12]. Consequently, waters already in transit to upwelling centers on the Pacific coast near Puget Sound carry increasing concentrations of anthropogenic CO_2 [12]. Although we are unable to stem the flow of CO_2-enriched marine waters into Puget Sound, understanding how human-caused impacts (e.g., eutrophication, pollution, and altered river flows) exacerbate or buffer against the effects of ocean acidification on biota may help identify actions regional and local governments can take to reduce the effects of OA. A key uncertainty facing coastal ecosystem management, however, lies in the response of organisms and communities to ocean acidification and its interactions with other stressors [2]. Fundamental to addressing this uncertainty is the design of manipulative experiments in which control water conditions are representative of the system under study. Our analyses offer insight into seasonal patterns of covariation in carbonate chemistry with temperature and oxygen, and we suggest that such considerations are needed in order to draw useful ecological inferences from manipulative experiments. Given the potential for synergistic interactions between temperature, oxygen and carbonate chemistry on organismal physiology [24], developing a baseline understanding of how these variables naturally covary can facilitate the development of experiments with more relevance to questions regarding ocean acidification.

Supporting Information

Table S1 Estimated intercept and slope coefficients and explained variance (R^2) corresponding to linear regression models relating $\log_{10}$-transformed pCO_2 and Ω_{ar} to temperature, oxygen, and salinity. Slope coefficients that differ from zero are indicated by asterisks. Slopes with p-values <0.05, <0.01, and <0.001 correspond to *, **, and ***, respectively.

Acknowledgments

The authors wish to thank the officers and crew of the R/V *Thomas G. Thompson*, EPA Ocean Survey Vessel *Bold*, and R/V *Robertson* for their assistance during the cruises and Cynthia Peacock, Geoff Lebon, Cathy Cosca, Corinne Bassin, Jill Coyle, Dana Greeley, Julia Bos, Kathy Krogslund, Amanda Gray, Megan Black, Sylvia Musielewicz, and Jennifer Nomura for their shipboard, laboratory, and graphics support of this research effort. Mariana Ribas-Ribas and an anonymous reviewer provided valuable comments on an earlier version of this work. This is PMEL contribution number 4031.

Author Contributions

Conceived and designed the experiments: JCPR SRA. Performed the experiments: JCPR SRA. Analyzed the data: JCPR SRA. Contributed reagents/materials/analysis tools: JCPR SRA RAF JN MW. Wrote the paper: JCPR SRA PM.

References

1. Orr JC, Fabry VJ, Aumont O, Bopp L, Doney SC, et al. (2005) Anthropogenic ocean acidification over the twenty-first century and its impact on calcifying organisms. Nature 437: 681–686.
2. Doney SC, Fabry VJ, Feely RA, Kleypas JA (2009) Ocean Acidification: The Other CO_2 Problem. Ann Rev Mar Sci 1: 169–192.
3. Caldeira K, Wickett ME (2003) Anthropogenic carbon and ocean pH. Nature 425: 365–365.
4. Feely RA, Sabine CL, Lee K, Berelson W, Kleypas J, et al. (2004) Impact of anthropogenic CO_2 on the $CaCO_3$ system in the oceans. Science 305: 362–366.
5. Orr J (2011) Recent and future changes in ocean carbonate chemistry. In: Gattuso J, Hansson L, editors. Ocean Acidification. New York, NY: Oxford University Press. pp. 41–66.
6. Kroeker KJ, Kordas RL, Crim R, Hendriks IE, Ramajo L, et al. (2013) Impacts of ocean acidification on marine organisms: quantifying sensitivities and interaction with warming. Global Change Biology 19: 1884–1898.
7. Koch M, Bowes G, Ross C, Zhang XH (2013) Climate change and ocean acidification effects on seagrasses and marine macroalgae. Global Change Biology 19: 103–132.
8. Fabry VJ, Seibel BA, Feely RA, Orr JC (2008) Impacts of ocean acidification on marine fauna and ecosystem processes. ICES J Mar Sci 65: 414–432.
9. Duarte CM, Hendriks IE, Moore TS, Olsen YS, Steckbauer A, et al. (2013) Is ocean acidification an open-ocean syndrome? Understanding anthropogenic impacts on seawater pH. Estuar Coasts: 1–16.
10. Hofmann GE, Smith JE, Johnson KS, Send U, Levin LA, et al. (2011) High-frequency dynamics of ocean pH: a multi-ecosystem comparison. PLoS ONE 6: e28983.
11. Andersson A, Mackenzie F (2012) Ocean acidification: setting the record straight. Biogeosciences 9: 893–905.

12. Feely RA, Sabine CL, Hernández-Ayón JM, Ianson D, Hales B (2008) Evidence for upwelling of corrosive "acidified" water onto the continental shelf. Science 320: 1490–1492.
13. Harris KE, DeGrandpre MD, Hales B (2013) Aragonite saturation state dynamics in a coastal upwelling zone. Geophys Res Lett 40: 2720–2725.
14. Hauri C, Gruber N, Vogt M, Doney SC, Feely RA, et al. (2013) Spatiotemporal variability and long-term trends of ocean acidification in the California Current System. Biogeosciences 10: 193–216.
15. Alin SR, Feely RA, Dickson AG, Hernández-Ayón JM, Juranek LW, et al. (2012) Robust empirical relationships for estimating the carbonate system in the southern California Current System and application to CalCOFI hydrographic cruise data (2005–2011). J Geophys Res 117: C05033.
16. Evans W, Hales B, Strutton PG (2011) Seasonal cycle of surface ocean pCO_2 on the Oregon shelf. J Geophys Res 116: C05012.
17. Frieder C, Nam S, Martz T, Levin L (2012) High temporal and spatial variability of dissolved oxygen and pH in a nearshore California kelp forest. Biogeosciences 9: 3917–3930.
18. McElhany P, Busch DS (2012) Appropriate pCO_2 treatments in ocean acidification experiments. Mar Biol 160: 1807–1812.
19. Andersson AJ, Mackenzie FT (2012) Revisiting four scientific debates in ocean acidification research. Biogeosciences 9: 893–905.
20. Waldbusser GG, Salisbury JE (2013) Ocean acidification in the coastal zone from an organism's perspective: multiple system parameters, frequency domains, and habitats. Ann Rev Mar Sci. DOI: 10.1146/annurev-marine-121211-172238
21. Yu PC, Matson PG, Martz TR, Hofmann GE (2011) The ocean acidification seascape and its relationship to the performance of calcifying marine invertebrates: Laboratory experiments on the development of urchin larvae framed by environmentally-relevant pCO_2 / pH. J Exp Mar Biol Ecol 400: 288–295.
22. Hettinger A, Sanford E, Hill T, Hosfelt J, Russell A, et al. (2013) The influence of food supply on the response of Olympia oyster larvae to ocean acidification. Biogeosciences 10: 6629–6638.
23. Juranek LW, Feely RA, Peterson WT, Alin SR, Hales B, et al. (2009) A novel method for determination of aragonite saturation state on the continental shelf of central Oregon using multi-parameter relationships with hydrographic data. Geophysl Res Lett 36: L24601.
24. Pörtner H-O (2010) Oxygen-and capacity-limitation of thermal tolerance: a matrix for integrating climate-related stressor effects in marine ecosystems. J Exp Biol 213: 881–893.
25. Pörtner HO, Farrell AP (2008) Physiology and climate change. Science 322: 690–692.
26. Feely RA, Alin SR, Newton J, Sabine CL, Warner M, et al. (2010) The combined effects of ocean acidification, mixing, and respiration on pH and carbonate saturation in an urbanized estuary. Estuar Coast Shelf Sci 88: 442–449.
27. PSP (2007) Puget Sound Conservation and Recovery Plan. Puget Sound Partnership, Olympia, WA. 90 p.
28. Mackas DL, Harrison PJ (1997) Nitrogenous nutrient sources and sinks in the Juan de Fuca Strait/Strait of Georgia/Puget Sound estuarine system: assessing the potential for eutrophication. Estuar Coast Shelf Sci 44: 1–21.
29. Newton J, Bassin C, Devol A, Kawase M, Ruef W, et al. (2008) Hypoxia in Hood Canal: an overview of status and contributing factors. In: Proceedings of the 2007 Georgia Basin Puget Sound Research Conference, Olympia, WA.
30. Carpenter JH (1965) The Chesapeake Bay Institute technique for the Winkler dissolved oxygen method. Limnol Oceanogr 10: 141–143.
31. Johnson KM, King AE, Sieburth JM (1985) Coulometric DIC analyses for marine studies: An introduction. Mar Chem: 61–82.
32. Johnson KM, Sieburth JM, Williams PJL, Brandstrom L (1987) Coulometric total carbon-dioxide analysis for marine studies e automation and calibration. Mar Chem 21: 117–133.
33. Dickson A, Sabine C, Christian J (2007) Guide to Best Practices for Ocean CO_2 Measurements. PICES Special Publication vol. 3, 191 p.
34. Lavigne H, Gattuso J (2010) Seacarb: seawater carbonate chemistry with R. R package version 2.4.3
35. Lueker TJ, Dickson AG, Keeling CD (2000) Ocean pCO_2 calculated from dissolved inorganic carbon, alkalinity, and equations for K1 and K2: validation based on laboratory measurements of CO_2 in gas and seawater at equilibrium. Mar Chem 70: 105–119.
36. Mucci A (1983) The solubility of calcite and aragonite in seawater at varioussalinities, temperatures, and one atmosphere total pressure. American Journal of Science 283: 780–799.
37. Millero FJ (1995) Thermodynamics of the carbon-dioxide system in the oceans.Geochim Cosmochim Ac 59: 661–677.
38. Kutner MH, Nachtsheim CJ, Neter J, Li W (2005) Applied linear statistical models. New York, NY: McGraw-Hill/Irwin. 1396 p.
39. Carrascal LM, Galvan I, Gordo O (2009) Partial least squares regression as an alternative to current regression methods used in ecology. Oikos 118: 681–690.
40. Mevik B-H, Wehrens R (2007) The pls package: principal component and partial least squares regression in R. J Stat Softw 18: 1–24.
41. Moore SK, Mantua NJ, Newton JA, Kawase M, Warner MJ, et al. (2008) A descriptive analysis of temporal and spatial patterns of variability in Puget Sound oceanographic properties. Estuar Coast Shelf Sci 80: 545–554.
42. Parker-Stetter SL, Horne JK, Langness MM (2009) The influence of midwater hypoxia on nekton vertical migration. ICES J Mar Sci 66: 1296–1302.
43. Frost BW (1988) Variability and possible adaptive significance of diel vertical migration in Calanus pacificus, a planktonic marine copepod. Bull Mar Sci 43: 675–694.
44. Bollens SM, Frost BW, Lin TS (1992) Recruitment, growth, and diel vertical migration of Euphausia pacifica in a temperate fjord. Mar Biol 114: 219–228.
45. Parker-Stetter SL, Horne JK (2009) Nekton distribution and midwater hypoxia: A seasonal, diel prey refuge? Estuar Coast Shelf Sci 81: 13–18.
46. Osgood KE, Frost BW (1994) Ontogenetic diel vertical migration behaviors of the marine planktonic copepods *Calanus pacificus* and *Metridia lucens*. Mar Ecol Prog Ser 104: 13–25.
47. Sutherland DA, MacCready P, Banas NS, Smedstad LF (2011) A model study of the Salish Sea estuarine circulation. J Phys Oceanogr 41: 1125–1143.
48. Steinberg PD, Brett MT, Bechtold JS, Richey JE, Porensky LM, et al. (2011) The influence of watershed characteristics on nitrogen export to and marine fate in Hood Canal, Washington, USA. Biogeochemistry 106: 415–433.
49. Brandenberger JM, Louchouarn P, Crecelius EA (2011) Natural and post-urbanization signatures of hypoxia in two basins of Puget Sound: historical reconstruction of redox sensitive metals and organic matter inputs. Aquat Geochem 17: 645–670.
50. Cannon GA, Holbrook JR, Pashinski DJ (1990) Variations in the onset of bottom-water intrusions over the entrance sill of a fjord. Estuaries 13: 31–42.
51. Lavelle J, Cokelet E, Cannon G (1991) A model study of density intrusions into and circulation within a deep, silled estuary: Puget Sound. J Geophys Res-Oceans 96: 16779–16800.
52. Borges AV, Abril G (2011) Carbon Dioxide and Methane Dynamics in Estuaries. In: Wolanski E, McLusky DS, editors. Treatise on Estuarine and Coastal Science. Waltham: Academic Press. pp. 119–161.
53. Portner HO, Gutowska MA, Ishimatsu A, Lucassen M, Melzner F, et al. (2011) Effects of ocean acidification on nektonic organisms. In: Gattuso JP, Hansson LJ, editors. Ocean Acidification. New York: Oxford University Press. pp. 154–175.
54. Wernberg T, Smale DA, Thomsen MS (2012) A decade of climate change experiments on marine organisms: procedures, patterns and problems. Global Change Biol 18: 1491–1498.
55. Shaw EC, McNeil BI, Tilbrook B (2012) Impacts of ocean acidification in naturally variable coral reef flat ecosystems. J Geophys Res-Oceans 117: 1978–2012.
56. Gray SE, DeGrandpre MD, Langdon C, Corredor JE (2012) Short-term and seasonal pH, pCO_2 and saturation state variability in a coral-reef ecosystem. Global Biogeochemical Cycles DOI: 10.1029/2011GB004114.
57. Delille B, Borges A, Delille D (2009) Influence of giant kelp beds (*Macrocystis pyrifera*) on diel cycles of pCO_2 and DIC in the Sub-Antarctic coastal area. Estuar Coast Shelf Sci 81: 114–122.
58. Wootton JT, Pfister CA, Forester JD (2008) Dynamic patterns and ecological impacts of declining ocean pH in a high-resolution multi-year dataset. Proc Nat Acad Sci 105: 18848–18853.
59. Barton A, Hales B, Waldbusser GG, Langdon C, Feely RA (2012) The Pacific oyster, *Crassostrea gigas*, shows negative correlation to naturally elevated carbon dioxide levels: Implications for near-term ocean acidification effects. Limnol Oceanogr 57: 698–710.
60. Rosa R, Seibel BA (2008) Synergistic effects of climate-related variables suggest future physiological impairment in a top oceanic predator. Proc Nat Acad Sci105: 20776–20780.
61. Dufault AM, Cumbo VR, Fan T-Y, Edmunds PJ (2012) Effects of diurnally oscillating pCO_2 on the calcification and survival of coral recruits. Proc Roy Soc B-Biol Sci 279: 2951–2958.
62. Sunda WG, Cai W-J (2012) Eutrophication induced CO_2-acidification of subsurface coastal waters: interactive effects of temperature, salinity, and atmospheric pCO_2. Environ Sci Technol 46: 10651–10659.
63. Gattuso J-P, Hansson L (2011) Ocean acidification: background and history. In: Gattuso J-P, Hansson L, editors. Ocean Acidification. Oxford: Oxford University Press. pp. 1–20.
64. Melzner F, Thomsen J, Koeve W, Oschlies A, Gutowska MA, et al. (2012) Future ocean acidification will be amplified by hypoxia in coastal habitats. Mar Biol: 1–14.
65. Shaw EC, McNeil BI, Tilbrook B, Matear R, Bates ML (2013) Anthropogenic changes to seawater buffer capacity combined with natural reef metabolism induce extreme future coral reef CO_2 conditions. Global Change Biol doi: 10.1111/gcb.12154.
66. Hassellöv IM, Turner DR, Lauer A, Corbett JJ (2013) Shipping contributes to ocean acidification. Geophys Res Lett 40: 2731–2736
67. Doney SC, Mahowald N, Lima I, Feely RA, Mackenzie FT, et al. (2007) Impact of anthropogenic atmospheric nitrogen and sulfur deposition on ocean acidification and the inorganic carbon system. Proc Nat Acad Sci 104: 14580–14585.
68. Gypens N, Lancelot C, Borges AV (2004) Carbon dynamics and CO_2 air-sea exchanges in the eutrophied coastal waters of the Southern Bight of the North Sea: a modelling study. Biogeosciences 1: 561–589.
69. Bianucci L, Denman K (2012) Carbon and oxygen cycles: Sensitivity to changes in environmental forcing in a coastal upwelling system. J Geophys Res: Biogeosciences 1117: 2005–2012.
70. Borges A, Gypens N (2010) Carbonate chemistry in the coastal zone responds more strongly to eutrophication than to ocean acidification. Limnol Oceanogr 55: 346–353.

19

Immediate Ecological Impacts of the 2011 Tohoku Earthquake Tsunami on Intertidal Flat Communities

Jotaro Urabe*, Takao Suzuki, Tatsuki Nishita, Wataru Makino

Division of Ecology and Evolutionary Biology, Graduate School of Life Sciences, Tohoku University, Sendai, Japan

Abstract

Following the Great East Japan Earthquake in 2011, a large tsunami developed and struck the Pacific coast of eastern Japan. To assess the immediate impacts of the tsunami on coastal communities, changes in taxon composition and richness of macrobenthic animals before and after the tsunami were examined at nine intertidal flats in Sendai Bay and the Sanriku Ria coast. The results showed that 30–80% of taxa indigenously inhabiting intertidal flats disappeared after the tsunami. Among animal types, endobenthic and sessile epibenthic animals were more vulnerable to the tsunami than mobile epibenthic animals like shore crabs and snails. For all the intertidal flats examined, animals that were originally dwellers in lower tidal zones and not recorded before the tsunami were also found right after the tsunami, indicating that the tsunami not only took away many benthic taxa from the intertidal flats but also brought in some taxa from elsewhere. However, overall changes in taxon community composition were greater for intertidal flats that experienced larger inundation heights. These results showed that the ecological impacts of the tsunami were proportional to the physical impacts as gauged by wave height and that mobile epibenthic animals were less vulnerable to the tsunami.

Editor: Jan Geert Hiddink, Bangor University, United Kingdom

Funding: This study was financially supported by Mitsui & Co., Ltd. Environment Fund (F11-F1-020), and research project funds, Tohoku Ecosystem-Associated Marine Science, from the Ministry of Education, Culture, Sports, Science and Technology, Japan. The funders had no role in study design, data collection and analysis, decision to publish, or preparation of the manuscript.

Competing Interests: Although this study was financially supported by Mitsui & Co., Ltd.

* E-mail: urabe@m.tohoku.ac.jp

Introduction

Rare and extreme events such as unusual floods, hurricanes, droughts and oil spills often reduce abundance of major native species and enhance colonization success of minor native species and even exotic species [1–5]. These events may, therefore, become unique opportunities causing drastic changes in community structures and providing insights on ecological processes shaping community structures. However, due to limited opportunities, little general knowledge has been accumulated on the ecological impacts of these events [1]. To assess ecological impacts of rare and extreme events, we need to examine changes in community composition immediately after an event as a first and essential step.

Tsunami is one type of a rare and extreme event. Since they are natural disturbances that have repeatedly affected coastal ecosystems over long time scales, tsunamis may play a role in the formation of coastal communities [6,7]. However, their ecological impacts are not, as yet, well understood. Using rare opportunities when tsunamis were generated by earthquakes, several studies have described subsequent structural changes of benthic communities in various coastal ecosystems including mangrove swamps [8], seagrass beds [9], sandy beaches [10] and subtidal soft bottoms [11]. These studies showed that the immediate impacts of a tsunami differed among communities at different sites and even among different species in the same ecosystem. However, it is not clear if the ecological impacts of a tsunami depend on the physical scale of the tsunami waves at a given site and what types of benthic animals are affected by tsunamis.

Following the Great East Japan Earthquake on 11 March 2011 registering 9.0 on the Moment Magnitude Scale (M_w), the resulting large tsunami (the 2011 Tohoku earthquake tsunami) struck the Pacific coastline of eastern Japan, including Sendai Bay and the Sanriku Ria coast [12]. In these areas, a number of intertidal flats were found in lagoons of river mouths, in inner parts of the bay and on shallow sheltered beaches. The physical, social and economic impacts of the tsunami were documented [12–14]. However, few studies have examined the ecological impacts of the tsunami on intertidal flat communities. The exception was a study by Kanaya et al. [15] who examined changes in macrobenthic fauna on the Gamo intertidal flat of Sendai Bay after the tsunami. They showed that 30 out of 79 benthic species recorded on the flat for 7 years before the tsunami were not present after the tsunami.

However, it was not clear that such a change in benthic fauna was also induced by the 2011 Tohoku earthquake tsunami at other intertidal flats. Fortunately, we had made benthic fauna census surveys of several of the intertidal flats in Sendai Bay and on the Sanriku Ria coast before the tsunami. If the tsunami seriously affected the intertidal communities, taxon composition and richness of benthic animals should have changed. Moreover, the differences that occurred in benthic communities were likely larger at sites that received the greater physical impacts. We examined these possibilities by performing census surveys of the intertidal flats right after the tsunami at locations where surveys were made before the tsunami. Using the data from these surveys, we also examined what types of benthic animals were more affected by the tsunami.

Methods

Census Sites and Ethics Statement

Pre-surveys of the areas subjected to the 2011 Tohoku earthquake tsunami were conducted in March to June between 2002 and 2010 and post-surveys in May to August 2011 at nine intertidal flats in Sendai Bay and on the Sanriku Ria coast (Table. 1, Fig. 1). All of these intertidal flats are so-called muddy flats and consisted of silt, clays and fine sands, that are exposed at low tide. The amount of exposed area differed among the intertidal flats and varied from 1 to 350 ha. Among these, Matsukawaura and Torinoumi, the two largest flats, and Gamo are located in coastal lagoons. Hitsugaura, Sokanzan and Hazutsuura are located on the shore of Matsushima Bay, while Katsura and Sabusawa are located on sheltered beaches of small islands in Matsushima Bay. Hosoura occurs on the Sanriku Ria coast. Because these intertidal flats are not privately owned or legislatively protected, no special permits were required to perform field surveys in these areas. No protected species were involved in this study.

Census Method

Taxon richness and composition of macrobenthos on intertidal flats were examined during low tide using intertidal census survey methods described below. Since surveys were done by the naked eye, animals recorded were those larger than ca. 5 mm in body size. Identification was generally made on-site using live animals. These animals were released after identification for nature conservation purpose. However, for animal taxa that could not be visually identified, several individuals were fixed with formalin and identified microscopically in the laboratory. Identification was generally made to species level. For animals that could not be identified to species level, identification was made to genus or the finest taxonomic level possible. Throughout the censuses, the same researcher confirmed taxon identifications to minimize artifacts caused by misidentification. The identified animals were categorized into three different types according to mobility and habitat: i.e., sessile epibenthic animals (SA) that lived by attaching to substrates at the sediment surface, mobile epibenthic animals (MA) that were able to move on the sediment surface, and endobenthic animas (EA) that lived in sediments. Because not all animals were identified to the species level, community characteristics were expressed as taxon richness and composition. In this study, taxon richness was defined simply as number of taxa found in a single survey.

Since the surveys before the tsunami were originally done for different objects, the census was made in different methods and date from spring to summer depending on sites. Fortunately, in some cases, different census methods were performed at the same date or site (Table 1). Therefore, census method and season were included in statistical analyses examining effects of tsunami on taxon richness. In sum, for the nine intertidal flats, total 11 census

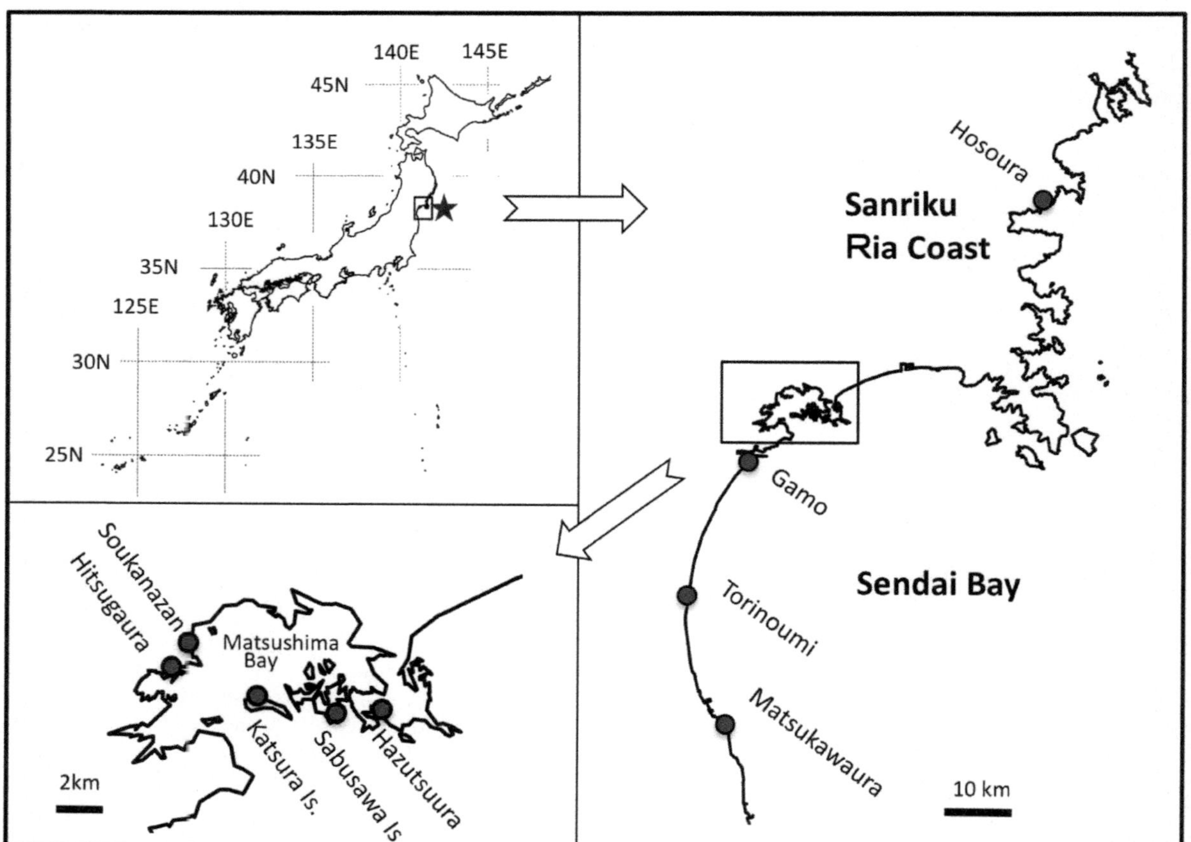

Figure 1. Maps of the Japanese Islands and the Pacific coastline of eastern Japan showing Sendai Bay and Matsushima Bay (an inner bay of Sendai Bay) with locations of the intertidal flats examined in this study (dark circles). Location of the seismic center of the Great East Japan Earthquake on 11 March 2011 is denoted by a star.

Table 1. Longitude and latitude (expressed in the decimal system) of intertidal flats examined, and date and census methods of the surveys before and after the 2011 Tohoku earthquake tsunami.

				Before the tsunami		After the tsunami	
Site	Latitude	Longitude	Surface area	Date	Census Method	Date	Census Method
Hosoura	38.69516	141.49884	2 ha	17-May-10	M1	11-Jun-11	M1
Hazutsuura	38.33216	141.14649	6 ha	28-May-02	M2	20-May-11	M1
Sabusawa Is.	38.33580	141.08732	1 ha	24-May-04	M1	18-Jun-11	M1
Katsura Is.	38.33001	141.12644	1 ha	16-Mar-09	M1	13-Aug-11	M3
Hitsugaura	38.35066	141.05203	4 ha	28-May-02	M2	28-May-11	M2
				28-Apr-09	M1	03-Jun-11	M3
Soukanzan	38.35253	141.05943	2 ha	08-Jun-09	M1	03-Jun-11	M2
						03-Jun-11	M3
Gamo	38.25574	141.01248	3 ha	22-May-04	M2	31-Jul-11	M3
						01-Aug-11	M2
Torinoumi	38.03096	140.91330	60 ha	05-Jun-08	M3	16-Jun-11	M3
				05-Jun-08	M3	16-Jun-11	M3
Matsukawaura	37.81933	140.98217	350 ha	04-Jun-08	M3	05-Jun-11	M3

surveys were done before the tsunami and 13 surveys after the tsunami. In each survey, one of the following three census methods was used.

In Method 1 (M1), one to three field researchers walked randomly for 30 to 90 min over whole area of a small intertidal flat or over an area <1 ha of large intertidal flats, and searched for epibenthic animals. Researchers often dug holes of approximately 30-cm in diameter and about 20 cm in depth with shovels and visually searched for endobenthic animals in muddy sediments. The animals found were either recorded to the finest taxonomic level and returned to the intertidal flat or were collected for later microscopic identification in the laboratory.

In Method 2 (M2), at a given intertidal flat, three transect lines, from the shore at low water level to the land, were first set. In each transect, a 5×5 m quadrant was placed near shore, midpoint and near land. Thus, a total of nine quadrants (total 0.023 ha) covering most of an intertidal flat was established. Animals found in the quadrants were collected and identified. In each quadrant, the substrate (about 20 cm in depth) was removed for 10 min by shovel and endobenthic animals were examined as in Method 1. The animals found were either recorded to the finest taxonomic level or collected for later identification as above.

In Method 3 (M3), eight to twelve researchers randomly walked for 15 min over the entire area of a given intertidal flat, and collected epibenthic animals they encountered. Each researcher also dug 15 holes of approximately 30-cm in diameter and about 20 cm in depth with shovels and visually searched endobenthic animals. The animals easily identified on-site were returned to the intertidal flat after recording, but those unidentified were taken to the laboratory for further examination as above. This method has been described in detail elsewhere [16].

Statistical Analyses

To examine differences in taxon richness before and after the tsunami, a generalized linear mixed model (GLMM) was used with the model selection procedure employing Akaike's information criteria (AIC) [17]. In the model for taxon richness, tsunami (pre-tsunami and post-tsunami surveys), season (surveys in March-May, June and July-August) and census methods (methods 1, 2 and 3) were treated as fixed effects, and intertidal flat as a random effect. The model assumed Poisson distribution and all of the explanatory variables were used as categorical data. Because the surveys were made only once in March and August, these were grouped with April-May and July, respectively. The intertidal flats examined differed greatly in surface area and location (Fig. 1, Table 1). In addition, texture, particle composition and nutrient condition of the sediments probably differed among the intertidal flats. To control for the possible influence of these site-specific conditions on taxon richness, intertidal flat was incorporated in the models as a random effect. Models were constructed separately for sessile epibenthic animals (SA), mobile epibenthic animals (MA), endobenthic animals (EA) and all animals (AA). For each animal category, a model with the lowest AIC value and models with an AIC not greater than 2 from the lowest AIC were selected as the best models [17].

Similarity (S) in taxon composition before and after the tsunami was measured by Jaccard index [18] as follows,

$$S = c/(a + b - c)$$

where a and b are taxon richness before and after the tsunami, respectively, and c is number of taxa commonly found both before and after the tsunami. Percent changes in number of taxa (T), percent taxa remaining (R) and percent new taxa (N) were calculated as follows:

$$T = b/a \times 100,$$

$$R = c/a \times 100,$$

$$N = (b - c)b \times 100.$$

If differences in taxon richness and taxonomic composition were caused mainly by the tsunami, values of these parameters were likely related to the physical impacts of the tsunami. To test this possibility, the relations of S, T, R and N with inundation height of the tsunami at each intertidal flat were examined by simple correlation analyses. The inundation height was obtained from data compiled by the 2011 Tohoku earthquake tsunami joint survey group [12] (http://www.coastal.jp/tsunami2011/). For intertidal flats where no inundation height was recorded, an average value of inundation heights (two to four) at coastal sites within 1 km radius was used.

Results and Discussion

Before the tsunami struck in 2011, a total of 172 benthic animal taxa (mainly at the species level) were recorded from nine intertidal flats (Table S1). After the tsunami, a total of 140 animal taxa were found at the same set of intertidal flats. Among these, the number of taxa commonly found both before and after the tsunami was only 106. Thus, 66 taxa were not found in the surveys immediately following the tsunami and 34 taxa were benthic animals newly recorded for the areas. On average, animal taxon richness per intertidal flat decreased by 22% after the tsunami (Fig. 2).

To clarify if the difference in number of benthic animal taxon richness before and after the tsunami was due to either differences in the census methods used or seasons when the surveys were performed or both, a GLMM was constructed with season, census method and tsunami as fixed factors and intertidal flat as a random factor. The best model for taxon richness of all benthic animals (AA) included only tsunami (timing of survey) as an explanatory variable (Table 2). The result suggested that the difference in taxon richness between pre-tsunami and post-tsunami surveys was mainly caused by the tsunami and not attributable to either census methods or seasons.

Among animals, sessile epibenthic animals and endobenthic animals decreased in taxon richness in surveys after the tsunami (Fig. 2). To explore whether these changes were attributable to the tsunami, a GLMM was also constructed for each of these types of animals. The best models for sessile and endobenthic animals included tsunami as an explanatory variable, although some of the changes were attributable to differences in census method, as shown by inclusion of census method in the best models (Table 2). For mobile epibenthic animals, differences in taxon richness were small between surveys before and after the tsunami, and only one of the two best models included tsunami as an explanatory variable. The results indicated that compared with sessile animals such as chitons (*Patelloida conulus*) and sponges (*Halichondria panicea*) and endobenthic animals such as bivalves (*Laternula marilina* and *Macoma incongrua*) and polychaetes (species of Nereididae), mobile epibenthic animals like shore crabs (*Hemigrapsus penicillatus, H. takanoi*) and hermit crabs (*Pagurus minutus*) were less affected by the tsunami.

Because pre-tsunami surveys were made in different years among the sites, one may suspect that changes in taxon composition and richness between pre-tsunami and post-tsunami surveys reflect a simple annual variation. However, if these changes were indeed caused by the 2011 Tohoku earthquake tsunami, it was most likely that the changes were greater for intertidal flats receiving larger tsunami waves. One method to gauge the magnitude of physical impacts brought on by the tsunami is to use inundation height. Due to the local geo-morphology, the height differed among sites and location on the Pacific coastline of eastern Japan [12]; it was <3 m for intertidal flats located at inner parts of Sendai Bay but >9 m for intertidal flats formed in lagoons that faced the Pacific Ocean. To examine effects of physical impacts by the tsunami, community composition similarity between pre- and post-tsunami at each flat are plotted against inundation height (Fig. 3a). It decreased significantly for intertidal flats having larger inundation heights. Significant correlation to inundation height was also detected for percent changes in number of taxa and percent taxa remaining. For some intertidal flats facing the Pacific Ocean (Matsukawaura, Hosoura and Sabusawa Is.), the similarity value was <0.2 and community composition changed drastically after the tsunami. At these intertidal flats, the number of taxa was halved by the tsunami and percent taxa remaining was <20%. However, more than 70% of taxa survived at intertidal flats in inner parts of Sendai Bay (Soukanzan and Hitsugaura) where the inundation height was <3 m (Fig 3b).

These results clearly indicated that taxon composition and richness of intertidal flat communities in Sendai Bay and on the Sanriku Ria coast were mainly changed by the 2011 Tohoku earthquake tsunami and that these changes were greater for intertidal flats that experienced tsunami waves with larger inundation heights. The fact implies that ecological impacts of the tsunami were proportional to the physical impacts as gauged by wave height. Because of their rare and extreme nature, only

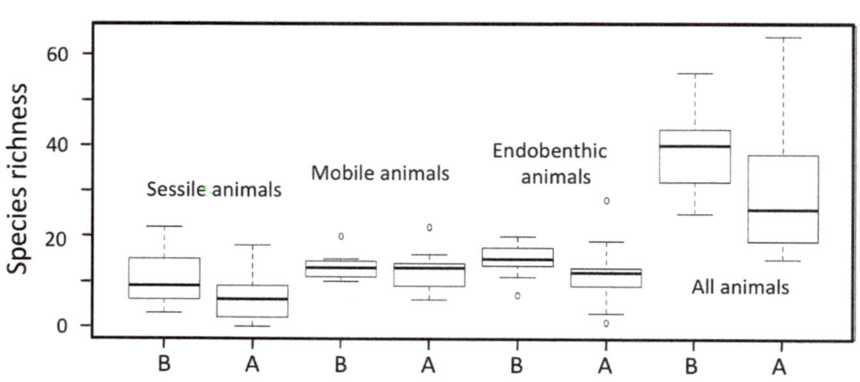

Figure 2. Box plots showing taxon richness of sessile epibenthic animals, mobile epibenthic animals, endobenthic animals and all animals before (B) and just after the tsunami (A).

Table 2. Results from the GLMM analyses showing the best models for taxon richness of sessile epibenthic animals (SA), mobile epibenthic animals (MA), endobenthic animals (EA) and all animals (AA) with fixed effects for census methods (M2 and M3), tsunami (T) and their interactions.

No.	Model		df	AIC	Chi-square	P-value
1	SA~M2+M3+T+(1|Site)		5	62.06	20.07	<0.001
2	MA~(1|Site)		2	30.05		
3	MA~T+(1|Site)		3	31.71	0.76	0.384
4	EA~M2+M3+T+M2*T+M3*T+(1|Site)		7	47.02	29.37	<0.001
5	AA~T+(1|Site)		3	69.39	12.16	<0.001
No.	**Intercept**	**Coefficient of M2**	**Coefficient of M3**	**Coefficient of T**	**Coefficient of T*M2**	**Coefficient of T*M3**
1	2.608+0.165	−0.687+0.226	−0.573+0.230	−0.314+0.160		
2	2.509+0.078					
3	2.563+0.098			−0.101+0.116		
4	2.602+0.149	0.227+0.217	0.217+0.242	−1.247+0.314	1.153+0.402	1.115+0.379
5	3.660+0.086			−0.245+0.070		

The models used site as a random effect (1|Site) controlling for the possible influence of site-specific conditions. Results of Chi-square tests against the null model and coefficients of the best models with the standard deviations are shown.

a limited number of studies have assessed immediate impacts of tsunamis on benthic communities [8–11]. Comparison of these studies suggests that ecological impacts of tsunamis vary depending

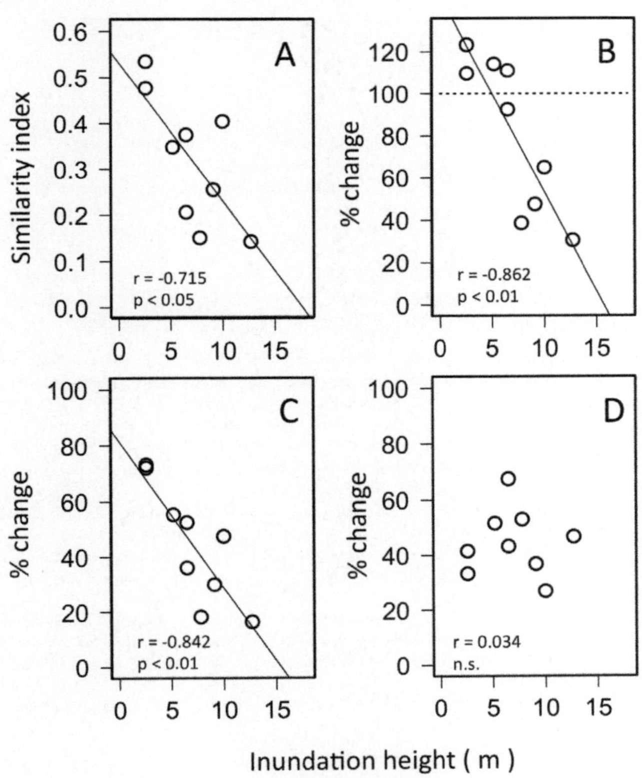

Figure 3. Jaccard similarity (see text) in taxon composition before and after the tsunami (A), percent changes in number of taxa (B), percent taxa remaining (C) and percent new taxa (D) plotted against inundation height of the tsunami. Correlation coefficients with significance levels are shown in each panel.

on physical impacts. For example, with the 2007 Peru earthquake (M_W 8.0), taxon composition of the benthic community of the soft bottom beach of Paracas Bay was largely unchanged after a tsunami with a run-up height of 2–3 m, although the abundance of several species were changed [11]. In general, run-up height of a tsunami is larger than the inundation height used in the present study. Thus, physical impacts of the tsunami reported for the beach of Paracas Bay would be much smaller than those observed in our study. Whanpetch et al. [9] examined impacts of the tsunami generated by the Sumatra–Andaman earthquake (M_w 9.1–9.3) in 2004 on seagrass beds and benthic communities therein and showed that changes in community structures after the tsunami were larger at the site facing open ocean, which likely experienced larger physical forces of the tsunami waves than more sheltered sites. This study showed that inundation or wave height is a useful gauge when immediate impacts of tsunamis on benthic communities are to be compared across different sites and events.

In the present study, taxon richness of mobile epibenthic animals was less affected by the tsunami compared to sessile and endobenthic animals. With the 2010 Chilean earthquake (M_W 8.8), abundance of mobile crustacean species on some sandy beaches was either unaffected or increased within two months of the tsunami [10]. Similarly, in mangrove swamps on the Indian Ocean coast of Thailand, abundance of endobenthic animals, like bivalves, decreased drastically after the tsunami generated by the Sumatra–Andaman earthquake (M_w 9.1–9.3) but those of mobile epibenthic animals, like crustaceans and snails, did not [8]. The results of these previous studies [8,10] are consistent with those of our study, implying that mobile epibenthic animals have abilities to persist against large physical disturbances or return promptly within a few months from unknown refuge habitats.

In estuaries with muddy bottoms, opportunistic polychaeta often increase in abundance right after large disturbances. In the Neuse River Estuary of North Carolina, for example, an opportunistic polychaete species showed a notable increase in abundance within two months following extensive damage by multiple hurricanes in 1999, although other native macrobenthic abundance declined. Similarly, Kanaya et al. [15] found that the abundance of endobenthic polycheates, such as *Hediste diadroma* and *Pseudopolydora* cf. *kempi*, increased on the Gamo intertidal flat five months after the 2011 Tohoku earthquake tsunami. These are known to be opportunistic species [19]. We also found these opportunistic polychaetes at several but not all intertidal flats examined after the tsunami. This implies that physical disturbance induced by the 2011 Tohoku erathquack tsunami favored opportunistic species at some intertidal flats, including Gamo.

Worth noting from these observation is that the effects of tsunamis are not limited to a species "take-away" action from intertidal flats. After the tsunami struck, animals that were not recorded before the tsunami were found at all the intertidal flats. For intertidal flats in the inner part of Sendai Bay, the number of taxa after the tsunami was larger than before the tsunami (>100%: Fig. 3B). The percentage of new taxa was not correlated with tsunami inundation height (Fig. 3D). These animals included species of sea cucumber (*Stichopus japonica*), sea star (*Asterina pectinifera*) and swimming crab (*Charybdis japonica*). Because most of these were originally dwellers in lower tidal zones, they were temporally brought onto the intertidal flats by the tsunami. We expect that these animal taxa will disappear from the intertidal flats shortly.

Recent studies suggest that large, rare and extreme events like unusual floods, hurricanes and tsunamis change the physical and chemical conditions of habitats [2,3,4,11,15,20] and may provide an opportunity for invasive species to colonize areas by transporting their propagules and depressing the abundance of native species [1]. Moreover, the events may push a community to an alternative state with a different taxon composition [21]. Large, rare and extreme events, therefore, are likely important processes in shaping and altering local community structures [1,2]. This implies that events like a tsunami generate not only immediate impacts but also long-lasting impacts on local communities. Among animal taxa recorded in the pre-tsunami surveys of the intertidal flats examined, 66 taxa were not found in the post-tsunami surveys. These animals might have been present in low numbers and been underway in re-establishing their populations. However, the tsunami may have created habitat conditions that impaired recovery of some native species. If this was the case, the 2011 Tohoku earthquake tsunami could result in long-lasting impacts on the intertidal flat communities of Sendai Bay and the Sanriku Ria coast. Long-term monitoring is essential to examine community-level recovery processes and thus to understand implications of tsunamis on coastal ecosystems.

Supporting Information

Table S1 Taxon list and results of census surveys used in this study.

Acknowledgments

We thank T. Uchino, G. Kanaya, K. Kinoshita, participants of the citizen research program by Erathwatch Japan and participants of the undergraduate field course of animal ecology, Tohoku University, for their field supports and M. Kyle and R. D. Roberts for suggestions on improving the manuscript.

Author Contributions

Conceived and designed the experiments: JU TS. Performed the experiments: JU TS TN WM. Analyzed the data: JU TS TN. Wrote the paper: JU.

References

1. Diez JM, D'Antonio CM, Dukes JS, Grosholz ED, Olden JD, et al. (2012) Will extreme climate events facilitate biological invasions? Front Ecol Environ 10: 249–257.
2. Thibault KM, Brown JH (2008) Impact of an extreme climatic event on community assembly. Proc. Natl Acad Sci 105: 3410–3415.
3. Mallin MA, Posey MH, Shank GC, McIver MR, Ensign SH, et al. (1999) Hurricane effects on water quality and benthos in the Cape Fear watershed: Natural and anthropogenic impacts. Ecol Appl 9: 350–362.
4. Balthis WL, Hyland JL, Bearden DW (2006) Ecosystem responses to extreme natural events: impacts of three sequential hurricanes in fall 1999 on sediment quality and condition of benthic fauna in the Neuse River Estuary, North Carolina. Environ Monit Assess 119: 367–389.
5. Schlacher TA, Holzheimer A, Stevents T, Rissik D (2011) Impacts of the 'Pacific Adventurer' oil spill on the macrobenthos of subtropical sandy beaches. Estuar Coast 34: 937–949.
6. Krishnankutty N (2006) Effects of 2004 tsunami on marine ecosystems – a perspective from the concept of disturbance. Current Biol 90: 772–773.
7. Bourrouilh-Le J, Beck C, Gorsline DS (2007) Catastrophic events (hurricanes, tsunami and others) and their sedimentary records: introductory notes and new concepts for shallow water deposits. Sediment Geol 199: 1–2.
8. Fujioka Y, Tabuchi R, Hirata Y, Yoneda R, Patanaponpaiboon P, et al. (2008) Disturbance and recovery of mangrove forests and macrobenthic communities in Andaman Sea, Thailand following the Indian Ocean tsunami. Proceedings of the 11th International Coral Reef Symposium, 1225–1229.
9. Whanpetch N, Nakaoka M, Mukai H, Suzuki T, Nojima S, et al. (2010) Temporal changes in benthic communities of seagrass beds impacted by a tsunami in the Andaman Sea, Thailand. Estuar Coastal Shelf Sci 87: 246–252.
10. Jaramillo E, Dugan JE, Hubbard DM, Melnick D, Manzano M, et al. (2012) Ecological implications of extreme events: footprints of the 2010 earthquake along the Chilean Coast. PLoS ONE 7(5): e35348. doi:10.1371/journal.pone.0035348.
11. Lomovasky BJ, Firstater FN, Salazar AG, Mendo J, Iribarne OO (2011) Macro benthic community assemblage before and after the 2007 tsunami and earthquake at Paracas Bay, Peru. J Sea Res 65: 205–212.
12. Mori N, Takahashi T, the 2011 Tohoku earthquake tsunami joint survey group (2012) Nationwide post event survey and analysis of the 2011 Tohoku Earthquake Tsunami. Coast Engng J. 54: 1250001. DOI 10.1142/S0578563412500015.
13. Cyranoski D (2012) Rebuilding Japan: After the deluge. Nature 483: 141–143.
14. Okada N, Tao Y, Kajitani Y, Shi P, Tatano H (2011) The 2011 eastern Japan great earthquake disaster: Overview and comments. Int J Disaster Risk Sci 2: 34–42.
15. Kanaya G, Suzuki T, Maki H, Nakamura Y, Miyajima Y, et al. (2012) Effects of the 2011 tsunami on topography, vegetation, and macrobenthic fauna in Gamo Lagoon, Japan [in Japanese with English Abstract]. Jpn J Benthology 67: 20–32.
16. Suzuki T, Sasaki M (2010) Civil procedure for researching benthic invertebrate animals inhabiting tidal flats in eastern Japan. Plankton Benthos Res 5: 221–230.
17. Burnham KP, Anderson DR (2002) Model Selection and Multimodel Inference. New York: Springer. 488p.
18. Legendre P, Legendre L (1998) Numerical ecology. Amsterdam: Elsevier. 853.
19. Kanaya G, Suzuki T, Kikuchi E (2011) Spatio-temporal variations in macrozoobenthic assemblage structures in a river-affected lagoon (Idoura Lagoon, Sendai Bay, Japan): influences of freshwater inflow. Estuar Coast Shelf Sci 92: 169–179.
20. Noda A, Katayama H, Sagayama T, Suga K, Uchida Y, et al. (2007) Evaluation of tsunami impacts on shallow marine sediments: an example from the tsunami caused by the 2003 Tokachi-oki earthquake, northern Japan. Sediment Geol 200: 314–327.
21. Scheffer M, Carpenter S, Foley JA, Folke C, Walker B (2001) Catastrophic shifts in ecosystems. Nature 413: 591–596.

20

Decadal Changes in the World's Coastal Latitudinal Temperature Gradients

Hannes Baumann[1]*, Owen Doherty[2]

1 School of Marine and Atmospheric Sciences, Stony Brook University, Stony Brook, New York, United States of America, 2 Scripps Institution of Oceanography, University of California San Diego, La Jolla, California, United States of America

Abstract

Most of the world's living marine resources inhabit coastal environments, where average thermal conditions change predictably with latitude. These coastal latitudinal temperature gradients (CLTG) coincide with important ecological clines,e.g., in marine species diversity or adaptive genetic variations, but how tightly thermal and ecological gradients are linked remains unclear. A first step is to consistently characterize the world's CLTGs. We extracted coastal cells from a global $1°\times1°$ dataset of weekly sea surface temperatures (SST, 1982–2012) to quantify spatial and temporal variability of the world's 11 major CLTGs. Gradient strength, i.e., the slope of the linear mean-SST/latitude relationship, varied 3-fold between the steepest (North-American Atlantic and Asian Pacific gradients: −0.91°C and −0.68°C lat^{-1}, respectively) and weakest CLTGs (African Indian Ocean and the South- and North-American Pacific gradients: −0.28, −0.29, −0.32°C lat^{-1}, respectively). Analyzing CLTG strength by year revealed that seven gradients have weakened by 3–10% over the past three decades due to increased warming at high compared to low latitudes. Almost the entire South-American Pacific gradient (6–47°S), however, has considerably cooled over the study period (−0.3 to −1.7°C, 31 years), and the substantial weakening of the North-American Atlantic gradient (−10%) was due to warming at high latitudes (42–60°N, +0.8 to +1.6°C,31 years) and significant mid-latitude cooling (Florida to Cape Hatteras 26–35°N, −0.5 to −2.2°C, 31 years). Average SST trends rarely resulted from uniform shifts throughout the year; instead individual seasonal warming or cooling patterns elicited the observed changes in annual means. This is consistent with our finding of increased seasonality (i.e., summer-winter SST amplitude) in three quarters of all coastal cells (331 of 433). Our study highlights the regionally variable footprint of global climate change, while emphasizing ecological implications of changing CLTGs, which are likely driving observed spatial and temporal clines in coastal marine life.

Editor: Inés Álvarez, University of Vigo, Spain

Funding: The authors have no support or funding to report.

Competing Interests: The authors have declared that no competing interests exist.

* E-mail: hannes.baumann@stonybrook.edu

Introduction

Coastal and continental shelf regions are vitally important to humans [1]. They constitute the most diverse and productive parts of the world's oceans and provide an estimated 43% of all ecosystem services [2]. Hence, understanding major drivers of change in coastal ecosystems, both in space and time, remains a long-standing scientific goal that has become more urgent due to unfolding global climate change [3]. Given that temperature is the central abiotic variable affecting marine life from the cell to the ecosystem level [4,5], studies of coastal ecological change are particularly interested in spatial and temporal temperature patterns along coastal areas.

One important large-scale pattern is the common decrease in temperature with latitude. Although ubiquitous on earth, latitudinal temperature gradients differ greatly between terrestrial and marine systems or between the open ocean and coastal areas as a result of regional wind and cloud patterns, alongshore currents and coastal upwelling [6–8]. Latitudinal temperature patterns coincide with – and potentially drive – important ecological clines such as the common latitudinal decrease in species diversity [9] or many morphological, physiological and life history traits that are linked directly or indirectly to temperature (e.g., body size [10,11], growth and development rates [12], life span [13], fecundity [14], recruitment [15], and vertebral number [16,17]).

However, despite their potential ecological significance, coastal latitudinal temperature gradients (CLTG) are rarely implicated in studies of large-scale ecological change, perhaps because they have yet to be consistently quantified. While temperature patterns and trends in average open oceans have been a scientific focus for decades [7,18–20], the world's coastal temperature patterns have received considerably less attention [21]. Recently, Lima and Wethey [6] used sea surface temperature (SST) data of the highest available spatial and temporal resolution (1/4 degree daily SST analysis) to document the heterogeneity in fine-scale coastal SST changes as well as trends in the number of extreme hot or cold days. Here we explore how regional SST patterns determine the diversity of CLTGs on larger spatial and temporal scales, which is useful for potential analyses of spatial trait patterns in the ocean or ecological consequences of temporal changes over the past three decades.

We used SSTs from a global data set as a proxy for ambient thermal conditions in the coastal waters. Selecting only the coastal cells along the world's major north-south oriented continental coastlines enabled us to define and comparatively characterize 11

CLTGs. Apart from SST itself, our spatio-temporal analyses included a second ecologically relevant variable, seasonality, which describes the magnitude of seasonal temperature fluctuations. For many coastal organisms, seasonality plays an important role in determining the length of growing and spawning seasons [12], the severity of winter mortality [22] or the degree of temperature-dependent sex determination [23].

The two goals of this study were (1) to quantify the regional variability in long-term average strength and seasonality among the world's major CLTGs and (2) to examine how these gradients have themselves changed over the past three decades. Specifically, we hypothesized that greater warming at high compared to low latitudes [3,6,20] will lead to a general weakening in gradient strengths, accompanied by trends towards more extreme seasons [24]. For clarity, the scope of these analyses was limited to large scale patterns in space and time as relevant to large-scale ecological clines, while purposely excluding analyses of gradient features on finer spatial and temporal scales.

Methods

CLTGs were extracted from the NOAA Optimum Interpolation (OI) SST product (V2), a 31 year dataset (1982–2012) of global SSTs with a $1° \times 1°$ spatial and one week temporal resolution[25]. SSTs comprise satellite AVHRR data (Advanced Very High Resolution Radiometer) blended with *in situ* observations (International Comprehensive Ocean-Atmosphere Data Set, ICOADS) using optimal interpolation algorithms described in Reynolds et al. [26] and Smith et al. [27]. From the global dataset we extracted SSTs only from coastal cells, defined as 'sea' cells immediately adjacent to 'land' in the datasets land-sea mask (Fig.1a). To be consistent, we excluded major islands from continental coastlines (e.g., British Isles, Japan, Madagascar) and omitted latitudes that would have led to inclusion of large embayments, gulf or marginal seas (e.g., Baltic, Gulf of Mexico, Red Sea,). No analysis was performed over land so that sea cells, even if partially extending over land, only contain SST data. Lima and Wethey [6] demonstrated that the degree of land 'contamination' of coastal SST cells did not bias SST trends in a similar dataset. We then further selected only those cells that belonged to the world's major north-south oriented continental coastlines (e.g., Fig.1b), yielding a total of 433 cells from 11 gradients: the North-American Pacific and Atlantic gradients (NAm_{Pac}: 27–60°N, NAm_{Atl}: 26–60°N), the South-American Pacific and Atlantic gradients (SAm_{Pac}: 8°N–55°S, SAm_{Atl}: 8°N–55°S), the North- and South-African Atlantic gradients (NAf_{Atl}: 5–35°N, SAf_{Atl}, 1°N–34°S), the African Indian Ocean gradient (Af_{Ind}: 23°N–34°S), the European Atlantic gradient (Eu_{Atl}: 37–70°N), the Asian Pacific gradient (As_{Pac}: 22–61°N), and the Australian Indian Ocean and Pacific gradients (Au_{Ind}: 12–35°S, Au_{Pac}: 11–39°S).

To characterize average gradient strengths, we calculated the long-term mean SST for each latitude (31 years $\times$52 weeks = 1,612 data points) and used the gradient-specific slope of the linear regression of mean SST against latitude. In addition, we calculated the long-term SST minimum and maximum (average of the 31 annual SST minima and maxima) for each latitude of each CLTG, and used the difference SST_{max} - SST_{min} as a measure of seasonality. To examine how gradient strength changed over time, annual gradient slopes (i.e., from linear regressions of annual mean SSTs against latitude) were linearly regressed against year. The total change in gradient strength was then expressed as the percentage difference between the predicted strength at the beginning (1982) and end (2012) of our 31 year time series. In addition, we characterized strength and temporal variability of partial sections of some gradients, where regional rates of latitudinal SST change differed markedly from the average gradient strength (Table 1).

Next we examined two ecologically relevant questions separately for each latitude and gradient: (1) did it get warmer, and (2) did seasons become more extreme over the past 31 years? For each latitude, thus, we linearly regressed (1) mean annual SST and (2) seasonality (annual $SST_{max-min}$) against year (Fig.2). In a second step, the slopes and *p*-values of these regressions were plotted against latitude to identify potential areas of significant ($p<0.05$) warming or cooling within each gradient (Fig.3). We then correlated the latitude-specific regression slopes of both variables to latitude in order to evaluate the hypotheses that higher latitudes have warmed more and increased more in seasonality relative to lower latitudes of a gradient (Fig.3).

Last, we examined how changes in average gradients over the past three decades (i.e., based on annual mean SST) might have resulted from seasonal changes in temperature patterns. For example, if a range of latitudes showed a consistent increase (or decrease) in mean annual SST, we asked whether this trend could have been the due to warming (or cooling) only during some seasons of the year. We therefore conducted separate linear regressions of SST on year for each gradient, latitude, and week (433 latitudes $\times$52 weeks = 22,516 regressions) and used the slopes and their *p*-values to construct contour plots of predicted latitude- and week-specific SST changes over the past 31 years (contours derived by Kriging, grid spacing for *latitude, week* = 0.5, Golden Software Surfer® V.8, Fig.4). All regression analyses were conducted using SPSS® statistical software (V19 IBM Inc.) and used unadjusted significance levels of $p = 0.05$. This is justifiable by our study's broad focus on consistent patterns over large spatial and temporal scales, which are unaffected by 5% of type I errors [i.e., chance detection of significant relationships, 28]. We assessed, however, the level of temporal autocorrelation in our time-series by calculating the lag-1 autocorrelation coefficients (SPSS® ACF procedure) of detrended time series of (i) gradient strength, latitude-specific trends in (ii) average SST, (iii) seasonality and (iv) weekly SSTs. Time series of gradient strength were not statistically significantly autocorrelated ($p>0.05$), while SST and seasonality time-series showed low levels of temporal autocorrelation ($p<0.05$ for 12% and 5% of the 433 cells, respectively) that did not necessitate a correction procedure.

Results

Average gradient characteristics

We found that average slopes vary more than 3-fold among the world's coastal latitudinal temperature gradients (Fig.3, Table 1). The steepest CLTGs are found along the east coasts of the North-American (NAmAtl = $-0.91°C\ lat^{-1}$) and Asian continents (As_{Pac} = $-0.60°C\ lat^{-1}$), while the weakest CLTGs occur along the South-African Indian Ocean ($Af_{Ind(1)}$ = $-0.28°C\ lat^{-1}$) and the South- and North-American Pacific coasts ($-0.29°C\ lat^{-1}$ and $-0.32°C\ lat^{-1}$, respectively). CLTGs of intermediate strength include the European, South-American and South-African Atlantic gradients ($-0.34°C\ lat^{-1}$, $-0.37°C\ lat^{-1}$ and $-0.42°C\ lat^{-1}$, respectively), as well as the Australian Indian Ocean gradient ($-0.46°C\ lat^{-1}$). Some CLTGs could reasonably be described by linear relationships across their entire latitudinal range (NAmPac, EuAtl, AuInd, AuPac), but most contained regions where the latitudinal SST decline differed notably from the overall average (Fig.3). Along the NAmAtl, for example, SST decreases with $-2.55°C\ lat^{-1}$ between 33–39°N (by far the steepest CLTG in the world), which is followed to the north by a rate of $-0.76°C\ lat^{-1}$

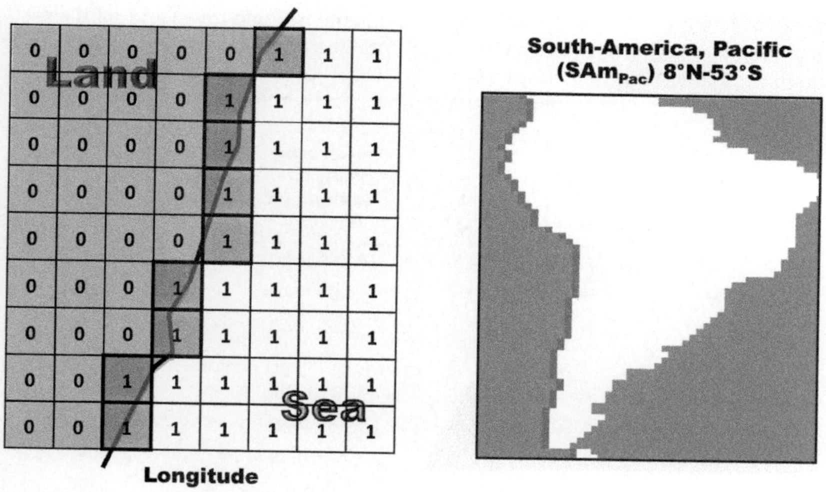

Figure 1. Extracting coastal latitudinal SST gradients. A: the 1°×1° land-see mask of the NOAA OISST (V2) dataset was used to select at each latitude and gradient the 'sea' cells ("1", white) closest to land ("0", dark grey). **B**: example of extracted cells (black) to characterize the South-American Pacific gradient (SAm$_{Pac}$).

(40–57°N). Similarly, sections of the SAmPac and the SAfAtl gradients have slopes twice as strong as their entire range average (SAmPac(1) 3°N-8°S = −0.82°C lat^{-1}; SAfAtl(1) 12-23°S = −0.98°C lat^{-1}). Furthermore, the partial SAmPac(1) comprises the world's strongest thermal decline in tropical latitudes (Table 1). In contrast, average SST along the African east coast (AfInd) is almost invariant along a more than 4,000 km long stretch of tropical coastline (SST23◦N-15◦S = 26.0−27.6°C), followed southward by the weak AfInd(1) gradient (15–34°S: −0.28°C lat^{-1}). Thus, partial CLTGs are even more variable than entire CLTGs and span almost a magnitude of slope differences (Table 1, Fig.3).

Average seasonality ($SST_{max-min}$) also varies strongly between CLTGsand latitudes (Fig.3). Lowest values generally occur in tropical and subpolar regions (<8°C), while seasonality maxima are located at subtropical and temperate latitudes (25–55°:10–26°C), with particularly high values in northern hemisphere gradients (NAm_{Atl} = 24.4$°C_{38.5°N}$; As_{Pac} = 25.6$°C_{39.5°N}$). The seasonality peak of the Eu_{Atl} gradient occurs at comparatively high latitudes (15.9$°C_{53.5°N}$), whereas the NAm_{Pac} gradient shows no peak at all but a depression at mid-latitudes (4.1$°C_{39.5°N}$) consistent with the Californian coastal upwelling system. Similar seasonality depressions are also apparent along two other major coastal upwelling areas, i.e., Canary (NAf_{Atl}) and Benguela coastal upwelling regions (SAf_{Atl})(Fig.3).

Spatio-temporal changes in gradient strength & seasonality

Temporal SST trends differed markedly between gradients, latitudes, and seasons of the year, thereby causing complex decadal changes in most CLTGs (Table 1, Fig.3). Out of 11 CLTGs, seven have weakened over the study period (i.e., their absolute latitudinal SST slopes decreased), while four have become stronger (i.e., slopes increased). All northern hemisphere CLTGs except the NAf_{Atl}, have decreased in strength by 3–10%, in particular the North-American Atlantic and Pacific gradients (NAm_{Atl}: −9.6%, $p<0.001$; NAm_{Pac}: −8.1%, $p=0.041$), the European Atlantic gradient (Eu_{Atl}: −5.3%, $p=0.057$), and the Asian Pacific gradient (As_{Pac}: −3.2%, $p=0.086$). In the southern hemisphere, both Australian gradients did not show statistically significant changes (Au_{Ind}: −5.7%, $p=0.226$ Au_{Pac}: −4.6%, $p=0.173$), whereas the South-African Indian Ocean gradient has lost 7% of its strength over the past 31 years ($Af_{Ind(1)}$ $p=0.022$). In contrast, significant increases in gradient strength were detected for the South-African and South-American Atlantic gradients (SAf_{Atl}: +6.2%, $p=0.030$; SAm_{Atl}: +3.3%, $p=0.007$). The steep latitudinal decline in tropical SSTs along the South-American Pacific gradient has strengthened further over the past three decades ($SAm_{Pac(1)}$: +22%, $p=0.016$).

Overall, two thirds of the examined 433 coastal cells showed positive SST trends over time (n = 286), approximately half of which were statistically significant ($p<0.05$, n = 141), while 147 cells had negative SST trends, 55 of which were significant (Fig.3). Regression slopes ranged from −0.07°C year^{-1} ($NAm_{Atl\ 31.5°N}$) to +0.09°C year^{-1} ($SAm_{Atl\ 35.5°S}$) corresponding to a 31 year change in local SST of −2.21°C to +2.67°C, respectively. Similarly, trends towards more extreme seasons (i.e., increasing annual SST_{max}-SST_{min}) were found at more than three quarters of all latitudes (n = 331, $p<0.05$ for 113 cells).

AMERICA (Figs.3A–C, 4A–D. Most cells of the ***North-American Pacific gradient*** (NAm_{Pac}) showed weak cooling trends (n = 27, $p<0.05$ for 4 cells) with the degree of cooling being latitude-dependent. Maximum cooling was observed in the gradient's lower latitudes, thereby causing a significant decrease in gradient strength over the study period (Fig.3B, Table 1). Seasonality trends, albeit insignificant, changed significantly with increasing latitude (Fig.3C). The seasonal decomposition of the average SST trends revealed that high latitudes (49–55°N) have actually warmed significantly throughout the second half of the year (weeks 35–52), but this effect has been partially offset by cooling trends affecting the late boreal winter to early summer weeks (Fig. 4A). Of the ***North-American Atlantic gradient***'s (NAm_{Atl}) 34 coastal cells, 22 showed SST increases ($p<0.05$ for 18 cells) and 12 showed SST decreases over the study period ($p<0.05$ for 9 cells). Warming predominantly occurred at high latitudes north of 42°N (42–60°N, +0.8 to +1.6°C in 31 years), whereas coastal cooling affected all lower latitudes south of 38°N (Florida to Cape Hatteras 26–35°N, −0.5 to −2.2°C in 31 years), therefore causing a 10% decrease in gradient strength over the study period (Fig.3B). Seasonality increased throughout the entire gradient and

Table 1. Strength and decadal changes of the world's coastal latitudinal temperature gradients (shaded rows) and partial sections (white rows), including range and number of analyzed latitudes (N), long-term mean SST minima and maxima across each gradient, average gradient strength and estimated change in gradient strength from 1982–2012 (%).

	Bordering ocean	Gradient	Latitude range[1]	N	Max/min SST (°C)	Mean (s.e.) gradient strength (°C °lat^{-1})	30 year change (%)	*p*
North America	Pacific	(1) NAm_{Pac}	27–60°N	33	22.5/5.1	**−0.32 (0.02)**	**−8.1**	**0.041***
	Atlantic	(2) NAm_{Atl}	26–60°N	34	29.7/−1.8	**−0.91 (0.03)**	**−9.6**	**<0.001***
		$NAm_{Atl(1)}$	33–39°N	6	28.8/0.1	**−2.55 (0.21)**	**−13.4**	**0.001***
		$NAm_{Atl(2)}$	40–57°N	17	23.5/−1.8	−0.76 (0.03)	−4.1	0.301
South America	Pacific	(3) SAm_{Pac}	8°N–55°S	63	29.0/5.2	−0.29 (0.01)	*+0.7*	*0.820*
		$SAm_{Pac(1)}$	3°N–8°S	11	28.4/16.6	**−0.82 (0.04)**	***+21.9***	***0.016****
	Atlantic	(4) SAm_{Atl}	8°N–55°S	63	29.6/4.6	**−0.37 (0.02)**	***+3.3***	***0.007****
		$SAm_{Atl(1)}$	17–55°S	38	28.4/4.6	−0.55 (0.01)	*+2.3*	*0.200*
Europe	Atlantic	(5) Eu_{Atl}	37–70°N	32	22.3/3.2	−0.34 (0.01)	−5.3	0.057
Asia	Pacific	(6) As_{Pac}	22–61°N	39	29.4/−1.7	−0.60 (0.02)	−3.2	0.086
		$As_{Pac(1)}$	22–55°N	33	29.4/−1.7	−0.68 (0.01)	−3.2	0.051
Africa	Atlantic	(7) NAf_{Atl}	5–35°N	30	29.6/16.3	−0.36 (0.02)	*+5.5*	*0.059*
		$NAf_{Atl(1)}$	7–23°N	16	29.5/17.6	−0.57 (0.02)	−0.5	0.865
		(8) SAf_{Atl}	1°N–34°S	35	29.5/13.1	**−0.42 (0.03)**	***+6.2***	***0.030****
		$SAf_{Atl(1)}$	12–23°S	11	29.3/13.1	**−0.98 (0.06)**	***+8.5***	***0.042****
	Indian Ocean	(9) Af_{Ind}	23°N–34°S	53	30.4/19.2			
		$Af_{Ind(1)}$	15–34°S	19	29.9/19.2	**−0.28 (0.02)**	**−7.0**	**0.022***
Australia	Indian Ocean	(10) AuInd	12–35°S	23	31.4/16.8	−0.46 (0.01)	−5.7	0.226
	Pacific	(11) AuPac	11–39°S	28	29.8/12.3	−0.35 (0.03)	−4.6	0.173

Significant changes ($p<0.05$) are in bold, weakening and strengthening gradients are in normal and italicized fonts, respectively.
[1]latitudes in the dataset are given as cell mid-points; hence a latitudinal range of e.g. 27–60°N includes 33 coastal cells from 27.5–59.5°N.

significantly from north to south (Fig.3C). The seasonal decomposition of SST trends revealed that low latitudes (26–35°N) cooled primarily because boreal winter and spring temperatures decreased strongly, whereas fall temperatures trended slightly upward. Conversely, high latitude warming along the NAm_{Atl} (>42°N) affected the late boreal summer, fall and winter weeks, but not those in spring (Fig.4B). The ***South-American Pacific gradient*** (SAm_{Pac}) encompassed 63 coastal cells, of which only 10 had positive, non-significant SST trends, whereas 53 latitudes showed pronounced cooling ($p<0.05$ for 34 cells) at maximum rates of approximately $-0.06°C$ year^{-1} (Fig. 3B). These cooling trends occurred consistently over a large contiguous coastline between 6–47°S (cooling rates of -0.3 to $-1.7°C$ in 31 years, Fig.3C), and resulted from relatively uniform SST decreases throughout the year with the strongest changes seen in austral spring and summer (i.e., weeks 1–6, 35–52, Fig. 4C). However, linear analyses cannot detect potential changes SST patterns due to cyclical changes in El Nino strength and frequency. Neither SST nor seasonality trends showed significant correlations to latitude along the SAm_{Pac} (Fig.3B,C). In contrast, 46 out of 63 cells of the ***South-American Atlantic gradient*** (SAm_{Atl}) showed warming trends ($p<0.05$ for 27 cells). Significant warming has occurred along two main clusters located between subtropical 34–41°S and equatorial 8°N–9°S latitudes, while coastal cells south of 45°S showed no or weak negative SST trends, explaining the weak but significant strengthening of the SAm_{Atl} over the study period (Table 1, Fig. 3B). Warming around the equatorial latitudes was caused by uniform trends across all seasons (Fig. 4D), whereas most latitudes south of 24°S showed increased seasonality (Fig. 3C) caused by warming patterns during weeks 5–35 (austral fall-winter) but cooling trends later in the year (austral spring-summer, Fig. 4D).

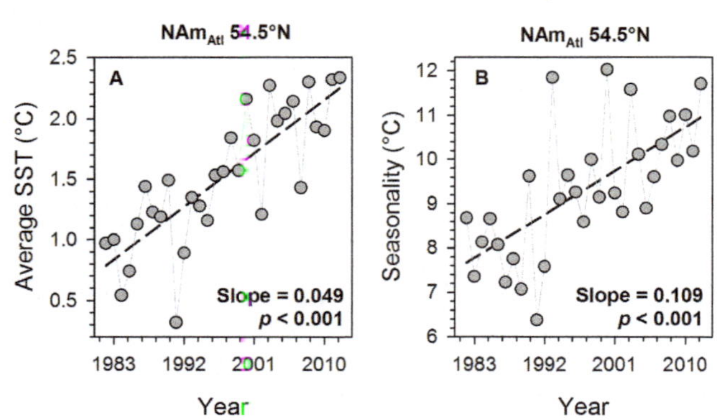

Figure 2. Example of the two linear regressions performed at each latitude, as shown here for 54.5°N of the North American Atlantic gradient (NAm_{Atl}). A: annual SST against year B: seasonality against year.

EURASIA (Figs.3D–F, 4H,I). All of the 32 coastal cells of the ***European Atlantic gradient*** (Eu_{Atl}) showed warming trends over the study period ($p<0.05$ for 29 cells). SST trends were significantly correlated to latitude; lower latitudes warmed about

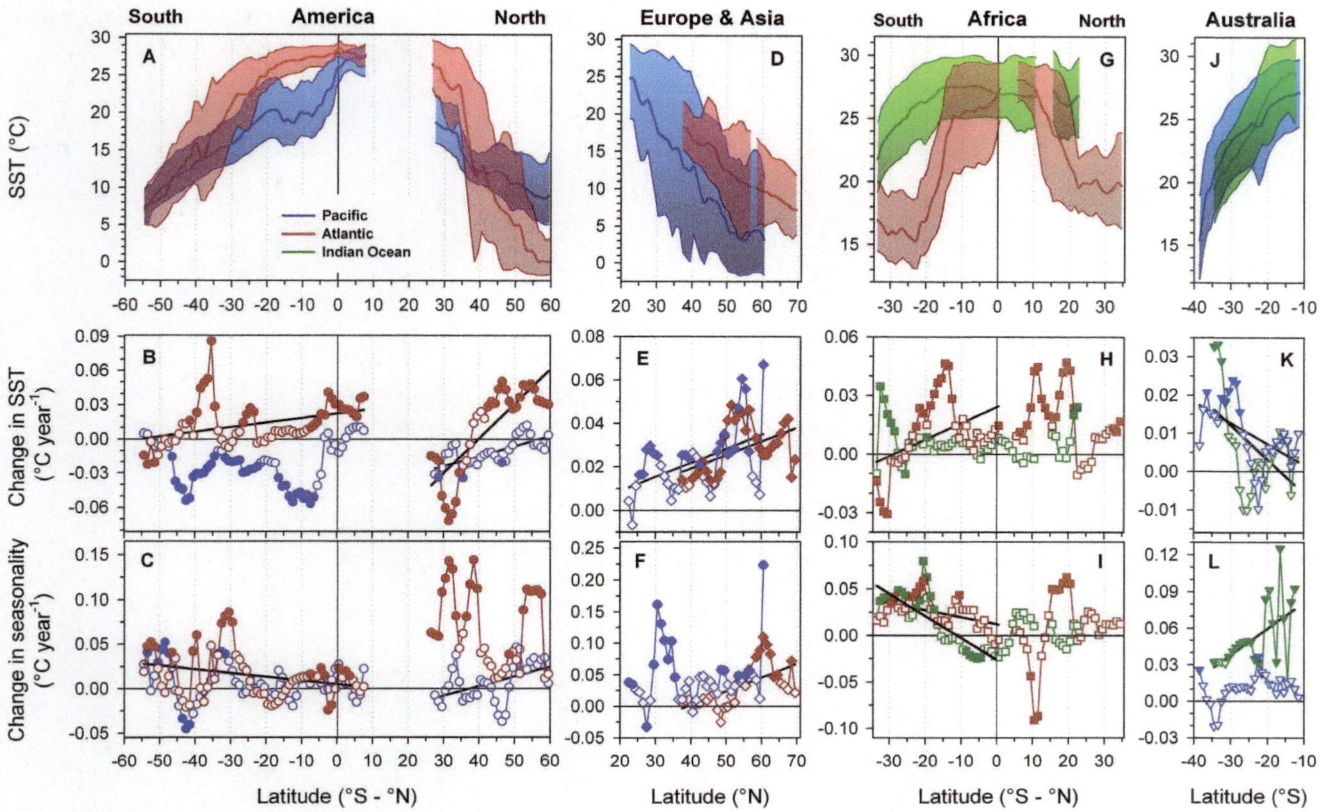

Figure 3.Spatio-temporal patterns of the world's major coastal latitudinal temperature gradients. Upper panel row (A, D, G, J): latitudinal patterns in long-term average SST (thick line), average SST minima and maxima (thin lines) and seasonality (shaded area). Middle panel row (B, E, H, K): latitudinal patterns in 31 year SST trends, with positive and negative slopes indicating warming and cooling, respectively, at specific latitudes.Bottom panel row (C, F, I, L): latitudinal patterns in 31 year trends in seasonality (SST_{max}-SST_{min}). Empty and filled symbols depict non-significant and significant linear regressions, respectively ($p<0.05$). Significant ($p<0.05$) latitudinal trends in middle and bottom panel rows are indicated by solid regression lines (black). Atlantic Ocean gradients: red lines, shades and symbols; Pacific Ocean gradients: blue lines, shades and symbols; Indian Ocean gradients: green lines, shades and symbols. Positive and negative latitudes refer to the northern and southern hemisphere respectively.

half as fast as higher latitudes (mean SST change 37–49°N = +0.015°C $year^{-1}$, 49–70°N = +0.035°C $year^{-1}$, Fig.3E), resulting in a loss of gradient strength. Seasonality has increased over the past 31 years mostly at the high latitudes of the Eu_{Atl} gradient (Fig.3F). These trends were attributable to seasonal differences in decadal SST change across the latitudinal range, with warming confined to late boreal summer, fall and winter weeks, partially offset by cooling trends during boreal spring and early summer (Fig.4H). Along the ***Asian Pacific gradient*** (As_{Pac}), 38 out of 39 coastal cells exhibited positive SST trends over the study period ($p<0.05$ for 21 cells). Similar to the Eu_{Atl} gradient, warming trends along the As_{Pac} gradient strengthened significantly with increasing latitude (Fig.3E). This, however, was mainly due to four latitudes that have warmed much faster (+0.06°C $year^{-1}$ for 53–56°N and 60–61°N) than all others (~ +0.02°C $year^{-1}$). Seasonality increased throughout the entire latitudinal range, but significantly mainly between 29–35°N and 53–61°N (Fig.3F) due to seasonal SST dynamics comparable to the Eu_{Atl} and NAm_{Atl} gradients, i.e., warming largely in boreal fall and early winter weeks, being partially offset by SST decreases throughout late boreal winter, spring and summer (Fig.4I).

AFRICA (Figs.3G–I, 4E–G). The ***African Indian Ocean gradient*** (Af_{Ind}) had 53 coastal cells, the majority of which showed weak and insignificant warming trends (n = 40, $p<0.05$ for 8 cells). However, the southernmost latitudes (28–33°S) warmed substantially over the past 30 years and also showed trends towards more extreme seasons (Fig.3H,I). The latter was a consequence of SST increases during the first half of the year (weeks 1–30, late austral summer to early winter) but no or weak negative SST trends during the second half of the year (Fig.4G). The ***South-African Atlantic gradient*** (SAf_{Atl}) was comprised of 35 cells that showed both significant warming and cooling patterns. The majority of latitudes experienced warming (n = 26, $p<0.05$ for 11 cells), which was most pronounced between 12–23°S and coincided with the steepest section of the SAf_{Atl} gradient (Fig.3G,H). Significant cooling trends, however, were observed at the gradients high latitude end (31–34°S, i.e., south-western tip of Africa), thus resulting in an overall strengthening of the SAf_{Atl} gradient. Seasonality changes were largely insignificant, with a few notable exceptions, including south of 20°S (Fig.3I). Warming and cooling trends were mostly uniform across seasons (Fig.4F). The 30 coastal cells of the ***North-African Atlantic gradient*** (NAf_{Atl}) showed mostly warming trends over the study period (n = 25, $p<0.05$ for 17 cells). Tropical latitudes saw greater SST increases than subtropical latitudes, thus explaining the overall strengthening of the NAf_{Atl} gradient (+5.5%). Similar to the SAf_{Atl}, the strongest SST trends occurred where the gradient was steepest, i.e., between 7–23°N Fig3H. Seasonality increased significantly

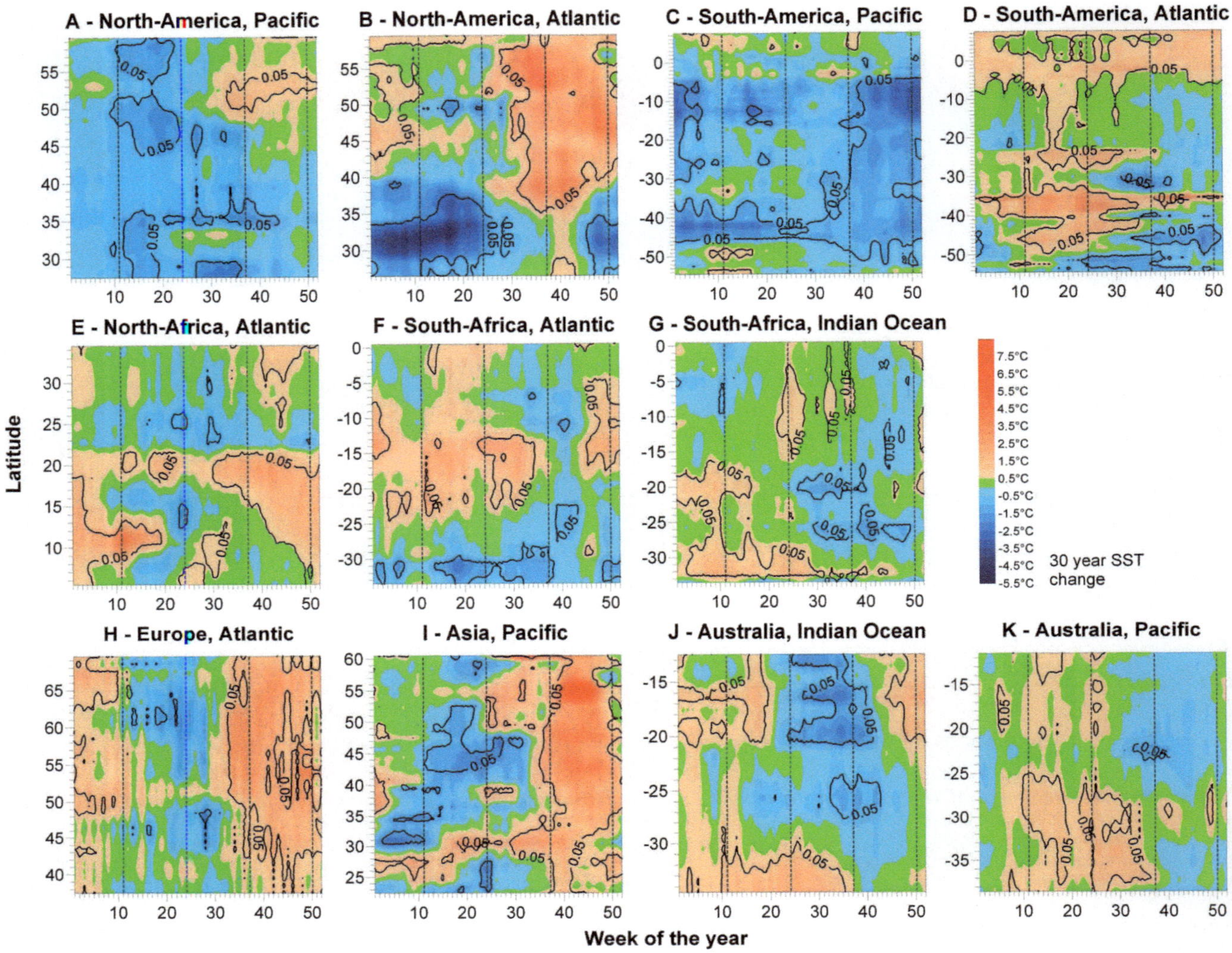

Figure 4. Seasonal decomposition of decadal SST trends in coastal latitudinal temperature gradients. Contours represent the absolute predicted change in SST for each latitude and week, derived by multiplying the linear regression slope *a* by 31 years ($SST_{ijk} = a$*year $+b$ with SST of gradient *i*, latitude *j* and week *k*). Isolines enclose areas of regressions with $p<0.05$ (overlaid contour). Dashed lines separate the 52 weeks (wks) into spring, summer, fall, and winter seasons in the northern (NH) and southern hemisphere (SH). Wks 12–24: $spring_{NH}/fall_{SH}$, wks 25–37: $summer_{NH}/winter_{SH}$, wks 38–50: $fall_{NH}/spring_{SH}$, wks 51–11: $winter_{NH}/summer_{SH}$.

between 15–21°N, but decreased between 9–12°N as a consequence of heterogeneous seasonal warming patterns (Figs 3I, 4E).

AUSTRALIA (Figs.3J–L, 4J,K). The ***Australian Indian Ocean gradient*** (Au_{Ind}) consisted of 23 coastal latitudes, of which 16 showed positive SST trends over the study period ($p<0.05$ for 3 cells). Significant warming was restricted to the highest three latitudes (Fig.3K). Increased seasonality was observed throughout the entire latitudinal range of the Au_{Ind} gradient, with stronger trends at lower latitudes (Fig.3L). This could be attributed to the seasonal decomposition of SST changes, revealing that low latitudes (11–21°S) warmed mainly during the first 20 and last 10 weeks of the year (late austral spring, summer, fall), but exhibited opposite SST trends during austral winter and early spring weeks (weeks 20–42, Fig.4J). In contrast, the highest Au_{Ind} latitudes south of 30°S primarily warmed in austral fall and winter (weeks 10–37). The ***Australian Pacific gradient*** (Au_{Pac}) encompassed 28 latitudes, most of which showed warming trends ($n=25$ $p<0.05$ for 7 cells) that were strongest at the gradients southern end (Fig.3K), thus slightly weakening the AuPac. Seasonality trends were non-significant, but consistently positive throughout the gradient range (Fig.3L). Seasonal decomposition indicated that significant warming affected all latitudes but only during austral fall and winter weeks (weeks 10–35, Fig.4K).

Discussion

Our study characterized how SST along the world's major coastlines changes with latitude, and how these large-scale thermal clines have themselves changed over the past three decades. This is of interest, because coastal and continental shelf regions are the most diverse, productive, and economically important parts of the ocean [1], and because temperature is the most important abiotic factor influencing marine life from the cell to the ecosystem level [4,5]. Hence, coastal ecological shifts in both space (i.e., latitude) and time (i.e., decade) are likely driven in part by the kind of latitudinal temperature patterns described here. We found considerable variation in gradient strength and seasonality between the world's coastal regions, while documenting complex and often opposing decadal trends in these metrics, consistent with a body of related previous works that have detailed the regional variability of unfolding global climate change [6,8,19,21,29–32].

SST was deemed a suitable proxy for average thermal conditions in shallow coastal areas. In addition, SST can be measured remotely and is therefore available at high data density and consistency over long periods of time [26,27]. The land-sea mask of the data set facilitated an objective algorithm for selecting coastal cells from the global grid, with the acknowledged trade-off of narrowing coastal representation with increasing latitude (e.g., one degree of longitude at 0° and 60° is 111.3 and 55.8 km wide, respectively). Our approach was to apply simple regression analyses over broad spatial and temporal scales to yield the desired general picture of gradients and their changes over past 30 years, thereby ignoring analyses of potential non-linear or cyclic patterns in global climate (e.g., ENSO, AMO).

Patterns & changes of coastal latitudinal temperature gradients

Latitudinal temperature patterns along coastlines are much more heterogeneous than in the open ocean [8,31,32], likely because of regional differences in well-understood oceanographic and climatic features such as alongshore surface currents, upwelling, thermocline depth, wind patterns, and cloudiness [7]. The world's strongest coastal gradients (i.e., NAm_{Atl} and As_{Pac}), for example, occur along the western boundaries of the Northern Hemisphere Atlantic and Pacific Oceans, where opposing warm and cold alongshore currents transport subarctic waters south- and tropical waters northward (NAm_{Atl}: Florida/Labrador Currents, As_{Pac}: Kuroshio/Oyashio Currents). The European gradient's (Eu_{Atl}) elevated thermal profile [7,33], with coastal latitudes being on average 5–10°C warmer than the same latitudes along the NAm_{Atl} or As_{Pac}, is consistent with the influence of the North-Atlantic Drift, i.e., the warm continuation of the Gulf Stream past its main branching point. Similarly, the world's weakest gradient was found along the South-African east coast ($Af_{Ind(1)}$ −0.28°C lat^{-1}), where another western boundary current - the warm Agulhas Current - transports heat southward [34]. Coastal upwelling, on the other hand, appears to be recognizable by distinct temperature depressions ('troughs') that cause substantially different regional gradient slopes. This signature pattern appeared in all four major coastal upwelling regions (Fig.3), i.e., (i) the California Current upwelling system (NAm_{Pac}), (ii) the Benguela Current upwelling system (SAf_{Atl}), (iii) the Canary Current upwelling region (NAf_{Atl}), and (iv) Humboldt Current upwelling system off Peru and Chile (SAm_{Pac}). In the latter, the observed depression in coastal temperatures (5–17°S) result in a uniquely steep tropical SST decline with latitude ($SAm_{Pac(1)\ 3°N\text{-}8°S}$ = −0.82°C lat^{-1}), which sets a potentially great natural contrast for large-scale ecological comparisons.

Temporal trends in CLTG patterns likely represent the complex net result of changes in atmospheric and oceanographic features [1,35], both of which are expected to differ between regions. However, we found that the NAm_{Atl}, Eu_{Atl} and As_{Pac} gradients share a common Northern Hemisphere warming pattern [21,30] that involves intensified warming at high compared to lower latitudes, generally as a consequence of precipitous SST increases during fall and winter (wks 35–11) that are partially offset by cooling trends in spring and summer (wks 12–34) [6]. This has caused the weakening in overall gradient strengths while increasing seasonality across latitudes. The latter is consistent with Lima & Wethey's [6] finding that the incidence of extreme SSTs (warm and cold) has increased in most coastal waters. However, common patterns may be evoked by different mechanisms [21], such as increased heat transport at high latitudes, stronger stratification due to warmer freshwater run-off [21] or changing ocean currents [36,37]. For example, the NAm_{Atl} weakened not only because of high latitude warming, but also because of strong cooling trends in coastal waters between 26–35°N [6]. SSTs in coastal waters between the tip of Florida (25°N) and Cape Hatteras (35°N) are influenced by the warm Florida Current, i.e., the beginning of the Gulf Stream [38]. Cooling of this coastal section is intriguing [39], given that it might signal an overall decrease in the Florida Current heat transport or a progressive diversion of warm Florida Current waters from the coast, both of which have been suggested by Ezer et al. [37] in a recent analysis. These SST changes have been substantial and therefore likely had profound ecological consequences for this region. Another example can be found along the Australian East coast (Au_{Pac}), where the warm East Australian Current has been shown to extend further and further southward [40], which is consistent with the significant warming trends observed between 36–39°S along the Au_{Pac} (mainly an austral fall and winter phenomenon) [6].

In general, SST trends at individual latitudes were consistent with those reported in studies of regional coastal temperature change [21]. For example, Nixon et al.'s [41] analysis of a 117-year SST record from the pier at Woods Hole, Cape Cod (located at ~41.5°N on the North-American Atlantic coast) showed average warming rates of 0.04°C year^{-1} between 1970–2002, which is comparable to rates estimated here ($NAm_{Atl,\ 41.5°N}$ = 0.02°C year^{-1} and $NAm_{Atl,\ 42.5°N}$ = 0.03°C year^{-1}). Similarly, Gomez-Gesteira et al. [7] found that SST in coastal European waters between 37–48°N increased at annual rates of 0.012 to 0.035°C year^{-1} (1985–2005), which is consistent with the present estimates of 0.014 ($Eu_{Atl,\ 37.5°N}$) to 0.034°C year^{-1} ($Eu_{Atl,\ 49.5°N}$).

Coastal upwelling has been predicted to intensify as a result of increasing wind-stress linked to global climate change . [1,35,42, but see 43]. Our analyses support this prediction by identifying a second remarkable pattern shared by some coastal latitudinal SST gradients: cooling [6]. This has been most evident along the South-American Pacific coast, where all latitudes between 6–47°S but particularly those between 6–14°S have consistently cooled (average of −1.58°C in 31 years), which is spatially consistent with the Humboldt Current upwelling system [44]. In addition, cooling affected all seasons of the year, but was most intense during austral spring and summer (wks 35–5). Hence, the substantial strengthening (+22%) of the partial $SAm_{Pac(1)}$ gradient, i.e., the unique decline in tropical SSTs along the South-American Pacific coast, may have largely been a consequence of upwelling intensification [21]. Similar patterns, albeit weaker, were also detected for the region of the NAm_{Pac} gradient corresponding to the California Current upwelling system, for which upwelling has been examined in more detail [21,45,46]. Intensification of the South African Benguela upwelling system, as suggested by Santos et al. [32] is consistent with the cooling patterns we detected at the southern edge of the SAf_{Atl} gradient. It may also explain why coastal latitudes of the Canary Upwelling system (22–30°N) showed no significant SST trends, contrary to substantial warming observed at all other coastal cells of the NAf_{Atl} gradient [6,31]. Conversely, several studies have detected a weakening of upwelling intensity along the western Iberian coast [47], which corresponds to small but significant warming trends along the lower latitudes of the Eu_{Atl} gradient in this study. In general, decade-scale cooling had likely as severe, if opposite, ecological consequences for marine life as decadal warming.

We conclude that the complex and regionally heterogeneous patterns imply that direct heat transfer from the atmosphere to ocean (i.e., global warming) cannot by itself explain the observed changes in CLTGs. Regional scale, wind-driven, indirect effects, via heat fluxes and upwelling, must drive changes in both the magnitude and seasonality of coastal SST gradients.

Ecological implications

Given temperature's fundamental role in governing marine life, long-term changes in CLTGs will likely precipitate shifts in ecological gradients. Adaptive latitudinal variations in intra-specific morphological, physiological and life history traits are well documented in marine taxa [reviewed in 12] and encompass traits such as body size [10,11], growth and development rates [48–50], predator vulnerability [51], life span [52], or vertebral number [16,17]. For example, adapting to decreasing temperatures with latitude, many estuarine fish populations from high latitudes have evolved higher genetic growth capacities than their lower latitude conspecifics, a form of genotype × environment interaction known as countergradient variation [53–55]. Baumann and Conover [56] found that the strength of countergradient growth variation in silverside fishes from the North-American Pacific versus Atlantic coasts mimicked the strength of their contrasting latitudinal SST gradients. Similarly, increases in vertebral number with latitude, a form of co-gradient variation [12], is twice as strong in silverside fishes along the North-American Atlantic (strong CLTG) versus the Pacific coast [weak CLTG, 57]. Hence, latitudinal temperature gradients likely determine genotype*environment interactions in coastal systems, and shifts e.g., towards weaker gradients may precipitate weaker genetic gradients by reducing the amount of genetic variance between populations.

These adaptive trait gradients along CLTGs may comprise useful spatial analogies for studying climate change adaptation [12,56]. How organisms will adapt to global climate change is uncertain, partly because of difficulties besetting the methods to study it (e.g., duration, replication, plastic vs. genetic responses). In contrast, adaptation principles can be rigorously examined across spatial climate gradients of different strength, such as the world's CLTGs. If countergradient variation, for example, is as ubiquitous an adaptation principle in time as it is in space, global warming may not lead to oft-anticipated increases in observed growth rates among ectotherms [1,12]. Instead, genetic decreases in growth capacity could offset the plastic, temperature-driven growth increases, analogous to the local growth adaptation in high versus low latitude fish populations [56].

CLTGs may also underlie the ubiquitous decline in species diversity from low to high latitudes [9,58,59]. Whether coastal diversity actually scales with the strength of the underlying temperature gradient has received less attention, but could be effectively tested using the full and partial thermal slopes described here. Paleo-biological analyses indicated that diversity-temperature relationships in planktic foraminifers have been surprisingly stable over the past three million years, suggesting that gradients in species richness rapidly reorganize in response to climatic changes [58]. Hence, weakening thermal gradients due to large-scale latitudinal temperature shifts could precipitate weaker diversity gradients, by increasing diversity at high latitudes and/or decreasing diversity at low latitudes [e.g., 60].

One already widely observed biotic reaction to global climate change [60–64], is the shift in geographic ranges in marine species, with the vast majority comprising poleward movements [e.g., 61: 75% of 129 marine species]. This is consistent with our observation of weakening coastal latitudinal temperature gradients, not only because high latitude warming increases potential habitat for lower latitude organisms, but also because decreases in thermal gradient strength would make habitats more similar across latitudes and thus even more suitable for some intermediately located genotypes. In the North Atlantic Ocean, for example, where both coastal temperature gradients (Eu_{Atl}, NAm_{Atl}) have weakened substantially, range shifts over the past decades are particularly evident, e.g., in the fish assemblages in the North Sea [65] and along the North-American continental shelf [66]. On the other hand, the considerable variability in geographic trends among regions of the world is consistent with the substantial differences in coastal thermal gradients we have described here. While focusing on temperature alone is likely too simplistic to predict shifting species distributions, 'bioclimate-envelope' approaches are still useful on large spatial and temporal scales [67,68] and may benefit from explicitly incorporating coastal latitudinal temperature gradients.

Lastly, shifts in coastal temperature patterns like the observed trends towards more extreme seasons may affect the connectivity between coastal marine populations. More extreme winters could increase overwinter mortalities in some organisms, while more extreme summers could both extend or curtail reproductive and growing seasons in others, thereby potentially turning continuous species distributions into more mosaic ones. Along the Australian Indian Ocean coast (Au_{Ind}), for example, seasonality increased significantly along the entire gradient, a trend that may challenge stenothermal tropical organisms (e.g., some corals or reef fish). Reduced connectivity may also result from opposing temperature trends between adjacent coastal regions. At the southern tip of the African continent, for example, the warm African Indian Ocean gradient and the bordering cold South-African Atlantic gradient have exhibited divergent trends over the past decades. Our analyses indicated significant warming on the Indian Ocean side (Af_{Ind}: $SST_{32.5°S} = 22.3°C_{1982} \rightarrow 23.3°C_{2012}$) but cooling on the Atlantic Ocean side (SAf_{Atl}: $SST_{32.5°S} = 17.0°C_{1982} \rightarrow 16.3°C_{2012}$), thereby exacerbating the temperature differential that migrating organisms must cross by more than 25% (5.3 to 7.0°C).

Conclusions

We found that the rate of latitudinal temperature change along the world's major coastlines varies more than threefold and has – in seven of 11 cases – decreased over the past three decades. SST and seasonality have increased in the majority of coastal cells, with the greatest changes observed in Northern Hemisphere gradients, but opposite trends, i.e., coastal cooling, have occurred particularly at upwelling influenced coastlines. As expected, decadal changes in gradient strength, latitude-specific SST and seasonality were found to be highly heterogeneous between regions, yet consistent with regionally differing consequences of global climate change. We suggest that past and future ecological changes in coastal marine ecosystems are linked to these regionally diverse CLTG patterns and that quantifying them is useful for future meta-analyses of large-scale ecological and thermal gradients. This would improve the current understanding about adaptive responses of marine life to climate gradients in space and, by inference, time. Valuable next steps include the identification of non-linear and cyclic SST trends, backward extension of the SST dataset, and forecasts of thermal gradient evolution under different IPCC scenarios [3] in general circulation climate models.

Acknowledgments

We are grateful to David Conover, Charles Flagg, and Sultan Hameed for their input and constructive criticisms. The NOAA_OI_SST_V2 data were provided by the National Oceanic and Atmospheric Administration (NOAA), Earth System Research Laboratory (ESRL), Physical Science Division (PSD), Boulder, Colorado, USA.

Author Contributions

Analyzed the data: HB OD. Wrote the paper: HB OD.

References

1. Harley CDG, Randall Hughes A, Hultgren KM, Miner BG, Sorte CJB, et al. (2006) The impacts of climate change in coastal marine systems. Ecology Letters 9: 228–241.
2. Costanza R, d'Arge R, de Groot R, Farber S, Grasso M, et al. (1997) The value of the world's ecosystem services and natural capital. Nature 387: 253–260.
3. IPCC (2007) Summary for Policymakers. Cambridge, United Kingdom and New York, NY, USA: Cambridge University Press.
4. Pauly D (1980) On the interrelationships between natural mortality, growth parameters, and mean environmental temperature in 175 fish stocks. ICES Journal of Marine Science 39: 175–192.
5. Clarke A (2003) Costs and consequences of evolutionary temperature adaptation. Trends in Ecology & Evolution 18: 573–581.
6. Lima FP, Wethey DS (2012) Three decades of high-resolution coastal sea surface temperatures reveal more than warming. Nature Communications 3: 704.
7. Gómez-Gesteira M, deCastro M, Alvarez I, Gómez-Gesteira JL (2008) Coastal sea surface temperature warming trend along the continental part of the Atlantic Arc (1985–2005). Journal of Geophysical Research: Oceans 113: C04010.
8. Santos F, Gomez Gesteira M, deCastro M (2011) Coastal and oceanic SST variability along the western Iberian Peninsula. Continental Shelf Research 31: 2012–2017.
9. Roy K, Jablonski D, Valentine JW, Rosenberg G (1998) Marine latitudinal diversity gradients: Tests of causal hypotheses. Proceedings of the National Academy of Sciences 95: 3699–3702.
10. Bergmann C (1847) Über die Verhältnisse der Wärmeökonomie der Thiere zu ihrer Grösse. Göttinger Studien 1: 595–708.
11. Blackburn TM, Gaston KJ, Loder N (1999) Geographic gradients in body size: a clarification of Bergmann's Rule. Diversity and Distributions 5: 165–174.
12. Conover DO, Duffy TA, Hice LA (2009) The covariance between genetic and environmental influences across ecological gradients - reassessing the evolutionary significance of countergradient and cogradient variation. Annals of the New York Academy of Sciences 1168: 100–129.
13. Munch SB, Salinas S (2009) Latitudinal variation in lifespan within species is explained by the metabolic theory of ecology. Proceedings of the National Academy of Sciences of the United States of America 106: 13860–13864.
14. Conover DO (1992) Seasonality and the scheduling of life history at different latitudes. Journal of Fish Biology 41: 161–178.
15. Connolly SR, Menge BA, Roughgarden J (2001) A latitudinal gradient in recruitment of intertidal invertebrates in the Northeast Pacific ocean. Ecology 82: 1799–1813.
16. McDowall RM (2008) Jordan's and other ecogeographical rules, and the vertebral number in fishes. Journal of Biogeography 35: 501–508.
17. Jordan DS (1891) Relations of temperature to vertebrae among fishes. Proceedings of the US National Museum 14: 107–120.
18. Paltridge G, Woodruff S (1981) Changes in global surface temperature from 1880 to 1977 derived from historical records of sea surface temperature. Monthly Weather Review 109: 2427–2434.
19. Burrows MT, Schoeman DS, Buckley LB, Moore P, Poloczanska ES, et al. (2011) The pace of shifting climate in marine and terrestrial ecosystems. Science 334: 652–655.
20. Folland CK, Karl TR, Vinnikov K (1990) Observed Climate Variations and Change. In: Houghton GT, Jenkins GJ, Ephraums JJ, Climate Change: The IPCC Scientific Assessment. Cambridge: Cambridge University Press. 195–238.
21. Belkin IM (2009) Rapid warming of large marine ecosystems. Progress in Oceanography 81: 207–213.
22. Munch SB, Mangel M, Conover DO (2003) Quantifying natural selection on body size from field data: winter mortality in *Menidia menidia*. Ecology 84: 2168–2177.
23. Conover DO, Heins SW (1987) Adaptive variation in environmental and genetic sex determination in a fish. Nature 326: 496–498.
24. Easterling DR, Meehl GA, Parmesan C, Changnon SA, Karl TR, et al. (2000) Climate extremes: observations, modeling, and impacts. Science 289: 2068–2074.
25. NOAA (2012) Earth System Research Laboratory: NOAA Optimum Interpolation (OI) Sea Surface Temperature (SST) V2. http://www.esrl.noaa.gov/psd/data/gridded/data.noaa.oisst.v2.html. Accessed 2013 May 27.
26. Reynolds RW, Rayner NA, Smith TM, Stokes DC, Wang W (2002) An improved in situ and satellite SST analysis for climate. Journal of Climate 15: 1609–1625.
27. Smith TM, Reynolds RW, Peterson TC, Lawrimore J (2008) Improvements to NOAA's Historical Merged Land–Ocean Surface Temperature Analysis (1880–2006). Journal of Climate 21: 2283–2296.
28. Beninger PG, Boldina I, Katsanevakis S (2012) Strengthening statistical usage in marine ecology. Journal of Experimental Marine Biology and Ecology 426–427: 97–108.
29. Goreau TJ, Hayes RL, McAllister D (2005) Regional patterns of sea surface temperature rise: implications for global ocean circulation change and the future of coral reefs and fisheries. World Resource Review 17: 350–370.
30. Strong AE, Kearns EJ, Gjovig KK (2000) Sea surface temperature signals from satellites—An update. Geophysical Research Letters 27: 1667–1670.
31. Santos F, deCastro M, Gómez-Gesteira M, Álvarez I (2012) Differences in coastal and oceanic SST warming rates along the Canary upwelling ecosystem from 1982 to 2010. Continental Shelf Research 47: 1–6.
32. Santos F, Gomez-Gesteira M, deCastro M, Alvarez I (2012) Differences in coastal and oceanic SST trends due to the strengthening of coastal upwelling along the Benguela current system. Continental Shelf Research 34: 79–86.
33. Rossby T (1996) The North Atlantic Current and surrounding waters: At the crossroads. Reviews of Geophysics 34: 463–481.
34. Bryden HL, Beal LM, Duncan LM (2005) Structure and transport of the Agulhas Current and its temporal variability. Journal of Oceanography 61: 479–492.
35. Bakun A (1990) Global climate change and intensification of coastal ocean upwelling. Science 247: 198.
36. Dengler M, Fischer J, Schott FA, Zantopp R (2006) Deep Labrador Current and its variability in 1996–2005. Geophysical Research Letters 33: L21S06.
37. Ezer T, Atkinson LP, Corlett WB, Blanco JL (2013) Gulf Stream's induced sea level rise and variability along the U.S. mid-Atlantic coast. Journal of Geophysical Research: Oceans 118: 685–697.
38. Schroeder EH (1966) Average surface temperatures of the western North Atlantic. Bulletin of Marine Science 16: 302–323.
39. Bryden HL, Longworth HR, Cunningham SA (2005) Slowing of the Atlantic meridional overturning circulation at 25°N. Nature 438: 655–657.
40. Roughan M, Middleton JH (2004) On the East Australian Current: variability, encroachment, and upwelling. Journal of Geophysical Research 109: C07003.
41. Nixon SW, Granger S, Buckley BA, Lamont M, Rowell B (2004) A one hundred and seventeen year coastal water temperature record from Woods Hole, Massachusetts. Estuaries 27: 397–404.
42. Bakun A, Field DB, Redondo-Rodriguez ANA, Weeks SJ (2010) Greenhouse gas, upwelling-favorable winds, and the future of coastal ocean upwelling ecosystems. Global Change Biology 16: 1213–1228.
43. Pardo PC, Padín XA, Gilcoto M, Farina-Busto L, Pérez FF (2011) Evolution of upwelling systems coupled to the long-term variability in sea surface temperature and Ekman transport. Climate Research 48: 231–246.
44. Bakun A, Nelson CS (1991) The seasonal cycle of wind-stress curl in subtropical eastern boundary current regions. Journal of Physical Oceanography 21: 1815–1834.
45. Bograd SJ, Schroeder I, Sarkar N, Qiu X, Sydeman WJ, et al. (2009) Phenology of coastal upwelling in the California Current. Geophysical Research Letters 36.
46. Snyder MA, Sloan LC, Diffenbaugh NS, Bell JL (2003) Future climate change and upwelling in the California Current. Geophysical Research Letters 30: 1823.
47. Lemos RT, Pires HO (2004) The upwelling regime off the West Portuguese Coast, 1941–2000. International Journal of Climatology 24: 511–524.
48. Parsons KE (1997) Contrasting patterns of heritable geographic variation in shell morphology and growth potential in the marine gastropod *Bembicium vittatum*: evidence from field experiments. Evolution 51: 784–796.
49. Sanford E, Holzman SB, Haney RA, Rand DM, Bertness MD (2006) Larval tolerance, gene flow, and the northern geographic range limit of fiddler crabs. Ecology 87: 2882–2894.
50. Jonassen TM, Imsland AK, Fitzgerald R, Bonga SW, Ham EV, et al. (2000) Geographic variation in growth and food conversion efficiency of juvenile Atlantic halibut related to latitude. Journal of Fish Biology 56: 279–294.
51. Munch SB, Conover DO (2003) Rapid growth results in increased susceptibility to predation in *Menidia menidia*. Evolution 57: 2119–2127.
52. Salinas S, Munch SB (2012) Thermal legacies: transgenerational effects of temperature on growth in a vertebrate. Ecology Letters 15: 159–163.
53. Schultz ET, Reynolds KE, Conover DO (1996) Countergradient variation in growth among newly hatched *Fundulus heteroclitus*: geographic differences revealed by common-environment experiments. Functional Ecology 10: 366–374.
54. Conover DO, Present TMC (1990) Countergradient variation in growth rate: compensation for length of the growing season among Atlantic silversides from different latitudes. Oecologia 83: 316–324.
55. Brown JJ, Ehtisham A, Conover DO (1998) Variation in larval growth rate among striped bass stocks from different latitudes. Transactions of the American Fisheries Society 127: 598–610.
56. Baumann H, Conover DO (2011) Adaptation to climate change: contrasting patterns of thermal-reaction-norm evolution in Pacific versus Atlantic silversides. Proceedings of the Royal Society B-Biological Sciences 278: 2265–2273.
57. Baumann H, Rosales-Casian JA, Conover DO (2012) Contrasting latitudinal variations in vertebral number and sex determination in Pacific vs. Atlantic silverside fishes. Copeia 2012: 341–350.
58. Yasuhara M, Hunt G, Dowsett HJ, Robinson MM, Stoll DK (2012) Latitudinal species diversity gradient of marine zooplankton for the last three million years. Ecology Letters 15: 1174–1179.
59. Hillebrand H (2004) On the generality of the latitudinal diversity gradient. The American Naturalist 163: 192–211.

60. Holbrook SJ, Schmitt RJ, Stephens JS, Jr. (1997) Changes in an assemblage of temperate reef fishes associated with a climate shift. Ecological Applications 7: 1299–1310.
61. Sorte CJB, Williams SL, Carlton JT (2010) Marine range shifts and species introductions: comparative spread rates and community impacts. Global Ecology and Biogeography 19: 303–316.
62. Parmesan C, Yohe G (2003) A globally coherent fingerprint of climate change impacts across natural systems. Nature 421: 37–42.
63. Precht WF, Aronson RB (2004) Climate flickers and range shifts of reef corals. Frontiers in Ecology and the Environment 2: 307–314.
64. Lima FP, Queiroz N, Ribeiro PA, Hawkins SJ, Santos AM (2006) Recent changes in the distribution of a marine gastropod, *Patella rustica* Linnaeus, 1758, and their relationship to unusual climatic events. Journal Of Biogeography 33: 812–822.
65. ter Hofstede R, Hiddink JG, Rijnsdorp AD (2010) Regional warming changes fish species richness in the eastern North Atlantic Ocean. Marine Ecology Progress Series 414: 1–9.
66. Nye JA, Link JS, Hare JA, Overholtz WJ (2009) Changing spatial distribution of fish stocks in relation to climate and population size on the Northeast United States continental shelf. Marine Ecology Progress Series 393: 111–129.
67. Pearson RG, Dawson TP (2003) Predicting the impacts of climate change on the distribution of species: are bioclimate envelope models useful? Global Ecology and Biogeography 12: 361–371.
68. Zacherl D, Gaines SD, Lonhart SI (2003) The limits to biogeographical distributions: insights from the northward range extension of the marine snail, *Kelletia kelletii* (Forbes, 1852). Journal of Biogeography 30: 913–924.

21

Sensitivity of Heterogeneous Marine Benthic Habitats to Subtle Stressors

Iván F. Rodil[1,2*¤], Andrew M. Lohrer[2], Simon F. Thrush[2]

1 Departamento de Ecología y Biología Animal, Universidad de Vigo, Vigo, Spain, 2 National Institute of Water and Atmospheric Research, Hamilton, New Zealand

Abstract

It is important to understand the consequences of low level disturbances on the functioning of ecological communities because of the pervasiveness and frequency of this type of environmental change. In this study we investigated the response of a heterogeneous, subtidal, soft-sediment habitat to small experimental additions of organic matter and calcium carbonate to examine the sensitivity of benthic ecosystem functioning to changes in sediment characteristics that relate to the environmental threats of coastal eutrophication and ocean acidification. Our results documented significant changes between key biogeochemical and sedimentary variables such as gross primary production, ammonium uptake and dissolved reactive phosphorus flux following treatment additions. Moreover, the application of treatments affected relationships between macrofauna communities, sediment characteristics (e.g., chlorophyll *a* content) and biogeochemical processes (oxygen and nutrient fluxes). In this experiment organic matter and calcium carbonate showed persistent opposing effects on sedimentary processes, and we demonstrated that highly heterogeneous sediment habitats can be surprisingly sensitive to subtle perturbations. Our results have important biological implications in a world with relentless anthropogenic inputs of atmospheric CO_2 and nutrients in coastal waters.

Editor: Sam Dupont, University of Gothenburg, Sweden

Funding: IFR acknowledges the financial support from the postdoctoral program Ángeles Alvariño (XUGA, Spain). This work was supported by funding from Ministry of Business, Innovation and Employment (MoBIE) (NIWA COMS1320 and COER1201). The funders had no role in study design, data collection and analysis, decision to publish, or preparation of the manuscript.

* E-mail: ibantxorodil@gmail.com

¤ Current address: Interdisciplinary Centre of Marine and Environmental Research, University of Porto, Porto, Portugal

Introduction

Major disturbance events are likely to affect the structure and function of ecological communities with large and persistent consequences [1–4]. However, recent studies have demonstrated that subtle sublethal stressors can also influence ecosystem functioning [5–7] and that small changes matter [8]. Thus, for example, poleward shifts in species distributions have been linked to small gradual increases in annually averaged temperatures globally, despite the comparatively greater frequency and magnitude of temperature fluctuations on short time scales [9,10].

As with temperature, a slow shift in seawater pH is considered to be a potentially pervasive long-term threat to marine communities [11,12]. There is concern at present about the gradually increasing acidity of our oceans driven by high, and increasing, levels of CO_2 in our atmosphere [13,14] and coastal eutrophication [15,16]. Furthermore, it is also accepted that environmental stressors such as ocean acidification, eutrophication, and elevated temperature, among others, may interact synergistically in natural systems to affect biodiversity and ecosystem function [17,18]. The biogeochemistry of coarse, permeable carbonate sands has been mainly investigated in tropical areas where carbonates from coral reefs comprise the majority of the sedimentary material [19,20]. However, permeable carbonate sediments are also widely distributed in non-tropical regions [21], with shell hash from bivalves contributing the majority of the carbonate material.

There is evidence that carbonate systems play a key role in buffering ocean acidification [22–24]. However, CO_2 exchange between the atmosphere and ocean is only partly controlled by carbonate mineral precipitation. Other major mechanisms in the system are production, respiration, remineralisation, and net sedimentary storage of organic matter in the ocean [25]. Therefore, to examine the potential for $CaCO_3$ to alter biogeochemistry in temperate sediments, we performed an orthogonally crossed experiment involving additions of organic matter (OM) and $CaCO_3$ (CC) to coarse, heterogeneous coastal sand (OM addition, CC addition, OM+CC addition as a mixture, and controls). Oxygen, dissolved inorganic carbon, and nutrient fluxes were measured in all treatments as response variables to understand effects on benthic respiration, net and gross primary production, and nutrient utilisation. Individually, these variables are key processes in benthic systems and commonly used indices of ecosystem functioning [26,27]. However, we also examined changes in relationships between processes. Analysing the strength and variability of coupled processes like ammonium uptake and primary production, has been shown to be useful for assessing the impacts of subtle stressors in soft sediments [5–7], and this was the approach that we used here. We hypothesised that additions of CC would counteract those of OM, and that the treatment types would drive relationships in opposite directions.

Materials and Methods

Study site and sampling set up description

The study was conducted at a shallow soft-sediment site (9 m deep) with coarse permeable shelly sands and strong tidal currents in Kawau Bay (36° 23′ 55″ S, 174° 49′ 42″ E), a large embayment on the NE coast of North Island, New Zealand. This bay has a diversity of benthic soft-sediment habitat types [28] and the study site was particularly heterogeneous; that is, all the environmental variables measured exhibited a high degree of variability across the studied ecosystem (see Supporting Information, Figures S1 and S2). The area was dominated by an assemblage of large macrofauna, such as dog cockles (*Glycymeris glycymeris*), hermit crabs, sponges, shrimps (*Heterosquilla* sp.), fan worms (Serpulidae) as well as a variety of smaller organisms. All necessary permits were obtained for the described field studies. No specific permissions were required to study these locations, as they were away from shore (subtidal marine) and not within specially protected areas. The studies did not involve endangered or protected species, and the collection of cores for the study of macroinvertebrates was authorised by Special permits 505 and 542 from the Ministry for Primary Industries.

In November 2011, 48 square plots (50 cm×50 cm) were established over two days by scuba divers in a 12 column by 4 row array on the seafloor (length 30 m, width 8 m). Three treatment additions plus a control (n = 12 each) were interspersed in three identical blocks in a balanced crossed design (see Supporting Information, Figure S3). The treatments were created by adding substances to the sediment surface of the plots (0.25 m^2): organic matter (OM; cornflour, 35 g.), calcium carbonate (CC; Aglime, 35 g.) or a mixture of both (Mix, 70 g total with equal parts OM and CC). The type and amount of OM used (food grade cornflour; aprox. 42% organic carbon, 0.1% organic nitrogen) was sufficient to elicit a benthic response but too little to cause sediment hypoxia or macrofaunal mortalities [5,6]. Agricultural lime (McDonald's Aglime©; 94% $CaCO_3$) is a natural product consisting of finely ground limestone. It is commonly used in agriculture to improve the quality of the soil by neutralising acid [29].

Each powdered substance was weighed into lidded containers and mixed with ambient seawater (100 mL) prior to application. This ensured that a slurry of material could be controllably added to the sediment surface. Square domes (50 cm×50 cm) were positioned atop the sediment surface and divers used syringes loaded with slurry to introduce the material to the plots (see Figure S3). The domes were left in place for 2–3 hours after treatment additions to allow the material to settle out of suspension and onto the sediment surface. After dome removal, a 1 mm mesh screen was positioned over the plots and ambient sediment was carefully sprinkled through the screen, thinly covering the treatment materials, reducing the likelihood of losses due to resuspension and ensuring incorporation into the sediment fabric. The experiment set up took over two days and left it for one week to allow the treatments to take effect.

Measurements of oxygen and nutrient fluxes

A week after treatment establishment, benthic incubation chambers were deployed in the experimental plots. This interval provided time for the materials to be processed and to begin to affect porewater nutrient concentrations and sediment biogeochemistry. In particular, we expected that most of the organic material added to the plots would be consumed and processed by the biota within a week, approximately, due to the relatively small amount of organic material added and the high activity rates in these types of ecosystems [5,6,30].

Three consecutive days were required to sample all 48 plots, with 16 chambers (4 replicates of each treatment) deployed per day. We were able to block the plots by establishment date (the eastern half of the array was established on day 1, the western half on day 2) or by sampling date (the eastern third was sampled on day 8, the middle third on day 9, the western third on day 10). The chambers were used to assess fluxes of oxygen and nutrients across the sediment-water interface. During deployment, a square chamber base (an aluminium border with sides 50×50×10 cm tall) was positioned around the perimeter of each experimental plot and pressed halfway down into the sediment. Clear acrylic plastic lids were fitted to the chamber bases and clamped in place, isolating a square patch of sediment and approximately 25 L of overlying seawater for incubation. Each chamber contained one D-Opto dissolved oxygen and water temperature logger (60 data points per hour) and a Seabird electronic pump (20 ml/sec flow rate). The pump ran for 5 seconds every 45 seconds to provide intermittent non-directional stirring of chamber waters, preventing stagnation and the build-up of boundary layer gradients.

The seawater enclosed within each chamber was collected through syringe activated sampling ports at the beginning and end of each incubation period (~4 hours). Collection times were recorded in all instances. Incubations were run during daytime: once with photosynthesis possible (clear chamber lids in the morning between 0800 and 1230 hrs); and once without photosynthesis (black chamber covers in the afternoons between 1230 and 1750). Prior to every incubation period (including in between each day's light and dark incubations), all of the chambers were flushed with ambient bottom water by briefly removing the chamber lids; this ensured that all chamber incubations were initialised (once lids were replaced) with ambient bottom water unaffected by previous runs. As biological reactions are often light and temperature dependent, we assessed underwater light levels and temperatures with submersible Hobo® loggers deployed to 10 of the 16 chambers (5 cm above the sediment surface, logging every 5 min) on each of the 3 days of sampling.

As soon as water samples (4 replicates per treatment, 48 samples total) were back aboard the boat, levels of dissolved oxygen (Hach® HQ40d dissolved oxygen sensor, accuracy ±0.1 mg/L DO) and pH (PreSense® needle-type microsensor, accuracy ±0.05 pH units) were measured. Water samples were then filtered (1.1 μm Whatman filter) and stored frozen and in the dark until analysis for dissolved inorganic nutrient concentrations. Samples for dissolved inorganic carbon (DIC) were filtered and preserved in saturated mercuric chloride in 100 mL glassware prior to analysis.

Sediment characteristics

Upon completion of the chamber incubations, the chamber lids were removed and two small cores of surface sediment (5 cm^2×2 cm deep) were collected from every plot for chlorophyll *a*, organic matter and grain size analyses. After collection, all samples were frozen and stored in dark. In addition, one macrofaunal core (133 cm^2×15 cm deep) from the middle of each plot was collected, sieved across a 0.5 mm mesh and preserved in 70% IPA. Lastly, the remaining area of sediment inside each plot was searched and data on the abundance of large organisms (e.g., dog cockles, burrowing urchins, sponges, shrimp burrows) was collected.

Laboratory analysis

Fluxes of dissolved oxygen and inorganic nutrients (*F*) were calculated in all light and dark chambers based on the magnitude of change in water chemistry during incubation,

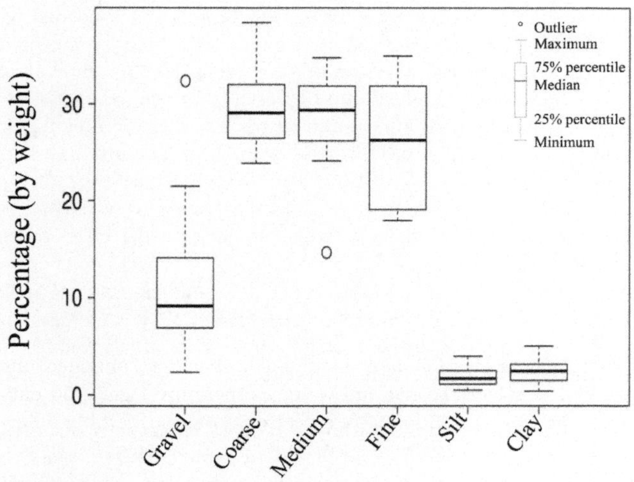

Figure 1. Boxplots showing the interquartile range (first quartile, median and third quartile) of the sediment data. Whiskers: the values that extend to 1.5 times the interquartile range. When there are no outliers the whiskers show the maximum and minimum values.

$$F = \frac{\{(Cend - Cinitial)xV\}}{(AxT)}$$

where C is nutrient or oxygen concentration (μmol l^{-1}) at the beginning or end of the incubation, V is the volume of seawater inside the chamber (25 l), A is the area of sediment enclosed by the chamber (0.25 m^2), and T is the elapsed time between initial and final samplings (hours).

Fluxes of DO in light chambers provided an estimate of net primary production (NPP); while fluxes of DO in dark chambers were used to estimate benthic community respiration (CR). Gross primary production (GPP) was calculated by subtracting CR from NPP. Analysis for ammonium nitrogen (NH_4^+- N), nitrate-plus-nitrite nitrogen (NO_X-N) and reactive phosphate phosphorus (HPO_4^{2-}-P) were performed using standard methods [31] on an Astoria Pacific International autoanalyzer (Clackamas, Oregon, USA) with detection limits of <1 μmol/L for N and P. Dissolved inorganic carbon (DIC) was obtained by means of a total organic carbon analyser (APHA5310C method) with detection limits of <0.2 μmol/L for C. NH_4^+- N uptake was calculated as dark NH_4^+-N flux minus light NH_4^+-N flux. NO_X-N and HPO_4^{2-}P uptakes were similarly calculated. Dissolved inorganic nitrogen (DIN) is sum of dissolved NH_4^+-N and the NO_X-N.

For sediment analysis, grain size samples were thawed and homogenized, and 5 g subsamples were digested in H_2O_2 until frothing ceased. The sediment was then wet sieved through 2000, 500, 250 and 63 μm mesh sieves. All fractions were dried at 60°C and weighed after 2 days. Sediment chlorophyll *a* (chla) content (μg.g^{-1} sediment) was used as an indicator of microphytobenthos abundance. Chla samples were freeze-dried prior to analysis (to standardize sample water content) and then homogenized. Subsamples (5 g) were boiled in 90% ethanol to extract the photopigments that were then analysed spectrophotometrically. An acidification step was used to separate degradation products from chla [32]. Organic matter content samples were thawed and homogenized, and 5 g subsamples were dried at 60°C. Dry masses were obtained after 2 days, and samples were moved to a 400°C combustion oven (6 hours). Organic matter content was assessed as the sediment's percentage mass lost during combustion.

Macrofauna were sorted and identified to the lowest taxonomic grouping practicable. From the raw data, we calculated the total abundance, taxonomic richness, and diversity of macrofauna in every plot. We later classified the macrofauna into main feeding guilds (deposit feeders, suspension feeders, predators/scavengers).

Data analysis

Patterns in the data were first assessed with boxplots, Cleveland dotplots, and pair plots to identify outliers and the potential for collinearity among explanatory variables. Sediment characteristics and macrofauna communities were then assessed by means of analysis of variance (ANOVA). Treatment and block were included in two-way ANOVA models; treatment was used as a fixed factor with four levels (organic matter, calcium carbonate, mixture and control) and "block" was included to test for the influence of either establishment date (2 levels: eastern and western halves of the sampling array) or sampling date (3 levels: eastern, middle and western thirds of the sampling array) (see Supporting Figure S3). Neither type of blocking had significant interaction effects with the treatments in the two-way ANOVA models (see

Table 1. The most parsimonious Generalized Linear Models (GLMs) relating different environmental and macrofauna community variables with and without treatment as a factor (terms that were not significant were dropped from the final models).

Model name	Model expression (without treatment)	AIC	pseudo-R^2	Model name	Model expression with treatment interaction (*Tr)	AIC	pseudo-R^2
Model 1a	GPP~DIC_{light}+NH_4^++NO_3^-+DF	733.9	0.609	Model 1b§	GPP~NH_4^++NH_4^+*Tr	682.1	0.655
Model 2a	DIN~GPP+DRP+DF	434.1	0.699	Model 2b	DIN~DRP+DRP*Tr	464.9	0.493
Model 3a	Chla~Abundance	135.1	0.100	Model 3b	Chla~Abundance+Abundance*Tr	134.12	0.334
Model 4a	$NO_3^-{}_{light}$~DF	387.8	0.100	Model 4b	$NO_3^-{}_{light}$~DF+DF*Tr	388.8	0.277

The Akaike's Information Criterion (AIC) and the proportional increase in explained deviance (pseudo-R^2) were used to evaluate each regression-based model fit and parsimony.
NH_4^+: ammonium uptake; NO_3^-: nitrate uptake; DIN: dissolved inorganic nitrogen (Σ NH_4^++NO_3^-); $NO_3^-{}_{light}$: nitrate flux during daylight.
DRP: dissolved reactive phosphorus (HPO_4^{2-}- P); DIC_{light}: dissolved inorganic carbon flux during daylight.
GPP: gross primary production; Chla: chlorophyll *a* concentration.
Abundance: total macrofauna abundance; DF: deposit feeders abundance.
pseudo-R^2 = (null deviance-residual deviance)/null deviance; (*sensu* [46]).
§There was no significant interaction between the response variable and treatment (NH_4^+*Tr). We dropped the interaction term from the analysis and ran the model again. Final Model 1b: GPP~NH_4^++Tr (AIC = 683.4, pseudo-R^2 = 0.593).

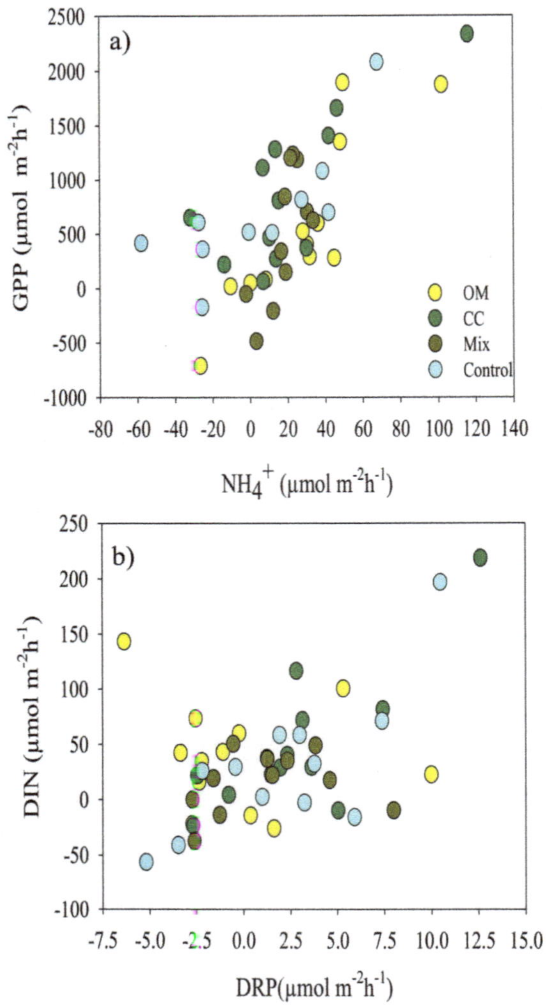

Figure 2. Relationship between main biogeochemical coupled processes (raw data). The top panel (a) shows the general relationship between gross primary production (GPP, µmol O_2 $m^{-2}h^{-1}$) and ammonium uptake (µmol NH_4^+-N $m^{-2}h^{-1}$) across the sampling area. The bottom panel (b) shows the relationship between dissolved inorganic nitrogen uptake (µmol DIN $m^{-2}h^{-1}$) and dissolved reactive phosphorus (µmol DRP $m^{-2}h^{-1}$) across the sampling area. Treatments: Organic Matter, Calcium carbonate, Mix, Control.

Supporting information, Table S1), so block was removed as a predictor variable from all subsequent analyses.

The response variables of interest (fluxes) and the explanatory co-variables (sediment characteristics, macrofauna, etc.) both exhibited a high degree of variability, reflecting the heterogeneity of the habitat (see Figures S1 and S2, Supporting Information). Also, as we were interested in relationships between processes, a continuous approach (e.g., regression based models or analysis of co-variance based) was more appropriate for this type of than a purely categorical approach. Therefore, we used generalized linear models (GLMs: Gaussian distribution, identity link function) to test for significant interactions between explanatory variables and treatments. Since three days were required to sample all the plots, the relationships between the main variables and treatment effects may have been influenced by the different sampling dates. Therefore, we included sampling date (3 days, one block each day) as a factor in the models. Unsurprisingly, there were some significant differences over time (see Supporting Information, Table S2). However, no significant interactions between sampling date and treatment and/or among sampling date, treatment and the main variables were found (Table S2), so time was dropped from the final models and treatment remained as the only factor. Finally, we ran the models first without treatment and later including treatment as a fixed factor to find those predictors mediated by the treatments. *A posteriori* contrasts were performed by reordering the levels of a factor to represent the effect of the treatments.

Akaike's Information Criterion (AIC) and the proportional increase in explained deviance (pseudo R^2; [33]) were used to evaluate each model fit and parsimony. In general, a variable was retained in the model only if it caused a significant increase in deviance when it was removed from the current model. Model assumptions were checked using (i) plots of residuals versus fitted values to verify homogeneity, (ii) quantile-quantile plots or histograms of the residuals for the assessment of normality, and (iii) residuals versus each explanatory variable to check for data independence. All the analyses were fitted in R 2.15.1 [34], glm package.

Results

Sediments at the sampling site were heterogeneous, with almost even mixtures of fine + very fine sand (63–250 µm, 26% by weight), medium sand (250–500 µm, 28.2%) and coarse + very coarse sand (500–2000 µm, 29.7%) (Figure 1). Gravel content (>2000 µm) was approximately 12%, with the large particles mostly comprised of bivalve shell fragments. Mud (<63 µm) was 4.2% of the total sedimentary composition (Figure 1).

There was a high degree of variability in the response variables that we measured, likely reflected the heterogeneity of the habitat (see Figures S1 and S2, Supporting Information). Sediment chlorophyll *a* (µg.g^{-1} sediment) and organic matter content (%) showed high variability (see Figure S1), and no significant differences were detected among the treatments (see Supporting Information, Table S1). Chla ranged between 2 to 4 µg.g^{-1} (IQR), and organic matter content between 2 to 3% (IQR) (Figure S1). Organic matter changed significantly ($p<0.01$) during the sampling dates, but no significant interaction was found between treatments (Table S1).

The abundance of large macrofaunal organisms was distributed haphazardly with respect to treatments at the site (Figure S1), with no significant differences detected among treatments (see Table S1 for the ANOVA summary). The rest of the macrofauna community components (total macrofauna abundance, species richness and the abundance of deposit and suspension feeders) also showed a variable distribution, again with no significant differences among treatments detected (see Table S1 and Figure S1, Supporting Information).

Water samples showed a pH range between 8.418 and 8.523 pH units (8.479±0.030). Statistical results indicated a significant change in pH during the sampling dates ($p<0.001$), but no significant interaction was found between treatments (see Table S1). Light intensity ($F_{2,24}=158.98$) and temperature ($F_{2,24}=48.3$) at the seabed changed significantly ($p<0.001$) during the 3 sampling days. However, light ($F_{3,23}=0.095$, $p=0.962$) and temperature ($F_{3,23}=0.614$, $p=0.613$) were not significantly different among treatments during this period.

The most parsimonious models (Models 1 to 4) relating different environmental and macrofauna community variables, first without treatment (a) and later including treatment (b) as a factor are presented in Table 1. Dissolved inorganic carbon in the presence

Table 2. ANOVA output from the GLM Model 1b from Table 1 (Gaussian distribution, identity link function) indicating the significance of ammonium uptake (NH_4^+) on gross primary production (GPP) among treatments (OM: Organic matter, CC: Calcium carbonate, Mix: OM+CC, and C: Control).

ANOVA output: GPP~NH_4^++ Treatment					**Treatment Contrasts (*p* results)**				
Source	**Rs df**	**Rs Dev**	**F**	***p***	**Tr**	**C**	**CC**	**Mix**	**OM**
Null	44	19561863			C	-			
NH_4^+	43	9960327	48.3	*<0.001****	CC	ns	-		
Treatment (Tr)	40	7951120	3.37	0.027*	Mix	***	*	-	
					OM	***	*	ns	-

No significant interaction was found (NH_4^+*Tr; $F_{1,3}=2.18$, $p=0.106$). The Treatment effect ($p<0.05$) sizes are showed as contrasts (see summary of the full regression-based model in Table S3).
ns: $p>0.1$,
$^{+}0.05<p<0.1$,
*$p<0.05$,
**$p<0.01$,
***$p<0.001$.
Rs df: residual degrees of freedom; Rs Dev: residual deviance.

of light (DIC_{light}), rates of NH_4^+ and NO_3^- uptake and the abundance of deposit feeders (DF) were significant predictors of benthic gross primary production (GPP) at our site in Kawau Bay (Model 1a), irrespective of treatment (Table 1). However, when treatment was included as a factor in the model, the only predictor variable showing a significant effect was NH_4^+ uptake (Table 1, Model 1b). NH_4^+ uptake showed a significant positive relationship with GPP (Figure 2a). There was no significant interaction between the response variable and treatment (NH_4^+*Tr) in this model (Model 1b; $F_{1,3}=2.18$, $p=0.106$). Therefore, we dropped the interaction term from the analysis and ran the model again (final Model 1b, Table 1). Our analysis revealed a significant effect of treatment on the form of the NH_4^+–GPP relationship (see Table 2). OM and Mix plots produced significantly stronger NH_4^+–GPP relationships than Control and CC plots (Table 2, Figure 2a). Significant differences were observed between treatment pairs, except for between Mix and OM, and C and CC. The GLM summary for all the treatments is presented as supporting information (see Table S3).

Gross primary production (GPP), dissolved reactive phosphorus flux (DRP) and the abundance of deposit feeders (DF) were significant predictors of DIN flux irrespective of treatment (Table 1, Model 2a). However, when treatment was included in the model, only DRP remained as a significant predictor variable (Table 1, Model 2b). The relationship between fluxes of DIN and DRP was positive overall (Figure 2b), though this relationship became negative as a result of the OM treatment (Figure 2b), signalling a significant interaction between DRP and treatment type in the prediction of DIN flux (Table 3). Judging from *post hoc* treatment contrasts, the form of the DRP–DIN flux relationship differed between OM plots and both Control and CC plots (Table 3). The GLM summary for all the treatments is presented as supporting information (see Table S4).

Macrofauna abundance was a significant predictor of chlorophyll *a* (Table 1, Model 3). The overall relationship, although negative and significant (Table 4), was not visually clear (Figure 3a). This relationship was mainly defined by the abundance*treatment interaction (Abund*Tr, Table 4). The chla vs. abundance relationship was significantly different in CC plots relative to OM and Control plots (Table 4). In the OM and Control plots this relationship was positive, whereas it was negative in the treatments containing calcium carbonate (Mix and CC) (Figure 3a). There

Table 3. ANOVA output from the GLM Model 2b (Gaussian distribution, identity link function) indicating the significance of dissolved reactive phosphorus (DRP) on dissolved inorganic nitrogen (DIN) among treatments (OM: Organic matter, CC: Calcium carbonate, Mix: OM+CC, and C: Control).

ANOVA output (GLM: DIN~DRP*Treatment)					**Contrasts (*p* results)**				
Source	**Rs df**	**Rs Dev**	**F**	***p***	**DRP*Tr**	**C**	**CC**	**Mix**	**OM**
Null	43	131192			C	-			
DRP	42	99762	17.01	*<0.001****	CC	ns	-		
Treatment (Tr)	39	95093	0.84	0.479	Mix	ns	ns	-	
DRP*Tr	36	66519	5.15	0.005**	OM	**	***	ns	-

The DRP*Treatment ($p<0.01$) interaction effects are showed as contrasts (see summary of the full regression-based model in Table S4).
ns: $p>0.1$,
$^{+}0.05<p<0.1$,
*$p<0.05$,
**$p<0.01$,
***$p<0.001$.
Rs df: residual degrees of freedom; Rs Dev: residual deviance.

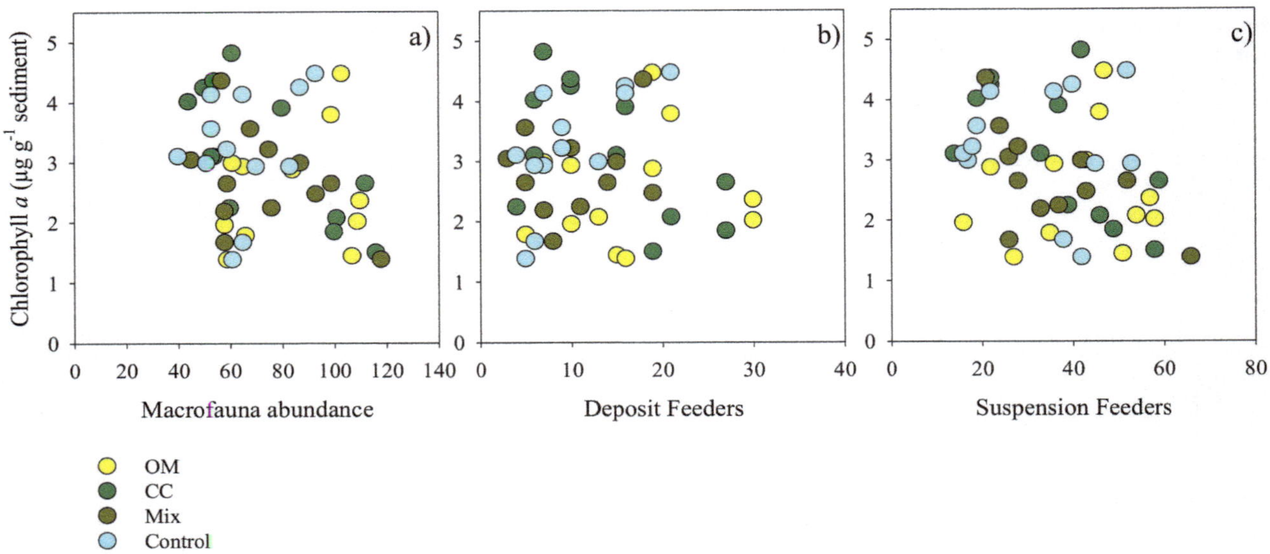

Figure 3. Biplots showing the relationship between chlorophyll *a* vs. macrofauna abundance, and the main feeding guilds. The panels show the general relationship across the sampling area. Chlorophyll *a* (µg g^{-1} sediment), total macrofauna abundance (number of individuals), main feeding guilds (abundance of deposit and suspension feeders; DF and SF, respectively). Plots are raw data. (Treatments: Organic Matter, Calcium carbonate, Mix, Control).

was also a significant and negative relationship between chla and the abundance of deposit feeders (DF, Figure 3b) mainly defined by the DF*treatment interaction (DF*Tr, Table 4). Again, this relationship differed in plots with and without calcium carbonate: negative in CC and Mix treatments, positive in OM and Control plots (Table 4, Figure 3b). The overall relationship between chla and the abundance of suspension feeders (SF) was also significant and negative ($F_{1,46} = 5.82$; $p = 0.021$), but no changes in the relationship according to treatment type were detected ($F_{3,40} = 1.76$; $p = 0.171$; Figure 3c). The GLM summary for all the treatments is shown as supporting information (see Table S5).

The abundance of deposit feeders was a significant predictor of light nitrate flux (Table 1, Model 4); the relationship was negative (Table 5, Figure 4). The relationship changed marginally ($p<0.10$) according to treatment, differing in the CC treatment relative to the OM and Mix treatments (Table 5, Contrasts). The estimated regression parameter was positive for CC and negative for OM and Mix (Figure 4). The GLM summary is given as supporting information (see Table S6).

Discussion

In marine soft-sediment systems, subtle changes to sediment characteristics have been shown to alter relationships between biogeochemical processes (e.g. nutrient utilisation and primary production [5,6]), suggesting that an analysis of coupled relationships may be appropriate for tracking the effects of subtle sublethal stressors.

In this study, we examined the response of a heterogeneous soft-sediment subtidal system to subtle, experimentally controlled, changes in sediment organic matter and calcium carbonate content. Organic matter occurs naturally in all sediments, with stocks of organic matter dependent upon the relative balance of supply (e.g. from the water column above) and degradation by organisms living in the sediment. The amount of organic matter in the sediment tends to be highest in eutrophic areas, with supply outpacing degradation. Organic enrichment of sediments has predictable consequences on sediment biogeochemistry and macrofaunal community composition [35]. For example, organic matter mineralization results in the formation of organic and inorganic acids, associated with O_2 consumption, which causes the pH of coastal bottom waters to decline [16,36]. However, most of the conceptual models for organic enrichment are designed to explain changes across very steep organic matter gradients.

Calcium carbonate is another naturally occurring substance in coastal sediments. Calcium carbonate concentrations are particularly high in tropical coral sands, but can be high in shell dominated sands in the middle and higher latitudes as well [21]. Whilst organic enrichment will result in increased rates of benthic respiration, elevated pCO_2 and increased porewater acidity, calcium carbonate is thought to counter these effects, acting as a buffer, though our knowledge of how (and whether) this occurs remains poorly understood [15,16].

In this study, we examined relationships between key biogeochemical and sedimentary variables to understand the individual and interactive effects of OM and CC addition. For example, we assessed the strength and variability of relationships between coupled processes such as gross primary production and ammonium release [6,7]. We expected primary production to be positively related to NH_4^+ uptake (with microphytes taking up this form of nitrogen in order to meet the demands of photosynthesis) and suggested the breakdown of this relationship in response to OM or CC additions to be a signal of impaired functioning. Furthermore, it is well known that stressors or responses such as abundance/diversity act in combination in natural systems, not in isolation. The interactive effects of multiple stressors on the ecosystem are more difficult to understand and predict [37–39]. For instance, in our study, the interaction between treatment and macrofauna abundance affected total chlorophyll *a* significantly; even though, according to our model (Table 4), treatment as a single variable was not significant. Therefore, the interactions may play a greater role in determining how an ecosystem responds than individual stressors, and at the same time those stressors may interact with other factors, such as abundance.

Table 4. ANOVA output from the GLM Model 3b (Gaussian distribution, identity link function) indicating the significance of chlorophyll *a* content on the total macrofauna abundance and the abundance of deposit feeders (DF) among treatments (OM: Organic matter, CC: Calcium carbonate, Mix: OM+CC, and C: Control).

ANOVA output (GLM: Chla~Abundance*Treatment)					**Abundance*Tr Contrasts (*p* results)**				
Source	**Rs df**	**Rs Dev**	**F**	***p***	**Tr**	**C**	**CC**	**Mix**	**OM**
Null	47	45.44			C	-			
Abundance	43	41.4	5.12	*0.03**	CC	*	-		
Treatment (Tr)	43	38.7	1.2	0.353	Mix	ns	ns	-	
Abund*Tr	40	31.6	3.02	*0.04**	OM	ns	*	ns	-
ANOVA output (GLM:Chla~DF*Treatment)					**DF*Tr Contrasts (*p* results)**				
Source	**Rs df**	**Rs Dev**	**F**	***p***	**Tr**	**C**	**CC**	**Mix**	**OM**
Null	46	43.1			C	-			
DF	45	42.9	0.19	0.658	CC	***	-		
Treatment (Tr)	42	39.1	1.71	0.18	Mix	ns	ns	-	
DF*Tr	39	29.3	4.37	*0.009***	OM	+	*	ns	-

The Abundance*Treatment ($p<0.05$) and DF*treatment ($p<0.01$) interaction effects are showed as contrasts (see summary of the full regression-based model in Table S5).
ns: $p>0.1$,
$^{+}0.05<p<0.1$,
$^{*}p<0.05$,
$^{**}p<0.01$,
$^{***}p<0.001$.
Rs df: residual degrees of freedom; Rs Dev: residual deviance.

Table 5. ANOVA output from the GLM Model 4b (Gaussian distribution, identity link function) indicating the significance of NO_3^- flux during daylight on the abundance of deposit feeders (DF) among treatments (OM: Organic matter, CC: Calcium carbonate, Mix: OM+CC, and C: Control).

ANOVA output: $NO_{3\ light}^-$~DF*Treatment					DF*Tr Contrasts (*p* results)				
Source	**Rs df**	**Rs Dev**	**F**	***p***	**Tr**	**C**	**CC**	**Mix**	**OM**
Null	46	10145.6			C	-			
DF	45	9271.6	4.65	*0.03**	CC	ns	-		
Treatment (Tr)	42	8716.1	0.98	0.411	Mix	ns	*	-	
DF*Tr	39	7343.5	2.43	0.07^+	OM	ns	*	ns	-

The DF*Treatment interaction (p = 0.07) effect is showed as contrasts (summary of the full regression-based model in Table S6).
ns: $p>0.1$,
$^+0.05<p<0.1$,
$^*p<0.05$,
$^{**}p<0.01$,
$^{***}p<0.001$.
Rs df: residual degrees of freedom; Rs Dev: residual deviance.

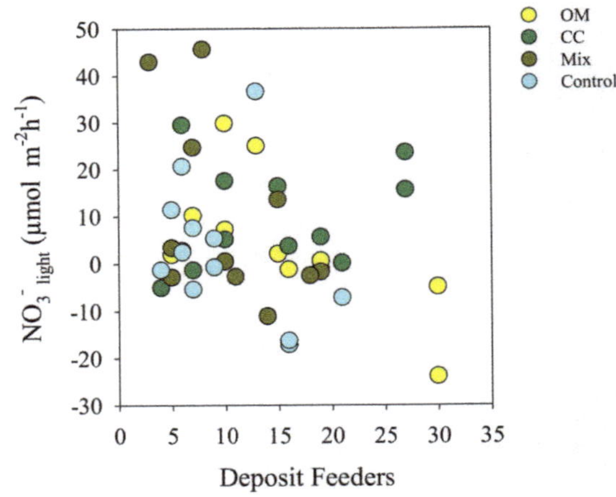

Figure 4. Relationship between nitrate flux during daylight (μmol $NO_3^-{}_{light}$ $m^{-2}h^{-1}$) and the abundance of deposit feeders. The panel shows the general relationship across the sampling area. Plots are raw data. (Treatments: Organic Matter, Calcium carbonate, Mix, Control).

The present experiment showed that the additions tended to act in opposite directions, as we hypothesized, and demonstrated that highly heterogeneous sedimentary habitats can be surprisingly sensitive to subtle perturbations. For instance, fluxes of NH_4^+ were affected by the experimental manipulation. It is known that the decomposition and remineralisation of OM in sediments releases nutrients to the overlying water [40], and that the first form of DIN to appear during OM remineralisation is NH_4^+. This essential nutrient is also readily utilised by bacteria and algae and has a positive effect on primary production in systems where nitrogen is limiting [41,42]. Gross primary production (GPP) was significantly higher in OM and Mix plots during this experiment when rates of NH_4^+ uptake were elevated, and this increase was significantly higher than in CC or Controls. We also documented changes in the relationship between N and P fluxes following treatment application, with a marked increase in nitrogen (DIN) to phosphorus (DRP) flux ratios in the OM plots. Regeneration of both N and P is expected to occur during the remineralisation of OM [43], generating the positive relationship between the two variables.

Although the application of organic matter and calcium carbonate material to the plots affected relationships between macrofauna and sediment biogeochemistry, no significant changes in macrofauna community characteristics (i.e., macrofauna abundance, diversity, number of species, feeding guilds) occurred among plots. However, macrofauna abundance was correlated with variables including nutrient fluxes and primary production. Specifically, the abundance of deposit feeders showed a negative relationship with NO_3^- flux during daylight in Control and OM plots, but the relationship changed to positive when calcium carbonate was added to the CC plots. Similarly, the relationship between macrofauna abundance (and deposit feeders) and chlorophyll *a* was positive in OM plots, but became negative in plots with both CC and Mix. The DIN that is regenerated during OM decomposition is likely available for use by microphytobenthos, the abundance of which can be inferred from chla. Moreover, an increase in the OM decomposition will tend to promote an increase in sediment oxygen demand, as bacteria and macrofauna utilise this source of carbon. A reduced availability of O_2, and an excess of CO_2, will reduce nitrification rates and therefore NO_x^- efflux in OM plots [44,45]. Thus, OM additions can conceivably account for an increase in NH_4^+ and chla in the treated plots and the negative effects on nitrate flux. CC additions would tend to counteract these effects.

Most of the seafloor is composed of unconsolidated sediments, and considerable biodiversity exists in these types of systems [46,47]. Most of the studies of carbonate sediments have focused on coral sands in tropical areas, and there is little information on the influence of carbonate materials on nutrient fluxes and biologically mediated ecosystem functions. The role of carbonates is increasingly important to define, given the steady rise in atmospheric CO_2 concentration and coastal eutrophication, with resultant acidification of seawater. CO_2 is sequestered in the form of $CaCO_3$ by calcifying species, and this remains in the hard shells in the sediment long after the organisms have died [48–50]. Understanding what the changes in sediment calcium carbonate concentration mean for marine ecosystem functions is pivotal in a world with rising atmospheric CO_2 concentrations [13].

The present contribution points out the high sensitivity of a highly heterogeneous soft-sediment ecosystem to small changes in biogeochemical drivers. Our study suggests that sediment carbonates may play a role in partially ameliorating coastal eutrophication effects by counteracting the effects of sediment organic matter enrichment. By analysing shifts in coupled relationships (*sensu* [6]) we were able to detect changes in the functioning of heterogeneous shallow marine carbonate sands at very low magnitudes of disturbance. This is useful, as predicting an ecosystem response prior to large and potentially irreversible changes is the key to the sustainability of our natural resources.

Supporting Information

Figure S1 Boxplots on sediment characteristics: chlorophyll a (μg.g-1 sediment), organic matter (%), and macrofauna community abundance (number of individuals) and Shannon's diversity index (H′) through the four treatments (Calcium carbonate, Control, Mix and Organic matter).

Figure S2 Boxplots showing the variability of the oxygen and nutrient fluxes through the four treatments (Calcium carbonate, Control, Mix and Organic matter). Gross primary production (GPP), net primary production (NPP), ammonium uptake (NH_4^+), dissolved reactive phosphorus (DRP), dissolved inorganic nitrogen uptake (DIN_{uptake}), dissolved inorganic nitrogen in daylight (DIN_{light}), nitrate uptake and during daylight ($NO_3^-{}_{uptake}$ and $NO_3^-{}_{light}$), dissolved inorganic carbon during daylight (DIC_{light}). Units: μmol m^{-2} h^{-1}.

Figure S3 Schematic diagram showing the layout of the plots with the treatments (Calcium carbonate, Organic matter, Mix: OM+CC, and Control) across the sampling area in a balanced orthogonal design with randomized interspersed treatment positions (3 blocks). Below a picture showing a diver introducing the treatment inside one of the plots through a syringe activated sampling port. Plots were made of two parts: a square chamber base (aluminum border with sides 50×50 cm×10 cm tall) pressed down into the sediment (~5 cm) and a clear acrylic plastic dome (50×50 cm) fitted to the chamber bases and clamped in place.

Table S1 Summary of the ANOVAs (Treatment and blocks as fixed factors) for sediment characteristics (chlorophyll a and organic matter), water pH and macrofauna communities (Abundance of large organisms, total abundance, number of species and the abundance of deposit and suspension feeders). Treatments: OM: Organic matter, CC: Calcium carbonate, Mix: OM+CC, Control. Significant results in italics.

Table S2 Summary of the ANOVA outputs from the Generalized Linear Models (Table 1), including Treatment and Day (3 sampling days) as fixed factors.

Table S3 Generalized Linear Model summary (regression-based models with Gaussian distribution and identity link function) indicating the significance of ammonium uptake (NH_4^+) on gross primary production (GPP) among treatments (OM: Organic matter, CC: Calcium carbonate, Mix: OM+CC, Control).

Table S4 Generalized Linear Model summary (regression-based models with Gaussian distribution and identity link function) indicating the significance of dissolved reactive phosphorus (DRP) on dissolved inorganic nitrogen (DIN) among treatments (OM: Organic matter, CC: Calcium carbonate, Mix: OM+CC, Control).

Table S5 Generalized Linear Model summary (regression-based models with Gaussian distribution and identity link function) indicating the significance of chlorophyll *a* content on the total macrofauna abundance and the abundance of deposit feeders among treatments (OM: Organic matter, CC: Calcium carbonate, Mix: OM+CC, Control). SE: Standard Error.

Table S6 Generalized Linear Model summary (regression-based models with Gaussian distribution and identity link function) indicating the significance of NO_3^- flux during daylight on the abundance of deposit feeders (DF) among treatments (OM: Organic matter, CC: Calcium carbonate, Mix: OM+CC, Control).

Acknowledgments

We thank our dive buddies and those who assisted in the field and in the lab, D. Buckthought, K. Cartner, S. Edhouse, B. Greenfield and J. Simpson. We appreciate discussions on carbonate chemistry with C. Depree. We thank J. Hewitt for critical reviewing an earlier version of the manuscript. We are very grateful to two anonymous reviewers and the Academic Editor who provided thorough constructive critiques and insightful comments on previous versions of the manuscript.

Author Contributions

Conceived and designed the experiments: IFR AML SFT. Performed the experiments: IFR AML. Analyzed the data: IFR. Contributed reagents/materials/analysis tools: IFR AML SFT. Wrote the paper: IFR AML SFT.

References

1. Dayton PK, Tegner MJ (1984) Catastrophic storms, El Niño, and patch stability in a southern California kelp forest community. Science 224:283–285.
2. Foster DR, Knight DH, Franklin JF (1998) Landscape patterns and legacies resulting from large, infrequent forest disturbances. Ecosystems 1: 497–510.
3. Peterson CH, Rice SD, Short JW, Esler D, Bodkin JL, et al. (2003) Long-Term Ecosystem Response to the Exxon Valdez Oil Spill. Science 302: 2082–2086.
4. Jaramillo E, Dugan JE, Hubbard DM, Melnick D, Manzano M, et al. (2012) Ecological Implications of Extreme Events: Footprints of the 2010 Earthquake along the Chilean Coast. PLoS ONE 7(5): e35348.
5. Lohrer AM, Hewitt JE, Hailes SF, Thrush SF, Ahrens M, et al. (2011) Contamination on sandflats and the decoupling of linked ecological functions. Austral Ecol 36l: 378–388.
6. Lohrer AM, Townsend M, Rodil IF, Hewitt JE, Thrush SF (2012) Detecting shifts in ecosystem functioning: The decoupling of fundamental relationships with increased pollutant stress on sandflats. Mar Pollut Bull 64: 2761–2769.
7. Rodil IF, Lohrer AM, Chiaroni LD, Hewitt JE, Thrush SF (2011) Disturbance of sandflats by thin terrigenous sediment deposits: consequences for primary production and nutrient cycling. Ecol Appl 21(2): 416–426.
8. Thrush SF, Hewitt JE, Lohrer AM, Chiaroni LD (2013) When small changes matter: the role of cross-scale interactions between habitat and ecological connectivity in recovery. Ecol Appl 23(1): 226–238.
9. Parmesan C, Yohe G (2003) A globally coherent fingerprint of climate change impacts across natural systems. Nature 421: 37–42.
10. Wethey DS, Woodin SA (2008) Ecological hindcasting of biogeographic responses to climate change in the European intertidal zone. Hydrobiologia 606: 139–151.
11. Widdicombe S, Dashfield SL, McNeill CL, Needham HR, Beesley A, et al. (2009) Effects of CO_2 induced seawater acidification on infaunal diversity and sediment nutrient fluxes. Mar Ecol Prog Ser 379: 59–75.
12. Hale R, Calosi P, McNeill L, Mieszkowska N, Widdicombe S (2011) Predicted levels of future ocean acidification and temperature rise could alter community structure and biodiversity in marine benthic communities. Oikos 120: 661–674.
13. Feely RA, Sabine CL, Lee K, Berelson W, Kleypas J, et al. (2004) Impact of Anthropogenic CO_2 on the $CaCO_3$ System in the Oceans. Science 305: 362–366.
14. Orr JC, Fabry VJ, Aumont O, Bopp L, Doney SC, et al. (2005) Anthropogenic ocean acidification over the twenty-first century and its impact on calcifying organisms. Nature 437: 681–686.
15. Borges AV, Gypens N (2010) Carbonate chemistry in the coastal zone responds more strongly to eutrophication than to ocean acidification. Limnol Oceanogr 55(1): 346–353.
16. Sunda WG, Cai W-J (2012) Eutrophication induced CO_2-acidification of subsurface coastal waters: Interactive effects of temperature, salinity, and atmospheric P_{CO2}. Environ Sci Technol 46: 10651–10659.
17. Crain CM, Kroeker K, Halpern BS (2008) Interactive and cumulative effects of multiple human stressors in marine systems. Ecol lett 11(12): 1304–1315.
18. Darling ES, Cote IM (2008) Quantifying the evidence for ecological synergies. Ecol lett 11(12): 1278–1286.
19. Alongi DM, Pfitzner J, Trott LA (2006) Deposition and cycling of carbon and nitrogen in carbonate mud of the lagoons of Arlington and Sudbury Reefs, Great Barrier Reef. Coral Reefs 25: 123–143.

20. Rao AMF, Polerecky L, Ionescu D, Meysman FJR, de Beer D (2012) The influence of pore-water advection, benthic photosynthesis, and respiration on calcium carbonate dynamics in reef sands. Limnol Oceanogr 57(3): 809–825.
21. Nelson CS (1988) An introductory perspective on non-tropical shelf carbonates. Sediment Geol 60: 3–12.
22. Morse JW, Andersson AJ, Mackenzie FT (2006) Initial response of carbonate-rich shelf sediments to rising atmospheric pCO_2 and "ocean acidification": Role of high Mg-calcites. Geochim Cosmochim Acta 70: 5814–5830.
23. Egleston ES, Sabine CL, Morel FMM (2010) Revelle revisited: Buffer factors that quantify the response of ocean chemistry to changes in DIC and alkalinity. Global Biogeochem Cycles 24, GB1002, doi:10.1029/2008GB003407.
24. Tynan S, Opdyke BN (2011) Effects of lower surface ocean pH upon the stability of shallow water carbonate sediments. Sci Total Environ 409: 1082–1086.
25. Lerman A, Mckenzie FT (2005) CO_2 air-sea exchange due to calcium carbonate and organic matter storage, and its implications for the global carbon cycle. Aquat Geochem 11: 345–390.
26. Lohrer AM, Thrush SF, Gibbs MM (2004) Bioturbators enhance ecosystem function through complex biogeochemical interactions. Nature 431:1092–1095.
27. Norling K, Rosenberg R, Hulth S, Grémare A, Bonsdorff E (2007) Importance of functional biodiversity and species-specific traits of benthic fauna for ecosystem functions in marine sediment. Mar Ecol Prog Ser 332: 11–23.
28. Thrush SF, Hewitt JE, Funnell GA, Cummings VJ, Ellis J, et al. (2001) Fishing disturbance and marine biodiversity: role of habitat structure in simple soft-sediment systems. Mar Ecol Prog Ser 221: 255–264.
29. Hamilton SK, Kurzman AL, Arango C, Jin L, Robertson P (2007) Evidence for carbon sequestration by agricultural liming. Global Biogeochem Cycles 21: GB2021.
30. Welsh DT (2003) It's a dirty job but someone has to do it: the role of marine benthic macrofauna in organic matter turnover and nutrient recycling to the water column. Chem Ecol 19: 321–342.
31. Grashoff K, Ehrhardt KM, Kremling F (1983) Methods of seawater analysis. Second edition. Verlag Chemie, Weinheim, Germany. 317 p.
32. Sartory DP (1982) Spectrophotometric analysis of chlorophyll a in freshwater phytoplankton. Technical Report TR 115, Hydrological Research Institute, Department of Environmental Affairs, Pretoria, South Africa.
33. Dobson AJ (2002) Introduction to Generalized Linear Models. Second Edition. Chapman & Hall/CRC Press. 225 p.
34. R Development Core Team (2012) R: A language and environment for statistical computing. R Foundation for Statistical Computing, Vienna, Austria. ISBN 3-900051-07-0, Available: http://www.R-project.org/.
35. Pearson TH, Rosenberg R (1978) Macrobenthic succession in relation to organic enrichment and pollution of the marine environment. Oceanogr Mar Biol: Ann Rev 16: 229–311.
36. Doney SC, Mahowald N, Lima I, Feely RA, Mackenzie FT, et al. (2007) Impact of anthropogenic atmospheric nitrogen and sulfur deposition on ocean acidification and the inorganic carbon system. Proc Natl Acad Sci 104 (37): 14580–14585.
37. Blake RE, Duffy JE (2010) Grazer diversity affects resistance to multiple stressors in an experimental seagrass ecosystem. Oikos 119(10): 1625–1635.
38. Hicks N, Bulling MT, Solan M, Raffaelli D, White PC, et al. (2011) Impact of biodiversity-climate futures on primary production and metabolism in a model benthic estuarine system. BMC Ecol 11: 7.
39. Wernberg T, Smale DA, Thomsen MS (2012) A decade of climate change experiments on marine organisms: procedures, patterns and problems. Glob Chang Biol 18(5): 1491–1498.
40. Engelsen A, Hulth S, Pihl L, Sundbäck K (2008) Benthic trophic status and nutrient fluxes in shallow-water sediments. Estuar Coast Shelf Sci 78: 783–795.
41. Sundbäck K, Miles A, Göransson E (2000) Nitrogen fluxes, denitrification and the role of microphytobenthos in microtidal shallow-water sediments: an annual study. Mar Ecol Prog Ser 200: 59–76.
42. Clavier J, Boucher G, Chauvaud L, Fichez R, Chifflet S (2005) Benthic response to ammonium pulses in a tropical lagoon: implications for coastal environmental processes. J Exp Mar Biol Ecol 316: 231–241.
43. Sundby B, Gobeil C, Silverberg N, Mucci A (1992) The phosphorus cycle in coastal marine sediments. Limnol Oceanogr 49:1095–1107.
44. Hueseman MH, Skillman AD, Crecelius EA (2002) The inhibition of marine nitrification by ocean disposal of carbon dioxide. Mar Pollut Bull 44: 142–148.
45. Hulth S, Aller RC, Canfield DE, Dalsgaard T, Engström P, et al. (2005) Nitrogen removal in marine environments: recent findings and future research challenges. Mar Chem 94: 125–145.
46. Gray JS (1997) Marine biodiversity: patterns, threats and conservation needs. Biodivers Conserv 6:153–175.
47. Snelgrove PVR (1999) Getting to the bottom of marine biodiversity: sedimentary habitats–ocean bottoms are the most widespread habitat on Earth and support high biodiversity and key ecosystem services. BioScience 49:129–138.
48. Miller AW, Reynolds AC, Sobrino C, Riedel GF (2009) Shellfish Face Uncertain Future in High CO_2 World: Influence of Acidification on Oyster Larvae Calcification and Growth in Estuaries. PLoS ONE 4(5): e5661.
49. Cebrián J (2002) Variability and control of carbon consumption, export, and accumulation in marine communities. Limnol Oceanogr 47(1): 11–22.
50. Tang Q, Zhang J, Fang J (2011) Shellfish and seaweed mariculture increase atmospheric CO_2 absorption by coastal ecosystems. Mar Ecol Prog Ser 424: 97–104.

Calcium Induces Long-Term Legacy Effects in a Subalpine Ecosystem

Urs Schaffner[1]*, Christine Alewell[2], René Eschen[1], Diethart Matthies[3], Thomas Spiegelberger[1,4], Otto Hegg[5]

1 CABI, Delémont, Switzerland, **2** Department of Environmental Sciences, University of Basel, Basel, Switzerland, **3** Department of Biology, University of Marburg, Marburg, Germany, **4** Ecosystèmes montagnards, Institut national de recherche en sciences et technologies pour l'environnement et l'agriculture, Saint-Martin-d'Hères, France, **5** Department of Biology, University of Bern, Bern, Switzerland

Abstract

Human activities have transformed a significant proportion of the world's land surface, with profound effects on ecosystem processes. Soil applications of macronutrients such as nitrate, phosphorus, potassium or calcium are routinely used in the management of croplands, grasslands and forests to improve plant health or increase productivity. However, while the effects of continuous fertilization and liming on terrestrial ecosystems are well documented, remarkably little is known about the legacy effect of historical fertilization and liming events in terrestrial ecosystems and of the mechanisms involved. Here, we show that more than 70 years after the last application of lime on a subalpine grassland, all major soil and plant calcium pools were still significantly larger in limed than in unlimed plots, and that the resulting shift in the soil calcium/aluminium ratio continues to affect ecosystem services such as primary production. The difference in the calcium content of the vegetation and the topmost 10 cm of the soil in limed vs. unlimed plots amounts to approximately 19.5 g m^{-2}, equivalent to 16.3% of the amount that was added to the plots some 70 years ago. In contrast, plots that were treated with nitrogen-phosphorus-potassium fertilizer in the 1930s did not differ from unfertilized plots in any of the soil and vegetation characteristics measured. Our findings suggest that the long-term legacy effect of historical liming is due to long-term storage of added calcium in stable soil pools, rather than a general increase in nutrient availability. Our results demonstrate that single applications of calcium in its carbonated form can profoundly and persistently alter ecosystem processes and services in mountain ecosystems.

Editor: Minna-Maarit Kytöviita, Jyväskylä University, Finland

Funding: The work was funded by the Bristol Stiftung (http://www.bristol-stiftung.ch). The funders had no role in study design, data collection and analysis, decision to publish, or preparation of the manuscript.

Competing Interests: The authors have declared that no competing interests exist.

* E-mail: U.Schaffner@cabi.org

Introduction

While intensive forms of land-use are essential for humans to obtain natural products such as food and fibre, they have contributed to the transformation of ecosystem patterns and processes [1]. In particular, human activities have altered the rate, pathways and efficiency of the movement of nutrients within and between the various biotic or abiotic ecosystem compartments. These changes in nutrient cycles do not only affect today's ecosystem functioning but may also result in long-term legacy effects on ecosystem processes, thereby changing the resilience of ecosystems and their adaptive capacity to sustain ecosystem services in the face of uncertainty and global change [2,3,4].

So far, most emphasis in terms of terrestrial nutrient cycling has been put on carbon (C), nitrogen (N) and phosphorus (P), because of their key roles in primary production and climate regulation [5,6]. For example, the effects of continuous fertilization of grasslands [7,8,9,10], and to some extent also the changes after cessation of fertilization [11,12,13], have been well documented. Less well understood are the short-term and long-term effects on ecosystem patterns and processes of the addition of elements other than C, N or P, although they are also known to affect ecosystem functioning. Calcium (Ca) is a key base cation in the soil influencing various physiological and structural processes both in animals and plants [14]. Ca is a component of the soil buffering system, as CaCO3– HCO3- is an important sink for protons with a high buffering rate. Furthermore, Ca as a bivalent cation with a rather strong sorption capacity is very effective in binding to soil particles, thereby increasing aggregate formation in soils and improving soil structure significantly, and an important early indicator of disturbance of nutrient cycling in terrestrial ecosystems [14]. It also affects soil pH and consequently all major soil processes that are modulated by soil pH. Large amounts of lime ($CaCO_3$) are applied to croplands, pastures and forests in many regions of the world [15,16,17], and they are expected to further increase globally due to the expansion of intensive land-use practices [18]. While liming can improve plant health in naturally or anthropogenically acidified soils [19,20], it may also contribute to the loss of rare and threatened species [11,16,21] and is discussed controversially as either decreasing [22] or increasing the CO_2 sink capacity of soils [23]. Processes that influence Ca availability are therefore highly relevant to ecosystem regulation and management [20,23,24].

In the early 1930s, a unique long-term experiment was set up at the Schynige Platte, Bernese Oberland (Switzerland), to improve

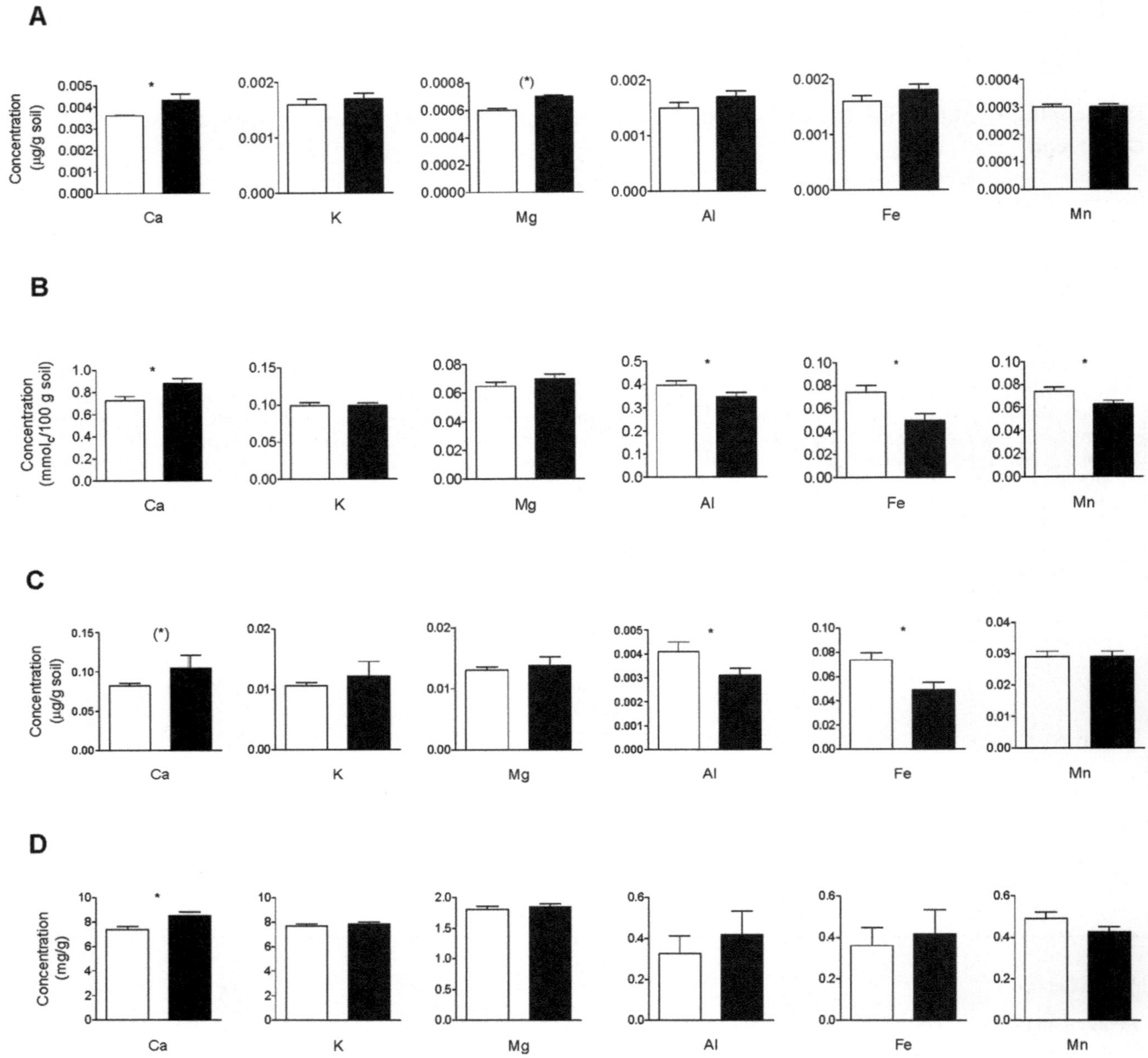

Figure 1. Cation concentrations in soil compartments and in above-ground vegetation of the subalpine grassland at Schynige Platte. A) Soil solution, B) soil cation exchange capacity, C) soil microbial community, and D) above-ground vegetation. Ca = calcium, K = potassium, Mg = magnesium, Al = aluminium, Fe = iron, Mn = manganese. Means +1 SE. Black bars indicate limed plots, white bars unlimed plots (*, $P<0.05$; (*), $P<0.1$).

site productivity and fodder quality and vegetational composition of subalpine grasslands [25]. Natural acidification of the top soil layers through leaching of cations is a common process in subalpine grasslands of the European Alps, since precipitation exceeds evapotranspiration considerably [14]. On the experimental site at the Schynige Platte, liming and NPK (nitrogen-phosphorus-potassium) fertilization were applied for a period of 2–4 years. Initially, both treatments had strong effects on vegetational composition and led to a significant increase in primary production [25]. Revisiting the experimental site more than 70 years after the last applications had been made, Spiegelberger et al. [11] found that limed plots continued to differ significantly from control plots in the composition of the soil microbial community and the vegetation, with a significant reduction in abundance of several rare plant species. In contrast, the initially strong effects of NPK fertilization on vegetation composition and primary production were not detectable any more [11].

Here we set out to compare the chemical composition of below- and above-ground ecosystem compartments of limed, fertilized and control plots. We focused our analyses on cations indicating increased nutrient availability, i.e. calcium (Ca), potassium (K) and magnesium (Mg), and on indicator cations of nutrient-poor acid soil conditions, i.e. aluminium (Al), iron (Fe) and manganese (Mn).

We also analysed total concentrations of Ca, N, and C as well as soil organic matter, and we assessed whether primary production is correlated with concentrations of Ca or N, soil pH or the Ca/Al ratio. The latter is important for soil processes, since free ionic Al is an important determinant of acidity stress because of its toxic effects on plant as well as microbial cell physiology, and Ca can compete with Al at cell membrane binding sites [14,26,27]. The concentration of Ca in the top 10 cm soil and in the above-ground vegetation was also used to estimate the difference in total Ca content between limed and control plots. We hypothesized that a) a significant amount of the Ca added more than 70 years ago had been retained in the soil-vegetation loop, despite the fact that cation leaching is a well-known process in subalpine grasslands, and that b) liming induced a shift towards either a generally higher nutritional level or a reduced acidity stress in the soil. Based on our results, we propose a possible mechanism for the long-term legacy effects of historical liming in this subalpine ecosystem.

Results

We found that limed and unlimed plots, 70 years after treatments were applied, still differed significantly in their Ca concentrations in the soil soluble fraction ($F_{1,73} = 4.028$; $P = 0.048$), the soil cation exchange capacity ($F_{1,73} = 6.064$; $P = 0.016$), as well as in total Ca concentration in the soil ($F_{1,73} = 7.431$; $P = 0.008$) and in the above-ground plant biomass ($F_{1,73} = 7.524$; $P = 0.008$). Ca concentration in the soil microbial fraction was marginally higher in limed plots ($F_{1,73} = 3.364$; $P = 0.071$) (Fig. 1; Table 1). Plots that had been treated for two years and plots that had been treated for four years did not differ significantly in Ca concentration in the different soil compartments or in the above-ground plant biomass (all $P>0.1$). Interestingly, with the exception of a marginally higher Mg concentration in the soil solution ($F_{1,73} = 3.302$; $P = 0.073$), the concentration of other nutrient elements that are indicators of improved nutrient availability under neutral or basic soil conditions, such as N, P and K, did not differ between limed and unlimed plots (Fig. 1; Table 1). In contrast, the percentage of exchangeable Al ($F_{1,73} = 5.113$; $P = 0.027$;), Fe ($F_{1,73} = 4.610$; $P = 0.035$) and Mn ($F_{1,73} = 5.757$; $P = 0.019$) were significantly lower, and the concentration of Al ($F_{1,73} = 4.427$; $P = 0.039$) and Fe ($F_{1,73} = 8.113$; $P = 0.006$) was also lower in the microbial fraction of limed plots. The soil C concentration did not differ between limed and unlimed plots (Table 1).

In contrast to limed plots, the plots treated with NPK fertilizer did not differ from the control plots in the concentration of Ca, Mg, K, Al, Fe or Mn cations in any of the ecosystem compartments analysed, or in soil total Ca, N or C concentrations (all $P>0.2$). Also, whether lime was added in the 1930s in combination with NPK fertilizer or alone did not affect cation concentrations in the different soil fractions or in the vegetation (for all interactions NPK×lime $P>0.2$).

In the early 1930s, each of the plots received either 80 g Ca m^{-2} (two years of application) or 160 g Ca m^{-2} (four years of application), respectively [26]. The results from total soil digestion and from vegetation analysis indicate that limed plots contained 143.5 g Ca m^{-2} in the top 10 cm of the soil and ca. 3.0 g Ca m^{-2} in the vegetation, while unlimed plots contained 124.6 g Ca m^{-2} in the soil and 2.4 g Ca m^{-2} in the vegetation. Hence, to date the average difference in Ca content in limed vs. unlimed plots in the top 10 cm of the soil amounts to 18.9 g m^{-2}, and that in the vegetation to 0.6 g m^{-2}. The sum (19.4 g Ca m^{-2}) is equivalent to 16.3% of the average amount of Ca that had been added to the soil in the early 1930s. Based on the differences in productivity and in Ca concentration of the above-ground biomass between limed and unlimed plots (Fig. 1; Table 1), annual removal of the above-ground biomass led to a 0.28 g higher loss of Ca in limed than in unlimed plots in 2006. Assuming that the differences in productivity and Ca concentration of the above-ground biomass between limed and unlimed plots have been relatively stable over the course of the 71 years between the last lime application and the year of soil and vegetation sampling, annual removal of the above-ground vegetation led to a higher loss of Ca on limed plots of c. 20.2 g Ca m^{-2}, equivalent to 16.8% of the Ca that had been added in the early 1930s. It should be noted, though, that these numbers are estimates based on a one-time sampling and should therefore be treated with caution; however, since the productivity of limed plots in the early years after the treatments were significantly higher than 70 years later [25], the numbers may be considered as conservative estimates.

Annual primary production increased significantly with the Ca/Al ratio (Fig. 2), but was not related to total soil Ca or N concentration or to soil pH (all $P>0.3$).

Discussion

Our results indicate that a significant amount of the added Ca is still present in the top soil layer and in the vegetation more than 70 years after the last liming treatment. Little is known about the migration velocity of Ca in the soil, but our findings are consistent

Table 1. Vegetation and soil characteristics in limed and unlimed plots.

Fraction	Limed	Unlimed	$F_{1,73}$	P
Above-ground biomass (g m^{-2})	117.0±0.90	95.8±0.57	4.250	0.043
Soil organic matter content (%)	16.0±0.4	17.2±0.9	3.591	0.062
Soil Ca concentration (‰)	2.05±0.08	1.78±0.07	7.344	0.008
Soil C concentration (%)	7.33±0.17	7.72±0.23	2.135	0.148
Soil H concentration (%)	1.43±0.03	1.48±0.03	1.244	0.268
Soil N concentration (%)	0.60±0.01	0.62±0.01	2.404	0.125

Mean values ±1 SE are shown.

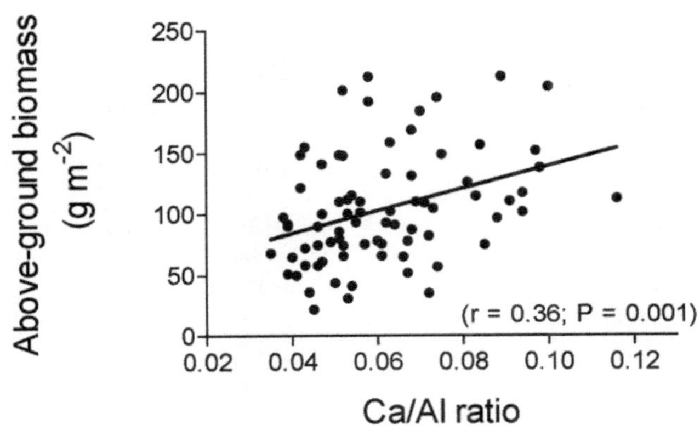

Figure 2. Relationship between the calcium to aluminium (Ca/Al) ratio in the soil and annual above-ground primary production in the subalpine grassland at the Schynige Platte.

with results from studies investigating the migration of ^{90}Strontium, an ion generally believed to behave very similar to Ca in soils and plants. For example, Bossew et al. [28] concluded that the migration velocity of ^{90}Strontium within the soil profile is between 0.1 and 2 mm yr^{-1} in a sandy soil. Assuming Ca to have a similar migration pattern, the major fraction of Ca should not have migrated more than 14 cm in depth. Since the soils at Schynige Platte have a higher content of silt, clay and organic matter than those studied by Bossew et al. [28] and thus have a higher sorption capacity and affinity for Ca, migration should be even slower. Of the added Ca still stored in the soil only 15% was found in the exchangeable Ca fraction. The latter is noteworthy, because this implies that the major part of the Ca is, either organically or as a mineral non-exchangeable fraction, bound in the soil. However, observed differences in vegetation between limed and unlimed plots are nevertheless likely to be related to differences in Ca.

There are two possible explanations for the long-term effects of liming: 1) Liming induced a general shift towards a higher nutritional level of the soil resulting in a relatively stable new state for the whole ecosystem (considering that the effects of 2–4 years of liming are still visible after more than 70 years), or 2) the added Ca is stored in stable "permanent" forms and thus not easily leached from the soil. Even though the soil pH is still enhanced in limed plots [11], no effect was observed on the other macro-nutrients (Mg, P, N, K). Thus, our data do not provide evidence that the whole system shifted towards a state of higher nutritional status. Rather, our findings suggest that long-term storage of the added Ca in relatively stable soil pools is responsible for the long-term legacy effect of single liming events on this subalpine ecosystem. As already observed by Lüdi [25], liming favoured plant species that have high rates of nutrient acquisition and high relative growth rates. We propose that a significant amount of lime that was applied in the 1930 was incorporated in plant or microbial biomass. The dead, Ca-enriched biomass that remained in the field was then incorporated into the soil organic matter pool, thereby increasing the Ca fraction that was more permanently bound. Since mountain grasslands can, despite the dominance of perennial species, undergo relatively rapid species turnover at small scales [29], we propose that the current plant species composition [11] is still affected by the altered soil Ca levels in limed plots. This implies that the more permanently bound Ca fractions are available to plants and/or microorganisms, or are affected by longer term weathering. The exchangeable cation pool in soils is a very old concept in soil science dating from the beginning of the 20th century [30]. Several studies have challenged the concept and asked for a redefinition considering (i) organically bound nutrient cations, especially in acid soils [30], and/or (ii) the existence of pools available to plant and microorganisms other than the 'exchangeable pool' [15,31]. It has been suggested, that Ca can be made directly available from rocks, most likely via mycorrhizal fungi, which are able to dissolve Ca from Ca feldspars and apatite [31,32]. Hence, plants with their symbiotic fungi in the limed plots might be able to take up Ca from non-exchangeable pools in the soil, thereby refilling the exchangeable Ca pool, or they use Ca from the exchangeable pool that is continuously refilled through weathering/decomposition of the Ca enriched 'non-exchangeable fraction'.

A very profound effect of the liming was an increase in soil pH from 4.5 to almost 7 within a few years after the treatments [25]. The soil pH has decreased considerably since, but is still higher than that of the unlimed plots (pH_{soil} 5.14±0.04 vs. 4.97±0.04) [11]. Soils at the Schynige Platte are deeply leached and strongly acidified overlying limestone. The predominant buffer mechanism in these soils is not the $CaCO_3$ system but dissolving Al oxides and hydroxides that release free ionic Al into the solution. The latter is potentially toxic to plants as well as microorganisms, most likely because of inter-ionic competition at cellular membranes. Thus, even if acidified soils have low Ca contents, a general Ca deficiency of plants is unlikely because of the low demand of plants for Ca [33]. However, since Ca can compete with Al at the cell membrane binding sites, the Ca/Al ratio in solution might be a better indicator of the Ca nutritional status of soils or the Al toxicity stress for plants than the Ca concentration [15,26,27]. In a recent review, Vanguelova et al. [27] concluded that the fine root Ca/Al ratio has been repeatedly found to be positively correlated with seedling growth and nutrient uptake in woody plant species. The fact that in our experimental site the Ca/Al ratio, but not other measures such as soil pH, explained a significant amount of the variation in annual primary production suggests that a low Ca/Al ratio might also be a suitable indicator of soil acidity stress for plants in acidified mountain grasslands.

While single applications of NPK fertilizer initially showed a strong effect on productivity and vegetation composition [25], there were no effects of historical fertilization on the chemical composition of any of the below-ground and above-ground compartments analysed 70 years after the last treatment. This suggests that the duration of a legacy effect of historical human activities strongly depends on the type of activity or disturbance and that Ca may, at least in acidic subalpine grassland soils and under addition rates comparable to those used in this experiment, cause a more long-term legacy effect than N and P. Despite the fact that P was added as superphosphate, which also includes Ca, the soil pH of fertilized plots only moderately and temporarily increased from 4.5 to 4.7 immediately following application of the treatment [25].

Our study provides unique experimental evidence of the long-term memory of subalpine soils to single events of liming. Despite the small size of the experimental plots and the continuous removal of the above-ground biomass, the Ca-concentration in all ecosystem compartments investigated differed between plots that were limed in the 1930s and unlimed plots, and these patterns continue to affect ecosystem processes. While the increase in primary production by the legacy of the liming treatment may be considered beneficial from an economic point of view, it has detrimental effects on biodiversity since it led to a replacement of rare species by ubiquitous species adapted to increased nutrient availability [11]. Thus, from a general perspective, our findings suggest that small-scale human and natural perturbations might cause long-lasting small-scale patterns in nutrient cycling and hence influence below-ground and above-ground community composition in acid mountain grasslands [11]. From the more specific view of sustainable ecosystem and soil management, the results indicate that the widely used practice of soil liming might have long-lasting effects on plant and soil microbial biodiversity as well as on the chemical composition of soils (pH_{soil}, Ca and Al content), at least for several decades or even centuries. Long-lasting legacies of human activities might be particularly common in mountain ecosystems, which are characterized by low rates of nutrient cycling [34]. The management of mountain ecosystems should therefore be based on a thorough understanding of the interrelationships between the legacy effects of historical land-use, ecosystem resilience and the sustainability of socio-ecological systems in order to sustain or restore desirable ecosystem patterns and processes in the face of global change [4].

Materials and Methods

Site Description

The experiment at the Schynige Platte near Interlaken (Bernese Oberland, Switzerland) was set up in 1930 by Werner Lüdi at 1925 m a.s.l. in a subalpine acid grassland (acid cambisol, pH at the beginning of the experiment: 4.5–5.0) [25] on a SSE-facing slope with an inclination of 20°. Mean annual precipitation is c. 1800 mm, and mean annual temperature is c. 1°C. Before 1930, the site had been used as a pasture for many centuries. Once the experimental plots had been set up and the treatments initiated, the whole site was fenced and was mown once a year at the end of the growing season in September. From 1958–1980 the site was grazed by cattle. Since 1980, the site has been fenced again and mown once a year. The vegetation of the site consisted mainly of grasses, in particular *Nardus stricta* L. and *Festuca rubra* L. with a mean cover of approximately 30% and 5%, respectively. Subordinate species with a mean cover of between 3% and 5% were *Arnica montana* L., *Crepis conyzifolia* (Gouan) Kerner, *Gentiana purpurea* L., and *Vaccinium myrtillus* L. [11].

In the early 1930s, a total of 340 plots of 2.56 m^2 each (1.6×1.6 m, separated by 40 cm wide access paths) were set up in blocks and were subjected to different treatments [35]. To assess the effect of liming and NPK fertilization on soil nutrient cycling, we analysed the 80 plots corresponding to experimental blocks 8–11.

Each block consisted of 20 plots which were arranged in two parallel rows of 10 plots along the main altitudinal gradient. The plots were subjected during two years (1932 and 1933) in a factorial design to the two treatments liming (yes/no) and NPK fertilizer (yes/no). Both treatments were applied once a year in early summer: N was applied as ammonium sulphate, P as superphosphate and K as 30% potassium chloride [35]. The fertilized plots also received a small amount of Ca as part of the P-fertilizer. In half of the blocks, the treatments were repeated in 1934 and 1935. In total, plots that were limed over a period of two years received 80 g m^{-2} of Ca, and plots that were fertilized received 1.4 g m^{-2} of N, 4.9 g m^{-2} of P, 9.7 g m^{-2} of K, and 18.7 g m^{-2} of Ca. Plots that were treated over a period of four years received twice the amount of nutrients. Within blocks, the four treatment combinations were allocated to the plots in an alternating order.

Soil and Vegetation Sampling

Soil samples were collected in June 2006 from each plot by taking one soil core (1.8 cm diameter and 10 cm deep) from each corner of the plot. The four subsamples from each plot were pooled for analysis, resulting in approximately 60 g of soil for each plot. Samples were kept for six months at 2°C until analysis. Prior to analysis, all samples were passed through a 2 mm sieve and the plant parts removed.

Above-ground biomass was sampled in early September from the central 1 m^2 of each plot using a bar mower of 1 m width. The biomass samples were dried at 60°C for 36 h and weighed. The exact surface of the cut area was determined by measuring the length of the area cut by the bar mower. The dried bulk samples were ground using a grass mill and two subsamples of c. 1 g each were taken from each sample. The C, H and N content of the two subsamples was analysed with a CHN analyser at 1050°C (Leco CHN-1000, Leco Corporation, St Joseph, MI, USA), and the results averaged.

Water-soluble cations were extracted by adding 10.0 ml millipore water to 5.0 g of soil and leaving this for 24 h, during which the mixture was manually stirred several times. After 24 h, the mixture was filtered through a 45 µm membrane filter (Whatman ME25, Schleicher and Schuell, Dassel, Germany). The amount of cations in the filtrate was determined at the day of filtration using an ICP Spectrometer (Spectro Ciros Vision, Spectro GmbH, Kleve, Germany). To extract cations from the microbial biomass, the soil was autoclaved (35 min at 126°C) in a steam sterilizer (All American 25X, Wisconsin Aluminium Foundry Co. Inc., Manitowoc, WI, USA). The amount of cations in the autoclaved soil was then determined in the same way as the water-soluble cations. The amount of cations in the microbial biomass was calculated by subtracting the amount of water-soluble cations in untreated soil from the amount of cations in the autoclaved soil. The total amount of cations in the soil was determined by ICP-OES (Ciros Vision; Spectro, Kleve, Germany) after fusion of the powdered samples with $Li_2B_4O_7$ and dissolution of the melt in 1M HNO_3. Chemical analysis of the solution was done with ICP-OES. To determine the cation exchange capacity, 2.5 g of soil was soaked overnight in 1 M NH_4Cl. The day after, 100 mL of 1 M NH_4Cl was allowed to percolate through the soil for at least 4 h. The resulting soil extract was filtered through a filter paper (LS 14, Schleicher and Schuell, Dassel, Germany) and 1 M NH_4Cl was added to make the volume of each sample 100 mL [30]. The amount of cations in the extract was analysed in an ICP spectrometer as described above. Water content of the sieved soils was determined gravimetrically by weighing fresh soil and reweighing the soil sample after drying it at 40°C for 36 h. Organic matter content of the dried soil was determined by weighing the soils again after burning at 550°C for 8 h. To determine the total C, H and N content of the soil (% of dry weight), the dried soils were ground and C, H and N content was determined in approximately 200 mg of ground soil in a CHN analyser as described above. Soil pH measures were taken from Spiegelberger et al. [11].

To calculate the Ca content of soil in limed and unlimed plots, we used average soil bulk density reported for subalpine grasslands in the same region of the Swiss Alps and at the same altitude (approximately 0.7 g cm^{-3}) [36]. To calculate total Ca content of the vegetation, the belowground vegetation biomass was estimated to be approximately three times as high as the above-ground biomass [36] and the Ca concentration in belowground vegetation biomass to be 20% lower than in above-ground biomass [37,38].

Statistical Analysis

Data from the 80 plots were analysed using ANOVA with block as a random factor and lime and fertilizer treatments as fixed factors. We did not apply a Bonferroni correction, because it is considered as overly conservative [39,40]. Out of 48 analyses 8 tests (17%) were significant at least at the 5% level (compared to 2.4 expected by chance alone), and three of these at the 1% level (compared to 0.48 expected by chance alone). All statistical analyses were done using SPSS version 16.0.

Acknowledgments

We thank the Geschäftsleitung of the Alpengarten Schynige Platte for allowing us to work on the Lüdi plots, Franz Conen and Heinz Hürlimann for technical support, Franz Conen and Richard Bardgett for valuable comments on an earlier version of the manuscript and the Bristol Foundation for financial support.

Author Contributions

Conceived and designed the experiments: US DM OH. Performed the experiments: CA RE TS OH. Analyzed the data: US DM. Contributed reagents/materials/analysis tools: CA. Wrote the paper: US CA DM OH.

References

1. Foley JA, DeFries RS, Asner GP, Barford C, Bonan G, et al. (2005) Global consequences of land use. Science 309: 570–574.
2. Carpenter SR, Mooney HA, Agard J, Capistrano D, DeFries RS, et al. (2009) Science for managing ecosystem services: Beyond the Millennium Ecosystem Assessment. Proceedings of the National Academy of Sciences of the United States of America 106: 1305–1312.
3. Elmqvist T, Folke C, Nyström M, Peterson G, Bengtsson J, et al. (2003) Response diversity, ecosystem change, and resilience. Frontiers in Ecology and the Environment 1: 488–494.
4. Folke C, Carpenter S, Elmqvist T, Gunderson L, Holling CS, et al. (2002) Resilience and sustainable development: building adaptive capacity in a world of transformations. AMBIO 31: 437–440.
5. Schlesinger WH (2003) Treatise on geochemistry, Volume 8. Amsterdam, Elsevier. 682 p.
6. Galloway JN, Townsend AR, Erisman JW, Bekunda M, Cai Z, et al. (2008) Transformation of the nitrogen cycle: Recent trends, questions, and potential solutions. Science 320: 889–892.
7. Wedin DA, Tilman D (1996) Influence of nitrogen loading and species composition on the carbon balance of grasslands. Science 274: 1720–1723.
8. Stevens CJ, Dise NB, Mountford JO, Gowing DJ (2004) Impact of nitrogen deposition on the species richness of grasslands. Science 303: 1876–1879.
9. Suding KN, Collins SL, Gough L, Clark C, Cleland EE, et al. (2005) Functional- and abundance-based mechanisms explain diversity loss due to N fertilization. Proceedings of the National Academy of Sciences of the United States of America 102: 4387–4392.
10. Silvertown J, Poulton P, Johnston E, Edwards G, Heard M, et al. (2006) The Park Grass Experiment 1856–2006: its contribution to ecology. Journal of Ecology 94: 801–814.
11. Spiegelberger T, Hegg O, Matthies D, Hedlund K, Schaffner U (2006) Long-term effects of short-term perturbation in a subalpine grassland. Ecology 87: 1939–1944.
12. Hrevušová Z, Hejcman M, Pavlů VV, Hakl J, Klaudisová M, et al. (2009) Long-term dynamics of biomass production, soil chemical properties and plant species composition of alluvial grassland after the cessation of fertilizer application in the Czech Republic. Agriculture, Ecosystems & Environment 130: 123–130.
13. Pierik M, Van Ruijven J, Bezemer TM, Geerts RHEM, Berendse B (2011) Recovery of plant species richness during long-term fertilization of a species-rich grassland. Ecology 92: 1393–1398.
14. McLaughlin SB, Wimmer R (1999) Calcium physiology and its role in terrestrial ecosystem processes. New Phytologist 142: 373–417.
15. Jandl R, Alewell C, Prietzel J (2004) Calcium loss in Central European forest soils. Soil Science Society of America Journal 68: 588–595.
16. Isselstein J, Jeangros B, Pavlu V (2005) Agronomic aspects of extensive grassland farming and biodiversity management. Grassland Science in Europe 10: 211–220.
17. West TO, McBride AC (2005) The contribution of agricultural lime to carbon dioxide emissions in the United States: dissolution, transport and net emissions. Agriculture, Ecosystems & Environment 108: 145–154.
18. Tilman D, Fargione J, Wolff B (2001) Forecasting agriculturally driven global environment change. Science 229: 281–284.
19. Alewell C, Manderscheid B, Meesenburg H, Bittersohl J (2000) Is acidification still an ecological threat? Nature 407: 856–857.
20. Shortle WC, Smith KT (1988) Aluminum-induced calcium deficiency syndrome in declining red spruce. Science 240: 1017–1018.
21. Tilman D, Dodd M E, Silvertown J, Poulton PR, Johnston AE, et al. (1994) The park grass experiment - insights from the most long-term ecological study. In: Leigh RA, Johnston AE, editors. Long-term experiments in agricultural and ecological sciences. Wallingford: CABI. 287–303.
22. Hamilton SK, Kurzman AL, Arango C, Jin L, Robertson GP (2007) Evidence for carbon sequestration by agricultural liming. Global Biogeochemical Cycles 21: 1–12.
23. Fornara DA, Steinbeiss S, McNamara NP, Gleixner G, Oakley S, et al. (2011) Increases in soil organic carbon sequestration can reduce the global warming potential of long-term liming to permanent grassland. Global Change Biology 17: 1925–1934.
24. Schaberg PG, DeHayes DH, Hawley GJ (2001) Anthropogenic calcium depletion: A unique threat to forest ecosystem health? Ecosystem Health 7: 214–228.
25. Lüdi W (1959) Versuche zur Alpweidenverbesserung auf der Schynigen Platte bei Interlaken (Alpengarten Schynige Platte, Interlaken).
26. Cronan CS, Grigal DF (1995) Use of calcium/aluminium ratios as indicators of stress in forest ecosystems. Journal of Environmental Quality 24: 209–226.
27. Vanguelova EI, Hirano Y, Eldhuset TD, Sas-Paszt L, Bakker MR, et al. (2007) Tree fine root Ca/Al molar ratio - Indicator of Al and acidity stress. Plant Biosystems 141: 460–480.
28. Bossew P, Gastberger M, Gohla H, Hofer P, Hubmer A (2004) Vertical distribution of radionuclides in soil of a grassland site in Chernobyl exclusion zone. Journal of Environmental Radioactivity 73: 87–99.
29. Herben T, Krahulec F, Handicová V, Skálová H (1993) Small-scale variability as a mechanism for large-scale stability in mountain grasslands. Journal of Vegetation Science 4: 163–170.
30. Ross DS, Matschonat G, Skyllberg U (2008) Cation exchange in forest soils: the need for a new perspective. European Journal of Soil Science 59: 1141–1159.
31. Van Breemen N, Lundström US, Jongmans AG (2000) Do plants drive podsolization via rock-eating mycorrhizal fungi? Geoderma 94: 163–171.
32. Blum JD, Klaue A, Nezat CA, Driscoll CT, Johnson CE, et al. (2002) Mycorrhizal weathering of apatite as an important calcium source in base-poor forest ecosystems. Nature 417: 729–731.
33. Rorison IH, Robinson D (1984) Calcium as an environmental variable. Plant, Cell & Environment 7: 381–390.
34. DeAngelis DL, Bartell SM, Brenkert AL (1989) Effects of nutrient recycling and food-chain length on resilience. American Naturalist 134: 778–805.
35. Dähler W (1993) Langfristige Auswirkungen menschlicher Eingriffe in alpine Vegetation. Geobotanica Helvetica 71: 1–143.
36. Leifeld J, Zimmermann M, Fuhrer J, Conen F (2009) Storage and turnover of carbon in grassland soils along an elevation gradient in the Swiss Alps. Global Change Biology 15: 668–679.
37. Whitehead DC (2000) Nutrient elements in grasslands: soil-plant-animal relationships. Wallingford: CABI. 369 pp.
38. Ray JG, George KJ (2010) Calcium accumulation in relation to their root cation exchange capacity. Journal of Agronomy 9: 70–74.
39. Moran MD (2003) Arguments for rejecting the sequential Bonferroni in ecological studies. Oikos 100: 403–405.
40. Garcia LV (2004) Escaping the Bonferroni iron claw in ecological studies. Oikos 105: 657–663.

23

Indigenous Knowledge and Science Unite to Reveal Spatial and Temporal Dimensions of Distributional Shift in Wildlife of Conservation Concern

Christina N. Service[1,2,3,4], Megan S. Adams[1,2,4], Kyle A. Artelle[2,4,5], Paul Paquet[1,2], Laura V. Grant[1,3], Chris T. Darimont[1,2,3,4]*

1 Department of Geography, University of Victoria, Victoria, BC, Canada, **2** Raincoast Conservation Foundation, Denny Island, BC, Canada, **3** Spirit Bear Research Foundation, Klemtu, BC, Canada, **4** Hakai Beach Institute, Heriot Bay, BC, Canada, **5** Earth to Ocean Research Group, Department of Biological Sciences, Simon Fraser University, Burnaby, BC, Canada

Abstract

Range shifts among wildlife can occur rapidly and impose cascading ecological, economic, and cultural consequences. However, occurrence data used to define distributional limits derived from scientific approaches are often outdated for wide ranging and elusive species, especially in remote environments. Accordingly, our aim was to amalgamate indigenous and western scientific evidence of grizzly bear (*Ursus arctos horribilis*) records and detail a potential range shift on the central coast of British Columbia, Canada. In addition, we test the hypothesis that data from each method yield similar results, as well as illustrate the complementary nature of this coupled approach. Combining information from traditional and local ecological knowledge (TEK/LEK) interviews with remote camera, genetic, and hunting data revealed that grizzly bears are now present on 10 islands outside their current management boundary. LEK interview data suggested this expansion has accelerated over the last 10 years. Both approaches provided complementary details and primarily affirmed one another: all islands with scientific evidence for occupation had consistent TEK/LEK evidence. Moreover, our complementary methods approach enabled a more spatially and temporally detailed account than either method would have afforded alone. In many cases, knowledge already held by local indigenous people could provide timely and inexpensive data about changing ecological processes. However, verifying the accuracy of scientific and experiential knowledge by pairing sources at the same spatial scale allows for increased confidence and detail. A similarly coupled approach may be useful across taxa in many regions.

Editor: Marco Festa-Bianchet, Université de Sherbrooke, Canada

Funding: The authors are grateful for support from the Habitat Conservation Trust Foundation, Hakai Network, Tides Canada, Tula Foundation, and Wilburforce Foundation. Additionally, the authors are thankful for the in-kind support provided by SEI Industries. CNS, KAA and MSA were supported by the Natural Sciences and Engineering Research Council of Canada [CNS and KAA by NSERC CGSA (Alexander Graham Bell Supplements) and MSA by an NSERC IPS]. CTD acknowledges a NSERC Discovery Grant (435683). Websites: http://www.hctf.ca/; http://hakai.org/; http://tidescanada.org/; http://tula.org/; http://www.wilburforce.org/; http://www.nserc-crsng.gc.ca/; http://www.sei-ind.com/. The funders had no role in study design, data collection and analysis, decision to publish, or preparation of the manuscript.

Competing Interests: The authors have declared that no competing interests exist.

* Email: darimont@uvic.ca

Introduction

Distributions of organisms are shaped and re-shaped over geological and ecological timescales. Broadly structured by a suite of natural processes and their interactions, changes to species distributions (hereafter 'range shifts'), can be driven by abiotic factors (*e.g.*, CO_2 enrichment, nitrogen deposition, climate; [1]), biotic processes (*e.g.*, competition and facilitation [2]), and dispersal capability [3,4]. In recent history, distributional shifts have often been rapid and associated with human-caused drivers. Causes include climate change, habitat modification, over-exploitation, persecution, introduction of exotic species, and re-introduction of native species (*e.g.*, [5–10]). The increased pace of range shifts caused by humans can impose ecological effects on other species, communities, and ecosystems by exposing recipients to novel predation pressure, competition, and diseases [11]. For example, the rapid, human-aided range expansion of the brown tree snake (*Boiga irregularis*) to Guam decimated native bird populations, which in turn reduced the reproductive success of vertebrate-pollinated native plant species [12].

Given such potential impacts, current and accurate knowledge of species distributions comprises a fundamental and important dimension in conservation management. For example, many policy-relevant processes, such as protected areas design, mapping of critical habitat, and land-use planning require distributional data [13,14]. Moreover, current distribution information can also inform proactive conservation intervention in the face of climate change and other stressors (*e.g.*, [10,15]). More generally, detecting shifts in species' distributions can signal underlying ecological changes within an ecosystem, providing managers with early insight that changes might be occurring in other species and communities.

Identifying contemporary ecological change requires knowledge of the past. A complementary methods approach that combines traditional and local ecological knowledge with conventional scientific methods can provide data that not only offer detailed occurrence data across large areas but also over long time periods [16]. Traditional ecological knowledge (TEK) of indigenous people is transmitted through generations and revolves around a cumulative body of knowledge, practice, and belief surrounding the relationships of living and nonliving beings with their environment and one another [17]. Local ecological knowledge (LEK), often but not exclusively associated with indigenous people, also provides information about ecosystem change, but is gained from observations over lifetimes and not via inter-generational transmission [18]. However, in practice the distinction between TEK and LEK is often imprecise as they may share many similarities [18,19]. In contrast, wildlife science uses a variety of empirical techniques that span differing temporal and spatial resolutions and can provide detailed and quantitative information on populations and individuals [18]. Such data, however, are often very expensive to acquire and temporally and spatially limited. Accordingly TEK/LEK data – potentially spanning decades or longer – can be summarized and analyzed to yield information on elusive species across large areas that are otherwise too expensive or difficult to monitor with conventional scientific tools (*e.g.*, [20,21]). Employing TEK/LEK and wildlife science approaches together might yield more comprehensive and detailed information about changes over time and space than either method alone [22,23]. Importantly, incorporating TEK/LEK into ecological research also can facilitate the engagement of communities [24,25]. Social research components of LEK and TEK necessarily include local people and affirms the importance of their contributions [[24]]; in turn, a collaborative methods approach can be a critical first step in establishing more collaborative management [17,19,25].

Grizzly bear (*Ursus arctos horribilis*) distribution on the remote and now sparsely populated central coast of British Columbia (BC), Canada, provides an ideal system to examine the temporal and spatial components of potential range shifts using a complementary methods approach. The provincial government's current management boundary, delineated as the western (*i.e.*, seaward) extent of the so-called Grizzly Bear Population Units (GBPUs), is the spatial scale at which grizzly bears are managed for hunting, habitat protection, and human-wildlife conflict [26]. Whereas formalized species accounts do not include details of coastal grizzly distribution at an appropriate spatial scale for this study [27], the current management boundary suggests grizzlies are functionally absent from all but five of the dozens of islands in the vast archipelago ([26,28–32]; T. Hamilton, BC Ministry of Environment, pers. comm; Figure 1). In contrast, local people in the area, including several First Nations communities that still rely heavily on subsistence activities and travel across their expansive Territories, now commonly observe grizzlies on islands. If present, subsistence hunters and fishers are likely to sight and remember large-bodied, diurnal and iconic wildlife, like grizzly bears.

Our overarching aim was to combine TEK, LEK, and western scientific methods (human-caused mortality records, non-invasive genetic sampling, and remote camera data) to record grizzlies on islands and detail a potential range shift. Understanding the potential colonization and occupancy of grizzly bears onto islands in the region has significant conservation implications because these animals possess tremendous ecological, cultural, and economic importance in this area and indeed, where they still exist throughout their global range (*e.g.*, [33–36]). More broadly applicable to other areas and taxa, we also test the hypothesis that data from each method yield similar spatial and temporal patterns. Finally, we explore how these approaches reveal complementary spatial and temporal dimensions of data and emerge with broadly applicable conclusions relevant to many systems.

Methods

Study area

The islands and nearby mainland of the central coast of BC (8800 km^2) occur within a nearly road-less and now sparsely populated region extending from its southern boundary of Calvert Island (51.58° N, 127.81° W) north to Princess Royal Island (53.21° N, 128.05° W) (Figures 1a & 1b). A complex matrix of landmasses, the central coast as a whole is composed of mainland valleys divided by extensive fjords and various sized islands (< 1 km^2 to >2200 km^2) separated by tidal waters [37]. Eighteen major islands (*i.e.*, greater than 45 km^2) were included in the study area. The closest major island to the mainland (Yeo) is separated by 230 m of tidal water. The Coastal Western Hemlock biogeoclimatic zone dominates low elevations of the region [38]. Potential foods for bears are well distributed across the coastal mainland and islands and include spring (sedges and forbs), summer (berries), and fall (spawning salmonids) resources [39,40]. Since colonization by Europeans, most people in the area now live in the communities of Bella Bella (Heiltsuk Nation, population ~2200) and Klemtu (Kitasoo/Xai'xais Nation, population ~400). Our study area comprised four islands (of five) now recognized by the current management limit and 14 islands beyond this management range (Figures 1 & 2; Table 1).

Data types

Interview data. Ethics Statement: The interview component of this study was approved by the Human Research Ethics Board at the University of Victoria (Victoria, BC, Canada - Protocol # 12-385), Heiltsuk Integrated Resource Management Department, and Kitasoo/Xai'xais Integrated Resource Authority. All participants provided written informed consent. Field sampling was approved by the Heiltsuk Integrated Resource Management Department, Kitasoo/Xai'xais Integrated Resource Authority, and BC Parks.

We conducted 22 LEK and seven TEK interviews using a "snowball sampling" method [41]. For TEK interviewees, the Kitasoo/Xai'xais Stewardship Department and the Heiltsuk Integrated Resource Management Department recommended initial participants, who hold traditional oral and observational history knowledge, and who in turn suggested additional experts to interview (University of Victoria Human Research Ethics Approval # 12-385) [22,42]. The LEK participant pool included mostly indigenous (19 of 22) candidates, including subsistence and commercial fishers, hunters, eco-tour operators, salmon counters, bear viewing guides, and biologists.

Using adapted general guidelines of TEK/LEK data collection, we guided participants through target questions while also recording additional comments [22]. To assess historical grizzly distributions, comprising occupancy over the past century and beyond, we asked TEK participants about their experience and cultural transmission of knowledge about bears. We asked them to depict on their Traditional Territory map where, based on their Nation's oral histories, they would historically expect to see grizzly bears (hereafter 'Pre 1992'). Given inter-generational transmission of knowledge, we estimate that this TEK 'temporal window' provides information about island occupancy that spans centuries or more.

We used a mapping approach to complement interview data. Specifically, LEK participants indicated on a map the islands on which they had observed a grizzly bear. For each observation,

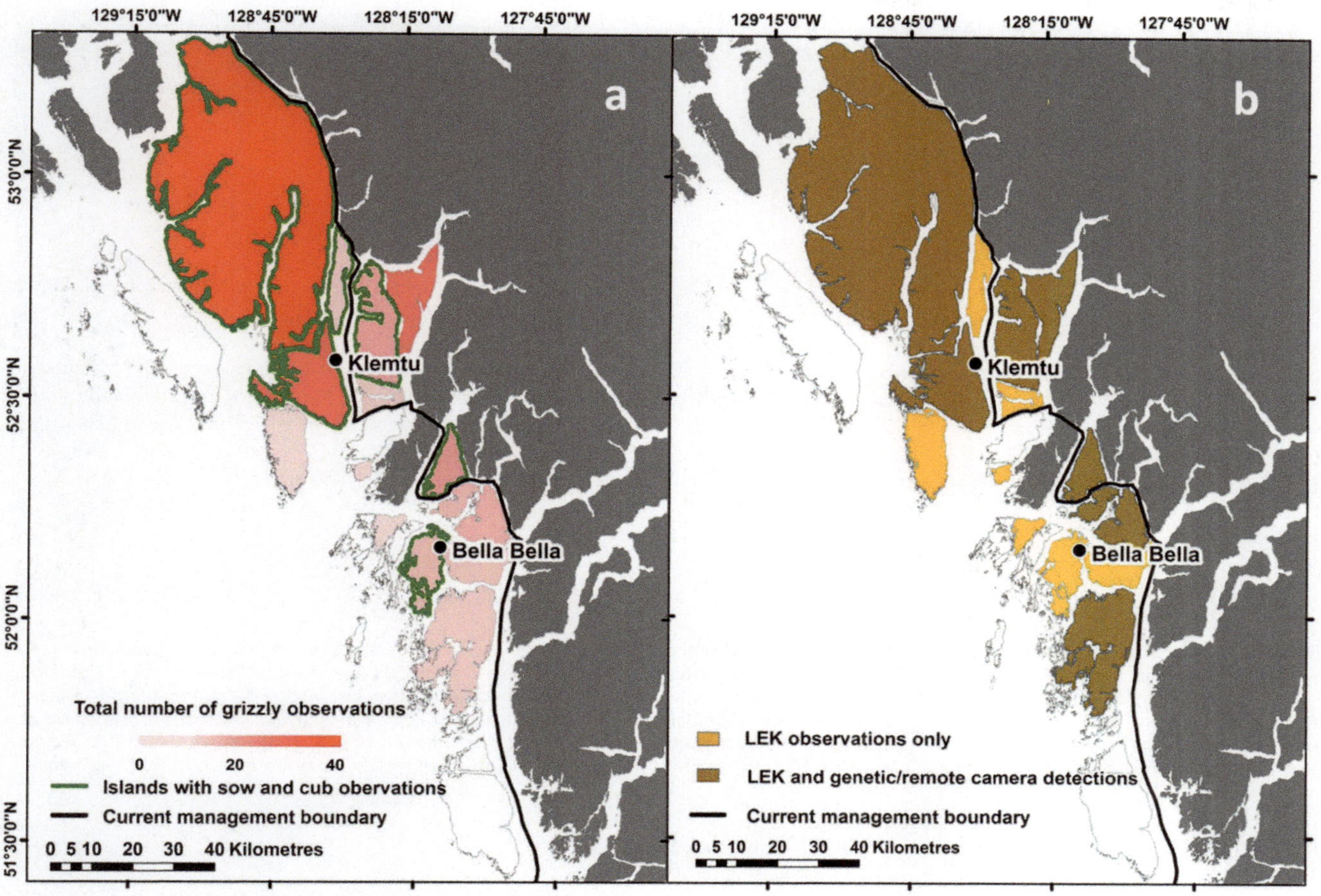

Figure 1. Spatial pattern of evidence of island occupancy and detection type. a) Weight of evidence of grizzly bear (*Ursus arctos horribilis*) occupancy at the island scale within Heiltsuk and Kitasoo/Xai'xais Territories in coastal British Columbia, Canada. Each data type (local ecological knowledge observation, mortality record, genetic ID and remote camera observation) was weighted equally to provide an indication of occupancy rather than bear density. Dark grey areas were not included in our study. The 'current management boundary' is the westward (i.e., seaward extent) of the Province of British Columbia's Grizzly Bear Population Units, the spatial scale at which grizzly bears are managed in the province. Spatial pattern in data types used to detect grizzly bears (*Ursus arctos horribilis*) in sampled areas within Heiltsuk and Kitasoo/Xai'xais Territories. Eighteen islands were included in the study.

participants provided supporting information including the island name, year, and season. We also asked LEK participants to draw on a map the areas they would define as coastal grizzly bear distribution during two timespans (1992–2002, 2003–2012). In addition, an estimate of survey effort for each LEK participant was quantified spatially – by indicating the extent of their area of expertise on a map – and temporally – as the estimated number of years and days per year the participant was in the field and potentially able to observe bears. All TEK/LEK data were verified with interview participants through follow up workshops. Once verified, interview data in transcript form were returned to both the participants and the Nations' Resource Stewardship offices in digital and print versions.

Genetic, remote camera, and mortality data. We used genetic data, remote camera images, and mortality records to identify location, date, and – in some cases sex, individual identity, and age class – of grizzly bears detected in the same area from 2009–2012. We sampled genetic data from non-invasive hair-snagging stations baited with a non-reward bait [43,44]. These stations (n = 33) were part of longitudinal carnivore monitoring programs across 10 islands in Kitasoo/Xai'xais (2012) and Heiltsuk (2009–2012) Territories (Table 1 & 2). In addition, 18 remote trail cameras (n = 1268 trap nights) were deployed on four islands in 2012 (Table 1 & 2). Finally, we queried island locations within the BC Ministry of Environment's kill records from their Compulsory Inspection Database [39,40], yielding dates, locations, and sexes of known human-killed grizzly bears. All island mortality records (n = 3) were hunted individuals rather than animal control (i.e., human-wildlife conflict) kills (Table 1 & 2).

Analyses

Occupancy. We summed the number of grizzly bear observations from all data sources on each island to assess the weight of evidence for contemporary occupancy. We accounted for observations of (genetically) undocumented bears from sightings and camera images equally with known, specific individuals from mortality and genetic data. Accordingly, it is possible that multiple observations were from the same individual bear. However, we did not expect bias across candidate islands in the spatial pattern where such multiple counting might have occurred. We calculated detection-per-unit-effort (DPUE) for all data types except mortality data (Table 1 and Table 2). For remote camera and genetic data, we quantified DPUE for each island by dividing the number of observations by the cumulative number of

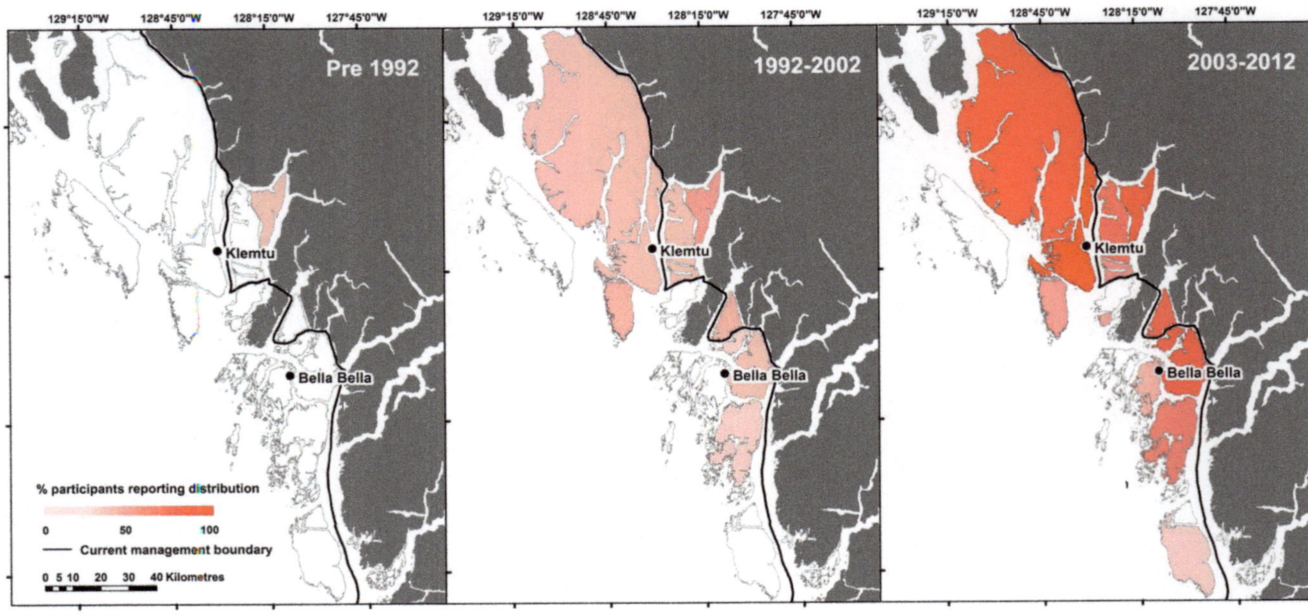

Figure 2. Time series of island grizzly bear occupation over three time periods. Local residents' perception of grizzly bear (*Ursus arctos horribilis*) distribution throughout Heiltsuk and Kitasoo/Xai'xais Territories in coastal British Columbia, Canada during the Pre 1992 (traditional ecological knowledge data), 1992–2002 (local ecological knowledge data) and 2003–2012 (local ecological knowledge data) time periods. Data are reported at the island scale (n = 18 islands) as the percentage of participants who indicated the island was within their area of observation and supported grizzly bears. Dark grey areas were not included in our study and the 'current management boundary' is the westward (i.e., seaward extent) of the Province of British Columbia's Grizzly Bear Population Units, the spatial scale at which grizzly bears are managed in the province.

days each camera (n = 18) or hair snag (n = 33) was employed (Table 2). We standardized LEK observations by dividing the total number of observations for each island by the total number of survey days each interviewee estimated s/he spent within their identified geographic area of expertise. Both survey effort and total number of observations were summed for all participants at the island scale. As level of experience could not be reliably estimated, we assumed the same observational abilities of all participants. To assess which islands contain reproducing females, we used both remote camera and LEK observations to identify all detections of sows with cubs.

Temporal Trends in Occupancy. We assessed temporal occupancy across three intervals: Pre 1992 (n = 7 interviewees), 1992–2002 (n = 27 interviewees), and 2003–2012 (n = 27 interviewees). We estimated 'Pre 1992' occupancy from TEK data, and 1992–2002 and 2003–2012 periods from LEK data. Because LEK observations are likely increasingly comprehensive with increasing proximity to communities of residence [18], for each island we only used interview data from the closest community. We indexed the total number of observations across all islands by dividing them by the total number of survey days in each year. Survey effort was calculated annually using Equation 1:

$$\text{Survey Effort} = \frac{\sum(\text{Survey days per participant})}{\#\text{ of participants} \times 365} \quad (1)$$

where # of participants is the number of LEK interviewees, 365 denotes the total possible number of survey days a year, and survey days per person indicates the total number of survey days for each participant. The number of island LEK observations per year was then divided by annual survey effort across all islands. This metric allows survey effort to be calculated per annum across all participants for all islands.

Results

We identified 149 grizzly bear observations across 15 major islands, including 10 islands outside the current grizzly bear management boundary (Figure 1a, Table 1). LEK yielded the largest number of data points relative to other sources, with 110 observations across 15 major islands (Table 1 & 2). Twelve islands hosted more than one observation (mean = 9.93, range = 1 to 39) (Table 1; Figure 1a). Across all LEK participants, 86% (19 of 22) reported at least one observation of an island grizzly (Table 1). Data on known individuals (genetic data and mortality records) indicated the presence of at least nine unique individuals (three female, five male and one of unknown gender) across 10 locations on seven islands. Remote cameras yielded a total of 28 observations of grizzlies on four islands (Table 1), two of which are outside of the current management boundary. Sow and cubs were detected in 23 LEK observations and seven remote camera observations across six islands (Table 1).

TEK/LEK data sources revealed a similar pattern of island occupation as did evidence from genetics, cameras, and hunting records; all newly occupied islands identified by scientific evidence were affirmed by complementary TEK/LEK observations (Figure 1b; Table 1). Of the seven islands solely associated with TEK/LEK evidence, all but one (Denny Island) lacked scientific evidence (Figure 1b).

Indigenous knowledge additionally yielded historical information suggesting that bear colonization of islands occurred recently, and that the process has increased in pace. Only one of seven TEK interviewees identified grizzly bears on an island before 1992. Accounts of island occurrence, within and outside the currently recognized distribution, were increasingly common and spatially widespread in 1992–2002 and 2003–2012 periods (encompassing 12 and 15 islands, respectively; Figure 2). The total island

Table 1. Detections of island grizzly bears (*Ursus arctos horribilis*) by data type (presented as raw observations and as detection-per-unit-effort (DPUE) values) within Heiltsuk and Kitasoo/Xai'xais Territories in coastal British Columbia, Canada (n = 18 islands).

Island	Total Remote Camera Observations	Remote Camera Detections Per Trap Night[a]	Total Hair Snag Observations	Hair Snag Year	Hair Snag Detections Per Trap Night[b,d]	LEK Observations	Total Sow/Cub Observations	Mortality Observation
Aristazabal	-	-	-	-	-	0	0	0
Calvert	-	-	-	-	-	0	0	0
Campbell	-	-	0	-	-	6	0	0
Chatfield	-	-	2	2009–2012	0.03	6	0	0
Cunningham	-	-	1	2009–2012	0.01	7	3	0
Denny	-	-	0	2009–2012	0	5	0	0
Dufferin	-	-	-	-	-	1	0	0
Hecate	-	-	-	-	-	0	0	0
Hunter	-	-	-	-	-	1	0	1
Yeo**	-	-	1	2009–2012	0.04	12	4	0
Lady Douglas	-	-	-	-	-	1	0	0
Pooley**	7	0.02	0	2012	0.01	15	0	0
Price	0	0	0	2012	0	1	0	0
Princess Royal	19	0.03	3	2012	0.02	16	17	1
Roderick**	1	0.01	0	2012	0.01	12	3	1
Sarah	-	-	-	-	-	9	3	0
Susan**	-	-	-	-	-	2	0	0
Swindle	1	0.03	1	2012	0.03	22	15	0

[a]For islands with remote cameras.
[b]For islands with hair snags.
[c]Indicates number of observations, not number of individual grizzly bears, across study period.
[d]Summed across seasons and years.
** Indicates islands within the current management boundary'.

Table 2. Sampling effort of island grizzly bears (*Ursus arctos horribilis*) by data type within Heiltsuk and Kitasoo/Xai'xais Territories in coastal British Columbia, Canada (n = 18 islands).

Island	Number of Remote Cameras	Total Remote Camera Nights[s]	Number of Hair Snag Sites	Hair Snag Year	Total Hair Snag Nights	Number of LEK Interviewees Who Reported on Island[b]	Average Number of Report Days Per Year[c]
Aristazabal	0	-	0	-	-	13	120
Calvert	0	-	0	-	-	9	120
Campbell	0	-	0	-	-	14	142
Chatfield	0	-	2	2009–2012	159	14	142
Cunningham	0	-	4	2009–2012	304	14	142
Denny	0	-	2	2009–2012	178	13	148
Dufferin	0	-	0	-	-	13	148
Hecate	0	-	0	-	-	9	120
Hunter	0	-	0	-	-	12	146
Yeo**	0	-	3	2009–2012	221	15	145
Lady Douglas	0	-	0	-	-	14	137
Pooley**	3	357	3	2012	416	10	136
Price	1	23	1	2012	23	13	140
Princess Royal	9	661	9	2012	1217	8	140
Roderick**	4	192	3	2012	398	11	139
Sarah	0	-	0	-	-	8	128
Susan**	0	-	0	-	-	14	145
Swindle	1	135	1	2012	121	11	140

[a]For islands with remote cameras.
[b]Across entire study period.
[c]Average number of 'report days' (i.e., days interviewees reported on for each island) per year from 1992–2012. Data derived from all participants who included. the island in their area of expertise.
** Indicates islands within the "current management boundary".

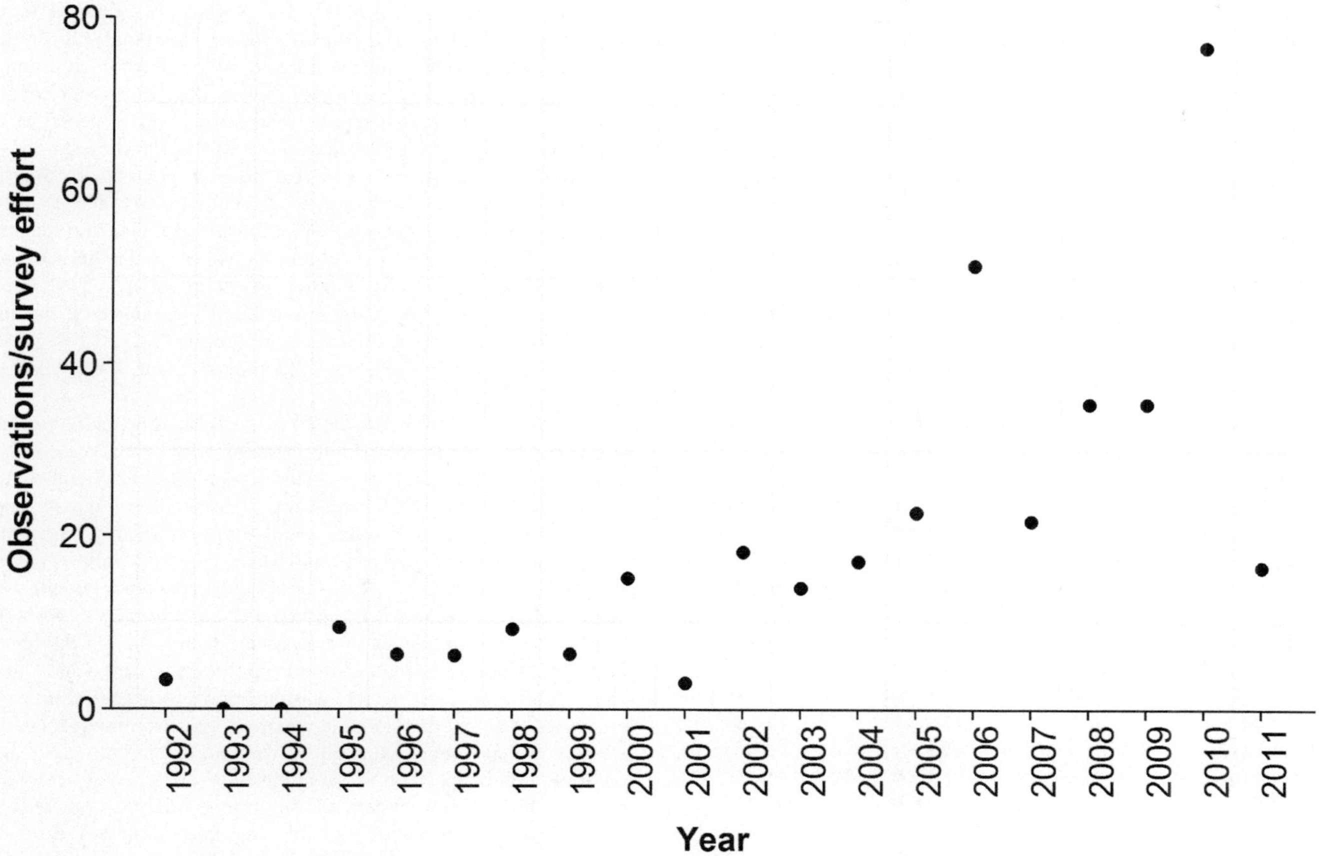

Figure 3. Trend of island grizzly occupation over time as derived from LEK data. Yearly number of island grizzly bear (*Ursus arctos horribilis*) observations per unit survey effort (a summation of all participants' observations across all islands; local ecological knowledge data only) between 1992–2011 in the Traditional Territories of the Heiltsuk and Kitasoo/Xai'xais First Nations in British Columbia, Canada. These observations encompass islands within and outside the current management boundary (n = 18 islands).

observations per year, indexed by survey effort, increased between 1992–2011 (Figure 3).

Discussion

Our approach coupling indigenous knowledge and western science offered multiple lines of evidence that grizzly bears of both sexes have recently colonized 10 islands outside of the currently accepted distribution. Moreover, camera and interview data provided strong evidence for reproductive individuals and their offspring on four islands. Collectively, this suggests a distributional process beyond any exploratory extra-range movements of mobile males. Such an abrupt distributional shift of an apex predator may invoke cascading ecological consequences (*e.g.*, [45]). More broadly, the drivers of such change may also act on other less iconic and recognizable species and processes at differing spatial and temporal scales [46,47].

Though trends in these data are apparent, our approach also presents some inherent limitations. Specifically, although highly useful, TEK/LEK data are not systematic in coverage [23]. Accordingly, use of islands by interviewees likely varies in their mode of use (hunting, fishing, ecotourism, etc.) and as a result the habitat types people frequent may differ. As such, some interviewees may be more likely to encounter grizzly bears than others. Additionally, it is possible that black bears may have been mistaken for grizzly bears by some interviewees. Furthermore, TEK/LEK may not always be appropriately responsive to detect re-colonization and abandonment events that could accompany dynamic distributional processes. Snowball sampling methods may have also excluded some knowledgeable participants who were outside of the social networks of interviewees [48]. Finally, we note that recent observations might be more likely to be remembered and reported. More likely, however, we postulate that older observations of grizzlies outside their known range would have been interpreted as surprising and, accordingly, equally or even more memorable. Despite these limitations, the inclusion of TEK/LEK data provides important process values through the engagement of local people and as a result this approach may facilitate collaborative rather than antagonistic conservation efforts (*e.g.*, [49–51]).

Conventional scientific tools, such as remote cameras and genetic identification through hair snagging, also present limitations. Although more systematic in their deployment, the temporal resolution of these data types are limited, both on an annual basis (*i.e.*, spring sampling) and over longer timeframes (most ecological monitoring programs span years rather than decades or centuries) [20]. Additionally, these tools are spatially static and can be impacted by poor placement. Indeed, in the nearly road-less landscape we study, most snag station are set up close to shorelines where they can be accessed by boat. Importantly, though all data sources have potential weaknesses, uniting these independent sources provides increased temporal and spatial detail of the range shift we describe.

Interviewee comments and relevant literature allow us to offer working hypotheses about potential drivers of this shift. Specifically, modifications to the abundance and distribution of food resources as well as changes to intra- and inter- specific competition might be relevant. Many interviewees cited the reduction of salmon (*Oncorhynchus* spp.), which has declined throughout coastal BC, especially since 2000 [52]. As a critical resource that influences ursid body condition, reproductive output, and population dynamics [36,53], salmon abundance could influence bear distribution; individuals within other carnivore species have expanded or abandoned their range following declines in prey and subsequent increased intraspecific competition (*e.g.*, [54–56]). Increasing berry abundance and accessibility, the result of recent logging on islands, might cause individuals on exploratory forays to islands to remain. Grizzly bears in other areas select recently logged habitat to exploit diverse food resources offered in early regeneration stages of disturbed habitat [57]. Alternatively, black bears (*Ursus americanus*) are thought to limit grizzlies via exploitative competition on coastal islands where food resources are more dispersed and more difficult to defend. Acting alone or synergistically with human-caused mortality from trophy hunting of grizzlies, such competition between species has been proposed as a mechanism for excluding grizzly bears from islands [31] or reducing their densities elsewhere [58]. Recent reductions of grizzly bear trophy hunting on the mainland (BC Ministry of Environment, unpublished data), and/or changes in competition from black bears, might have reduced the demographic constraints on grizzly bears, thereby allowing dispersal to nearby islands. Whatever the cause(s) in our system, range shifts observed in closely monitored species may signal changes to underlying ecological conditions and provide resource managers the opportunity to consider and plan for wider implications.

The ecological, cultural, and economic consequences of this distribution shift in coastal grizzly populations might be considerable. As a larger species, grizzly bears outcompete co-occurring black bears (*U. americanus*) for salmon through interference competition [59] and also kill juvenile and adult black bears [60]. In addition, 'spirit bears' (*Ursus americanus kermodei*), a rare and geographically restricted white morph of black bears that occur with relatively high frequency on several islands in the area [61,62], are not only revered culturally but are also economically prized as the foundation of wildlife eco-tourism. Consequently, any mortality or increased avoidance behaviour due to new grizzly bear presence might impact cultural and economic values. Finally, forest management plans for grizzlies – a 'species of conservation concern' in BC [63] – carries legally-binding measures to protect high quality habitat wherever grizzly bears occur [30]. Such multifaceted implications would also be likely for range shifts in other high-profile fauna. Indeed, reintroductions of wolves (*Canis lupus*) to the Yellowstone Ecosystem provide a flagship example of the tremendous ecological, cultural, and economic ramifications associated with a change in distribution [64,65]. Similarly, expansions to sea otter (*Enhydra lutris*) ranges, a system well-suited to TEK/LEK study, carry comparably broad implications for people and ecosystems [66].

Beyond its regional relevance, this study illustrates the synergistic benefits of combining science with TEK/LEK over concurrent and complementary spatial and temporal scales. In our study, both data types predominately affirmed one another; islands without scientific evidence for occupation also lacked concurrent TEK/LEK observations. The co-affirmation of data sources in locations where they overlapped added confidence in patterns observed by LEK where they did not overlap [16]. Genetic and camera data provided precise information on individuals captured, including individual identities, gender, age class, location, and time of visit. These data were captured in all weather and at all hours each day. Costs of these scientific methods, however, necessarily limit their spatial and temporal coverage. Moreover, they can only be employed to capture data in real time. In contrast, although yielding less detailed information on individual bears, TEK/LEK data afforded broader information across a larger area and longer timespan. Finally, despite the inherent value of TEK/LEK, it remains largely ignored as a source of ecological data [19].This case study provides one of the few examples we detected in the literature that illustrates the value in uniting TEK/LEK with scientific methods to provide meaningful input into wildlife management [18,24,25,67].

With careful consideration a TEK/LEK method alone could perform well in other wildlife systems. Such an approach, however, requires careful assessment. Specifically, LEK information might be most useful for questions of distribution but may lack the detail required for other population parameters, such as absolute abundance, without calibration from scientific sources [18,68]. The case for a TEK/LEK approach alone might be particularly compelling with conspicuous and culturally important focal species in data-deficient regions for which species distributions impose serious management implications and funds for science are scarce. Moreover, as many indigenous governments and societies across the globe play increasingly prominent roles in resource management again [69–71], methods that emphasize an integration of, or focus on, existing local knowledge might emerge as default approaches. Such a transition might help overcome the dual common barriers of lack of conservation action due to inadequate data and data deficiency resulting from scarcity of funds. In this way existing indigenous knowledge can be proactively incorporated into management. Such consideration may support a transition to implement management action more rapidly, the timescale in which conservation action is often required.

Supporting Information

Document S1 Interview instruments used in Local and Traditional Ecological Knowledge surveys of island grizzly bears (*Ursus arctos horribilis*) within Heiltsuk and Kitasoo/Xai'xais Territories in coastal British Columbia, Canada.

Acknowledgments

We thank interview participants, the Kitasoo/Xai'xais Integrated Resources Authority and the Heiltsuk Integrated Resource Management Department, and are grateful to Jennifer Carpenter, Tony Hamilton, William Housty, Wayne McCrory, Douglas Neasloss, and Laurie Whitehead for comments on this manuscript. We also acknowledge and thank our colleagues in the field: Harvey Brown Sr, Vernon Brown, Heather Bryan, Collin Reid, Ian Reid, Brandon Robinson, Krista Robinson, Leonard Robinson, Medrick Robinson, Shawn Robinson, Howard Humchitt, and Whitney Starr.

Author Contributions

Conceived and designed the experiments: CNS CTD PP. Performed the experiments: CNS MSA LVG. Analyzed the data: CNS MSA KAA LVG. Wrote the paper: CNS CTD PP KAA MSA LVG. GIS analysis support: KAA. Interview data collection: CNS LVG MSA. Field collection of remote camera and genetic data: CTD CNS LVG MSA KAA.

References

1. Tylianakis JM, Didham RK, Bascompte J, Wardle DA. (2008) Global change and species interactions in terrestrial ecosystems. Ecol Lett 11: 1351–1363.
2. Boulangeat I, Gravel D, Thuiller W (2012) Accounting for dispersal and biotic interactions to disentangle the drivers of species distributions and their abundances. Ecol Lett 15: 584–593.
3. MacArthur RH, Wilson EO (1967) The theory of island biogeography: monographs in population biology. Princeton, NJ, USA: Princeton University Press.
4. Pulliam HR (2000) On the relationship between niche and distribution. Ecol Lett 3: 349–361.
5. Robinson JG, Bennett EL (2000) Hunting for sustainability in tropical forests. New York, NY, USA: Columbia University Press.
6. Warren M, Chikin Y, Danilov I, King C, Morris N, et al. (2001) Rapid responses of British butterflies to opposing forces of climate and habitat change. Nature 414: 65–69.
7. Ripple WJ, Beschta RL (2004) Wolves and the ecology of fear: Can predation risk structure ecosystems? Bioscience 54: 755–766.
8. Whitfield D, Fielding AH, Mcleod DRA, Haworth PF (2004) The effects of persecution on age of breeding and territory occupation in golden eagles in Scotland. Biol Conserv 118: 249–259.
9. Chen I-C, Hill JK, Ohlemüller R, Roy DB, Thomas CD (2011) Rapid range shifts of species associated with high levels of climate warming. Science 333: 1024–1026.
10. Bennie J, Hogdgson JA, Lawson,CR, Holloway, CTR, Roy, DB,Brereton,T., Thomas, CD, and Wilson RJ (2013) Range expansion through fragmented landscapes under a variable climate. Ecol Lett 16: 921–929.
11. Vermeij GJ (1991) When biotas meet: Understanding biotic interchange. Science 253: 1099–1104.
12. Mortensen HS, Dupont YL, Olesen JM (2008) A snake in paradise: Disturbance of plant reproduction following extirpation of bird flower-visitors on Guam. Biol Conserv 141: 2146–2154.
13. Araújo MB, Williams PH (2000) Selecting areas for species persistence using occurrence data. Biol Conserv 96: 331–345.
14. Laliberte AS, Ripple WJ (2004) Range contractions of North American carnivores and ungulates. Bioscience 54: 123–138.
15. Guisan A, Thuiller W (2005) Predicting species distribution: offering more than simple habitat models. Ecol Lett 8: 993–1009.
16. Huntington HP, Suydam RS, Rosenberg DH (2004) Traditional knowledge and satellite tracking as complementary approaches to ecological understanding. Environ Conserv 31: 177–180.
17. Berkes F, Colding J, Folke C (2000) Rediscovery of traditional ecological knowledge as adaptive management. Ecol Appl 10: 1251–1262.
18. Gilchrist G, Mallory M, Merkel F (2005) Can local ecological knowledge contribute to wildlife management? Case studies of migratory birds. Ecol Soc 10.
19. Brook RK, McLachlan SM (2008) Trends and prospects for local knowledge in ecological and conservation research and monitoring. Biodivers Conserv 17: 3501–3512.
20. Anadón JD, Giménez A, Ballestar R, Pérez I (2009) Evaluation of local ecological knowledge as a method for collecting extensive data on animal abundance. Conserv Biol 23: 617–625.
21. Carter BTG, Nielsen EA (2011) Exploring ecological changes in Cook Inlet beluga whale habitat though traditional and local ecological knowledge of contributing factors for population decline. Mar Policy 35: 299–308.
22. Huntington HP (2000) Using traditional ecological knowledge in science: Methods and applications. Ecography (Cop) 10: 1270–1274.
23. Moller H, Berkes F, Lyver POB, Kislalioglu M (2004) Combining science and traditional ecological knowledge: Monitoring populations for co-management. Ecol Soc 9.
24. Brook RK, Mclachlan SM (2005) On using expert-based science to "test " Local Ecological Knowledge. Hum Ecol 10: 2–4.
25. Berkes F (2004) Rethinking community-based conservation. Conserv Biol 18: 621–630.
26. Ministry of Environment (2012) British Columbia Grizzly Bear Population Estimates for 2012. Available: http://www.env.gov.bc.ca/fw/wildlife/big_game_info.html#GrizzlyBear.
27. Committee on the Status of Endangered Wildlife in Canada (2012) COSEWIC assessment and status report on the Grizzly Bear, western population, in Canada. Ottawa, Canada.
28. McCrory WP, Bergdah J, Paquet PC, Cross B (2001) A conservation area design for protection of the white black bear (Ursus americanus kermodei) on the central coast of British Columbia: 25.
29. Schwartz CC, Miller SD, Haroldson M (2003) Grizzly bear. In: Feldhamer G, Thompson B, Chapman J, editors. Wild Mammals of North America. Baltimore, MD, USA: The Johns Hopkins University Press. pp. 556–586.
30. Ministry of Forests, Lands and NRO (2013) Ministerial Order: Central and North Coast Order Schedule 2 Grizzly Bear Habitat April 2013 Amendments.
31. Mattson DJ, Herrero S, Merrill T (2005) Are black bears a factor in the restoration of North American grizzly bear populations? Are black bears a factor in the restoration of North American grizzly bear populations? Ursus 16: 11–30.
32. Servheen C (1990) The Status and Conservation of the Bears of the World. International Conference of Bear Research and Management. Victoria, Canada. pp. 11–20.
33. Rockwell D (1991) Giving voice to bear: North American indian rituals, myths, and images of the bear. Lanham, Maryland, USA: Rowman & Littlefield.
34. Nevin OT, Gilbert BK (2005) Perceived risk, displacement and refuging in brown bears: positive impacts of ecotourism? Biol Conserv 121: 611–622.
35. Helfield JM, Naiman RJ (2006) Keystone interactions: Salmon and bear in riparian forests of Alaska. Ecosystems 9: 167–180.
36. Levi T, Darimont CT, Macduffee M, Mangel M, Paquet P, et al. (2012) Using grizzly bears to assess harvest-ecosystem tradeoffs in salmon fisheries. PLoS Biol 10: e1001303.
37. Darimont, C.T & Paquet PC (2002) The gray wolves, Canis lupus, of British Columbia's central and north coast: distribution and conservation assessment. Can F Nat 116: 416–422.
38. Krajina VJ (1965) Biogeoclimatic Zones and Classification of British Columbia. Ecology of Western North America. Ecol 1: 1: 1–17.
39. Pojar J, McKinnon A (1994) Plants of the Pacific Northwest Coast: Washington, Oregon, British Columbia, and Alaska. Auburn, WA, USA: Lone Pine Publishing.
40. Darimont CT, Bryan HM, Carlson SM, Hocking MD, MacDuffee M, et al. (2010) Salmon for terrestrial protected areas. Conserv Lett 3: 379–389.
41. Davis A, Wagner JR (2003) Who knows? On the importance of identifying " experts" when researching local ecological knowledge. Hum Ecol 31: 463–489.
42. Noy C (2008) Sampling knowledge: The hermeneutics of snowball sampling in qualitative research. Int J Soc Res Methodol 11: 327–344.
43. Press A, Woods JG, Paetkau D, Lewis D, Mclellan BN, et al. (1999) Genetic tagging of black bears. Wildl Soceity Bull 27: 616–627.
44. Bryan HM, Darimont CT, Paquet PC, Wynne-Edwards KE, Smits JE (2013) Stress and reproductive hormones in grizzly bears reflect nutritional benefits and social consequences of a salmon niche. PLoS One: In press
45. Beschta RL, Ripple WJ (2009) Large predators and trophic cascades in terrestrial ecosystems of the western United States. Biol Conserv 142: 2401–2414.
46. Carroll C, Noss RF, Paquet PC (2001) Carnivores as focal species for conservation planning in the Rocky Mountain region. Ecol Appl 11: 961–980.
47. Niemi GJ, McDonald ME (2004) Application of Ecological Indicators. Annu Rev Ecol Evol Syst 35: 89–111.
48. Biernacki P, Waldorf D, Methods S (1981) Snowball sampling: Problems and techniques of chain referral sampling. Sociol Methods Res 10. doi:10.1177/004912418101000205
49. Adams M, Carpenter J, Housty J, Neasloss D, Paquet P, et al. (2014) Towards increased engagement between academic and indigenous community partners in ecological research. Ecol Soc: In press.
50. Dowie M (2009) Conservation refugees: the hundred-year conflict between global conservation and native peoples. Cambridge, Massachusetts, USA: MIT Press.
51. Grove R (1995) Green imperialism: colonial expansion, tropical island edens and the origins of environmentalism. Cambridge, England: Cambridge University Press.
52. Price MHH, Darimont CT, Temple NF, MacDuffee SM (2008) Ghost runs: management and status assessment of Pacific salmon (Oncorhynchus spp.) returning to British Columbia's central and north coasts. Can J Fish Aquat Sci 65: 2712–2718.
53. Hilderbrand GV, Schwartz CC, Robbins CT, Jacoby ME, Hanley TA, et al.(1999) The importance of meat, particularly salmon, to body size, population productivity, and conservation of North American brown bears. Can J Zool 7: 132–138.
54. Ward RMP, Krebs CJ (1985) Behavioural responses of lynx to declining snowshoe hare abundance. Can J Zool 63: 2817–2824.
55. Knick ST (1990) Ecology of bobcats relative to exploitation and a prey decline in Southeastern Idaho. Wildl Monogr 108: 3–42.
56. Schmidt K (2008) Behavioural and spatial adaptation of the Eurasian lynx to a decline in prey availability. Acta Theriol (Warsz) 53: 1–16.
57. Nielsen SE, Boyce MS, Stenhouse GB (2004) Grizzly bears and forestry. For Ecol Manage 199: 51–65.
58. Mowat G, Heard DC, Schwarz CJ (2013) Predicting grizzly bear density in western North America. PLoS One 8: e82757.
59. Belant JL, Kielland K, Follmann EH, Adams LG (2006) Interspecific resource partitioning in sympatric ursids. Ecol Appl 16: 2333–2343.
60. Mattson DJ, Knight RR, Blanchard BM (1992) Cannilbalism and predation on black bears by grizzly bears in the Yellowstone ecosystem, 1975–1990. J Mammal 73: 422–425.
61. Ritland K, Newton C, Marshall HD (2001) Inheritance and population structure of the white-phased "Kermode" black bear. Curr Biol 11: 1468–1472.
62. Klinka DR, Reimchen TE (2009) Adaptive coat colour polymorphism in the Kermode bear of coastal British Columbia. Biol J Linn Soc 98: 479–488.
63. Austin MA, Heard DC, Hamilton AN (2004) Grizzly bear (Ursus arctos) harvest management in British Columbia. Victoria, BC, Canada: Ministry of Water, Land and Air Protection. Available: www.env.gov.bc.ca/wld/documents/gb_harvest_mgmt.pdf.

64. Morell V (2008) Wolves at the door of a more dangerous world. Science (80) 319: 890–892.
65. Wilson MA (1997) The wolf in Yellowstone: Science, symbol, or politics? Deconstructing the conflict between environmentalism and wise use. Soc Nat Resour 10: 453–468.
66. Riedman ML, Estes JA (1988) A Review of the History, Distribution and Foraging Ecology of Sea Otters. In: VanBlaricom G, Estes J, editors. The Community Ecology of Sea Otters. Berlin, Germany: Springer. pp. 4–21.
67. Abraão MB, Nelson BW, Baniwa JC, Yu DW, Shepard Jr GH (2008) Ethnobotanical ground-truthing: indigenous knowledge, floristic inventories and satellite imagery in the upper Rio Negro, Brazil. J Biogeogr 35: 2237–2248.
68. Anadón JD, Giménez A, Ballestar R (2010) Linking local ecological knowledge and habitat modelling to predict absolute species abundance on large scales. Biodivers Conserv 19: 1443–1454.
69. Folke C, Carpenter S, Elmqvist T, Gunderson L, Holling CS, et al. (2002) Resilience and sustainable development: building adaptive capacity in a world of transformations. Ambio 31: 437–440.
70. Houde N (2007) The six faces of traditional ecological knowledge: challenges and opportunities for Canadian co-management arrangements. Ecol Soc 12: 34.
71. Wyatt S (2008) First Nations, forest lands, and "aboriginal forestry" in Canada: from exclusion to co management and beyond. Can J For Res 38: 171–180.

Permissions

All chapters in this book were first published in PLOS ONE, by The Public Library of Science; hereby published with permission under the Creative Commons Attribution License or equivalent. Every chapter published in this book has been scrutinized by our experts. Their significance has been extensively debated. The topics covered herein carry significant findings which will fuel the growth of the discipline. They may even be implemented as practical applications or may be referred to as a beginning point for another development.

The contributors of this book come from diverse backgrounds, making this book a truly international effort. This book will bring forth new frontiers with its revolutionizing research information and detailed analysis of the nascent developments around the world.

We would like to thank all the contributing authors for lending their expertise to make the book truly unique. They have played a crucial role in the development of this book. Without their invaluable contributions this book wouldn't have been possible. They have made vital efforts to compile up to date information on the varied aspects of this subject to make this book a valuable addition to the collection of many professionals and students.

This book was conceptualized with the vision of imparting up-to-date information and advanced data in this field. To ensure the same, a matchless editorial board was set up. Every individual on the board went through rigorous rounds of assessment to prove their worth. After which they invested a large part of their time researching and compiling the most relevant data for our readers.

The editorial board has been involved in producing this book since its inception. They have spent rigorous hours researching and exploring the diverse topics which have resulted in the successful publishing of this book. They have passed on their knowledge of decades through this book. To expedite this challenging task, the publisher supported the team at every step. A small team of assistant editors was also appointed to further simplify the editing procedure and attain best results for the readers.

Apart from the editorial board, the designing team has also invested a significant amount of their time in understanding the subject and creating the most relevant covers. They scrutinized every image to scout for the most suitable representation of the subject and create an appropriate cover for the book.

The publishing team has been an ardent support to the editorial, designing and production team. Their endless efforts to recruit the best for this project, has resulted in the accomplishment of this book. They are a veteran in the field of academics and their pool of knowledge is as vast as their experience in printing. Their expertise and guidance has proved useful at every step. Their uncompromising quality standards have made this book an exceptional effort. Their encouragement from time to time has been an inspiration for everyone.

The publisher and the editorial board hope that this book will prove to be a valuable piece of knowledge for researchers, students, practitioners and scholars across the globe.

List of Contributors

Claire J. Santorelli and Colleen M. Schaffner
Department of Psychology, University of Chester, Chester, United Kingdom

Christina J. Campbell
Department of Anthropology, California State University, Northridge, California, United States of America

Hugh Notman
Anthropology, Centre for Work and Community Studies, Athabasca University, Athabasca, Alberta, Canada

Mary S. Pavelka
Department of Anthropology, University of Calgary, Calgary, Alberta, Canada

Jennifer A. Weghorst
Natural History Museum and Biodiversity Research Center, University of Kansas, Lawrence, Kansas, United States of America

Filippo Aureli
Research Centre in Evolutionary Anthropology and Palaeoecology, School of Natural Sciences and Psychology, Liverpool John Moores University, Liverpool, United Kingdom

Martinho Marta-Almeida, Rosa Reboreda, Carlos Rocha, Jesus Dubert, Rita Nolasco, Nuno Cordeiro, Alfredo Rocha and João D. Lencart e Silva
CESAM-Dep Física, Universidade de Aveiro, Aveiro, Portugal

Tiago Luna
Instituto do Ambiente e Desenvolvimento, Universidade de Aveiro, Aveiro, Portugal

Henrique Queiroga
CESAM-Dep Biologia, Universidade de Aveiro, Aveiro, Portugal

Alvaro Peliz
Centro de Oceanografia, Universidade de Lisboa, Lisboa, Portugal

Manuel Ruiz-Villarreal
Instituto Español de Oceanografía, Centro Oceanográfico A Coruña, Galicia, Spain

Kelly Ortega Cisneros
Leibnitz Center for Marine Tropical Ecology, Bremen, Germany
School of Biological and Conservation Sciences, University of KwaZulu-Natal, Durban, South Africa

Albertus J. Smit
School of Biological and Conservation Sciences, University of KwaZulu-Natal, Durban, South Africa

Jürgen Laudien
Alfred Wegener Institute for Polar and Marine Research, Bremerhaven, Germany

David S. Schoeman
School of Biological and Conservation Sciences, University of KwaZulu-Natal, Durban, South Africa
School of Environmental Science, University of Ulster, Coleraine, Northern Ireland

Jay V. Gedir
Centre for Biodiversity and Restoration Ecology, School of Biological Sciences, Victoria University of Wellington, Wellington, New Zealand

Wayne L. Linklater
Centre for Biodiversity and Restoration Ecology, School of Biological Sciences, Victoria University of Wellington, Wellington, New Zealand
Department of Zoology, Centre for African Conservation Ecology, Nelson Mandela Metropolitan University, Port Elizabeth, South Africa

Graham I. H. Kerley
Department of Zoology, Centre for African Conservation Ecology, Nelson Mandela Metropolitan University, Port Elizabeth, South Africa

Peter R. Law
PRLDB Modeling, Monroe, New York, United States of America

Ron R. Swaisgood
San Diego Zoo Global Institute for Conservation Research, Escondido, California, United States of America

Keryn Adcock
IUCN African Rhino Specialist Group, Hilton, KwaZulu-Natal, South Africa

Pierre du Preez
Directorate of Scientific Services, Ministry of Environment and Tourism, Windhoek, Namibia

Michael H. Knight
Department of Zoology, Centre for African Conservation Ecology, Nelson Mandela Metropolitan University, Port Elizabeth, South Africa
Park Planning and Development, South African National Parks, Port Elizabeth, South Africa

Elodie Vercken
Institut Sophia Agrobiotech, UMR 1355 ISA, Institut National de la Recherche Agronomique, Sophia-Antipolis, France

Maren Wellenreuther and Erik I. Svensson
Department of Biology, Lund University, Lund, Sweden

Benjamin Mauroy
Laboratoire J.A. Dieudonné, UMR CNRS 7351, Université de Nice-Sophia Antipolis, Nice, France

Charlotte Noyer and Mikel A. Becerro
Center for Advanced Studies of Blanes (CEAB, CSIC), Blanes, Spain

Olivier P. Thomas
Université de Nice - Sophia Antipolis, Laboratoire de Chimie des Molécules Bioactives et des Arômes, LCMBA-UMR 6001 CNRS, Nice, France

Tuomas Leinonen, Gabor Herczeg and Juha Merilä
Ecological Genetic Research Unit, Department of Biosciences, University of Helsinki, Helsinki, Finland

John Loehr
Ecological Genetic Research Unit, Department of Biosciences, University of Helsinki, Helsinki, Finland
University of Helsinki, Lammi Biological Station, Lammi, Finland

Robert B. O'Hara
Department of Mathematics and Statistics, University of Helsinki, Helsinki, Finland
Biodiversity and Climate Research Centre, Frankfurt am Main, Germany

Leonardo Ariel Saravia and Fernando Momo
Instituto de Ciencias, Universidad Nacional de General Sarmiento, Los Polvorines, Buenos Aires, Argentina

Adonis Giorgi
CONICET, Departamento de Ciencias Básicas, Universidad Nacional de Luján, Luján, Buenos Aires, Argentina

Lars O. Hedin and Stephen W. Pacala
Department of Ecology and Evolutionary Biology, Princeton University, Princeton, New Jersey, United States of America

Duncan N. L. Menge
Department of Ecology and Evolutionary Biology, Princeton University, Princeton, New Jersey, United States of America
National Center for Ecological Analysis and Synthesis, Santa Barbara, California, United States of America
Department of Biology, Stanford University, Stanford, California, United States of America

Nelson Valdivia
Instituto de Ciencias Marinas y Limnológicas, Facultad de Ciencias, Universidad Austral de Chile, Valdivia, Chile
Centro de Estudios Avanzados de Zonas Áridas (CEAZA), Facultad de Ciencias del Mar, Universidad Católica del Norte, Coquimbo, Chile

Ricardo A. Scrosati and Amanda S. Knox
Saint Francis Xavier University, Department of Biology, Antigonish, Nova Scotia, Canada

Markus Molis
Section Functional Ecology, Biologische Anstalt Helgoland, Alfred Wegener Institute for Polar and Marine Research, Helgoland, Germany

Yves Desdevises, Vincent Maillet, Michael Fuentes and Hector Escriva
CNRS UMR 7232, UPMC Univ Paris 06, Observatoire Océanologique, Banyuls-sur-Mer, France
3Ji Luo and Yanhong Wu
Institute of Mountain Hazards and Environment, Chinese Academy of Sciences, Chengdu, China
3Youchao Chen, Jia She and Peng Zhou
Institute of Mountain Hazards and Environment, Chinese Academy of Sciences, Chengdu, China
Graduate University of the Chinese Academy of Sciences, Chinese Academy of Sciences, Beijing, China

Peili Shi
Institute of Geographic Sciences and Natural Resources Research, Chinese Academy of Sciences, Beijing, China

Robert W. Li, Ransom L. Baldwin VI and Congjun Li
United States Department of Agriculture- Agricultural Research Service, Bovine Functional Genomics Laboratory, Beltsville, Maryland, United States of America

Weizhong Li and Sitao Wu
Center for Research in Biological Systems, University of California, San Diego, California, United States of America

Steven B. Scyphers, Sean P. Powers and Kenneth L. Heck Jr.
Department of Marine Sciences, University of South Alabama, Mobile, Alabama, United States of America
Dauphin Island Sea Lab, Dauphin Island, Alabama, United States of America

Dorothy Byron
Dauphin Island Sea Lab, Dauphin Island, Alabama, United States of America

Eduardo Jaramillo, Mario Manzano, Cesar Campos and Roland Sanchez
Instituto de Ciencias Ambientales y Evolutivas, Universidad Austral de Chile, Valdivia, Chile

Jenifer E. Dugan and David M. Hubbard
Marine Science Institute, University of California Santa Barbara, Santa Barbara, California, United States of America

Daniel Melnick
Institut für Geowissenschaften, Universität Potsdam, Potsdam, Germany

Cristian Duarte
Instituto de Ciencias Marinas y Limnológicas, Universidad Austral de Chile, Valdivia, Chile

Laura Airoldi
Dipartimento di Biologia Evoluzionistica Sperimentale and Centro Interdipartimentale di Ricerca per le Scienze Ambientali in Ravenna, Universitá di Bologna, Ravenna, Italy
Department of Earth and Environmental Sciences, Wesleyan University, Middletown, Connecticut, United States of America

Fabio Bulleri
Dipartimento di Biologia, Universitá di Pisa, CoNISMa, Pisa, Italy
Dipartimento di Science Botaniche, Ecologiche e Geologiche, Universitá di Sassari, Sassari, Italy

Sally A. Bracewell, Matthew Spencer, Rob H. Marrs, Matthew Iles and Leonie A. Robinson
School of Environmental Sciences, University of Liverpool, Liverpool, Merseyside, United Kingdom

Jonathan C. P. Reum and Paul McElhany
Conservation Biology Division, Northwest Fisheries Science Center, National Marine Fisheries Service, National Oceanic and Atmospheric Administration, Seattle, Washington, United States of America

Simone R. Alin and Richard A. Feely
Pacific Marine Environmental Laboratory, National Oceanic and Atmospheric Administration, Seattle, Washington, United States of America

Jan Newton
Applied Physics Laboratory, University of Washington, Seattle, Washington, United States of America

Mark Warner
School of Oceanography, University of Washington, Seattle, Washington, United States of America

Jotaro Urabe, Takao Suzuki, Tatsuki Nishita and Wataru Makino
Division of Ecology and Evolutionary Biology, Graduate School of Life Sciences, Tohoku University, Sendai, Japan

Hannes Baumann
School of Marine and Atmospheric Sciences, Stony Brook University, Stony Brook, New York, United States of America

Owen Doherty
Scripps Institution of Oceanography, University of California San Diego, La Jolla, California, United States of America

Iván F. Rodil
Departamento de Ecologı´a y Biologı´a Animal, Universidad de Vigo, Vigo, Spain
National Institute of Water and Atmospheric Research, Hamilton, New Zealand
Interdisciplinary Centre of Marine and Environmental Research, University of Porto, Porto, Portugal

Andrew M. Lohrer and Simon F. Thrush
National Institute of Water and Atmospheric Research, Hamilton, New Zealand

Urs Schaffner and René Eschen
CABI, Delémont, Switzerland

Christine Alewell
Department of Environmental Sciences, University of Basel, Basel, Switzerland

Diethart Matthies
Department of Biology, University of Marburg, Marburg, Germany

Thomas Spiegelberger
CABI, Delémont, Switzerland
Ecosystémes montagnards, Institut national de recherche en sciences et technologies pour l'environnement et l'agriculture, Saint-Martin-d'Héres, France

Otto Hegg
Department of Biology, University of Bern, Bern, Switzerland

Paul Paquet
Department of Geography, University of Victoria, Victoria, BC, Canada
Raincoast Conservation Foundation, Denny Island, BC, Canada

Laura V. Grant
Department of Geography, University of Victoria, Victoria, BC, Canada
Spirit Bear Research Foundation, Klemtu, BC, Canada

Megan S. Adams
Department of Geography, University of Victoria, Victoria, BC, Canada
Raincoast Conservation Foundation, Denny Island, BC, Canada
Hakai Beach Institute, Heriot Bay, BC, Canada

Christina N. Service and Chris T. Darimont
Department of Geography, University of Victoria, Victoria, BC, Canada
Raincoast Conservation Foundation, Denny Island, BC, Canada
Spirit Bear Research Foundation, Klemtu, BC, Canada
Hakai Beach Institute, Heriot Bay, BC, Canada

Kyle A. Artelle
Raincoast Conservation Foundation, Denny Island, BC, Canada
Hakai Beach Institute, Heriot Bay, BC, Canada
Earth to Ocean Research Group, Department of Biological Sciences, Simon Fraser University, Burnaby, BC, Canada

Index